Diversity of Fungi and Fungus-Like Organisms in Tropical Forests

Tropical forests cover approximately 6% of the land surface of the earth and more than 30% of all forested areas on the entire planet. These forests are characterized by a very high level of diversity for the assemblages of organisms present. The overall diversity of tropical forests is higher than that of any other type of terrestrial ecosystem, yet the fungi and fungus-like organisms (basidiomycetes, ascomycetes, myxomycetes, and dictyostelids) associated with tropical forests remain understudied.

Diversity of Fungi and Fungus-Like Organisms in Tropical Forests brings together information on these organisms obtained by scientists who are actively carrying out research in every region of the world where tropical forests occur. Among the major subject areas considered are ecology, biogeography, and taxonomy. Furthermore, information is provided on the use of certain fungi as food for humans and the potential of some fungi as sources of products of economic, pharmaceutical, or medical value.

This book will be of great interest to students, teachers, and all who wish to learn more about the fungi and fungus-like organisms associated with tropical forests around the world.

Diversity of Fungi and Fungus-Like Organisms in Tropical Forests

Edited by

Steven L. Stephenson, Samantha Chandranath Karunarathna, Changlin Zhao, and Arun Kumar Dutta

CRC Press
Taylor & Francis Group
Boca Raton London New York

CRC Press is an imprint of the
Taylor & Francis Group, an **informa** business

Designed cover image: Noah Siegel

First edition published 2027
by CRC Press
2385 NW Executive Center Drive, Suite 320, Boca Raton FL 33431

and by CRC Press
4 Park Square, Milton Park, Abingdon, Oxon, OX14 4RN

CRC Press is an imprint of Taylor & Francis Group, LLC

Library of Congress Cataloging-in-Publication Data
Names: Stephenson, Steven L. editor | Karunarathna, Samantha C. editor | Zhao, Changlin editor | Dutta, Arun Kumar editor
Title: Diversity of fungi and fungus-like organisms in tropical forests / edited by Steven L. Stephenson, Samantha Chandranath Karunarathna, Changlin Zhao, Arun Kumar Dutta.
Description: First edition. | Boca Raton, FL : CRC Press, 2026. | Includes bibliographical references and index.
Identifiers: LCCN 2026010163 (print) | LCCN 2026010164 (ebook) | ISBN 9781041088301 hardback | ISBN 1041088302 hardback | ISBN 9781041085904 paperback | ISBN 1041085907 paperback | ISBN 9781003647140 ebook
Subjects: LCSH: Forest fungi—Tropics | Forest fungi—Tropics—Classification | Forest fungi—Therapeutic use—Tropics | Forest fungi—Economic aspects—Tropics
Classification: LCC QK603.2 .D58 2026 (print) | LCC QK603.2 (ebook)
LC record available at https://lccn.loc.gov/2026010163
LC ebook record available at https://lccn.loc.gov/2026010164

ISBN: 978-1-041-08830-1 (hbk)
ISBN: 978-1-041-08590-4 (pbk)
ISBN: 978-1-003-64714-0 (ebk)

DOI: 10.1201/9781003647140

Typeset in Times
by Apex CoVantage, LLC

Contents

Preface

Tropical forests cover approximately 6% of the land surface of the earth and slightly more than 30% of all forested areas on the entire planet. Tropical rainforests, perhaps the best-known type of tropical forest to most people, are characterized by the presence of tall, broadleaf trees that form a continuous canopy. Except for rare instances of monodominance by a single tree species, diversity of trees and other vascular and nonvascular plants in a tropical rainforest is very high. The richness of this plant diversity is indicative of an increased number of fungi and fungus-like organisms compared to any other terrestrial ecosystem.

However, studies of fungi and fungus-like organisms in tropical rainforests and other types of tropical forests were significantly sparse until recently. It is well known that both groups of these organisms play essential roles in all types of terrestrial ecosystems. Some of these roles, such as the decomposition of dead plant material by fungi, are readily apparent, since some of the fungi involved produce fruiting bodies that are visible to the naked eye. Most trees depend on mycorrhizal associations with fungi, while parasitic and pathogenic fungi diminish the fitness of their plant hosts, including trees, potentially leading to host mortality. Macroscopic fruiting bodies of fungi are fed upon by various animals, and certain species are harvested and consumed by humans. However, the vast majority of fungi associated with tropical forests are microfungi, which do not produce conspicuous fruiting bodies and often remain unnoticed.

Much less obvious is the role that the fungus-like eumycetozoans (dictyostelids and myxomycetes) play in tropical forests. These two groups (commonly referred to as slime molds) are not true fungi, although the myxomycetes were long thought to be. Eumycetozoans are generally abundant in soils and (along with nematodes) are involved in what is known as the "microbial loop" by microbiologists. In brief, the roots of plants (including trees) exude nutrients into the soil. These nutrients are then taken up by soil bacteria, resulting in an increase in bacterial biomass. If it were not for the presence of bacteria-consuming organisms in the soil, nutrients would be locked up in this bacterial biomass and thus no longer available for use at higher trophic levels. Although not generally appreciated, eumycetozoans are among the most important of the organisms that feed upon bacteria, thereby maintaining a natural balance in soils. The myxomycetes also occur in association with decaying wood and other types of dead plant material. Many species produce fruiting bodies that are visible to the naked eye and can be collected and curated in the same manner used for macrofungi.

The purpose of this book is to provide a comprehensive overview of fungi and fungus-like organisms in tropical forests throughout the world. Collectively, the contributing authors of the chapters in the book have actively carried out research in every region of the world where tropical forests occur. The countries represented by these authors include Australia, Brazil, Colombia, Costa Rica, Guyana, India, Kenya, Paraguay, Sri Lanka, Thailand, the Philippines, and the United States. The organisms discussed include a variety of fungi and fungus-like organisms, ranging from microscopic forms such as dictyostelids to macrofungi in the divisions Ascomycota and Basidiomycota.

About the Editors

Steven L. Stephenson retired from the University of Arkansas in 2023, where he was a research professor in the Department of Biological Sciences. He received his Ph.D. from Virginia Polytechnic Institute and State University and taught at Fairmont State University for 27 years before moving to the University of Arkansas. During his academic career, he has been a senior Fulbright scholar at Himachal University in India, a visiting scientist at the Australian Antarctic Division, and the William Evans visiting fellow at the University of Otago in New Zealand. He also has received Fulbright Specialist awards to India and Vietnam. Stephenson has carried out research on eumycetozoans and fungi in the forests of India, Australia, New Zealand, Thailand, Peru, Ecuador, Puerto Rico, and Costa Rica. He is the author or coauthor of more than 500 papers and 20 books on eumycetozoans, fungi, and various aspects of ecology.

Samantha Chandranath Karunarathna is a professor at the Center for Yunnan Plateau Biological Resources Protection and Utilization in the College of Biological Resource and Food Engineering of Qujing Normal University, Qujing, Yunnan, People's Republic of China. Karunarathna has vast experience in fungal taxonomy, phylogeny, and their economic importance. He has published over 300 SCI papers related to fungal taxonomy and phylogeny. He also received Yunnan Talent and Friendship Awards in 2022 and was recognized as one of the world's top 2% scientists in 2022, 2023, 2024, and 2025, according to rankings by Stanford University.

Changlin Zhao works in the College of Forestry at the Southwest Forestry University of the People's Republic of China. Currently, his research is focused on the taxonomy and phylogeny of wood-decaying fungi (Basidiomycota). He received a Ph.D. degree from Beijing Forestry University (2016), studied evolutionary biology at Harvard University as a joint Ph.D. during 2014 and 2015, and worked at Tsinghua University as a senior visiting scholar from 2021 to 2022. He is the first author or corresponding author of more than 150 publications and five books (as first editor) in mycology.

Arun Kumar Dutta is currently working as an assistant professor of botany at Gauhati University, Assam, India. Prior to this position, he worked at West Bengal State University as a DST-Inspire Faculty Fellow. Dutta has received numerous prestigious awards, including the ISCA Young Scientist Award for 2017–2018 given by the Indian Science Congress Association (Section: Plant Sciences) and the DST-Inspire Faculty Award of the Department of Science and Technology, India, in the fields of plant, animal, and agricultural sciences (2018). He earned his Ph.D. in botany with a specialization in mycology from the University of Calcutta, West Bengal, India, in 2017. Dutta began his research career by investigating the diversity of marasmioid and gymnopoid fungi of West Bengal, utilizing both morphological and molecular characteristics. Currently, his research focuses on the molecular systematics of wild edible macrofungi in Eastern India and examining their potential as nutraceuticals. To date, he is the author of 100 peer-reviewed papers in national and international journals.

About the Editors

Contributors

Krishnendu Acharya
Department of Botany
University of Calcutta, Kolkata
West Bengal, India

Carlos Rojas Alvarado
Universidad de Costa Rica
San Pedro de Montes de Oca
Costa Rica

Denis Cristina Benjumea-Aristizábal
Instituto de Biología
Facultad de Ciencias Exactas y Naturales
Universidad de Antioquia-UdeA
Medellín, Antioquia, Colombia

Michelle G. Campi
Universidad Nacional de Asunción
Laboratorio de Recursos Vegetales, area Micología
San Lorenzo, Paraguay

Laise de Holanda Cavalcanti
Department of Mycology
Federal University of Pernambuco
Recife, Pernambuco, Brazil

Ilannya A. Chamorro-Martínez
Universidad de Sucre
Sincelejo, Sucre, Colombia

Pinaki Chattopadhyay
Department of Botany
Gauhati University
Guwahati, Assam, India

Nikki Heherson A. Dagamac
Department of Biological Sciences
University of Santo Tomas
Manila, Philippines

Diogo Henrique Costa de Rezende
Cell Biology Laboratory
Federal University of Ceara
Fortaleza, Ceará, Brazil

Thomas Edison E. dela Cruz
Department of Biological Sciences
University of Santo Tomas
Manila, Philippines

Buddhika P. Dharmasena
Department of Biosystems Technology
Sabaragamuwa University of Sri Lanka
Belihuloya, Sri Lanka

Elisandro Ricardo Drechsler-Santos
Laboratory of Mycology
Federal University of Santa Catarina, Florianópolis
Santa Catarina, Brazil

Arun Kumar Dutta
Department of Botany
Gauhati University
Guwahati, Assam, India

Ana Esperanza Franco-Molano
Instituto de Biología
Facultad de Ciencias Exactas y Naturales
Universidad de Antioquia-UdeA
Medellín, Antioquia, Colombia

Alvaro Alexander Dávila Giraldo
Universidad del Tolima
Ibagué, Tolima, Colombia

Lina Rocío Dávila Giraldo
Instituto de Biología
Facultad de Ciencias Exactas y Naturales
Universidad de Antioquia-UdeA
Medellín, Antioquia, Colombia

Salomé Gómez-Gómez
Universidad Nacional de Colombia
Bogotá D.C., Colombia

Nataly Gómez-Montoya
Universidad de Antioquia
Campus El Carmen de Viboral-Kilometro 6, vía Rionegro la Ceja
El Carmen de Viboral, Antioquia, Colombia

Terry W. Henkel
Department of Biological Sciences
California State Polytechnic University-Humboldt,
Arcata, California, USA

Melissa Mardones Hidalgo
Universidad de Costa Rica
San Pedro de Montes de Oca
Costa Rica

Laura Vélez Jaramillo
Instituto de Biología
Facultad de Ciencias Exactas y Naturales
Universidad de Antioquia-UdeA
Medellín, Antioquia, Colombia

Samantha Chandranath Karunarathna
College of Biology and Food Engineering
Qujing Normal University
Qujing, Yunnan, China

Jaturong Kumla
Office of Research Administration
Chiang Mai University
Chiang Mai, Thailand

Lucas Leonardo-Silva
Laboratory of Basic and Applied Mycology and Scientific Dissemination
University of Goiás State
Anápolis, Goiás, Brazil

Yu Li
Engineering Research Center of Edible and Medicinal Fungi
Ministry of Education, Jilin Agricultural University
Changchun, Jilin Province, China

Pu Liu
Engineering Research Center of Edible and Medicinal Fungi
Ministry of Education, Jilin Agricultural University
Changchun, Jilin Province, China

Sittie Aisha B. Macabago
Department of Biological Sciences
University of Santo Tomas
Manila, Philippines

Garumuni Dilrukshi Nadeeshani Menike
Faculty of Science
Chiang Mai University
Chiang Mai, Thailand

Yanine E. Maubet
Universidad Nacional de Asunción
Facultad de Ciencias Exactas y Naturales
San Lorenzo, Paraguay

Viviana Motato-Vásquez
Universidad Nacional de Colombia
Instituto de Ciencias Naturales
Ciudad Universitaria, Bogotá, Colombia

Juliana Moura-Mendes
Universidad Nacional de Asunción
Centro Multidisciplinario de Investigaciones Tecnológicas
San Lorenzo, Paraguay

Mary Nyawira Muchane
Botany Department
National Museums of Kenya
Nairobi, Kenya

Mudugama Vidanage Sachini Sithma Mudugama
Department of Plant Science
Faculty of Agriculture, Rajarata University of Sri Lanka
Anuradhapura, Sri Lanka

Yulena S. Osorio-Navarro
Universidad de Sucre, Sincelejo
Sucre, Colombia

Paola Andrea Zapata Ocampo
Universidad CES
Medellín, Antioquia, Colombia

Melissa H. Pecundo
Research Center for the Natural and Applied Sciences
University of Santo Tomas
Manila, Philippines

Prakash Pradhan
West Bengal Biodiversity Board, Salt Lake
Kolkata, West Bengal, India

Bhagya M. Premarathne
National Institute of Fundamental Studies (NIFS)
Hanthana, Kandy, Sri Lanka

Tikka Dewage Chamarika Priyadarshani
Department of Plant Science
Rajarata University of Sri Lanka
Anuradhapura, Sri Lanka

A. K. Hasith Priyashantha
Department of Biology
Faculty of Science, Chiang Mai University
Chiang Mai, Thailand

Leela Maya Rizal
University of Queensland
CSIRO, Australia

Ana Cristina Bolaños Rojas
Universidad del Valle
Ciudad Universitaria Meléndez
Valle del Cauca, Colombia

Anirban Roy
West Bengal Biodiversity Board, Salt Lake
Kolkata, West Bengal, India

Carolina Ríos Sarmiento
Universidad de Antioquia-UdeA
Medellín, Antioquia, Colombia

Steven L. Stephenson
Department of Biological Sciences
University of Arkansas
Fayetteville, AR, USA

Nakarin Suwannarach
Office of Research Administration
Chiang Mai University
Chiang Mai, Thailand

Entaj Tarafder
Qujing Normal University
Qujing, Yunnan
People's Republic of China

Kasun M. Thambugala
Genetics and Molecular Biology Unit
Faculty of Applied Sciences
University of Sri Jayewardenepura
Nugegoda, Sri Lanka

Isuru Sakbo Uyangoda
Centre for Plant Materials and Herbal Products Research
University of Sri Jayewardenepura
Nugegoda, Sri Lanka

Julieta Carranza Velázquez
Universidad de Costa Rica
San Pedro de Montes de Oca
Costa Rica

Solange Xavier-Santos
Laboratory of Basic and Applied Mycology and Scientific Dissemination
University of Goiás State
Anápolis, Goiás, Brazil

Cristian Zambrano-Forero
Universidad del Tolima
Ibagué, Tolima, Colombia

Zhaojuan Zhang
Engineering Research Center of Edible and Medicinal Fungi
Ministry of Education, Jilin Agricultural University
Changchun, Jilin Province, China

Changlin Zhao
College of Forestry
Southwest Forestry University
Kunming, China

Jing Zhao
Engineering Research Center of Edible and Medicinal Fungi
Ministry of Education, Jilin Agricultural University
Changchun, Jilin Province, China

Yue Zou
Engineering Research Center of Edible and Medicinal Fungi
Ministry of Education, Jilin Agricultural University
Changchun, Jilin Province, China

1 Introduction to Tropical Forests and Their Associated Fungi and Fungus-Like Organisms

Steven L. Stephenson and Entaj Tarafder

1.1 INTRODUCTION

Tropical rainforests are among the most biologically diverse ecosystems in the world. They have generally moist conditions and occur between approximately 23.5 degrees north and south latitude, between the Tropic of Cancer and the Tropic of Capricorn. The high volume of annual rainfall is regularly distributed throughout the year in tropical rainforests (Artaxo et al. 2022; Aguirre-Gutiérrez et al. 2025). In contrast, monsoon forests are dominated by strongly differentiated wet and dry seasons, and dry tropical forests receive much less rainfall per year (He et al. 2024; Muehleisen et al. 2024). Lowland rainforests have been the most extensively researched tropical forests by mycologists historically, but there are other ecologically significant types, such as the montane tropical rainforests and cloud forests that occur on tropical mountain peaks and ridgetops (Abera et al. 2024; Wang and Huang 2024).

Tropical forests blanket approximately 6% of the terrestrial land surface of the earth (Figure 1.1). They are found in the southernmost region of North America, the Caribbean, Central America, northwestern South America, the central region of Africa, the island of Madagascar, South Asia, the East Indies, the northern tip of Australia, and numerous islands in the Pacific Ocean (Casais et al. 2024; Chang et al. 2025). The structure and composition of the vegetation of tropical forests also vary tremendously from place to place. Mainland tropical forests are characterized by high diversity of plants, animals, and microbes and high tree diversity, a trend that is generally less pronounced on oceanic islands (Srinivasan 2024).

Tropical forests are old and have an ancient evolutionary history. The earliest universal tropical forests were present 300 million years ago and thus were in existence during the late Carboniferous Period and survived well into the subsequent Permian Period. They were the "coal swamp" forests, and their legacy is the coal, oil, and natural gas deposits of today. The most striking and largest plants that could be found in such swamp forests were tree-sized lycopods, which grew to heights of 50 meters and had basal trunk diameters of 1 meter on a regular basis (Li et al. 2023; Pal et al. 2023; Stephenson et al. 2024; Ragon et al. 2025).

Members of the genus *Lepidodendron* are the best-known tree-sized lycopods. *Lepidodendron* had a straight, tall, and often unbranched trunk with a crown of branches at the top that bore long grass-like leaves more than half a meter in length. The trunk was overlaid with diamond-shaped leaf scars on its surface, and the "scale tree" common name arises because of this. It is thought that *Lepidodendron* lived in the poorly drained areas of the coal swamp forests. Also present in the coal swamp forests were members of the genus *Calamites*, tree-form plants 30 meters tall (Thomas and Cleal 2018; Wachtler 2023).

DOI: 10.1201/9781003647140-1

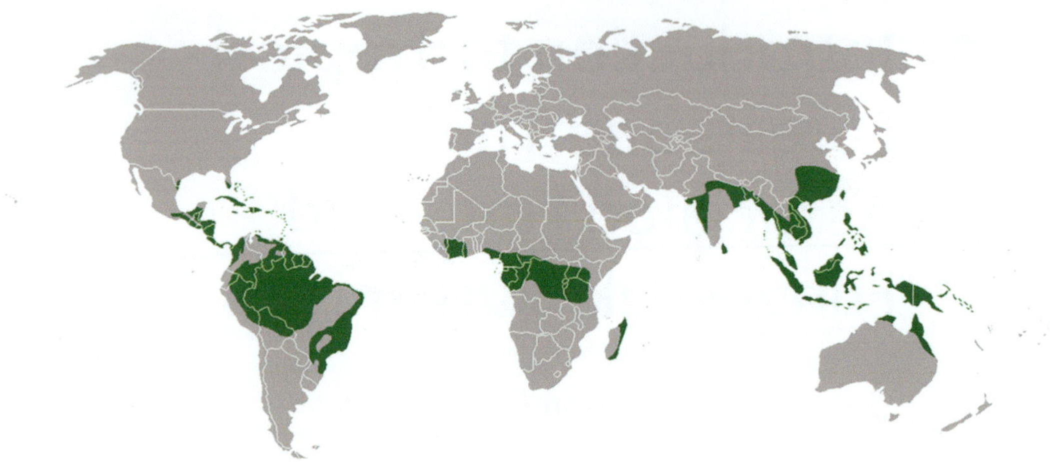

FIGURE 1.1 Distribution of tropical forests throughout the world.

Calamites belonged to the same plant family as extant *Equisetum*. Like *Lepidodendron*, *Calamites* apparently occupied poorly drained habitats. In addition to *Lepidodendron*, other characteristic flora of those ancient tropical forests included the seed ferns (the first plants to evolve seeds), true ferns, and the earliest of the primitive gymnosperms (Azevedo-Schmidt et al. 2024; Sun et al. 2024; Wang et al. 2024). Although the tree-sized plants are typically of prime interest, there were also plants in these ancient forests that had the type of growth forms today that would be recognized to represent herbaceous plants and shrubs and even some liana-like ones, such as members of the family *Sphenophyllum* (De Felice et al. 2019; D'Antonio et al. 2025). The overall image that emerges is that these early forests would have supplied much of the same microhabitats and substrates that fungi and fungus-like organisms still inhabit today. By the close of the Permian Period 250 million years ago, the tree-sized plants that dominated these ancient forests had been mostly replaced by gymnosperms (D'Antonio et al. 2025).

Recent conifers, the widespread and most abundant gymnosperms of today, appear for the first time in the Permian Period. Early Cretaceous Period fossil angiosperms are known from about 133 to 140 million years ago. For the rest of the Cretaceous (to about 66 million years ago), angiosperms diversified very rapidly, and by about 100 million years ago, numerous woody angiosperms existed. Late Cretaceous fossil wood is fairly common.

The angiosperms are the most diverse and widespread seed plants on our planet today. They are most diversified in the tropical rainforests, and in conjunction with the non-vascular plants, they are the dominant vegetation biota of such rainforests (Zuntini et al. 2024; Ding et al. 2025; Lin et al. 2025). Thus, the stage was set for the complex fungal communities of the present day. The diversity and ecological value of the fungi and fungus-like organisms that have co-evolved with these plants and underpin the functioning of tropical forests today will be discussed in this chapter.

1.2 FUNGUS-LIKE ORGANISMS IN TROPICAL FORESTS

The myxomycetes (also called plasmodial slime molds or myxogastrids) and dictyostelids (also called "cellular slime molds) are two groups of eumycetozoans found in all types of terrestrial ecosystems examined to date (Liu et al. 2025). Both groups are especially common in forests. The primary habitat for dictyostelids is the soil-litter interface of soils, but some species are known to occur in the soil-like material (canopy soil) that develops beneath vascular and nonvascular epiphytes on the trunks and branches of trees (Zhang et al. 2024). Myxomycetes are associated with numerous habitats. These include decaying coarse woody debris, ground litter, the bark of living trees, lianas, inflorescences of large herbaceous plants, aerial litter, and soil (Stephenson and Stempen 1994). In both dictyostelids and myxomycetes, the vegetative amoeboid stages in the life cycle feed mostly upon bacteria. Myxomycetes are capable of feeding upon other eukaryotic microbes, and a few species are known to feed upon the mycelia and fruiting bodies of macrofungi (Stephenson 2023).

Most studies of myxomycetes and dictyostelids have been carried out in temperate forests, especially those of eastern North America and western Europe (Chen et al. 2024). Dictyostelids were unknown to science until 1859, and there are no fossil records of this group. Tropical forests remain understudied, and limited data for dictyostelids were available until Cavender and Raper (1968) published the results of a survey they carried out in tropical and subtropical areas of North and Central America. Nine species were recovered from tropical rainforests in Costa Rica. The three most abundant dictyostelids were *Dictyostelium mucoroides*, *D. purpureum*, and *Heterostelium pallidum*. All three species are common and widespread throughout the world and are not restricted to tropical forests. The numbers of clones of dictyostelids per gram of soil (which ranged from 50 to 600) were low for samples collected in tropical forests when compared to values commonly reported for temperate forests. Since this initial publication, a number of additional studies have been carried out in tropical forests throughout the world. These studies will be reviewed in Chapter 9 of this book.

In contrast to dictyostelids, myxomycetes do possess a fossil record, albeit a very limited one, derived from specimens preserved in amber. The oldest of these is a fossil of *Stemonitis* from the mid-Cretaceous, about 100 million years ago. Except for its age, this fossil could easily have been assigned to a still extant species. This would suggest that myxomycetes may very well have been present in even the most ancient of forests (Rikkinen et al. 2019). The first specimens of myxomycetes for which we have published records were collected from the early eighteenth to the early nineteenth centuries by Europeans, whose primary focus was on other groups of organisms, particularly fungi (Waterfield 2004; Stephenson 2023). Since myxomycetes were considered to be fungi at the time, it is not surprising that the records of myxomycetes appeared along with fungi in species lists and publications. The first published records of myxomycetes from tropical forests appear to be those of Montagne (1837) from Brazil, Chile, Cuba, and Peru. However, it is likely that myxomycetes were included within lists of fungi at some undetermined earlier date (Rojas and Doss 2013; Stephenson 2023).

Due to their perceived lack of economic impact, non-pathogenic nature, and general obscurity, myxomycetes have historically attracted limited research funding, particularly for field-based surveys (Moroz et al. 2024). However, that changed in 1997, when the National Science Foundation of the United States funded a three-year project titled "Studies of Neotropical Myxomycetes." Funding for the project was extended for an additional three years in 2001. The overall objective of the project was to carry out surveys for myxomycetes in tropical forests at a number of localities throughout the Neotropics (i.e., the New World tropics) of Puerto Rico, Mexico, Costa Rica, and Ecuador (Stephenson 2023). These localities encompassed a wide range of tropical forest types, ranging from tropical dry forests to high-elevation cloud forests. Since this project ended, additional surveys have been carried out, the most important of which was the worldwide "Global Biodiversity of Eumycetozoans" project funded by the National Science Foundation. As a result of these efforts, the myxomycetes of tropical forests are now relatively well known, as will be described in Chapter 7.

1.3 FUNGI IN TROPICAL FORESTS

High levels of biodiversity, including the kingdom Fungi, are well known in tropical forests (Adnan et al. 2022). A nearly incomprehensible number of ecological niches are created for fungi by the intricate vertical stratification, numerous microhabitats, and wide variety of plant hosts (Stallman et al. 2024) (Figure 1.2). Given that many fungal guilds have essential, complementary roles in maintaining the environment, this diversity is both taxonomic and functional (Díaz 2024). These

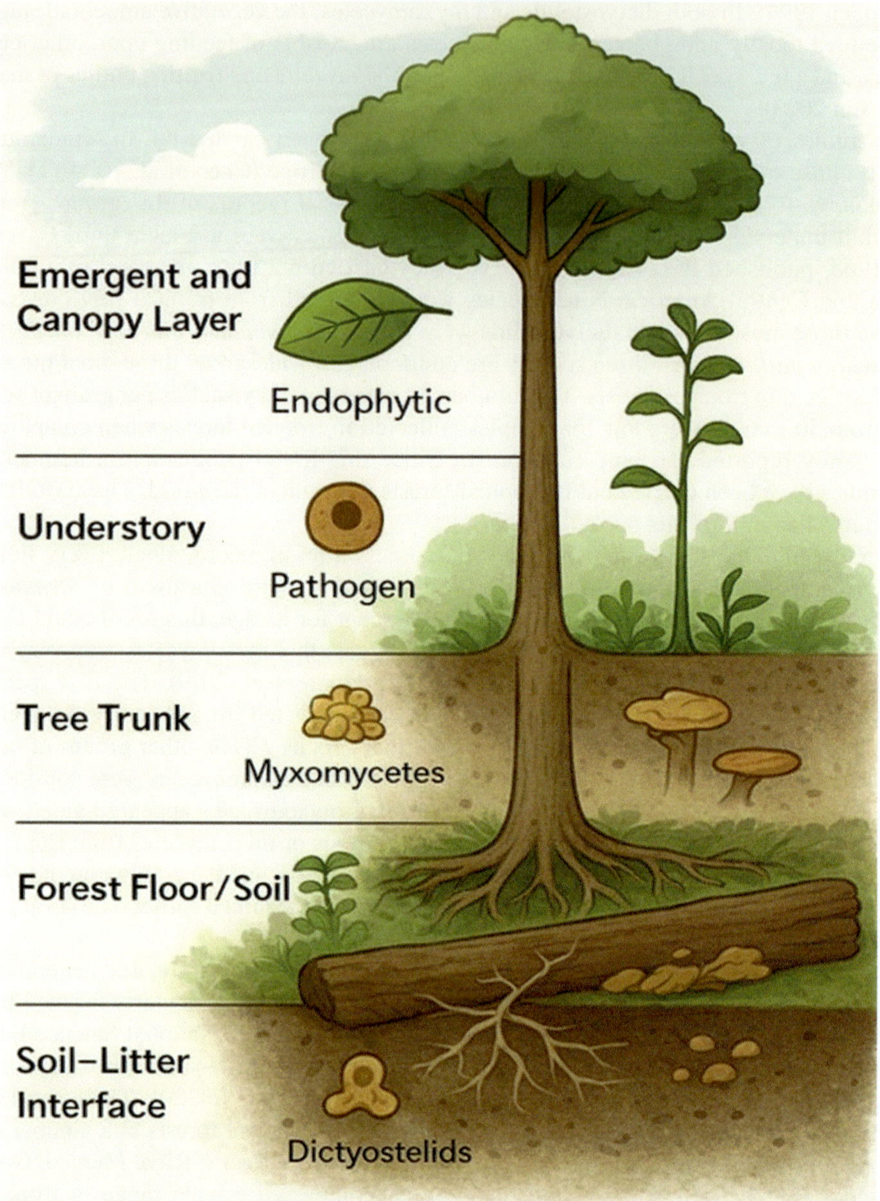

FIGURE 1.2 The complex structure of the forest creates a multitude of microhabitats, supporting an immense diversity of fungi and fungus-like organisms.

fungal interactions are not only beneficial but also frequently essential for ecosystem stability and plant survival due to the difficult edaphic conditions found in the tropics, which frequently include acidic, nutrient-poor, and heavily leached soils (Adnan et al. 2022).

1.3.1 Mycorrhizal Fungi

A vast majority of tropical plants are associated with mycorrhizal fungi that colonize the root systems of these plants in different ways to achieve a critical root extension for nutrient uptake (Favre-Godal et al. 2020). This symbiosis forms the framework of the tropical nutrient cycle by enabling plants to thrive in soils deficient in available phosphorus and nitrogen (Li et al. 2021). By increasing the absorptive surface area of root systems, these fungi expertly facilitate the uptake of limiting nutrients, for which they trade carbohydrates of plant origin (Li et al. 2021). The primary symbiosis in tropical lowland forests is represented by the arbuscular mycorrhizal fungi (AMF). AMF penetrate root cortical cells and then form tree-like structures called "arbuscules" as the interface for nutrient exchange with their plant host (Favre-Godal et al. 2020). This symbiosis influences plant fitness very significantly through enhancement of seedling survival, increased growth rates, and greater resilience to environmental stress such as drought (Li et al. 2021).

More prevalent in the temperate zone are ectomycorrhizal fungi (ECM), which also form ecologically crucial roles in certain tropical forest ecosystems. These typically form the dominant tree-fungus associations in the Dipterocarpaceae-dominated rainforests of Southeast Asia and with many important leguminous trees (Favre-Godal et al. 2020). Instead of penetrating root cells, ECM fungi form a dense sheath (or mantle) around root tips and a network of hyphae among the cells, creating a different but equally effective interface for nutrient exchange (Li et al. 2021). Among these is a further extraordinary and extremely specialized type of symbiosis called "orchid mycorrhizae." The extremely diverse orchid family of the tropics relies entirely on these fungal partners for propagation. Orchid seeds are very small, have no energy reserves, and proceed to germination and initial seedling development almost completely dependent on carbon and nutrients supplied by their specialized mycorrhizal fungi (Li et al. 2021; Favre-Godal et al. 2020). This central phenomenon seriously highlights the extent of interdependency that must exist at that level in order to support the establishment of these quite iconic tropical plants (Chauhan and Attri 2024).

1.3.2 Pathogenic Fungi

Mutualistic fungi are the ultimate nutrient suppliers for food for plants, while pathogenic fungi are quite antagonistic but equally important factors in the structuring and biodiverse scenarios of tropical forests (Peña and Tibbett 2024). Pathogenic fungi are like natural regulators on a large scale for plant community dynamics, as they often target particular host species. Their exclusive infection of a competitive host limits dominance by a common species and thus provides a competitive opportunity for those species that are rare (Nemetschek et al. 2024). Among the most important factors that regulate this are the types of plant pathogens involved as well as how and when they affect the host plant. For example, disease-causing fungi such as *Colletotrichum* cause anthracnose (Zakaria 2021). The water mold *Phytophthora*, a devastating oomycete, and others in the deep and long list of rust fungi belonging to the genus *Puccinia* exert their most influence during the very vulnerable seedling stage, where mortalities commonly occur (WHO 2022). These fungi have uniform density-dependent mortality patterns, as was presumed long before the Janzen–Connell hypothesis was proposed. The latter suggested that seeds and seedlings in close proximity to a parent tree or at high densities are more likely to die from disease, thus preventing any one species from monopolizing the landscape (Nemetschek et al. 2024).

Pathogen impact goes beyond plants and can heavily influence the population dynamics of animals. The very detrimental impact of the pathogen *Batrachochytrium dendrobatidis* speaks volumes. This pathogen infected the host through the skin and has thus been implicated as a major factor driving drastic declines and extinctions of numerous amphibians in Central and South America (Scheele et al. 2019; Casalini et al. 2024). The seriousness of the impact of this fungus reinforces the idea of how far-reaching and sometimes catastrophic the effects of fungal pathogens can be on the stability of tropical ecosystems and global biodiversity (Zakaria 2021).

1.3.3 Endophytic Fungi

Beyond the roots and the visible signs of disease, a hidden world of fungal diversity exists within the healthy tissues of the tropical plants themselves. This hidden diversity is represented by the endophytic fungi that live symbiotically within leaves, stems, and other plant parts without producing any visible disease symptoms (Majhi et al. 2025). The diversity of these cryptic communities in tropical forests is staggering; a single leaf often contains several dozen distinct fungal taxa, making an entire forest a great reserve in which there is tremendous undocumented microbial diversity (Salehi and Safaie 2024). Far from being passive residents, endophytes form a major line of defense for their plants. They also significantly enhance the tolerance of the host to a range of both biotic and abiotic stresses, including herbivory, drought, and pathogen attack, usually by providing a cocktail of bioactive compounds (Majhi et al. 2025). This chemical arsenal, evolved from the constant "arms race" in the tropics, has also made endophytes prolific sources of entirely new secondary metabolites with great pharmaceutical and industrial potential.

1.4 ECOLOGICAL ROLES AND INTERACTIONS

It is not always readily apparent, but fungi play a major role in the functioning of terrestrial ecosystems, and such things as nutrient cycles could not operate without fungi. Their importance in carbon sequestration cannot be overstated. Moreover, numerous different types of interactions occur between fungi and other organisms, and the products that fungi produce remain understudied.

1.4.1 Nutrient Cycling

Fungi in tropical ecosystems occupy the role of major decomposers and carry out the decomposition of recalcitrant compounds, such as lignin and cellulose, that cannot be processed by other organisms. By secreting potent extracellular enzymes, fungi convert litter and woody debris from complex organic matter to simple forms, releasing important nutrients into the soil. This critical biochemical process keeps the nutrient resources in a continuously available state for plant uptake, thus sustaining directly the primary productivity characteristic of tropical forests (Xu et al. 2025; Dou et al. 2025).

1.4.2 Plant-Fungi Mutualisms

These mutualistic associations between fungi and plants underpin the structural and functional vitality of forests. The most widespread of them, AMF, extend far beyond the root zone of their host plants and create a vast hyphal network, thus substantially increasing the soil volume explored for water and important nutrients, such as phosphorus and nitrogen. This symbiosis not only enhances plant nutrition but also confers considerable resilience on the host against environmental stresses, particularly drought and pathogen attack (Umer et al. 2025; Torres et al. 2024).

1.4.3 Role in Carbon Sequestration

As a result of their role in decomposition and soil organic matter formation, the activities of fungi affect the global carbon cycle. While decomposition releases carbon dioxide into the atmosphere, fungi contribute to long-term storage of carbon through the transformation of plant detritus into stable soil organic matter or the formation of fungal residues that endure in the soil (Figure 1.3). The balance between such processes helps to determine the capability of tropical forests to sequester atmospheric carbon, thereby making fungi essential regulators of carbon storage within those ecosystems (Prayudyaningsih et al. 2025; Kumari et al. 2025).

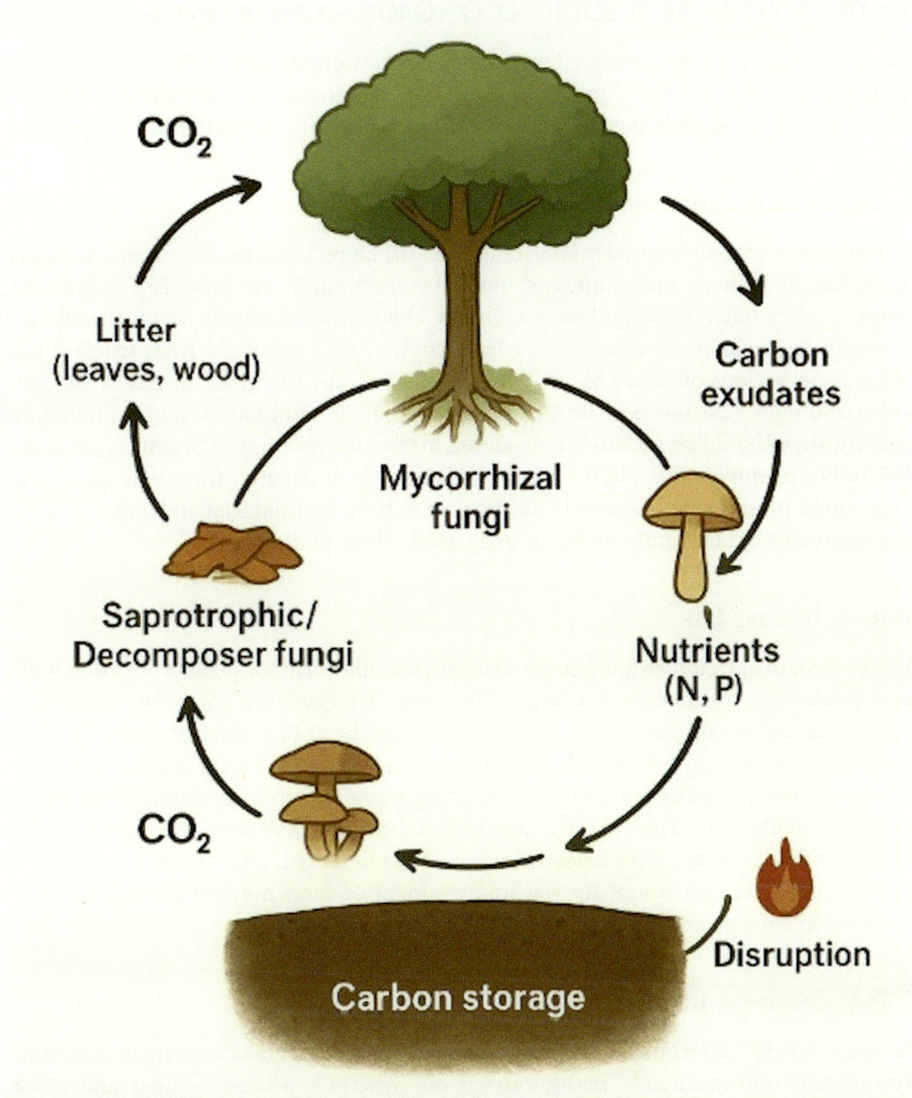

FIGURE 1.3 The central role of fungi in tropical forest nutrient and carbon cycles. This schematic illustrates how different functional groups of fungi drive essential ecosystem processes.

1.4.4 Learning From Community Dynamics

Pathogenic fungi provide an essential ecological function by helping to maintain tropical forest biodiversity through density-dependent population control. By attacking to a greater extent seeds and seedlings that occur in great densities close to parent trees, host-specific pathogens prevent the dominance of any one plant species, thereby providing a mechanism, expressed in the Janzen–Connell hypothesis, for the establishment of less common species, thereby further sustaining the remarkable species richness that characterizes tropical forests (Bonanomi et al. 2023; Bruijning et al. 2024).

1.5 BIOTECHNICAL AND SOCIO-ECONOMIC SIGNIFICANCE

Fungi have long been used in traditional medicine, and new compounds with pharmaceutical potential continue to be isolated. Other applications of fungi in agriculture and biotechnology are known to exist but have yet to be fully developed. Some of these are discussed in this section.

1.5.1 Medicinal Benefits

Tropical fungi represent a potentially enormous but still unexplored treasure house of pharmaceutical compounds. Secondary metabolites of bioactive compounds are produced overabundantly by these fungi, as they have developed very complex chemistries for their survival and competition in exceedingly lush and biodiverse forests. Endophytic fungi extracted from tropical plants have produced a wide variety of novel bioactive molecules useful for many diseases, including those that involve antibiotic resistance as well as potent anticancer compounds and immunomodulators. Consequently, they should be seriously considered in the process of drug discovery (Bao et al. 2025; Chen 2025). For example, endophytic fungi producing taxol isolated from yew (*Taxus* spp.) trees and various other promising compounds that appear to have antimalarial and antiviral activity have been investigated or are currently under investigation (Hyde et al. 2024).

1.5.2 Agricultural Use

Biological control in agriculture has gained a sustainable platform for plant production with reduced environmental impact from beneficial fungi. The genus *Trichoderma* has been shown to be widely effective against soil-borne pathogens at the commercial level (Tarafder et al. 2024). These fungi persist on plant roots, acting through a multiplicity of mechanisms at once (e.g., antagonism through mycoparasitism and competition) and also inducing plant systemic resistance, which suppresses many soil-borne pathogens. This, in turn, diminishes the reliance of the agricultural sector on heavily polluting synthetic chemical pesticides (Saldaña-Mendoza et al. 2023; Cortez Lázaro et al. 2025). The use of mycorrhizae inoculum for the improvement of crop nutrient acquisition and drought resistance in tropical agricultural systems is gaining momentum.

1.5.3 Environmental Biotechnology

Fungi have extremely important impacts on environmental protection with their metabolic capacities. Mycoremediation is already being carried out whereby white-rot fungi and other species are used to fix a host of environmental contaminants that resist other avenues for decomposition. These contaminants include such things as hydrocarbons, pesticides, or industrial colors. All these white-rot fungi possess strong ligninolytic enzyme systems capable of converting several forms of recalcitrant compounds to much less hazardous forms, making it a true boon and a crucial bioremediation tool for cleaning up soil across the tropical regions already affected by industrial pollution (Ningombam et al. 2025; Thirumalaivasan et al. 2024).

1.5.4 Food Security and Nutrition

Highly nutritious mushrooms from tropical forests also represent a primary means of improving food security while enriching the diet of people in rural areas. Mushrooms such as *Lentinula edodes* (shiitake), *Volvariella volvacea* (paddy straw mushroom), and many wild species are natural sources of proteins, vitamins, and minerals while also yielding profit in relation to local markets. Cultivating these mushrooms with agricultural waste offers a double boost in preventing malnutrition caused by poverty and reducing the hunger issue in the poor tropical regions (Hait et al. 2025; Itubochi et al. 2025).

1.5.5 Industrial Applications

Saprophytic fungi are power-packed industrial enzyme biofactories. They have superior capability of decomposing plant biomass. As a result, they secrete extracellular enzymes such as laccases, cellulases, and xylanases, all of which are crucial for lignocellulose breakdown. These extracellular enzymes have revolutionized areas of industry such as biofuels, pulp and paper, textiles, and agro-processing. Species of fungi under tropical conditions are especially suitable for producing the robustly operational thermos-stable enzyme variants needed for more efficient industrial bioprocesses (Khalil 2023; Jafarzadeh et al. 2025).

1.5.6 Socio-Economic Effects and Bioprospection

There are wider economic aspects to the potential contribution from tropical fungi beyond direct utilization of these fungi. Bioprospecting is an important activity for the discovery of new species of fungi and the compounds they produce, but concerns remain about issues such as benefit-sharing and intellectual property rights. An array of such provisions was formulated within the ethical framework of the Nagoya Protocol to allow for the equitable sharing of benefits derived from genetic resources and to determine that source countries and local communities do receive fair compensation from all their biological resources and traditional knowledge (Van Vooren 2024; CBD 2025).

1.6 THREATS TO FUNGAL DIVERSITY IN TROPICAL FORESTS

The most immediate and severe threat to fungal diversity is the rampant deforestation and changes in land use in tropical regions of the world. The conversion of primary forests into farmlands or woodlots or urban settlements actually destroys the natural habitat of all fungi, which includes a specific set of host plants, their soil networks, and decaying wood. Habitats are fragmented and destroyed, ending the fungal generations directly involved while at the same time disrupting a critical symbiotic relationship with plants that will cascade down to become biodiversity loss and ecosystem dysfunction (Global Forest Review 2025; Kamnitzer 2025).

1.6.1 Effects of Climate Change

Anthropogenic climate change represents a tremendous challenge to tropical fungi in terms of their temperature regimes, changing patterns of precipitation, and the ever-increasing frequency of extreme weather events. The altered temperature and moisture conditions, by themselves, could affect the physiology, reproduction, and survival of tropical fungi. Temperature and moisture changes may displace fungal phenology (timing of sporocarp formation) and species distributions and desynchronize the already tenuous timing of their fundamental interactions with plant hosts and animal dispersers that bind the ecological networks together (Salazar-Hamm and Torres-Cruz 2024; Williams et al. 2024).

1.6.2 Pollution and Environmental Contaminants

Increasing rates of pollution of air and soil, including nitrogen deposition from agricultural runoff and industrial activities, have emerged as the next defining threats for fungal communities. At elevated nitrogen levels, ecological balance by slow-growing species of nitrophilic fungi is disrupted, while mycorrhizal fungi, which importantly support tree health, are suppressed. Heavy metals from mining and industrial activities are known to pollute the environment. Fungal mycelia accumulate these heavy elements, with negative ramifications for reproductive success and shifts in community composition by undermining their ecosystem functioning (Singh and Singh 2024; Ningombam et al. 2025).

1.6.3 Overexploitation and Unsustainable Harvesting

The exploding demands of commerce for edible and medicinal mushrooms have led to "unsustainable harvesting" in most parts of the tropics. Among the targets for harvesting that have gained immense popularity are such fungi as chanterelles, truffles, and some medicinal polypores. The depressing truth is that these mushrooms are harvested at rates that wipe out the natural regeneration capability of the species involved. This overharvesting not only poses a threat to the target species but also causes disruption in patterns of spore dispersal and the processes that are involved in cycling of nutrients, thus taking center stage in their impact upon forest regeneration and health (Jacob et al. 2025; Premarathne et al. 2025).

1.6.4 Invaders and Pathogens

Prime examples of non-native invading species are fungal and oomycete pathogens, which often act together in disrupting a natural tropical ecosystem. An example is introduced root rot pathogen *Phytophthora cinnamomi*, which potentially causes an infection that destabilizes native soil fungal communities within a very short period. Often, their introduced host plants lack natural checks in their new habitats. Consequently, *Phytophthora cinnamomi*, with overwhelming vengeance, kills vulnerable native plants, dismantles established plant-fungal mutualisms, and triggers a loss in biodiversity that imperils the whole forest ecosystem (Del Castillo-González et al. 2024; Paap et al. 2025).

1.7 CONSERVATION AND FUTURE DIRECTIONS

Conservation efforts in tropical forests rarely involve fungi. In large part, this is due to the biology of most fungi. Unless they produce fruiting bodies easily visible to the naked eye, they are difficult to detect in nature. Moreover, the presence and relative abundance of fruiting bodies reveal little about the component of the ecosystem that fungi represent or, without study, the extent of the role they are playing. However, because of their ecological importance, fungi need to be considered in management plans.

1.7.1 Introducing Fungi in Conservation

Even having the most basic ecological roles, fungi are almost never included in conservation policies or in protected area management plans. This mycological blindness constitutes a major gap in the efforts of people around the globe to protect biodiversity. There is every reason why fungi should be included in the conservation framework. There should be holistic ecosystem protection, as plant and animal diversity preservation are closely linked to fungi communities that support them. Such an approach requires recognizing fungi as explicit conservation targets, mapping key fungal areas, and training a new generation of conservation biologists in fungal identification and ecology (Stallman et al. 2024; Arévalo et al. 2025).

1.7.2 Research Gaps and Priorities

Our knowledge of tropical fungi and their biodiversity is still at a very infant stage. While high-throughput methods such as metabarcoding are now revealing tremendous biodiversity that has previously been hidden, more than 90% of all tropical fungi are yet to be formally described. A huge gap exists between the detection of these taxa and biological understanding of what they represent in the context of the overall ecosystem. A critical research priority is moving beyond inventories toward functional studies that clarify their ecologies and activities in ecosystem processes, the resilience of those fungi against environmental changes, and tracking the complicated web of interactions that allow tropical forests to exist (Niskanen et al. 2023; Mittelstrass et al. 2025).

1.7.3 Fungi Inventories and Culture Collections

Strong and accessible biological repositories serve as a foundation for both effective conservation and applied research. Systematic fungal inventories, along with well-maintained herbaria and living culture collections, hold tremendous importance. These collections play an important role for accurate taxonomic work, long-term ecological monitoring, and genetic diversity conservation. There is an immediate need to strengthen such resource bases, especially in tropical countries, through capacity building and international collaboration. Such collections also can be a tremendous resource treasure for bioprospecting for new discoveries in medicine, agriculture, and industry (Thiers 2025).

1.7.4 Policy, Education, and Citizen Science

To close the gap between what we know scientifically and what is done in society to improve the conservation of mushrooms, demand-side and supply-side policy changes must occur along with public engagement. New useful policies for the development of fungal red lists, the integration of fungi into EIA (environmental impact assessment) approaches, and legal measures for protecting rare and endangered fungi badly need to be developed. In addition, education and citizen science initiatives, offering mushroom forays into communities and digital identification platforms, would engage the public, generate valuable data on distribution, and grow a constituency for fungal conservation (Haelewaters et al. 2024; Mueller and Furci 2025; Canteiro et al. 2025).

1.8 CONCLUSIONS

Tropical forests are major centers of biodiversity in the world, and the fungi and fungus-like organisms present are major factors in maintaining the ecological complexity of this biodiversity. In this chapter, the roles these organisms have in tropical ecosystems have been discussed. These include being the primary decomposers, serving as important symbionts, their impact as strong pathogenic agents, and the fact that they are hidden engineers of biogeochemical cycles. Their indispensable roles in the cycling of nutrients, regulation of plant community composition and structure, sequestration of carbon, and maintenance of the overall species diversity all contribute to the health and resilience of tropical forests.

The fungi and fungus-like organisms that occur in tropical forests also have tremendous biotechnological and socioeconomic promise with respect to representing a source of significant drugs, sustainable agricultural solutions, and novel industrial enzymes. However, this promise dwindles each day because of human activity. Deforestation, climate change, pollution, and invasive species all pose threats to these organisms, with repercussions that extend throughout the entire ecosystem. Conserving tropical forest ecosystems should engage all disciplines of science as well as all sectors of society. Biodiversity strategies and protected area management for fungi and fungus-like organisms would represent one urgent step. The further development of systematic and functional research, the enhancement of fungal collections and inventories, and efforts to raise public awareness and the development of ethical frameworks for bioprospecting also are important. All of these are necessary for ensuring future sustainability. Conservation of tropical fungi and fungus-like organisms is therefore not just specialized knowledge but rather a fundamental prerequisite for the conservation of tropical forests themselves. Through developing a better understanding, addressing key threats, and sustainably tapping their potential, these vital organisms will be in a good position to thrive for all time, ensuring that the most irreplaceable ecosystems in the world remain healthy, biodiverse, and productive.

REFERENCES

Abera, T.A., J. Heiskanen, E.E. Maeda, M.A. Muhammed, N. Bhandari, V. Vakkari, B.T. Hailu, P.K. Pellikka, A. Hemp, P.G. van Zyl & D. Zeuss. 2024. Deforestation amplifies climate change effects on warming and cloud level rise in African montane forests. *Nature Communications* 15:6992. https://doi.org/10.1038/s41467-024-51324-7

Adnan, M., W. Islam, L. Gang & H.Y. Chen. 2022. Advanced research tools for fungal diversity and its impact on forest ecosystem. *Environmental Science and Pollution Research* 29:45044–45062. https://doi.org/10.1007/s11356-022-20317-8

Aguirre-Gutiérrez, J., et al. 2025. *Tropical Forests in the Americas Are Struggling to Keep Pace with Climate Change*. University of Oxford, Environmental Change Institute. https://www.ox.ac.uk/news/2025-03-07-tropical-forests-americas-are-struggling-keep-pace-climate-change

Arévalo, Y., M.E. Avila-Salem, P. Loján, N. Urgiles-Gómez, D. Pucha-Cofrep, N. Aguirre & C. Benavidez-Silva. 2025. Arbuscular mycorrhizal fungi in the ecological restoration of tropical forests: A bibliometric review. *Forests* 16:1266. https://doi.org/10.3390/f16081266

Artaxo, P., H.C. Hansson, L.A.T. Machado & L.V. Rizzo. 2022. Tropical forests are crucial in regulating the climate on Earth. *PLOS Climate* 1:e0000054. https://doi.org/10.1371/journal.pclm.0000054

Azevedo-Schmidt, L., E.D. Currano, R.E. Dunn, E. Gjieli, J. Pittermann, E. Sessa & J.L. Gill. 2024. Ferns as facilitators of community recovery following biotic upheaval. *BioScience* 74:322–332. https://doi.org/10.1093/biosci/biae0222

Bao, M., Y. Shi, X. Gong, Y. Guo, J. Wang, X. Chen & L. Liu. 2025. New bioactive secondary metabolites from fungi: 2024. *Mycology* 16:961–987. https://doi.org/10.1080/21501203.2025.2526772

Bonanomi, G., A. Bobrovskikh, F. Cartenì, S. Mazzoleni & F. Giannino. 2023. Adult conspecific density affects Janzen–Connell patterns by modulating the recruitment exclusion zones. *Frontiers in Plant Science* 14:1079975. https://doi.org/10.3389/fpls.2023.1079975

Bruijning, M., C.J. Metcalf & M.D. Visser. 2024. Closing the gap in the Janzen-Connell hypothesis: What determines pathogen diversity? *Ecology Letters* 27:e14316. https://doi.org/10.1111/ele.14316

Canteiro, C., A. Corrales, T. Kiers & K. Walther. 2025. Integrating fungi in policy for a sustainable future. *Current Conservation* 19(3). https://www.currentconservation.org

Casais, G., N.S. Guimarães, T.R. Cortes, J. Pescarini, P.R. de Magalhães, V. Wells, J.F. de Sousa Filho, D.J. Neves, M. Shimonovich, J.R. Olsen & E.M. de Carvalho Neto. 2024. Wildfire, deforestation and health in tropical rainforest areas: A scoping review protocol. *BMJ Open* 14:e082381. https://doi.org/10.1136/bmjopen-2023-082381

Casalini, G., A. Giacomelli & S. Antinori. 2024. The WHO fungal priority pathogens list: A crucial reappraisal to review the prioritisation. *The Lancet Microbe* 5:717–724.

Cavender, J.C. & K.B. Raper. 1968. The occurrence and distribution of Acrasiae in forests of subtropical and tropical America. *American Journal of Botany* 55:504–513. https://doi.org/10.2307/2440581

Chang, C.H., J.T. Erbaugh, P. Fajardo, L. Lu, I. Molnár, D. Papp, B.E. Robinson, K.G. Austin, M. Castro, S.H. Cheng & S. Cook-Patton. 2025. Global evidence of human well-being and biodiversity impacts of natural climate solutions. *Nature Sustainability* 8:75–85. https://doi.org/10.1038/s41893-024-01454-z

Chauhan, P. & L. Attri. 2024. Mycorrhizal associations in orchids: A review. *Asian Journal of Biological and Life Sciences* 13:278–286. https://doi.org/10.5530/ajbls.2024.13.36

Chen, J. 2025. How fungi are reshaping drug discovery: 907 novel compounds described in 2024. *MycoStories*. https://www.mycostories.com

Chen, Y., W. Song, G. Rao, Q. Dai, S. Yan & S. Chen. 2024. The occurrence of myxomycete communities in *Taxodium distichum* forests: Influence of substrates and seasons. *Microorganisms* 12:2156. https://doi.org/10.3390/microorganisms12112156

Convention on Biological Diversity (CBD). 2025. *Nagoya Protocol on Access to Genetic Resources and the Fair and Equitable Sharing of Benefits Arising from their Utilization to the Convention on Biological Diversity*. Secretariat of the Convention on Biological Diversity, Montreal, Canada. https://www.cbd.int/abs/

Cortez Lázaro, A.A., P.J. Vázquez-Medina, J.V. García-Evangelista, R.A. Cortez-Lázaro, J.L. Rojas-Paz, C.R. Pesantes-Rojas, R.W. Ocrospoma-Dueñas, S.M. Oliva-Cruz, G.A. Manes-Cangana, J.L. Romero Bozzetta & S.T. Leiva-Espinoza. 2025. Global trends in the research of *Trichoderma* secondary metabolites in sustainable agricultural bioprotection. *Frontiers in Microbiology* 16:1595946. http://doi.org/10.3389/fmicb.2025.1595946

D'Antonio, M.P., P.R. Crane, C.L. Hotton, J. Wittry & F. Herrera. 2025. Sphenophyllales from the Mazon Creek flora (Upper Moscovian: Illinois, USA). *Botanical Journal of the Linnean Society* boaf043. http://doi.org/10.1093/botlinnean/boaf043

De Felice, H.T., A.L. Decombeix & J. Galtier. 2019. Anatomy, affinities, and evolutionary implications of new silicified stems of *Sphenophyllum* Brongniart, 1828 from the early Carboniferous (Mississippian) of France and Germany. *Geodiversitas* 41:587–599. http://doi.org/10.5252/geodiversitas2019v41a14

Del Castillo-González, L., S. Soudani, N. De La Cruz-Gómez, J.A. Manzanera & M. Berrocal-Lobo. 2024. An improved method to study *Phytophthora cinnamomi* Rands zoospores interactions with host. *BMC Plant Biology* 24:508. http://doi.org/10.1186/s12870-024-05205-2

Díaz, A. 2024. Effect of forest fragmentation on fungal diversity in tropical rainforests in Argentina. *International Journal of Biology* 4:13–22.

Ding, W., D. Silvestro, R.E. Onstein, M. Wu, Z. Zhou & Y. Xing. 2025. The stepwise rise of angiosperm-dominated terrestrial ecosystems. *Biological Reviews*. http://doi.org/10.1111/brv.70039

Dou, T., K. Zhang, X. Shi, W. Liu, F. Yu & D. Liu. 2025. Crop–mushroom rotation: A comprehensive review of its multifaceted impacts on soil quality, agricultural sustainability, and ecosystem health. *Agronomy* 15:563. http://doi.org/10.3390/agronomy15030563

Favre-Godal, Q., L. Gourguillon, S. Lordel-Madeleine, K. Gindro & P. Choisy. 2020. Orchids and their mycorrhizal fungi: An insufficiently explored relationship. *Mycorrhiza* 30:5–22. http://doi.org/10.1007/s00572-020-00934-2

Global Forest Review. 2025. *Fires Drove Record-Breaking Tropical Forest Loss in 2024*. World Resources Institute, Washington, DC. https://research.wri.org/gfr/latest-analysis-deforestation-trends

Haelewaters, D., C.A. Quandt, L. Bartrop, J. Cazabonne, M.E. Crockatt, S.P. Cunha, R. De Lange, L. Dominici, B. Douglas, E.R. Drechsler-Santos & J. Heilmann-Clausen. 2024. The power of citizen science to advance fungal conservation. *Conservation Letters* 17:e13013. http://doi.org/10.1111/conl.13013

Hait, M., A.K. Bhardwaj & N.K. Kashyap. 2025. Role of edible mushroom in addressing malnutrition and food security. In S.C. Izah, M.C. Ogwu & M. Akram (eds.), *Bioactive Compounds in Edible Mushrooms*. Reference Series in Phytochemistry. Springer, Cham. http://doi.org/10.1007/978-3-031-80050-4_15

He, J., W. Li, Z. Zhao, L. Zhu, X. Du, Y. Xu, M. Sun, J. Zhou, P. Ciais, J.P. Wigneron & R. Liu. 2024. Recent advances and challenges in monitoring and modeling of disturbances in tropical moist forests. *Frontiers in Remote Sensing* 5:1332728. http://doi.org/10.3389/frsen.2024.1332728

Hyde, K.D., P. Baldrian, Y. Chen, et al. 2024. Current trends, limitations and future research in the fungi? *Fungal Diversity* 125:1–71. http://doi.org/10.1007/s13225-023-00532-5

Itubochi, C.O., J.O. Ugwuanyi & I.B. Ezea. 2025. Advancing sustainable mushroom production in developing countries: A pathway to nutritional security and economic growth. *EAS Journal of Nutrition and Food Sciences* 7:48–55. http://doi.org/10.36349/easjnfs.2025.v07i01.006

Jacob, D.E., S.C. Izah, I.D. Jacob, K.S. Daniel & U.A.I. Udoakpan. 2025. Threats, harvesting, conservation, and sustainable harvesting practices of edible mushroom. In S.C. Izah, M.C. Ogwu & M. Akram (eds.), *Bioactive Compounds in Edible Mushrooms*. Reference Series in Phytochemistry. Springer, Cham. http://doi.org/10.1007/978-3-031-80050-4_7

Jafarzadeh, S.A., K.F. Bakhshaliyeva, E.O. Iskender, P.Z. Muradov, A. Leyla & S.M. Muradova. 2025. Exploration of novel industrial enzymes from extremophilic communities for biotechnological applications. *International Journal of Agriculture and Biosciences* 14:944–954. http://doi.org/10.47278/journal.ijab/2025.066

Kamnitzer, R. 2025. Tropical forest loss hit new heights in 2024; fire a major driver in Latin America. *Mongabay*, 21 May 2025. https://news.mongabay.com/2025/05/tropical-forest-loss-hit-new-heights-in-2024-fire-a-major-driver-in-latin-america

Khalil, N.M. 2023. Recent progress on fungal enzymes. In Y.M. Rashad, Z.A.M. Baka & T.A.A. Moussa (eds.), *Plant Mycobiome*. Springer, Cham. http://doi.org/10.1007/978-3-031-28307-9_13

Kumari, S., S. Chowdhury, S. Kumar & A. Chauhan. 2025. Arbuscular mycorrhizal fungi-mediated carbon sequestration: Mechanisms, influencing factors, and future directions. *Journal of Advances in Biology & Biotechnology* 28:144–162. http://doi.org/10.9734/jabb/2025/v28i32077

Li, T., W. Yang, S. Wu, M.-A. Selosse & J. Gao. 2021. Progress and prospects of mycorrhizal fungal diversity in orchids. *Frontiers in Plant Science* 12:646325. http://doi.org/10.3389/fpls.2021.646325

Li, Y., Y. Song, X. Li, D.G. Kaskaoutis, H. Gholami & Y. Li. 2023. Disentangling variations of dust concentration in Greenland ice cores over the last glaciation: An overview of current knowledge and new initiative. *Earth-Science Reviews* 242:104451. http://doi.org/10.1016/j.earscirev.2023.104451

Lin, D., B. Shao, Z. Gao, J. Li, Z. Li, T. Li, W. Huang, X. Zhong, C. Xu, M.W. Chase & X. Jin. 2025. Phylogenomics of angiosperms based on mitochondrial genes: Insights into deep node relationships. *BMC Biology* 23:45. http://doi.org/10.1186/s12915-025-02135-9

Liu, P., Z.J. Zhang, J.H. Ge, Y. Zou, Z. Li, S. Li, Y. Li & S.L. Stephenson. 2025. Dictyostelids: The second major group of slime molds. *Mycosphere* 16:423–516. http://doi.org/10.5943/mycosphere/16/1

Majhi, P., U. Pradhan, A. Toppo & A.K. Shukla. 2025. Fungal endophytes: An insight into diversity, stress tolerance, biocontrol and plant growth-promoting potentials. *Current Microbiology* 82:283. http://doi.org/10.1007/s00284-025-04266-2

Mittelstrass, J., R. Heinzelmann, R. Eschen, M. Hartmann, Q. Kupper, S. Schneider, S. Prospero & I. Franić. 2025. Metabarcoding with Illumina and Oxford Nanopore Technologies provides complementary insights into tree seed mycobiota. *Environmental Microbiome* 20:53. http://doi.org/10.1186/s40793-025-00712-7

Montagne, C. 1837. Centurie de plantes cellulaires exotiques nouvelles. Ser. II. *Annales des Sciences Naturelles. Paris* 8:345–370.

Moroz, E.L., V.I. Gmoshinskiy, O.N. Shchepin & Y.K. Novozhilov. 2024. The systematics and phylogeny of myxomycetes: Yesterday, today, and tomorrow. *Doklady Biological Sciences* 519:356–369. http://doi.org/10.1134/S0012496624701242

Muehleisen, A.J., N.B. Schwartz, S.M. Stump & A.C. Staver. 2024. Rainfall variability and deciduous-evergreen coexistence in tropical forests. *Theoretical Ecology* 17:255–268. http://doi.org/10.1007/s12080-024-00588-2

Mueller, G. & G. Furci. 2025. *2024–2025 report of the Fungal Conservation Committee. IUCN SSC and Secretariat. Species: Annual Report of the IUCN Species Survival Commission and Secretariat 2024–2025*. Gland, Switzerland: IUCN. https://www.iucn-fungi.org

Nemetschek, D., G. Derroire, E. Marcon, M. Aubry-Kientz, J. Auer, V. Badouard, C. Baraloto, D. Bauman, Q. Le Blaye, M. Boisseaux & D. Bonal. 2024. Climate anomalies and neighbourhood crowding interact in shaping tree growth in old-growth and selectively logged tropical forests. *Journal of Ecology* 112:590–612. https://doi.org/10.1111/1365-2745.14256

Ningombam, L., T. Mana, S. Pradhan, G. Apum & Y.D. Singh. 2025. Fungal bioremediation in environmental pollution and recent strategies. *Discover Environment* 3:100. https://doi.org/10.1007/s44274-025-00267-x

Niskanen, T., R. Lücking, A. Dahlberg, E. Gaya, L.M. Suz, V. Mikryukov, K. Liimatainen, I. Druzhinina, J.R. Westrip, G.M. Mueller & K. Martins-Cunha. 2023. Pushing the frontiers of biodiversity research: Unveiling the global diversity, distribution, and conservation of fungi. *Annual Review of Environment and Resources* 48:149–176. https://doi.org/10.1146/annurev-environ-112621-090937

Paap, T., F. Balocchi & M.J. Wingfield. 2025. The root rot pathogen *Phytophthora cinnamomi*: A long-overlooked threat to the Cape Floristic Region of South Africa. *Biological Invasions* 27:110. https://doi.org/10.1007/s10530-025-03570-z

Pal, S., J.P. Shrivastava & M.S. Kalpana. 2023. Fire activity across the Cretaceous/Paleogene transition: Evidence from pyrogenic biomarkers preserved in the Mahadeo-Cherrapunji section, Meghalaya, India. *Geobiology* 21:612–628. https://doi.org/10.1111/gbi.12567

Peña, R. & M. Tibbett. 2024. Mycorrhizal symbiosis and the nitrogen nutrition of forest trees. *Applied Microbiology and Biotechnology* 108:461. https://doi.org/10.1007/s00253-024-13298-w

Prayudyaningsih, R., M. Turjaman, M. Christita, N.E. Lelana, R.S.B. Irianto, S. Antonius, S.S. Hakim, A.I. Putri, H.H. Rachmat, V.B. Arifanti, et al. 2025. Tropical fungi and LULUCF: Synergies for climate mitigation through nature-based culture (NbC). *Climate* 13:208. https://doi.org/10.3390/cli13100208

Premarathne, B.M., M.C.A. Galappaththi, N.M. Patabendige, S.C. Karunarathna, N.N. Wijayawardene, Y.A.P.K. Dayasena, K.L.W. Kumara, D.S.A. Wijesundara, A. Ediriweera & S. Madawala. 2025. A review of wild edible and medicinal mushrooms in Sri Lanka: Systematic exploration and industrial applications. *MycoAsia* 2025/01. https://doi.org/10.59265/mycoasia.2025-01

Ragon, C., C. Vérard, J. Kasparian, H. Nowak, E. Kustatscher & M. Brunetti. 2025. Comparison between plant fossil assemblages and simulated biomes across the Permian-Triassic boundary. *Frontiers in Earth Science* 13:1520846. https://doi.org/10.3389/feart.2025.1520846

Rikkinen, J., D.A. Grimaldi & A.R. Schmidt. 2019. Morphological stasis in the first myxomycete from the Mesozoic, and the likely role of cryptobiosis. *Scientific Reports* 9:19730. https://doi.org/10.1038/s41598-019-55622-9

Rojas, C. & R.G. Doss. 2013. Brief research history and status of myxomycete conservation in the Neotropics. *Brenesia* 79:37–43.

Salazar-Hamm, P. & T.J. Torres-Cruz. 2024. The impact of climate change on human fungal pathogen distribution and disease incidence. *Current Clinical Microbiology Reports* 11:140–152. https://doi.org/10.1007/s40588-024-00224-x

Saldaña-Mendoza, S.A., S. Pacios-Michelena, A.S. Palacios-Ponce, M.L. Chávez-González & C.N. Aguilar. 2023. *Trichoderma* as a biological control agent: Mechanisms of action, benefits for crops and development of formulations. *World Journal of Microbiology and Biotechnology* 39:269. https://doi.org/10.1007/s11274-023-03695-0

Salehi, M. & N. Safaie. 2024. Endophytic fungi: Secondary metabolites and plant biotic and abiotic stress management. *Frontiers in Microbiology* 15:1345210. https://doi.org/10.3389/fmicb.2024.1345210

Scheele, B.C., F. Pasmans, L.F. Skerratt, L. Berger, A.N. Martel, W. Beukema, A.A. Acevedo, P.A. Burrowes, T. Carvalho, A. Catenazzi & I. De la Riva. 2019. Amphibian fungal panzootic causes catastrophic and ongoing loss of biodiversity. *Science* 363:1459–1463. https://doi.org/10.1126/science.aav0379

Singh, V.K. & R. Singh. 2024. Role of white rot fungi in sustainable remediation of heavy metals from the contaminated environment. *Mycology* 15:585–601. https://doi.org/10.1080/21501203.2024.2389290

Srinivasan, U. 2024. Forest-degradation thresholds shape tropical biodiversity. *Nature* 631:741–742. https://doi.org/10.1038/d41586-024-02155-5

Stallman, J.K., D. Haelewaters, R.A. Koch Bach, M. Brann, S. Fatemi, P. Gomez-Zapata, D.R. Husbands, B. Jumbam, P.J. Kaishian, A. Moffitt & M.C. Aime. 2024. The contribution of tropical long-term studies to mycology. *IMA Fungus* 15:35. https://doi.org/10.1186/s43008-024-00166-5

Stephenson, M.H., B. Bomfleur, P.E. Jardine, H. Kerp, P. Blomenkemper, R. Bäumer & J.W. Schneider. 2024. Palynological variability within the Permian (Changhsingian) Umm Irna Formation (Jordan): Implications for biostratigraphy and fluid-flow character in alluvial formations. *Palynology* 48. https://doi.org/10.1080/01916122.2024.2388134

Stephenson, S.L. 2023. Past and ongoing field-based studies of myxomycetes. *Microorganisms* 11:2283. https://doi.org/10.3390/microorganisms11092283

Stephenson, S.L. & H. Stempen. 1994. *Myxomycetes: A Handbook of Slime Molds*. Timber Press, Portland, Oregon.

Sun, W., Z. Wei, Y. Gu, T. Wang, B. Liu & Y. Yan. 2024. Chloroplast genome structure analysis of *Equisetum* unveils phylogenetic relationships to ferns and mutational hotspot region. *Frontiers in Plant Science* 15:1328080. https://doi.org/10.3389/fpls.2024.1328080

Tarafder, E., W. Zeng, S.C. Karunarathna, A.M. Elgorban, H. Ma, N. Wang, X. Zeng, W. Yong & F.-H. Tian. 2024. Unveiling two new species of *Trichoderma* (Hypocreales, Hypocreaceae) that cause green mold disease on *Stropharia rugosoannulata* from Guizhou Province, China. *MycoKeys* 110:361–383. http://doi.org/10.3897/mycokeys.110.134154

Thiers, B.M. 2025. *The World's Herbaria 2023: A Summary Report Based on Data from Index Herbariorum, issue 6.0*. New York Botanical Garden, New York, USA. http://sweetgum.nybg.org/science/ih/

Thirumalaivasan, N., L. Gnanasekaran, S. Kumar, R. Durvasulu, T. Sundaram, S. Rajendran, S. Nangan & K. Kanagaraj. 2024. Utilization of fungal and bacterial bioremediation techniques for the treatment of toxic waste and biowaste. *Frontiers in Materials* 11:1416445. https://doi.org/10.3389/fmats.2024.1416445

Thomas, B.A. & C.J. Cleal. 2018. Arborescent lycophyte growth in the late Carboniferous coal swamps. *New Phytologist* 218:885–890.

Torres, M.J., G. Moreira, J.H. Bhadha & E.S. McLamore. 2024. Arbuscular mycorrhizal fungi as inspiration for sustainable technology. *Encyclopedia* 4:1188–1200. https://doi.org/10.3390/encyclopedia4030077

Umer, M., N. Anwar, M. Mubeen, Y. Li, A. Ali, M.O. Alshaharni & P. Liu. 2025. Roles of arbuscular mycorrhizal fungi in plant growth and disease management for sustainable agriculture. *Frontiers in Microbiology* 16:1616273. https://doi.org/10.3389/fmicb.2025.1616273

Van Vooren, B. 2024. The Nagoya Protocol at its 10th anniversary: Lessons learned and new challenges from "access and benefit-sharing." *Covington & Burling LLP Client Alert*, 16 October 2024. https://www.cov.com/en/news-and-insights/insights/2024/10/the-nagoya-protocol-at-its-10th-anniversary-lessons-learned-and-new-challenges-from-access-and-benefit-sharing

Wachtler, M. 2023. Calamites horsetails of the Alps in the Carboniferous. In *Carboniferous Fossil Floras from the Eastern Alps*. Dolomythos-Museum, Innichen, Italy.

Wang, H.C. & C.Y. Huang. 2024. Cross-scale assessments of the impacts and resilience of subtropical montane cloud forests to chronic seasonal droughts and episodic typhoons. *Global Change Biology* 30:e17000. https://doi.org/10.1111/gcb.17000

Wang, X., Y. Miao, Y. Ji & B. Sun. 2024. New discovery of Calamitaceae from the Cisuralian in Northwest China: Morphological evolution of strobilus. *Biology* 13:347. https://doi.org/10.3390/biology13050347

Waterfield, A. 2004. The listers and the myxomycete collections at the Natural History Museum (BM), London. *Systematics and Geography of Plants* 74:289–291. http://www.jstor.org/stable/3668497

Williams, S.L., M. Toda, T. Chiller, J.M. Brunkard & A.P. Litvintseva. 2024. Effects of climate change on fungal infections. *PLoS Pathogens* 20:e1012219. https://doi.org/10.1371/journal.ppat.1012219

World Health Organization. 2022. *WHO Fungal Priority Pathogens List to Guide Research, Development and Public Health Action*. Geneva: World Health Organization. ISBN 978-92-4-006024-1.

Xu, Y., K. Cui, X. Zhang, G. Diwu, Y. Zhu, L. Deng, Y. Zhong & W. Yan. 2025. Shifts in fungal communities drive soil profile nutrient cycling during grassland restoration. *mBio* 16:e02834-24. https://doi.org/10.1128/mbio.02834-24

Zakaria, L. 2021. Diversity of *Colletotrichum* species associated with anthracnose disease in tropical fruit crops: A review. *Agriculture* 11:297. https://doi.org/10.3390/agriculture11040297

Zhang, Z., C. Zhai, Y. Li, S.L. Stephenson & P. Liu. 2024. Slime molds (Myxomycetes) causing a "disease" in crop plants and cultivated mushrooms. *Frontiers in Plant Science* 15:1411231. https://doi.org/10.3389/fpls.2024.1411231

Zuntini, A.R., T. Carruthers, O. Maurin, P.C. Bailey, K. Leempoel, G.E. Brewer, N. Epitawalage, E. Françoso, B. Gallego-Paramo, C. McGinnie & R. Negrão. 2024. Phylogenomics and the rise of the angiosperms. *Nature* 629:843–850. https://doi.org/10.1038/s41586-024-07324-0

2 Diversity and Ecology of Macrofungi of Tropical India

Arun Kumar Dutta, Prakash Pradhan, Pinaki Chattopadhyay, Krishnendu Acharya, and Anirban Roy

2.1 INTRODUCTION

Tropical India is a highly varied region in terms of climatic and geographical characteristics, with a vast flora and fauna. Macrofungi, which represent the visible, spore-producing elements of this kingdom, remain a crucial but underrated element of ecosystem function. Macrofungi represent a type of organism that produces structures visible to the human eye (>1 cm) (Arnolds 1992). They play a crucial role as saprotrophic, mutually beneficial mycorrhizal, and parasitic organisms in forests (Arnolds 1992; Watling 1997). Macrofungi also play a crucial role in providing food, medical, and economic support to local communities (Pradhan et al. 2010; Chatterjee et al. 2012). Despite estimates of millions of species of fungi on this planet (Hawksworth 2001; Mora et al. 2011), there remains a striking deficiency of data on macrofungi in Indian tropical ecosystems. The riveting richness of this biodiversity is, above all, not diligently described, particularly within regions of high diversity of plants and animals, such as the Eastern Himalayas, which encompass 3,160 plants, all of which can be found only within this region, and this area remains inadequately surveyed from a conservation perspective (Pradhan et al. 2016). Recent studies undertaken in specific ecoregions have started to unmask this hidden diversity. For example, fungal surveys in the Darjeeling Himalayan Region recorded 98 species of macrofungi belonging to 72 genera and 47 families, of which the Russulaceae, Marasmiaceae, and Coprinaceae turned out to be the most speciose families (Pradhan et al. 2016). Saprotrophic fungi dominated with a relative abundance of 58.16%, ectomycorrhizal fungi contributed 17.34%, and parasitic fungi accounted for 10.2%. These data clearly reflect not only functional diversity but potential forest disturbance as well. In the lateritic forests of West Bengal with acidic soils, ectomycorrhizal fungi such as *A. vaginata*, other species of *Amanita*, and species of *Russula* showed very tight associations only with natural Sal (*Shorea robusta*) forests, while *Pisolithus arhizus* was associated only with exotic plantations (Pradhan et al. 2012, 2013). In the coastal belt that included Sundarbans mangrove forests, dominated by saprotrophs, different species of fungi played significant roles as key agents of nutrient cycles for similar salty, tidal habitats (Dutta et al. 2013).

Recent progress in ecological theory and analytical methods enables fungal studies to transcend mere inventories. Hierarchical environmental filtering, a well-established concept in plant and animal ecology, asserts that biodiversity is organized by a series of nested filters, ranging from regional climates to local biotic interactions (Mondal et al. 2022a). For fungi, the most important are macroclimatic filters (precipitation and temperature) that define possible species pools, edaphic and topographic filters (soil pH, texture, and elevation) that divide niches, and micro-environmental or biotic filters (host availability, wind dispersal, and UV radiation) that determine actual occurrences. Historically constrained by data scarcity, modern methodologies employing molecular identification, standardized surveys, and computational ecology now facilitate the rigorous testing of these models within assemblages of tropical fungi.

This chapter consolidates existing knowledge regarding the diversity and ecology of macrofungi within the diverse tropical ecosystems of India, with a specific focus on West Bengal. This area has a

DOI: 10.1201/9781003647140-2

short latitudinal and elevational gradient, going from the saline Sundarbans mangroves to the fertile alluvial plains, ancient lateritic plateaus, and temperate Himalayas. This gradient made it a wonderful place to test ideas about hierarchical filtering, functional trait-environment matching, and disturbance-diversity relationships. Using data from 312 transects surveyed over four consecutive monsoon seasons (2008–2025) and including 208 species divided into 12 functional groups, we used multivariate ordination, occupancy-based frequency metrics, and machine learning (LASSO and Random Forest) to measure environmental drivers from microhabitats to ecoregions.

This chapter makes an effort to offer an integrated view through morphological, ecological, and computational analyses that provide a comprehensive overview of tropical Indian macrofungi. This includes discussions on tropical Indian macrofungi from an ecological perspective, functional diversity, and spatial distribution and formulating a predictive framework based on frequency that is significant to conservation strategies. More significantly, it underlines that persistent research and conservation strategies are critical for these ecologically important but very sensitive organisms in light of ongoing habitat degradation and climate change.

2.2 MATERIALS AND METHODS

The state of West Bengal, India, lies between 27°13′15″N and 21°25′24″N latitude and 85°48′20″E and 89°53′04″E longitude and has a total area of 88,752 km² (34,267 square miles). It shares varied edaphic and climatic conditions with most other Indian ecoregions, with the exception of arid desert conditions (Pradhan et al. 2010). This makes the state suitable for a high level of macrofungal diversity, along with diversification of their ecological functions due to varied climates, soils, vegetation, and organic matter turnover (Pradhan et al. 2010).

To systematically analyze this diversity, the landscape was classified into four ecoregions that are ecologically very distinct. The coastal region, which included habitats in and around the Sundarbans, was the most southern part. It included habitats that are mudflats in some areas, estuarine channels in some areas, and dense halophytic mangrove forests that are supported by saline, silty-clay alluvial soils (Dutta et al. 2013). The soils found in this region are geologically very young, freshwater-loving, and salty, which imposes a great deal of stress upon organisms. The region has a tropical, humid, wet climatic condition that is moderated by the sea, which makes it neither too cold nor too hot (Dutta et al. 2013). The region has halotolerant lignicolous fungi that resist high moisture and the high-water level that covers large areas in a normal tide, which inundates this place frequently (Dutta et al. 2013).

The Gangetic Plains form a more northern part of this region that is very fertile, having alluvial soils that are either neutral or very mildly alkaline (Dutta et al. 2013). The region was primarily moist deciduous forests in some distant past but is a landmass of intensive agriculture, scattered orchards, and semi-natural forests at present. This area has a tropical monsoon climate with distinct seasons, scorching summers before the monsoon, and soils that can hold a lot of water, which makes it very productive. The western Lateritic region, which is the same as the Rarh Plateau, is very old geologically (Mondal et al. 2022b). It is mostly made up of acidic, ferruginous alkaline substances and oxides that don't hold water or nutrients well. This ecoregion is mostly made up of dry deciduous forests of *Shorea robusta*, which are home to obligate ectomycorrhizal fungi (Mondal et al. 2022b). Darjeeling and Kalimpong are in the northernmost Himalayan Hills ecoregion. The area has rough terrain with changes in elevation, temperate to subtropical montane forests, brown forest soils that are high in organic matter, and complex microclimates resulting from varied aspect, slope, and elevation. Thermal extremes, frost, and slower decomposition rates create unique environments for cold-adapted ectomycorrhizal and wood-decay fungi to thrive.

Field surveys were conducted from 2008 to 2025, encompassing both monsoon and post-monsoon seasons, to capture the complete phenology of sporocarp emergence. There were 312 transects, each 12 m × 1000 m, set up across ecoregions in a way that was considered adequate (Figure 2.1). They were placed along forest trails that were easy to navigate, in the middle of forest patches, and at transitional ecotones to minimize bias while still capturing both undisturbed core and moderately

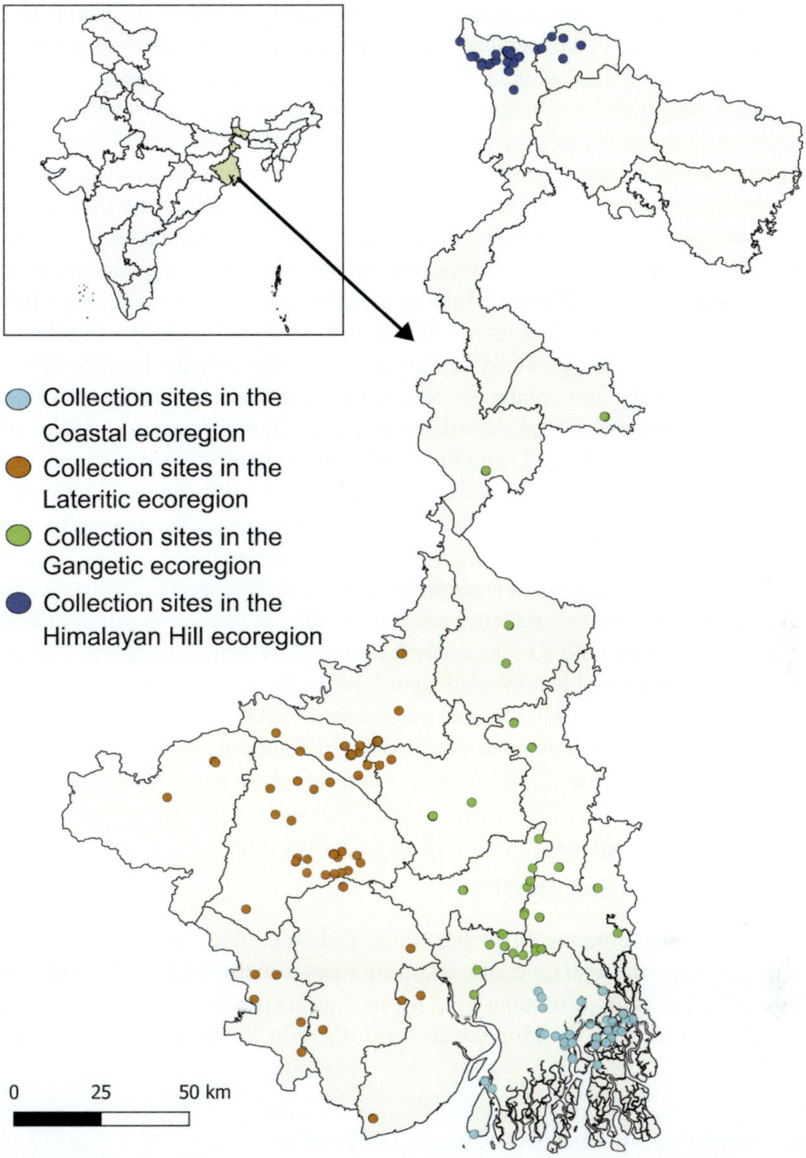

FIGURE 2.1 Collection sites of macrofungi in the different ecoregions of the study area (Prepared using QGIS software version 3.28.7).

disturbed edge habitats. Transects were regarded as independent sampling units, and all discernible macrofungal sporocarps exceeding 1 cm in diameter were documented. To avoid overestimating, only mature and morphologically distinct fruiting bodies were included. Young or degraded specimens were not included. For each observation, metadata were collected that included the type of substrate, the pattern of clustering, and the plants that were adjacent. High-resolution field photography was also used to record brief features, such as the color of the pileus, hymenial features, or the release of latex.

We brought the specimens to the laboratory to observe their appearance. The macroscopic features that we identified and described included the shape and color of the pileus, the attachment of the lamellae, the properties of the stipe, the existence of the volva and annulus, and the chemical spot reactions. On the other hand, the features observed under the microscope included the size and ornamentation of the spores, the basidia, the cystidia, and the hyphal systems (Chattopadhyay et al. 2025).

The species were assigned into functional groups according to observed habitat associations and documented nutritional modes in the literature. Functional groups consisted of saprotrophs in litter, soil, and/or fine wood; ectomycorrhizal associations in trees such as *Shorea*, *Quercus*, and *Pinus*; parasitic on living plant tissues; coprophilous on animal feces; entomopathogenic associated with insects; termitophilous associations on termite hills; bryophilous on mosses; and facultative associations involving saprotroph-ectomycorrhizal, saprotroph-coprophilous, saprotroph-parasitic, and saprotroph-biotroph.

For every transect, data on the environment were collected to determine the variables that drive community composition. There were 130 variables, including climatic factors, irradiation, soil, and topography. Climate variables included 19 bioclim variables, water vapor pressure, wind speed, solar radiation variables from the WorldClim v2.1 dataset (Fick and Hijmans 2017), and ultraviolet-B radiation variables from http://www.ufz.de/gluv/ (Beckmann et al. 2014). Soil attributes were obtained from SoilGrids (http://soilgrids.org/) at six depths, covering bulk density, coarse fragments, textures, pH, cation exchange capacity, and organic carbon density and mass, following the protocol in ISRIC (2022). To provide topographic data at the microclimate level, topographic variables, such as elevation, slope, and aspects, were obtained from the SRTM DEM. Data analysis was performed in R version 4.2.1, where all variables were preprocessed to standardize all variables in units of z-scores, and missing data were estimated with the median in order to standardize all variables in terms of units.

We used species richness, occupancy (the percentage of the transect occupied by each species), and diversity indices such as Shannon-Wiener and Pielou's evenness to measure community structure. To determine continuous occurrence patterns, we used inverse distance weighting to interpolate functional groups on a 20 km grid with a 60 km search radius. A principal component analysis (PCA) of standardized environmental variables revealed significant gradients that organize fungal assemblages. A hierarchical modeling framework combined species-level LASSO logistic regression to assess the impact of high-dimensional environmental predictors and Random Forest classification at the functional group, genus, and ecoregion levels to identify non-linear interactions and threshold effects.

2.3 AN UP-TO-DATE INVENTORY OF MACROFUNGI FOUND IN TROPICAL INDIA

The present study conducted across diverse tropical Indian ecoregions has significantly enhanced our understanding of macrofungal diversity, resulting in a more thorough and current inventory. The following subsections focus on ecoregion-wise, up-to-date accounts on the details of the macrofungi across various ecoregions, like the Himalayan foothills, lateritic landscapes, coastal Sundarbans, and Gangetic plains.

2.3.1 Macrofungi in the Himalayan Hills Ecoregion

As indicated in Table 2.1, a preliminary survey of macrofungi within the Himalayan Hills ecoregion documented a total of 74 distinct species (Figures 2.2, 2.3, 2.4, and 2.5). This diversity spans 60 genera, distributed across 45 families and 15 orders. The ratios from this dataset show a wide range of taxonomic groups, which is typical of a rich mycobiota. The species-to-genus ratio is about 1.23:1 (74:60), which means that there is a moderate amount of generic diversification. Meanwhile, the species-to-family and species-to-order ratios are approximately 1.64:1 (74:45) and 4.93:1 (74:15), respectively, reflecting a significant concentration of species within certain higher taxonomic groups. The order Agaricales was the most dominant, accounting for 32 of the 74 recorded species (43.24%), highlighting its ecological prominence in the study area. The orders Polyporales (11 species) and Russulales (7 species) were also well represented, reflecting the strong presence of wood-decaying fungi and ectomycorrhizal communities within the ecoregion.

2.3.2 Macrofungi of the Lateritic Ecoregion

As shown in Table 2.2, the macrofungi recorded for the Lateritic Ecoregion represented a cumulative count of 47 species (Figures 2.2, 2.3, 2.4, and 2.5). The species comprise 29 genera, 21 families,

TABLE 2.1
List of Macrofungi of the Himalayan Hills Ecoregion along with their representative Order and Families.

Sl. No.	Name of the Taxa	Order	Family
1	*Agaricus xanthodermus* Genev.	Agaricales	Agaricaceae
2	*Aleuria aurantia* (Pers.) Fuckel	Pezizales	Pyronemataceae
3	*Aleuria cestrica* (Ellis & Everh.) Seaver	Pezizales	Pyronemataceae
4	*Amanita vaginata* (Bull.) Lam.	Agaricales	Amanitaceae
5	*Apioperdon pyriforme* (Schaeff.) Vizzini	Agaricales	Lycoperdaceae
6	*Armillaria vara* (Berk.) Sacc.	Agaricales	Physalacriaceae
7	*Auricularia auricula-judae* (Bull.) Quél.	Auriculariales	Auriculariaceae
8	*Auricularia mesenterica* (Dicks.) Pers.	Auriculariales	Auriculariaceae
9	*Bjerkandera adusta* (Willd.) P. Karst.	Polyporales	Meruliaceae
10	*Boletellus ananas* (M.A. Curtis) Murrill	Boletales	Boletaceae
11	*Calocera cornea* (Batsch) Fr.	Dacrymycetales	Dacrymycetaceae
12	*Calocera viscosa* (Pers.) Fr.	Dacrymycetales	Dacrymycetaceae
13	*Clavaria fumosa* Pers.	Agaricales	Clavariaceae
14	*Clavaria vermicularis* Sw.	Agaricales	Clavariaceae
15	*Clavulina cristata* (Holmsk.) J. Schröt.	Cantharellales	Clavulinaceae
16	*Clavulinopsis fusiformis* (Sowerby) Corner	Agaricales	Clavariaceae
17	*Coltricia cinnamomea* (Jacq.) Murrill	Hymenochaetales	Hymenochaetaceae
18	*Coprinellus disseminatus* (Pers.) J.E. Lange	Agaricales	Psathyrellaceae
19	*Coprinus comatus* (O.F. Müll.) Pers.	Agaricales	Psathyrellaceae
20	*Cordyceps militaris* (L.) Link	Hypocreales	Cordycipitaceae
21	*Cortinarius iodes* Berk. & M.A. Curtis	Agaricales	Cortinariaceae
22	*Crepidotus applanatus* (Pers.) P. Kumm.	Agaricales	Crepidotaceae
23	*Cyathus striatus* (Huds.) Willd.	Agaricales	Nidulariaceae
24	*Dacryopinax spathularia* (Schwein.) G.W. Martin	Dacrymycetales	Dacrymycetaceae
25	*Daldinia concentrica* (Bolton) Ces. & De Not.	Xylariales	Xylariaceae
26	*Deflexula subsimplex* (Henn.) Corner	Gomphales	Lentariaceae
27	*Entoloma lividoalbum* (Kühner & Romagn.) Kubicka	Agaricales	Entolomataceae
28	*Favolaschia tonkinensis* (Pat.) Kuntze	Agaricales	Mycenaceae
29	*Favolus manipularis* Berk.	Polyporales	Polyporaceae
30	*Fistulina hepatica* (Schaeff.) With.	Agaricales	Fistulinaceae
31	*Flammulina velutipes* (Curtis) Singer	Agaricales	Physalacriaceae
32	*Fomes fomentarius* (L.) Fr.	Polyporales	Polyporaceae
33	*Ganoderma lucidum* (Curtis) P. Karst.	Polyporales	Ganodermataceae
34	*Geastrum saccatum* Fr.	Geastrales	Geastraceae
35	*Geastrum triplex* Jungh.	Geastrales	Geastraceae
36	*Grifola frondosa* (Dicks.) Gray	Polyporales	Meripilaceae
37	*Gymnopus confluens* (Pers.) Antonín, Halling & Noordel.	Agaricales	Omphalotaceae
38	*Gymnopus polygrammus* (Mont.) J.L. Mata	Agaricales	Omphalotaceae
39	*Helvella crispa* (Scop.) Fr.	Pezizales	Helvellaceae
40	*Hericium coralloides* (Scop.) Pers.	Russulales	Hericiaceae
41	*Hygrocybe cantharellus* (Schwein.) Murrill	Agaricales	Hygrophoraceae
42	*Hymenoscyphus scutula* (Pers.) W. Phillips	Helotiales	Helotiaceae
43	*Hypholoma fasciculare* (Huds.) P. Kumm.	Agaricales	Strophariaceae
44	*Laccaria laccata* (Scop.) Cooke	Agaricales	Hydnangiaceae
45	*Laetiporus sulphureus* (Bull.) Murrill	Polyporales	Fomitopsidaceae
46	*Lentinus squarrosulus* Mont.	Polyporales	Polyporaceae

(Continued)

TABLE 2.1 (*Continued*)
List of Macrofungi of the Himalayan Hills Ecoregion along with their representative Order and Families.

Sl. No.	Name of the Taxa	Order	Family
47	*Leotia lubrica* (Scop.) Pers.	Helotiales	Leotiaceae
48	*Lycoperdon perlatum* Pers.	Agaricales	Lycoperdaceae
49	*Marasmius vladimirii* A.K. Dutta & K. Acharya	Agaricales	Marasmiaceae
50	*Meripilus giganteus* (Pers.) P. Karst.	Polyporales	Meripilaceae
51	*Oudemansiella radicata* (Relhan) Singer	Agaricales	Physalacriaceae
52	*Parasola plicatilis* (Curtis) Redhead, Vilgalys & Hopple	Agaricales	Psathyrellaceae
53	*Pholiota squarrosa* (Oeder) P. Kumm.	Agaricales	Strophariaceae
54	*Phylloporus rhodoxanthus* (Schwein.) Bres.	Boletales	Boletaceae
55	*Pleurotus flabellatus* Sacc.	Agaricales	Pleurotaceae
56	*Psathyrella piluliformis* (Bull.) P.D. Orton	Agaricales	Psathyrellaceae
57	*Pseudohydnum gelatinosum* (Scop.) P. Karst.	Auriculariales	Pseudohydnaceae
58	*Russula brunneoaurantiaca*	Russulales	Russulaceae
59	*Russula buyckii* K. Acharya, Paloi & A.K. Dutta	Russulales	Russulaceae
60	*Russula cyanoxantha* (Schaeff.) Fr.	Russulales	Russulaceae
61	*Russula hookeri* Paloi, A.K. Dutta & K. Acharya	Russulales	Russulaceae
62	*Russula rosea* Pers.	Russulales	Russulaceae
63	*Russula senecis* S. Imai	Russulales	Russulaceae
64	*Schizophyllum commune* Fr.	Agaricales	Schizophyllaceae
65	*Scleroderma areolatum* Ehrenb.	Boletales	Sclerodermataceae
66	*Scutellinia scutellata* (L.) Lambotte	Pezizales	Pyronemataceae
67	*Sparassis crispa* (Wulfen) Fr.	Polyporales	Sparassidaceae
68	*Termitomyces heimii* Natarajan	Agaricales	Lyophyllaceae
69	*Trametes coccinea* (Fr.) Hai J. Li & S.H. He	Polyporales	Polyporaceae
70	*Trametes versicolor* (L.) Lloyd	Polyporales	Polyporaceae
71	*Tremella fuciformis* Berk.	Tremellales	Tremellaceae
72	*Trichocoma paradoxa* Jungh	Agaricales	Tricholomataceae
73	*Xerula radicata* (Relhan) Dörfelt	Agaricales	Physalacriaceae
74	*Xylaria hypoxylon* (L.) Grev.	Xylariales	Xylariaceae

TABLE 2.2
List of Macrofungi of the Lateritic Ecoregion along with their representative Order and Families.

Sl. No.	Name of the Taxa	Order	Family
1	*Agaricus endoxanthus* Berk. & Broome	Agaricales	Agaricaceae
2	*Amanita hemibapha* (Berk. & Broome) Sacc.	Agaricales	Amanitaceae
3	*Amanita princeps* Corner & Bas	Agaricales	Amanitaceae
4	*Amanita spreta* (Peck) Sacc.	Agaricales	Amanitaceae
5	*Amanita vaginata* (Bull.) Lam.	Agaricales	Amanitaceae
6	*Astraeus hygrometricus* (Pers.) Morgan	Hysterangiales	Diplocystidiaceae
7	*Boletellus ananas* (M.A. Curtis) Murrill	Boletales	Boletaceae
8	*Clathrus ruber* P. Micheli ex Pers.	Phallales	Phallaceae
9	*Clavaria straminea* Cotton	Agaricales	Clavariaceae
10	*Coltricia cinnamomea* (Jacq.) Murrill	Hymenochaetales	Hymenochaetaceae
11	*Daedalea flavida* Lév.	Polyporales	Polyporaceae

TABLE 2.2 (*Continued*)

List of Macrofungi of the Lateritic Ecoregion along with their representative Order and Families.

Sl. No.	Name of the Taxa	Order	Family
12	*Flavodon flavus* (Klotzsch) Ryvarden	Polyporales	Irpicaceae
13	*Geastrum saccatum* Fr.	Geastrales	Geastaceae
14	*Hygrocybe angustifolia* (Murrill) Candusso	Agaricales	Hygrocybaceae
15	*Inocybe umbonata* Quél.	Agaricales	Inocybaceae
16	*Laccaria laccata* (Scop.) Cooke	Agaricales	Hydnangiaceae
17	*Lactarius gerardii* Peck	Russulales	Russulaceae
18	*Lactarius zonarius* (Bull.) Fr.	Russulales	Russulaceae
19	*Lamprospora carbonicola* Boud.	Pezizales	Pyronemataceae
20	*Lentinus squarrosulus* Mont.	Polyporales	Polyporaceae
21	*Leucocoprinus fragilissimus* (Berk. & M.A. Curtis) Pat.	Agaricales	Agaricaceae
22	*Marasmius haematocephalus* f. *haematocephalus* N. Wannathes	Agaricales	Marasmiaceae
23	*Marasmius siccus* (Schwein.) Fr.	Agaricales	Marasmiaceae
24	*Marasmius suthepensis* Wannathes, Desjardin & Lumyong	Agaricales	Marasmiaceae
25	*Microporus flabelliformis* (Klotzsch) Kuntze	Polyporales	Polyporaceae
26	*Microporus xanthopus* (Fr.) Kuntze	Polyporales	Polyporaceae
27	*Pleurotus ostreatus* (Jacq.) P. Kumm.	Agaricales	Pleurotaceae
28	*Polyporus brumalis* (Pers.) Fr.	Polyporales	Polyporaceae
29	*Pseudochaete tabacina* (Sowerby) T. Wagner & M. Fisch.	Hymenochaetales	Hymenochaetaceae
30	*Pulveroboletus icterinus* (Pat. & C.F. Baker) Watling	Boletales	Boletaceae
31	Russula alatoreticula K. Acharya, S. Khatua, A.K. Dutta & Paloi	Russulales	Russulaceae
32	*Russula albonigra* (Krombh.) Fr.	Russulales	Russulaceae
33	*Russula brevipes* Peck	Russulales	Russulaceae
34	*Russula kanadii* A.K. Dutta & K. Acharya	Russulales	Russulaceae
35	*Russula pseudocyanoxantha* Paloi, K. Acharya & S. Khatua	Russulales	Russulaceae
36	*Russula senecis* S. Imai	Russulales	Russulaceae
37	*Russula xerampelina* (Schaeff.) Fr.	Russulales	Russulaceae
38	*Scleroderma cepa* Pers.	Boletales	Sclerodermataceae
39	*Scleroderma macrorhizon* Wallr.	Boletales	Sclerodermataceae
40	*Scytinopogon angulisporus* (Pat. & Gaillard) Corner	Cantharellales	Hydnaceae
41	*Spathularia flavida* Pers.	Helotiales	Cudoniaceae
42	*Termitomyces clypeatus* R. Heim	Agaricales	Lyophyllaceae
43	*Termitomyces heimii* Natarajan	Agaricales	Lyophyllaceae
44	*Termitomyces microcarpus* (Berk. & Broome) R. Heim	Agaricales	Lyophyllaceae
45	*Trametes cingulata* Berk.	Polyporales	Polyporaceae
46	*Trametes coccinea* (Fr.) Hai J. Li & S.H. He	Polyporales	Polyporaceae
47	*Trametes hirsuta* (Wulfen) Lloyd	Polyporales	Polyporaceae

and 11 orders, denoting a moderately diverse macrofungal species community. The calculated taxa ratio denotes a rational pattern of diversification of macrofungi in this ecoregion. The species to genus ratio is about 1.62:1 (47:29), denoting a lack of redundancy of genera; on the other hand, species to family and species to order ratio are about 2.24:1 (47:21) and 4.27:1 (47:11), respectively, denoting an accumulation of species at particular family and order levels.

The Agaricales showed the highest dominance, represented by 18 species that made up 38.3% of the total macrofungal diversity. The second highest was the Russulales, which was represented by 9 species that made up 19.1%, and the third highest was the Polyporales, with 8 species that made up 17.02%. In families, the Russulaceae showed the highest richness, represented by

9 species that made up 19.1% of the total, while the Polyporaceae was represented by 7 species that made up 14.9%. Genus-wise, the highest richness is represented by the genus *Russula*, which had seven recorded species, which indicates its importance in terms of ecological point of view in the lateritic landscape. Overall, the observed indices are suggestive of the fact that the macrofungal community is one that has wide distributions at higher taxa, though some groups show definitive domination, represented by agaric and polypore taxa exigencies fulfilled in the lateritic region.

2.3.3 Macrofungi of the Coastal Ecoregion (Indian Part of Sundarbans)

Table 2.3 shows the diversity of the macrofungal species in the Coastal ecoregion of the Indian Sundarbans, for which a total of 58 species of macrofungi were recorded (Figures 2.2, 2.3, 2.4, and 2.5). The total number of species of macrofungi was found to be about 45 genera, 27 families, and nine orders, which thus represents a diverse composition of the taxonomic groups of these organisms. The total number of the taxonomic ratios in the above data shows moderate

TABLE 2.3
List of Macrofungi of the Coastal Ecoregion (Indian Part of Sundarbans) along with their representative Order and Families.

Sl. No.	Name of the Taxa	Order	Family
1	*Agaricus placomyces* Peck	Agaricales	Agaricaceae
2	*Amauroderma rugosum* (Blume & T. Nees) Torrend	Polyporales	Ganodermataceae
3	*Amylosporus campbellii* (Berk.) Ryvarden	Polyporales	Polyporaceae
4	*Apioperdon pyriforme* (Schaeff.) Vizzini	Agaricales	Lycoperdaceae
5	*Ascobolus scatigenus* (Berk. & M.A. Curtis) Brumm	Pezizales	Ascobolaceae
6	*Auricularia auricula-judae* (Bull.) Quél.	Auriculariales	Auriculariaceae
7	*Calocybe indica* Purkay. & A. Chandra	Agaricales	Lyophyllaceae
8	*Conocybe tenera* (Schaeff.) Kühner	Agaricales	Bolbitiaceae
9	*Coprinellus disseminatus* (Pers.) J.E. Lange	Agaricales	Psathyrellaceae
10	*Coprinopsis lagopus* (Fr.) Redhead, Vilgalys & Moncalvo	Agaricales	Psathyrellaceae
11	*Coriolopsis badia* (Berk.) Murrill	Polyporales	Polyporaceae
12	*Coriolopsis occidentalis* (Klotzsch) Murrill	Polyporales	Polyporaceae
13	*Cubamyces flavidus* (Lév.) Lücking	Polyporales	Fomitopsidaceae
14	*Cyathus striatus* Willd	Agaricales	Nidulariaceae
15	*Dacryopinax spathularia* (Schwein.) G.W. Martin	Dacrymycetales	Dacrymycetaceae
16	*Daldinia childiae* J.D. Rogers & Y.M. Ju	Xylariales	Xylariaceae
17	*Earliella scabrosa* (Pers.) Gilb. & Ryvarden	Polyporales	Polyporaceae
18	*Fabisporus sanguineus* (L.) Zmitr.	Polyporales	Insertae sedis
19	*Flavodon flavus* (Klotzsch) Ryvarden	Polyporales	Phanerochaetaceae
20	*Funalia leonina* (Klotzsch) Pat.	Polyporales	Polyporaceae
21	*Ganoderma applanatum* (Pers.) Pat.	Polyporales	Ganodermataceae
22	*Ganoderma lucidum* (Curtis) P. Karst.	Polyporales	Ganodermataceae
23	*Geastrum saccatum* Fr.	Geastrales	Geastraceae
24	*Gymnopilus dilepis* (Berk. & Broome) Singer	Agaricales	Hymenogastraceae
25	*Lentinus squarrosulus* Mont.	Polyporales	Polyporaceae
26	*Leucocoprinus birnbaumii* (Corda) Singer	Agaricales	Agaricaceae
27	*Leucocoprinus cepistipes* (Sowerby) Pat.	Agaricales	Agaricaceae
28	*Macrocybe crassa* (Sacc.) Pegler & Lodge	Agaricales	Callistosporiaceae
29	*Macrocybe gigantea* (Massee) Pegler & Lodge	Agaricales	Callistosporiaceae
30	*Macrocybe lobayensis* (R. Heim) Pegler & Lodge	Agaricales	Callistosporiaceae
31	*Marasmius haematocephalus* f. *haematocephalus* N. Wannathes	Agaricales	Marasmiaceae

TABLE 2.3 (*Continued*)

List of Macrofungi of the Coastal Ecoregion (Indian Part of Sundarbans) along with their representative Order and Families.

Sl. No.	Name of the Taxa	Order	Family
32	*Marasmius imitarius* Wannathes, Desjardin & Lumyong	Agaricales	Marasmiaceae
33	*Marasmius nigrobrunneus* (Pat.) Sacc.	Agaricales	Marasmiaceae
34	*Marasmius rotula* (Scop.) Fr.	Agaricales	Marasmiaceae
35	*Marasmius siccus* (Schwein.) Fr.	Agaricales	Marasmiaceae
36	*Meripilus lineatus* (Pers.) Westph. & Rajchenb.	Polyporales	Meripilaceae
37	*Microporus xanthopus* (Fr.) Kuntze	Polyporales	Polyporaceae
38	*Mutinus caninus* (Schaeff.) Fr.	Phallales	Phallaceae
39	*Parasola plicatilis* (Curtis) Redhead, Vilgalys & Hopple	Agaricales	Psathyrellaceae
40	*Phallus indusiatus* Vent.	Phallales	Phallaceae
41	*Pisolithus arhizus* (Scop.) Rauschert	Boletales	Sclerodermataceae
42	*Pleurotus ostreatus* (Jacq.) P. Kumm.	Agaricales	Pleurotaceae
43	*Pluteus ephebeus* (Fr.) Gillet	Agaricales	Pluteaceae
44	*Podoscypha petalodes* (Berk.) Boidin	Polyporales	Podoscyphaceae
45	*Schizophyllum commune* Fr.	Agaricales	Schizophyllaceae
46	*Scleroderma cepa* Pers.	Boletales	Sclerodermataceae
47	*Sphaerobolus stellatus* Tode	Geastrales	Geastraceae
48	*Termitomyces clypeatus* R. Heim	Agaricales	Lyophyllaceae
49	*Tomophagus colossus* (Fr.) Murrill	Polyporales	Polyporaceae
50	*Trametes apiaria* (Pers.) Zmitr., Wasser & Ezhov	Polyporales	Polyporaceae
51	*Trametes cingulata* Berk.	Polyporales	Polyporaceae
52	*Trametes coccinea* (Fr.) Hai J. Li & S.H. He	Polyporales	Polyporaceae
53	*Trametes elegans* (Spreng.) Fr.	Polyporales	Polyporaceae
54	*Tulostoma chudaei* Pat.	Agaricales	Agaricaceae
55	*Volvariella pusilla* (Pers.) Singer	Agaricales	Volvariellaceae
56	*Volvariella volvacea* (Bull.) Singer	Agaricales	Volvariellaceae
57	*Volvopluteus gloiocephalus* (DC.) Vizzini, Contu & Justo	Agaricales	Volvariellaceae
58	*Xylaria hypoxylon* (L.) Grev.	Xylariales	Xylariaceae

diversification with a ratio of 1.29:1 and 2.15:1 in the ratio of the number of species to genera and the number of species to families, respectively.

At the ordinal rank, the Agaricales proved to be the most dominant, with a total of 27 species, which accounted for a whopping 46.55% of the entire macrofungal diversity, evidencing the dominance of gilled mushrooms in this ecosystem. The Polyporales came in as the second most common, with a total of 20 species (34.48%), which reflects wood-rotting polypores, whose importance in this mangrove ecosystem, as far as nutrient regeneration is concerned, is substantial. At the family level, the family Polyporaceae topped the list as far as species richness was considered, with a total of 12 species, while diversity at the ordinal level of Agaricales comprised a range of families, namely the Agaricaceae, Bolbitiaceae, Callistosporiaceae, Hymenogastraceae, Lycoperdaceae, Lyophyllaceae, Marasmiaceae, Nidulariaceae, and Pleurotaceae. In general, these data on taxonomic structure reveal a macro-fungal assemblage marked by high familial representation but moderate generic diversification. These are clear pointers to the various specialized habitats, which are created due to the salinity, humidity, and richness associated with the Sundarbans coastline.

2.3.4 Macrofungi of the Gangetic Ecoregion

When the species of macrofungi recorded from the Gangetic ecoregion are considered, as depicted in Table 2.4 and Figures 2.2, 2.3, 2.4, and 2.5, there is a rich diversity of species, which includes a

TABLE 2.4

List of Macrofungi of the Gangetic Ecoregion along with their representative Order and Families.

Sl. No.	Name of the Taxa	Order	Family
1	*Agaricus albostipitatus* Tarafder, A.K. Dutta & K. Acharya	Agaricales	Agaricaceae
2	*Agaricus duplocingulatoides* Tarafder, A.K. Dutta & K. Acharya	Agaricales	Agaricaceae
3	*Agaricus endoxanthus* Berk. & Broome	Agaricales	Agaricaceae
4	*Agaricus glabriusculus* S. Hussain	Agaricales	Agaricaceae
5	*Agaricus trisulphuratus* Berk.	Agaricales	Agaricaceae
6	*Agrocybe pediades* (Fr.) Fayod	Agaricales	Strophariaceae
7	*Amylosporus campbellii* (Berk.) Ryvarden	Russulales	Bondarzewiaceae
8	*Aspidella aureofloccosa* (Bas) Vizzini & Contu	Agaricales	Amanitaceae
9	*Auricularia auricula-judae* (Bull.) Quél.	Auriculariales	Auriculariaceae
10	*Calocybella swetha* K.N.A. Raj & Manim	Agaricales	Lyophyllaceae
11	*Candolleomyces candolleanus* (Fr.) D. Wächt. & A. Melzer	Agaricales	Psathyrellaceae
12	*Candolleomyces efflorescens* (Sacc.) D. Wächt. & A. Melzer	Agaricales	Psathyrellaceae
13	*Cantharocybe virosa* (Manim. & K.B. Vrinda) T.K.A. Kumar	Agaricales	Hygrophoraceae
14	*Chlorophyllum molybdites* (G. Mey.) Massee	Agaricales	Agaricaceae
15	*Chlorophyllum squamulosum* A.K. Dutta, Soum. Bera & K. Acharya	Agaricales	Agaricaceae
16	*Clathrus delicatus* Berk. & Broome	Phallales	Phallaceae
17	*Clavaria vermicularis* Batsch	Agaricales	Clavariaceae
18	*Clitocybula albida* A.K. Dutta, K. Acharya & Antonín	Agaricales	*Incertae sedis*
19	*Collybiopsis melanopus* (A.W. Wilson, Desjardin & E. Horak) R.H. Petersen	Agaricales	Omphalotaceae
20	*Collybiopsis menehune* (Desjardin, Halling & Hemmes) R.H. Petersen	Agaricales	Omphalotaceae
21	*Coprinellus disseminatus* (Pers.) J.E. Lange	Agaricales	Psathyrellaceae
22	*Coprinopsis lagopides* (P. Karst.) Redhead, Vilgalys & Moncalvo	Agaricales	Psathyrellaceae
23	*Cubamyces flavidus* (Lév.) Lücking	Polyporales	Polyporaceae
24	*Cyathus striatus* Willd	Agaricales	*Incertae sedis*
25	*Dacryopinax spathularia* (Schwein.) G.W. Martin	Dacrymycetales	Dacrymycetaceae
26	*Daldinia childiae* J.D. Rogers & Y.M. Ju	Xylariales	Hypoxylaceae
27	*Earliella scabrosa* (Pers.) Gilb. & Ryvarden	Polyporales	Polyporaceae
28	*Entoloma albotomentosum* Noordel. & Hauskn	Agaricales	Entolomataceae
29	*Entoloma brunneoapplanatum* K.N.A. Raj & Manim	Agaricales	Entolomataceae
30	*Exidia glandulosa* (Bull.) Fr.	Auriculariales	Auriculariaceae
31	*Fabisporus sanguineus* (L.) Zmitr.	Polyporales	*Incertae sedis*
32	*Favolus grammocephalus* (Berk.) Imazeki	Polyporales	Polyporaceae
33	*Flavodon flavus* (Klotzsch) Ryvarden	Polyporales	Irpicaceae
34	*Ganoderma lucidum* (Curtis) P. Karst.	Polyporales	Polyporaceae
35	*Geastrum saccatum* Fr.	Geastrales	Geastraceae
36	*Gymnopilus dilepis* (Berk. & Broome) Singer	Agaricales	Hymenogastraceae
37	*Gymnopilus purpureosquamulosus* Høil.	Agaricales	Hymenogastraceae
38	*Hygrocybe lucida* K. Acharya & A.K. Dutta	Agaricales	Hygrocybaceae
39	*Laccaria vinaceoavellanea* Hongo	Agaricales	Hydnangiaceae
40	*Laetiporus sulphureus* (Bull.) Murrill	Polyporales	Laetiporaceae
41	*Lentinus polychrous* Lév.	Polyporales	Polyporaceae
42	*Lentinus sajor-caju* (Fr.) Fr.	Polyporales	Polyporaceae
43	*Lentinus squarrosulus* Mont.	Polyporales	Polyporaceae
44	*Lepiota bengalensis* Hosen & T.H. Li, in Hosen, Li, Ge & Vellinga	Agaricales	Agaricaceae
45	*Lepiota brunneodisca* A.K. Dutta & N. Roy	Agaricales	Agaricaceae
46	*Lepista sordida* (Schumach.) Singer	Agaricales	*Incertae sedis*
47	*Leucocoprinus birnbaumii* (Corda) Singer	Agaricales	Agaricaceae

TABLE 2.4 (*Continued*)

List of Macrofungi of the Gangetic Ecoregion along with their representative Order and Families.

Sl. No.	Name of the Taxa	Order	Family
48	*Leucocoprinus brunneodiscus* (A.K. Dutta & K. Acharya) Kun L. Yang, Jia Y. Lin & Zhu L. Yang	Agaricales	Agaricaceae
49	*Leucocoprinus cepistipes* (Sowerby) Pat.	Agaricales	Agaricaceae
50	*Leucocoprinus cretaceus* (Bull.) Locq.	Agaricales	Agaricaceae
51	*Macrocybe gigantea* (Massee) Pegler & Lodge	Agaricales	Callistosporiaceae
52	*Macropsalliota tropica* (A.K. Dutta, Stallman & K. Acharya) Kun L. Yang, Jia Y. Lin & Zhu L. Yang	Agaricales	Agaricaceae
53	*Marasmius bambusiniformis* Singer	Agaricales	Marasmiaceae
54	*Marasmius beniensis* Singer	Agaricales	Marasmiaceae
55	*Marasmius haematocephalus* f. *haematocephalus* N. Wannathes	Agaricales	Marasmiaceae
56	*Marasmius hypochroides* Berk. & Broome	Agaricales	Marasmiaceae
57	*Marasmius imitarius* Wannathes, Desjardin & Lumyong	Agaricales	Marasmiaceae
58	*Marasmius indojasminodorus* A.K. Dutta, K. Acharya & K. Das	Agaricales	Marasmiaceae
59	*Marasmius leucorotalis* Singer	Agaricales	Marasmiaceae
60	*Marasmius leveilleanus* (Berk.) Pat	Agaricales	Marasmiaceae
61	*Marasmius luculentus* A.K. Dutta, K. Acharya & Antonín	Agaricales	Marasmiaceae
62	*Marasmius midnapurensis* A.K. Dutta, P. Pradhan & K. Acharya	Agaricales	Marasmiaceae
63	*Marasmius nigrobrunneus* (Pat.) Sacc.	Agaricales	Marasmiaceae
64	*Marasmius pellucidus* Berk. & Broome.	Agaricales	Marasmiaceae
65	*Marasmius rotula* (Scop.) Fr.	Agaricales	Marasmiaceae
66	*Marasmius siccus* (Schwein.) Fr.	Agaricales	Marasmiaceae
67	*Marasmius tenuissimus* (Sacc.) Singer	Agaricales	Marasmiaceae
68	*Marasmius trichotus* Corner	Agaricales	Marasmiaceae
69	*Micropsalliota lateritia* Heinem.	Agaricales	Agaricaceae
70	*Micropsalliota xanthorubescens* Heinem.	Agaricales	Agaricaceae
71	*Oudemansiella canarii* (Jungh.) Höhn.	Agaricales	Physalacriaceae
72	*Panus neostrigosus* Drechsler-Santos & Wartchow	Polyporales	Polyporaceae
73	*Paramarasmius palmivorus* (Sharples) Antonín & Kolařík	Agaricales	Marasmiaceae
74	*Parasola plicatilis* (Curtis) Redhead, Vilgalys & Hopple	Agaricales	Psathyrellaceae
75	*Phallus indusiatus* Vent.	Phallales	Phallaceae
76	*Phellinopsis conchata* (Pers.) Y.C. Dai	Hymenochaetales	Hymenochaetaceae
77	*Phellinus rimosus* (Berk.) Pilát	Hymenochaetales	Hymenochaetaceae
78	*Pleurotus ostreatus* (Jacq.) P. Kumm.	Agaricales	Pleurotaceae
79	*Podoscypha petalodes* (Berk.) Boidin	Polyporales	Podoscyphaceae
80	*Rhodocybe brunneoaurantiaca* A.K. Dutta, G.M. Gates & K. Acharya	Agaricales	Entolomataceae
81	*Russula alatoreticula* K. Acharya, S. Khatua, A.K. Dutta & Paloi	Russulales	Russulaceae
82	*Russula arunii* Paloi, A.K. Dutta & K. Acharya	Russulales	Russulaceae
83	*Russula indoeburnea* N. Roy, P. Chattopadhyay & A.K. Dutta	Russulales	Russulaceae
84	*Russula kanadii* A.K. Dutta & K. Acharya	Russulales	Russulaceae
85	*Schizophyllum commune* Fr.	Agaricales	Schizophyllaceae
86	*Termitomyces clypeatus* R. Heim	Agaricales	Lyophyllaceae
87	*Termitomyces heimii* Natarajan	Agaricales	Lyophyllaceae
88	*Termitomyces medius* R. Heim & Grassé	Agaricales	Lyophyllaceae
89	*Termitomyces microcarpus* (Berk. & Broome) R. Heim	Agaricales	Lyophyllaceae
90	*Tremella fuciformis* Berk.	Tremellales	Tremellaceae
91	*Tremella mesenterica* (Schaeff.) Pers.	Tremellales	Tremellaceae
92	*Trogia benghalensis* K. Acharya & A.K. Dutta	Agaricales	*Incertae sedis*

(*Continued*)

TABLE 2.4 (*Continued*)

List of Macrofungi of the Gangetic Ecoregion along with their representative Order and Families.

Sl. No.	Name of the Taxa	Order	Family
93	*Volvariella bilobata* A.K. Dutta & Chattopadhyay	Agaricales	Volvariellaceae
94	*Volvariella volvacea* (Bull.) Singer	Agaricales	Volvariellaceae
95	*Volvopluteus gloiocephalus* (DC.) Vizzini, Contu & Justo	Agaricales	Volvariellaceae
96	*Xanthagaricus subepipastus* (Heinem. & Little Flower) Little Flower, Hosag. & T.K. Abraham	Agaricales	Agaricaceae
97	*Xylaria hypoxylon* (L.) Grev.	Xylariales	Xylariaceae

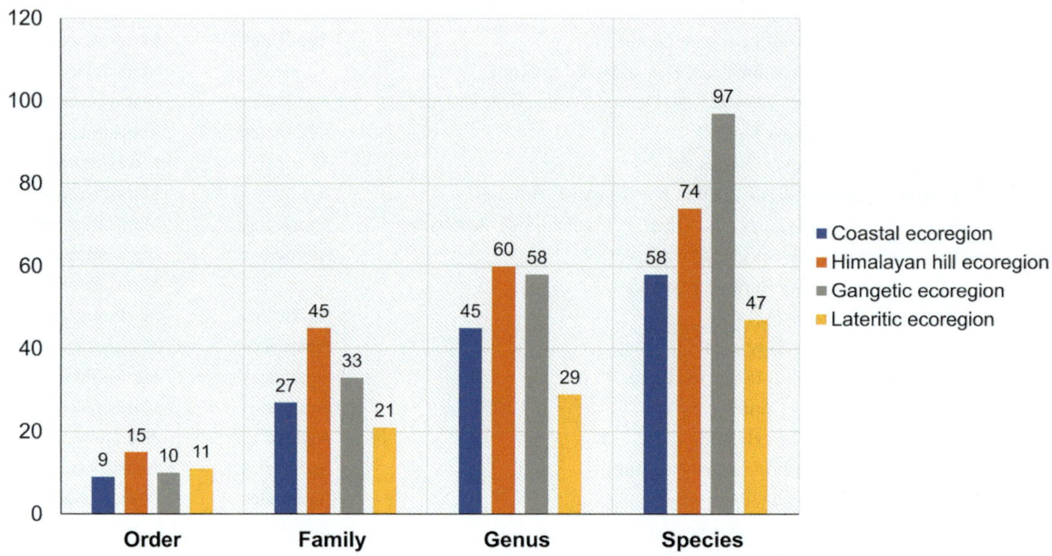

FIGURE 2.2 Representation of the total number of orders, families, genera, and species across the Coastal, Himalayan Hills, Gangetic, and Lateritic Ecoregions of West Bengal, tropical India.

total of 97 species belonging to a staggering 58 genera. These, in their turn, are further categorized under 33 families and 10 orders. The most diverse assemblage of species is found in the Agaricales. What is noteworthy is that there is a substantial family level diversity, which includes a total of 17 species of the Agaricaceae and Marasmiaceae. It is further interesting to note that there is a substantial spread of species throughout various taxa, which is reflected in the ratio of species, where on an average, there are approximately 2.94 species in a family, with about 1.67 species in a genus, and about 9.70 species represented by an order.

2.4 SPECIES DIVERSITY, FUNCTIONAL COMPOSITION, FREQUENCY DISTRIBUTION, AND HABITAT SPECIFICITY ACROSS ECOREGIONS

The large-scale macrofungal inventory carried out in the four main ecoregions in West Bengal, viz., Coastal, Gangetic Plains, Lateritic uplands, and Himalayan Hills, revealed a total of 208 species, which not only revealed a large degree of spatial structure in terms of climatic, edaphic, and vegetation gradients but also demonstrated that macrofungal communities are divided into various ecoregions, rather than a collective homogeneous group or a regional species pool. This

FIGURE 2.3 Field photographs of some of the striking macrofungi collected from tropical India: (A) *Agaricus albostipitatus*, (B) *Agaricus endoxanthus*, (C) *Agaricus glabriusculus*, (D) *Agaricus trisulphuratus*, (E) *Agaricus xanthodermus*, (F) *Agrocybe pediades*, (G) *Aleuria aurantia*, (H) *Amanita hemibapha*, (I) *Amanita princeps*, (J) *Amanita spreta*, (K) *Amanita vaginata*, (L) *Amauroderma rugosum*, (M) *Amylosporus campbellii*, (N) *Apioperdon pyriforme*, (O) *Ascobolus scatigenus*, (P) *Aspidella aureofloccosa*, (Q) *Astraeus hygrometricus*, (R) *Auricularia auricula-judae*, (S) *Calocybella swetha*, (T) *Cantharocybe virosa*, (U) *Chlorophyllum molybdites*, (V) *Chlorophyllum squamulosum*, (W) *Clathrus delicatus*, (X) *Clathrus ruber*, (Y) *Clavaria fumosa*, (Z) *Clavaria vermicularis*, (AA) *Clavulinopsis fusiformis*, (AB) *Clitocybula albida*, (AC) *Collybiopsis melanopus*, (AD) *Collybiopsis menehune*, (AE) *Conocybe tenera*, (AF) *Coprinellus disseminatus*, (AG) *Coprinopsis lagopides*, (AH) *Coprinopsis lagopus*, (AI) *Coriolopsis badia*, (AJ) *Coriolopsis occidentalis*, (AK) *Cortinarius iodes*, (AL) *Crepidotus applanatus*, (AM) *Cyathus striatus*, (AN) *Dacryopinax spathularia*, (AO) *Daedalea flavida*, (AP) *Daldinia childiae*, (AQ) *Daldinia concentrica*, (AR) *Deflexula subsimplex*, and (AS) *Earliella scabrosa*.

Source: Photos by Arun Kumar Dutta

FIGURE 2.4 Field photographs of some more intriguing macrofungi collected from various ecozones in tropical India: (A) *Entoloma albotomentosum*, (B) *Entoloma brunneoapplanatum*, (C) *Entoloma lividoalbum*, (D) *Exidia glandulosa*, (E) *Fabisporus sanguineus*, (F) *Favolus grammocephalus*, (G) *Fistulina hepatica*, (H) *Flammulina velutipes*, (I) *Flavodon flavus*, (J) *Funalia leonina*, (K) *Ganoderma applanatum*, (L) *Ganoderma lucidum*, (M) *Geastrum saccatum*, (N) *Geastrum triplex*, (O) *Grifola frondosa*, (P) *Gymnopilus dilepis*, (Q) *Gymnopilus purpureosquamulosus*, (R) *Hygrocybe lucida*, (S) *Hymenoscyphus scutula*, (T) *Laccaria vinaceoavellanea*, (U) *Lactarius zonarius*, (V) *Laetiporus sulphureus*, (W) *Lentinus polychrous*, (X) *Lentinus sajor-caju*, (Y) *Lentinus squarrosulus*, (Z) *Leotia lubrica*, (AA) *Lepista sordida*, (AB) *Leucocoprinus birnbaumii*, (AC) *Leucocoprinus cepistipes*, (AD) *Leucocoprinus cretaceus*, (AE) *Leucocoprinus fragilissimus*, (AF) *Macrocybe crassa*, (AG) *Macrocybe gigantea*, (AH) *Marasmius bambusiniformis*, (AI) *Marasmius beniensis*, (AJ) *Marasmius haematocephalus* f. *haematocephalus*, (AK) *Marasmius hypochroides*, (AL) *Marasmius imitarius*, (AM) *Marasmius indojasminodorus*, (AN) *Marasmius leucorotalis*, (AO) *Marasmius leveilleanus*, (AP) *Marasmius luculentus*, (AQ) *Marasmius midnapurensis*, (AR) *Marasmius nigrobrunneus*, and (AS) *Marasmius pellucidus*.

Source: Photos by Arun Kumar Dutta

FIGURE 2.5 Field photographs of some other intriguing macrofungi collected from various ecozones in tropical India. (A) *Marasmius rotula*, (B) *Marasmius tenuissimus*, (C) *Marasmius trichotus*, (D) *Meripilus giganteus*, (E) *Microporus xanthopus*, (F) *Micropsalliota lateritia*, (G) *Micropsalliota xanthorubescens*, (H) *Mutinus caninus*, (I) *Oudemansiella canarii*, (J) *Panus neostrigosus*, (K) *Paramarasmius palmivorus*, (L) *Parasola plicatilis*, (M) *Phallus indusiatus*, (N) *Phellinopsis conchata*, (O) *Phellinus rimosus*, (P) *Phylloporus rhodoxanthus*, (Q) *Pisolithus arhizus*, (R) *Pleurotus ostreatus*, (S) *Pluteus ephebeus*, (T) *Podoscypha petalodes*, (U) *Psathyrella piluliformis*, (V) *Russula alatoreticula*, (W) *Russula arunii*, (X) *Russula kanadii*, (Y) *Russula pseudocyanoxantha*, (Z) *Schizophyllum commune*, (AA) *Scleroderma cepa*, (AB) *Scleroderma macrorhizon*, (AC) *Scutellinia scutellata*, (AD) *Scytinopogon angulisporus*, (AE) *Sparassis crispa*, (AF) *Sphaerobolus stellatus*, (AG) *Termitomyces clypeatus*, (AH) *Termitomyces heimii*, (AI) *Termitomyces medius*, (AJ) *Termitomyces microcarpus*, (AK) *Trametes coccinea*, (AL) *Tremella fuciformis*, (AM) *Tremella mesenterica*, (AN) *Trichocoma paradoxa*, (AO) *Trogia benghalensis*, (AP) *Tulostoma chudaei*, (AQ) *Volvariella volvacea*, (AR) *Xerula radicata*, and (AS) *Xylaria hypoxylon*.

Source: Photos by Arun Kumar Dutta

has been facilitated by a combination of various environmental factors such as substrate, presence of host trees, nutrient status, disturbance patterns, and microclimatic stability. All these findings suggest that, in tropical eastern India, the diversity of macrofungal communities is regulated by a scale-dependent combination of environmental filters at both the landscape and local scales simultaneously.

Across all ecoregions, saprotrophic macrofungi formed the largest proportion of the regional assemblage, reflecting the central role of decomposition processes in these ecosystems. This shows how important they are for breaking down litter, mining nutrients, and turning over organic matter. Nonetheless, despite their prevalence, saprotrophic richness, spatial persistence, and dominance exhibited significant variation across ecoregions, indicating a high sensitivity of decomposer communities to variations in substrate availability, moisture regimes, and environmental stress. The dominance of saprotrophs at the regional level minimizes significant regional variations in ecological performance and community structure.

Species richness provides a list of the different organisms, while frequency metrics show realized dominance, spatial persistence, and functional performance. These are all important for fungi, whose mycelial networks extend far beyond the visible sporocarp stage. In all ecoregions, frequency distributions showed clear dominance-rarity gradients with long-tailed distributions. A small number of taxa had high occupancy, while the rest remained spatially rare. Coefficients of variation in frequency often surpassed 100%, and median frequencies consistently fell below mean values, indicating right-skewed distributions influenced by a limited number of highly persistent species. Dominance-rarity structures are prevalent in fungal communities globally, yet their occurrence is context-dependent. In certain systems, dominance signifies superior competition, whereas in others, it signifies strong environmental filtering.

Frequency analyses showed a sense of dissociation between richness and dominance, where the ecoregions showing high species richness did not always demonstrate high dominance and vice versa. Cross-ecoregional analyses of dominance gradients show the contrasting factors dictating the role of dominant species. In the Coastal ecoregion, dominance is a result of abiotic stress and physiological tolerance; in the Gangetic ecoregion, it is depressed by disturbance and soil turbulence; in the Lateritic Ecoregion, it is enhanced by symbiotic monopolization by ectomycorrhizal fungi; and, finally, in the Himalayan ecoregion, it is depressed by climatic limitations and microhabitat fragmentation. These differences reveal that richness patterns don't indicate how macrofungal communities are organized. Frequency structure shows which taxa are actively involved in decomposition, nutrient cycling, and symbiotic exchange.

Species distribution analysis indicated the presence of strong habitat association and low inter-ecoregional overlaps, suggesting harsh environmental filters over the surveyed landscapes. The Venn diagram revealed that most species were restricted to certain ecoregions, and the highest proportions were recorded in the Gangetic Plains (28.7%) and the Himalayan Hills regions (25.8%), followed by the Lateritic plateaus (12.0%) and the Coastal regions (10.5%) (Figures 2.6 and 2.7). A small number of species were shared by different ecoregions due to the strong species turnover along the environmental gradients.

Cases of co-occurring species in three ecoregions were highly exceptional, as each combination was composed of no more than 0.5–1.4% of the entire species pool, while only one was shared among all four ecoregions (0.5%). This lack of widely distributed species points to the preponderance of habitat specialists and suggests that true generalists in the region-wide macrofungal assemblages are actually absent.

2.4.1 Functional Group Distribution Patterns and Ecoregional Structuring

The interpolated surfaces showed clear spatial separation between functional groups—that is, the macrofungal community assembly is strongly influenced by ecoregions (Figure 2.8). Saprotrophic fungi, encompassing pure saprotrophs and various saprotrophic combinations (saprotroph/biotrophic,

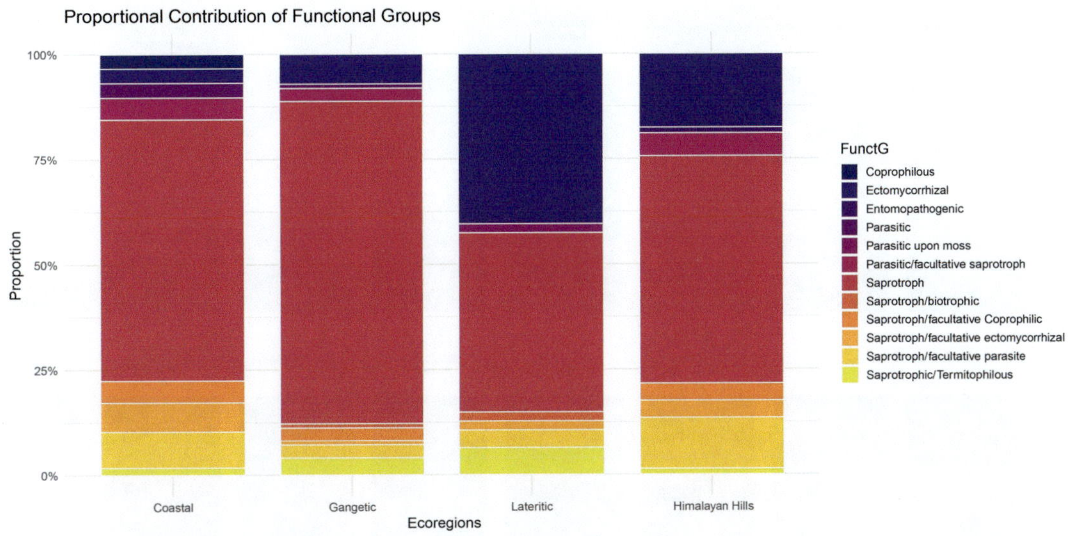

FIGURE 2.6 Proportion of functional groups across different ecoregions.

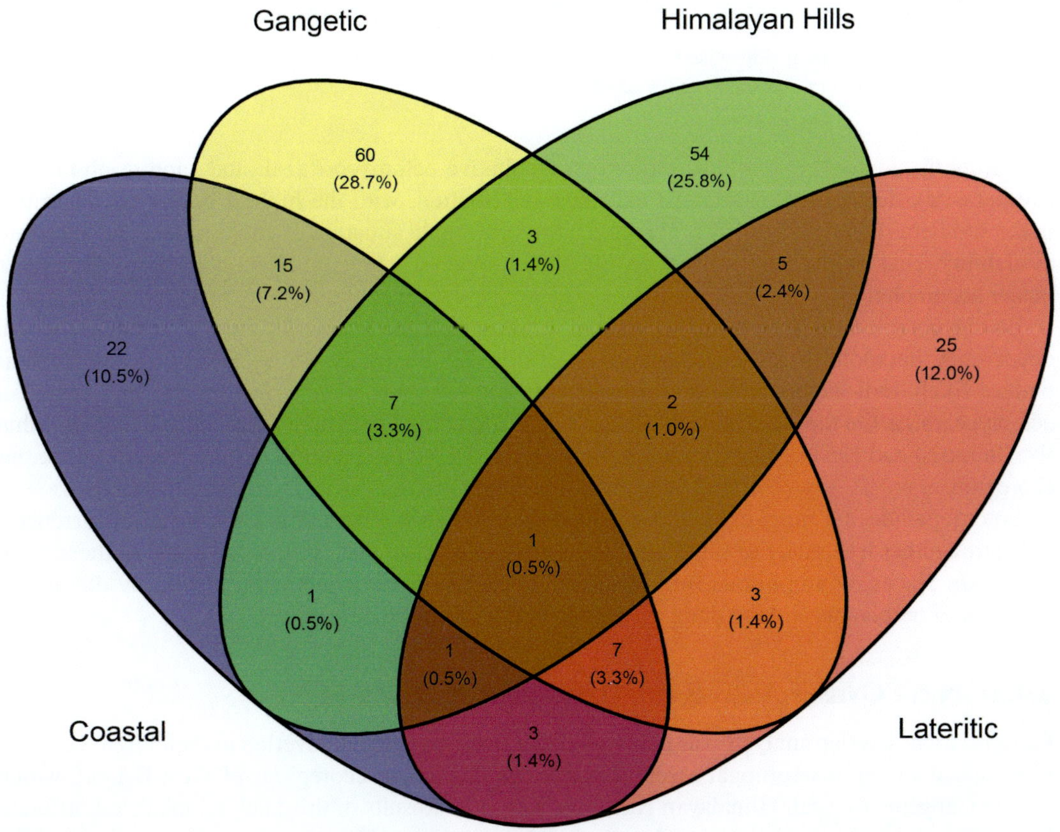

FIGURE 2.7 Venn diagram of species assemblage across four ecoregions.

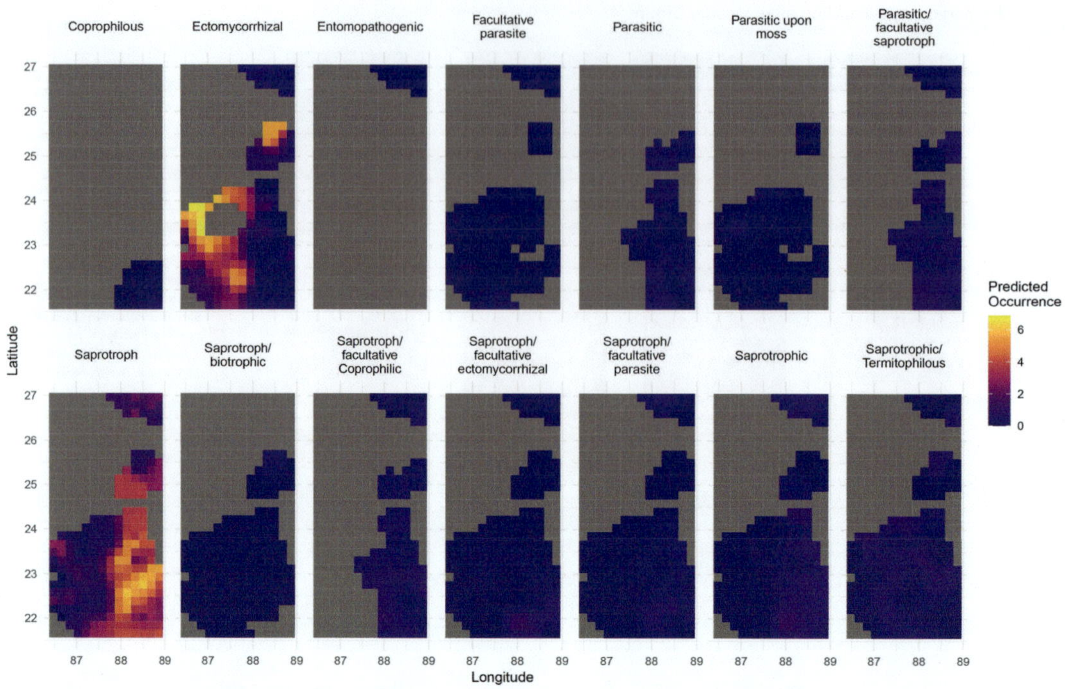

FIGURE 2.8 Functional group-wise inverse distance weighted (IDW) interpolation maps with 20 km grid and 60 km IDW search radius showing levels of predicted occurrence.

saprotroph/facultative coprophilic, saprotroph/facultative ectomycorrhizal, and saprotroph/facultative parasite), demonstrated the most extensive distribution, with the highest occurrence densities in the southern regions (22–24°N). This indicates a favorable situation of environment and substrate availability, characterized by warmer temperatures, increased precipitation, and enhanced organic matter accumulation.

Ectomycorrhizal and parasitic functional groups exhibited more limited distributions, with occurrence hotspots in the central-western sector (87–88°E, 23–25°N). The ectomycorrhizal pattern likely indicates that the host plants were present, particularly tree species that establish ectomycorrhiza. On the other hand, parasitic fungi (including facultative parasites and parasites that live on moss) had clustered distributions that coincided with the patterns of host plants in different ecoregions.

Entomopathogenic fungi exhibited minimal spatial distribution and low predicted occurrences, indicating either true rarity or sampling bias toward substrates that are less favorable to these taxa. In certain places, coprophilous fungi were found at moderate levels, which is likely due to the presence of herbivores and the dung they leave behind in those ecoregions.

2.4.2 Niche Overlap Analysis and Ecoregional Segregation

Pairwise niche overlap analysis was performed to assess the extent of overlap in their environmental niche space among macrofungal assemblages in the four major ecoregions of West Bengal, which are the Gangetic, Coastal, Himalayan Hills, and Lateritic. Results of this analysis indicated different levels of ecological segregation based on their environment gradients.

The niches of the Gangetic and the Lateritic regions showed the maximal level of similarity in the entire study (top-right in Figure 2.9). There was a significant overlap not only in the marginal regions but also in the core density areas, which indicates a sharp transition where the

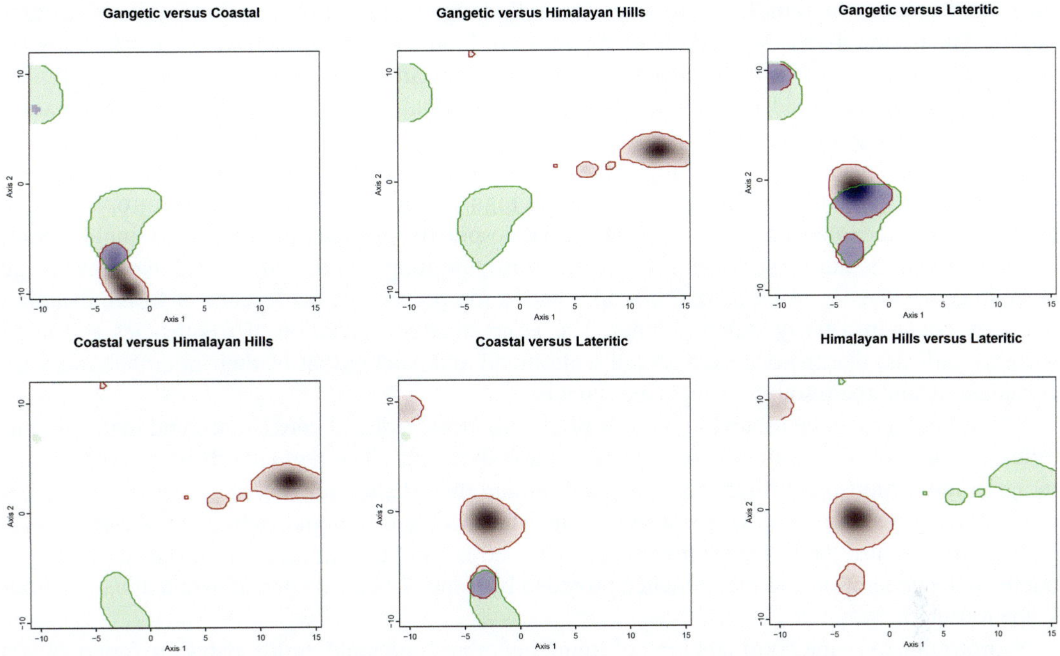

FIGURE 2.9 Pairwise comparisons of environmental niches among four ecoregions.

alluvial plains and the lateritic plateaus have a considerable level of climatic and edaphic factors to support a similar functional fungal group. On the other hand, the niches of the Gangetic and the Coastal regions, although geographically close, showed a moderate level of similarity. The core density contour of the two regions remained largely separate; thus, although they are geographically close, the saline and hydrological factors of the Coastal region play a major role in their niches:

Niche separation was most extreme between the Himalayan Hills ecoregion and the other regions. While the pair Gangetic Plains and Himalayan Hills had nearly total niche separation with density contours at opposite ends of the niche space, this is consistent with the vastly different ecologies of the tropical alluvial plain and the temperate mountain belt. Likewise, the pair Coastal and Himalayan Hills had the most extreme niche separation with no overlap in niche space, so it is clear that the niches of tropical coastal and temperate mountain fungi are fundamentally different.

The Lateritic niche is clearly distinct from the Coastal niche, with very little areas of intersection allowed only at the boundaries. This is also likely because of the substantial difference in substrate levels of dampness and salinity. Next, Himalayan Hills versus Lateritic niches presented a clear difference in core densities, with no intersection at all. However, the distance between these two niches seems a little less compared to that of Coastal and Himalayan niches. This is almost because of similar values of terrestrial substrate (non-saline soils) and not because of similarity in climate; lateritic regions are known to be hot and dry, unlike the cold and wet Himalayan climate.

2.4.3 Gangetic Plains

The Gangetic Plains came out as one of the main locations for species of macrofungi in these ecoregions, with a cumulative 97 species recorded, which formed the highest record for species of macrofungi surveyed for this study. Out of these, 75 species belonged to the saprotrophic functional group, which indicates to what extent this region can support various types of decomposers. The region displays a very diverse pattern of land use, which includes orchards, bamboo stands, agroforestry patterns, roadside vegetation, riparian zones, and crop residues. Climate patterns are also affected to a great extent by monsoonal patterns, which are a major factor influencing these patterns.

Diversity of species of macrofungi across various ecoregions, with their major categorized descriptions, is given below. These conditions are highly supportive of flexible saprotrophs, which can establish and exploit a new resource. However, despite the diversity of species, no single species dominated a particular geographical region. Saprotrophic fungi, on the other hand, had an average percentage occurrence of 0.076 (± 0.084 SD) with a top percentage occurrence of 0.385, meaning that they occur in a fairly stable manner. The large standard deviation of frequencies and large variance indicate that the occurrences are widespread and patchy; that is, their microhabitats vary considerably and the resource is replaced quickly.

These findings can be related to models of disturbance-mediated coexistence and metacommunities that account for mass effects and source-sink dynamics. The regular substrate turnover due to agriculture, monsoonal litter addition, and human disturbance prevents any one species from monopolizing a resource patch for a long time. The local and regional richness is therefore kept high. In this regard, the richness of fungi is sustained not by the stabilization of the environment, but by niche generation and maintenance provided by long-distance dispersal over a dynamic landscape mosaic.

Saprotrophs were the dominant type of fungi, but other functional guilds also were found. Seven species of ectomycorrhizal fungi, 4 species of parasitic fungi, 5 species of coprophilous fungi, and 3 species of termitophilous fungi were recorded. It is rather surprising to note that ectomycorrhizal species are found in this degraded environment. Although the ECM species were moderately diverse, the associations did not tend to remain in one place for too long. The frequency was only 0.0286, implying that the organism is very limited. This limited distribution is because the host trees are grouped together only in small numbers. For instance, only a few *Shorea robusta* are left, along with mixed moist deciduous forests along riverine tracts, marked as sacred groves, and protected forest patches. These act as ecological refugia harboring symbiotic fungal populations in a disturbed environment; however, they do not have the continuity required for the development of large ETM networks.

The Facultative Functional groups were much better in the Gangetic Plains than expected, and this is because they have the ability to shift their functional diets to cope with different landscape conditions. The functional group with the highest average frequency was the saprotroph-coprophilous fungi with a frequency of 0.138, which clearly shows that plant dung is abundant because of grazing and agricultural pasture use. The frequency is also higher for saprotroph-facultative parasites with a frequency of 0.062, which suggests there is sufficient plant distress because of agricultural disturbance.

The termitophilous fungi were of medium spatial persistence (mean frequency = 0.094), which corresponds to the wide range of termite hills found in the fine-textured alluvial soil. The termite hills act as nutrient-rich and temperature-buffered habitats that favor the local persistence of fungi and function to partially shield termitophilous species from the spatially large-scale variability. The termitophilous assemblages have an uneven spatial distribution and are not capable of filling all their spaces despite the many spaces being in one place.

This region is home to several unique species, including multiple species of *Agaricus* (*A. albostipitatus, A. duplocingulatoides, A. endoxanthus, A. glabriusculus,* and *A. trisulphuratus*), *Agrocybe pediades, Amanita aureofloccosa, Calocybella swetha, Cantharocybe virosa, Chlorophyllum hortense, C. molybdites, Clathrus delicatus, Clitocybula albida, Coprinopsis lagopides*, and several species of *Entoloma. Exidifda glandulosa, Gymnopilus* spp., *Hygrocybe lucida, Hymenoagaricus subepipestus, Laccaria vinaceoavellanea, Lentinus polychrous, L. sajorcaju, L. strigosus, Lepiota bengalensis, Lepista sordida, Leucoagaricus* spp., *Leucocoprinus cretaceous*, and a wide range of species of *Marasmius, Micropsalliota* spp., *Oudemansiella canarii, Phellinus conchatus, P. rimosus, Polyporus grammocephalus, Psathyrella* spp., *Rhodocybe brunneoaurantiaca, Russula* spp., *Termitomyces medius, Trametes flavida, Tremella mesenterica, Trogia benghalensis*, and *Volvariella bilobata* (Pradhan et al. 2010, 2012, 2013, 2016; Dutta et al. 2013; Chattopadhyay et al. 2022; Talukdar et al. 2023; Roy et al. 2025; Chattopadhyay et al. 2025) The prevalence of soil-dwelling saprotrophs, fungi associated with termites, and disturbance-tolerant lignicolous taxa highlight the ecological adaptability and niche diversification inherent to the Gangetic Plains.

Ecologically, the Gangetic Plains exist as an open, dynamically unstable environment for macrofungi. The region's fungal diversity is maintained not through the long-term coexistence of a given set of species, but rather as a constantly changing process influenced by fluctuations of environmental factors. As a consequence, a habitat characterized by a balance of strong abiotic factors and competitive exclusion, rather than a composition determined by habitat heterogeneity and habitat regeneration, comes into existence. In other words, the Gangetic Plains experience a constant process of maintaining a dynamically changing pool of fungal diversity, although a majority of those species exist patchily, lacking a broad geographic range.

2.4.4 Coastal Ecoregion

The Coastal ecoregion supported a mean of 58 species of macrofungi, with saprotrophic species being the most frequent (36). The ecoregion species composition indicates that abiotic factors are exceptional due to factors such as salinity intrusion, tides, high temperatures, moisture variation, and soil conditions, which are mainly composed of sandy to sandy-clay material. It is biologically restrictive, making it favorable to those that can tolerate conditions such as osmotic stress, periodic submergence, and high and low temperatures, particularly those associated with lignicolous and litter-fungal species.

The shoreline habitats supported a less functionally diverse assemblage than those of inland areas. The majority of fungi were saprotrophs, and ectomycorrhizal fungi were rare, with only two species recorded. However, this is not surprising because few obligate ECM host trees are present in mangrove-influenced and salt-tolerant shorelines. Only a few coprophilous and termitophilous fungi were present, as would be expected due to low grazing pressure and termite mound density. Parasitic fungi, although with few species (only three), formed a stable component of assemblages, possibly because their host plants are stressed and more susceptible to infection. Overall, abiotic stress factors rather than lack of substrate or biological interactions drove the macrofungal community composition.

The frequency analysis also added strength to the irregular occupancy pattern observed in this ecoregion. Frequency analysis showed that the mean occupancy frequency for the saprotrophic fungus was 0.092 (±0.105 SD), ranging from 0 to 0.468, and the median occupancy frequency was 0.052. Thus, it could be inferred that species occupied very rare locations, except for very abundant ones. *Ganoderma lucidum* and *Auricularia auricula-judae* had an occupancy value of 0.468, which meant they occurred under hyper-dominant conditions, as they could occupy very hostile environmental conditions. Thus, due to the presence of rare species, low evenness index, and lack of redundancy in species functionality in the community, the abiotic filter exerted on the species in this ecosystem is quite strong.

There was a reasonably high mean frequency with low variation registered in the parasite fungi (0.130) community, and this was due to host constancy in coastal conditions under physiological stress. The facultative saprotroph-ectomycorrhizal species recorded a higher mean frequency (0.133) due to adaptability to conditions where wood and roots are mixed together to create suitable conditions for a few symbiotic interactions to exist. Ectomycorrhizal species still recorded a lower frequency (0.039) and confirmed the secondary function of the obligates.

These data analysis revealed that the Coastal ecoregion is physiologically stressed, having limited biodiversity, yet better representing the prevalence of species that are physiologically more robust.

The Coastal ecoregion contained 22 exclusive species (10.5%), the fewest of any of the four regions. This ecoregion was home to a unique group of macrofungi that could live in salty environments, sandy to alluvial soil, and high humidity. *Agaricus placomyces, Amauroderma rugosum, Ascobolus scatigenus, Calocybe indica, Conocybe tenera, Coprinopsis lagopus*, and *Coriolopsis occidentalis* are some of the species that are found only in this area. Other species recorded were *Ganoderma applanatum, G. colossum, Hexagonia apiaria, H. badia, Lenzites elegans, Macrocybe crassa, M. lobayensis, Marasmius rotula, Mutinus caninus* var. *caninus*, *Pisolithus arhizus, Rigidoporus lineatus, Sphaerobolus stellatus, Trametes leonina, Tulostoma chudaei*, and *Volvariella pusilla*. The prevalence of lignicolous polypores and saprotrophic agarics highlights the significance of decomposing woody debris, coastal vegetation detritus, and moisture-laden microhabitats in shaping fungal communities.

2.4.5 Himalayan Hills

There were records of 74 species of macrofungi from the Himalayan Hills ecoregion, ecologically ascribed to the quick transitions between elevational zones, the mostly rugged terrain, and varied microhabitats for habitation. However, most of those species were saprotrophic fungi. This is indicative of a rather high decomposition potential for montane ecosystems as a whole. Fungi are optimal for environments with high rainfall, high cloud cover, and the presence of coarse woody material. There were 13 species of ectomycorrhizal fungi, with moderate abundance, ascribed to the availability of ECT tree host ranges including *Quercus*, *Castanopsis*, and *Lithocarpus*, with *Pinus* higher up, but due to uneven dispersal of tree ranges themselves, ECT fungal distribution is fragmentary, thereby reducing the chance of a community of fungi being geographically aggregated for mutual symbiotic advantage and exclusion of exotic fungi. There were rare and endemic ECT fungi such as *Cordyceps militaris*, a pathogen of insects, thereby showing the importance of a functional insect community along with a stable and cool micro-ecology for maintaining rare fungal ecologies themselves.

Most taxa displayed very spatial fragmented distributions. The macrofungal sporocarp is adapted to temporal and spatial conduciveness for its survival, including such things as low temperature, fog, steep slopes, or short fruiting seasons. Hence, communities that are diverse in one place but fragmented in another can be observed as per their adaptation or availability of the substrate/hosts. A large turnover of species along elevational, slope, or moisture gradients was observed. The beta diversity was observed to be high, and microhabitats were highly specialized in this manner.

Such fragmentation is revealed by the frequency analysis. The average frequency of the saprotrophic fungi was 0.042 (± 0.053 SD), with a maximum of 0.263. On the other hand, the average frequency of the ectomycorrhizal fungi was very low at 0.0231. However, all the functional groups maintained low values of average frequency, indicating that most of the species in this case were rare. The Shannon diversity index values of the saprotrophic fungi were high, at about 3.30. However, this diversity was associated with rarity rather than dominance, which implies niche partitioning (Figure 2.10).

This ecoregion had the second-highest number of species able to thrive in particular habitats, with 54 species (25.8%) occurring only in this area. This group was mostly made up of ectomycorrhizal fungi (*Russula, Cortinarius*, and *Amanita* allies), cold-adapted saprotrophs, and lignicolous polypores. This shows how the structure of the montane forest, with cooler temperatures, better moisture retention, and less human disturbance affected them. Some of the exclusive taxa that were representative were *Agaricus xanthodermus, Aleuria aurantia, Armillaria vara, Auricularia auricula-judae, Bjerkandera adusta, Calocera cornea, Clavulina cristata, Coprinus comatus,*

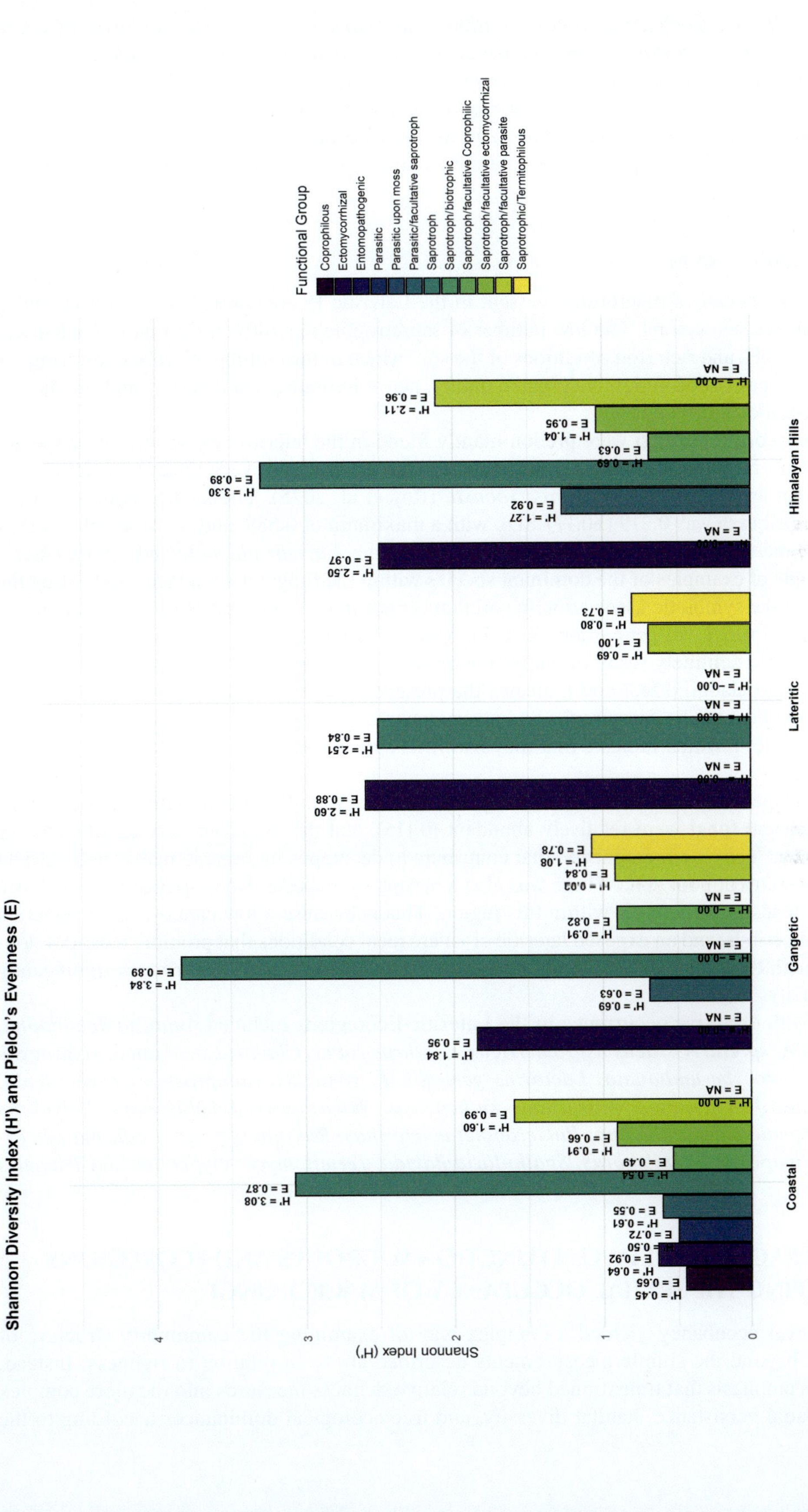

FIGURE 2.10 Shannon diversity index (H′) and Pielou's evenness (E) of macrofungal functional groups across four ecoregions (coastal, Gangetic, lateritic, and Himalayan Hills). Bars represent Shannon diversity (H′) for each functional group within an ecoregion, with corresponding evenness (E) values indicated above bars.

Cordyceps militaris, Cortinarius iodes, Daldinia concentrica, Favolus manipularis, Fistulina hepatica, Flammulina velutipes, Fomes fomentarius, Geastrum triplex, Grifola frondosa, Helvella crispa, Hericium coralloides, Hypholoma fasciculare, Meripilus giganteus, Pholiota squarrosa, Pleurotus flabellatus, Pseudohydnum gelatinosum, several *Russula* species, *Scleroderma areolatum, Sparassis crispa, Trametes versicolor*, and *Xerula radicata*. The fact that there are so many unique species shows how important it is to protect Himalayan forest ecosystems because they are so different from other ecosystems.

2.4.6 Lateritic Region

The number of species of macrofungi present in the Lateritic Ecoregion (47 species) was among the lowest in four ecoregions. The low number of saprotrophic macrofungi (20) indicates nutrient deficiency, low pH, and rich iron conditions of the soil, which in turn inhibit microbes, lowering the accessibility of easily decomposable organic matter, hence inhibiting the diversity and function of the non-symbiotic macrofungi.

The ectomycorrhizal fungi were predominantly found in the lateritic region, where 19 species were recorded, the highest among all the regions. The species were very much associated with sal forests that are dominated by *Shorea robusta* (Roy et al. 2025). The mean frequency of the ectomycorrhizal taxa was 0.219 (±0.177 SD), with a maximum of 0.589, and the assemblages they formed were massive and persisted for a long period of time. *Laccaria laccata* and *Russula kanadii* are a couple of examples of the dominant species within the fungal assemblages, indicating the importance of the symbiotic association in soil that is poor in nutrients and rich in iron, where the ectomycorrhizal fungi can get carbon from the host and control the organic nitrogen pools. The dominance would definitely result in the inhibition of saprotrophic fungal activity; this is in consonance with the Gadgil effect and indicates the presence of substantial functional trade-offs in the soil carbon fluxes. The saprotrophs had a very low mean frequency (0.041) and a low median occupancy (0.0167), which indicate that they had little spread, although there were many species among them.

Functional specialization of the lateritic region extends beyond ECM domination. For example, termite-associated fungi were relatively abundant (0.115), and this indicates how termite species also function as 'ecosystem engineers' that contribute to decomposing organic matter and nutrient movement in nutrient-poor soils. There was also a bryophilous species (moss-parasitic), which is a species that tends to thrive only within this region. This is because a low canopy, water retention during moist periods, and an exposed lateritic crust are ideal conditions that promote moss development. Although these species are not abundant, their functional diversity contributes significantly to beta diversity.

Species with exclusive occurrence in the Lateritic Ecoregions included *Amanita hemibapha, A. princeps, A. spreta, Astraeus hygrometricus, Clathrus ruber, Clavaria straminea, Hygrocybe angustifolia, Inocybe umbonata, Lactarius gerardii, L. zonarius, Lamprospora carbonicola, Leucocoprinus fragilissimus, Marasmius suthepensis, Microporus flabelliformis, Polyporus brumalis, Pseudochaete tabacina, Pulveroboletus icterinus, Russula* spp., *Scleroderma macrorhizon, Scytinopogon angulisporus, Spathularia flavida, Termitomyces clypeatus*, and *Trametes hirsuta*.

2.5 INTERACTIONS BETWEEN FUNCTIONAL GROUPS AND ECOREGIONS SHAPING THE SPATIAL OCCUPANCY OF MACROFUNGI

A species-level occupancy yielded a complex way of examining the community structure of macrofungi beyond the simple measurements described above in relation to richness. Instead, there was an emphasis that transitioned beyond relative richness measures into the more complex realm of spatial persistence, habitat diversity, and true ecological dominances according to the largely different concept of landscapes (Pradhan et al. 2013). A useful aspect of occupancy was that the presence of sporocarps was informative in relation to fungi, since their presence indicated the range and duration of the mycelium beneath the soil's surface rather than their relative abundance at any point in time. Here, occupancy subsumed a multiplicity of ecological factors and perspectives all at once—such as the continuity of substrate support, the availability of hosts, microclimatic suitability, effective dispersal, and the persistence of resource functionality both within and between ecoregions.

To quantify the effect of macrofungal occupancy as influenced by the intersection of the two factors mentioned above, the general linear model described above was applied to calculate the proportionate occupancy of the macrofungi represented. Therefore, the general linear model applied, which described the occupancy represented by the proportionate occupancy of the macro-fungi, took the form $Occ_i = \beta_0 + \Sigma\beta\_FG(FunctionalGroup) + \Sigma\beta\_Eco(Ecoregion) + \Sigma\beta\_FG{\times}Eco(FunctionalGroup \times Ecoregion) + \varepsilon_i$. There was a general significance of the model's proportionate effect on the occupancy represented, which accounted for a statistically significant proportion of the observed variation ($R^2 = 0.257$; Adjusted $R^2 = 0.160$; $F_{32,244} = 2.64$; $p = 1.44 \times 10^{-5}$). A larger R^2 indicated that the model accounted for more than half the observed proportionate effect, while the low value given by the residual standard error indicated that the model accounted for ecologically significant changes when the coefficients changed by very small amounts, indicated by the residual standard error of the model ($\sigma = 0.0908$).

The intercept ($\beta_0 = 0.039$, $p = 0.544$) represented the presence of a saprotrophic species at 3.9% of the transect in the Coastal ecoregion. The presence ratio was close to what actually existed on the planet. It is practical evidence that even in such open conditions, there existed limited ranges of every species of macrofungus.

A lower baseline highlighted the importance of microhabitat specificity, the lack of continuity in available substrates, and the effect of dispersal limitation on fungal communities. It supported that even if richness is high, conditions are unlikely to be so.

However, the leading effects of functional group identity were mostly at or near zero, suggesting the presence of significant intragroup variation and strongly context-dependent effects. Yet, distinct ecological cues were evident. Facultative parasitic-saprotrophs had the strongest positive main effect ($\beta = 0.156$, $p = 0.061$), suggesting greater occupancy on average within the capabilities of species that efficiently exploit decomposed organic material along with their live hosts. This conformed with their reported patterns of occurrence within ecoregions featuring the geographical co-occurrence of host tissue and decomposable organic matters in fine-scale spatial mosaics, specifically in the Coastal and Gangetic regions. Facultative parasitic fungi had the least positive main effect ($\beta = 0.091$, $p = 0.318$), with suggestions that the geographical distributions of these fungi were largely regulated by the patchiness of host matters. Biotroph-saprotrophs with intermediate properties had negative main effects ($\beta = -0.104$, $p = 0.498$), suggesting the importance of geographical microsites having naturally more restricted spatial scales. For moss-parasitic fungi, the main effects also were negative ($\beta = -0.059$, $p = 0.699$). This suggested the importance of geographical sites having naturally restricted spatial scales. Ectomycorrhizal had negligible main effects ($\beta \approx -2.08\times10^{-15}$, $p = 1.000$), suggesting the absence of natural occupancy advantages regardless of host geographical distributions. Termitophilous fungi had the same negligible main effects ($\beta \approx 3.96\times10^{-15}$, $p = 1.000$), validating the absence of any necessary connection with extensive geographical factors.

By contrast, the overall effect of ecoregions was also small and nonsignificant (Gangetic Plains: $\beta = 0.042$, $p = 0.681$; Lateritic Plateau: $\beta = 0.076$, $p = 0.470$; Himalayan Hills: $\beta = 0.061$, $p = 0.635$). This indicates that overall, ecoregions did not weaken and strengthen fungal occupancy in a similar manner when considered alone. Rather, overall, group by ecoregion interactions contributed most to ecological differentiation, confirming the importance of the interaction model and confirming that geographic location alone was insufficient and matched by trait-environment pairings concerning fungal spatial distributions.

However, the interaction values shed the most light on the ecological aspects. A positive ectomycorrhizal-lateritic interaction ($\beta = 0.104$, $p = 0.405$) raised the predicted ECM presence by about 14.3% over and above the main and interaction effects. This was very close to the observed mean ECM presence of 0.219 in lateritic forests. This interaction indicated the very close spatial association between the presence of ectomycorrhizal fungi and lateritic forests that harbor a lot of salt, where the continuous host root tissue and nutrient-poor soil allow the fungi to absorb nutrients derived from their host. On the other hand, the negative saprotroph-lateritic interaction ($\beta = -0.128$, $p = 0.238$) reduced the predicted saprotroph presence to be about 4%, which was very close to the observed suppression of decomposers in the system. On the other hand, in the Himalayan Hill forests, all the interaction terms were significantly negative, but the strongest was that of the facultative parasitic-saprotrophs ($\beta = -0.228$, $p = 0.120$), suggesting the very close effect of the environment, the large climatic gradients, and the fragmentation of the landscape as a whole. However, many of the rare group interactions were not estimable, primarily because of the sparse data, as the occurrence of highly specialized species was very difficult.

The occupancy distributions were highly right-skewed in all four ecoregions. Species that occupied very low numbers (<0.05) accounted for 68% of the total number of species, while those that demonstrated high spatial persistence (>0.25) accounted for just 7%. However, these dominant species made up 43% of all occupied instances in their ecological niches, while their strategies demonstrated a high degree of differences in their ecological niches, despite having similar statistical structures in all four ecoregions (Figure 2.11).

Stress-resisting saprotrophic fungi represented the bulk of the fungal assemblages in the Coastal ecoregion. *Auricularia auricula-judae* and *Ganoderma lucidum* represented around 42–48% of the total occupancies in the sampled transects, revealing their capacity to infect a wide range of lignocellulosic materials with higher temperatures, salinity, and moisture. Moderate-range saprotrophs such as *Schizophyllum commune* (0.28), *Dacryopinax spathularia* (0.19), and *Trametes hirsuta* (0.17) demonstrated the mechanism of fine-scale niche partitioning with respect to the vertical and successful scales of wooden debris. Ectomycorrhizal fungi, on the other hand, were exceedingly sparse in the studied region, with a mean occupancy of 0.021%. The reason behind this low prevalence might be the limited range of suitable host trees owing to frequent land use changes in many areas.

The Gangetic Plains exhibited a mixture of occupancy, moderate dominance, and the highest functional diversity. Saprotrophic fungi such as *Marasmius nigrobrunneus* (0.32%), *Auricularia auricula-judae* (0.29%), *Schizophyllum commune* (0.27%), and *Chlorophyllum molybdites* (0.26%) were recorded over large areas, including crop fields, orchards, riversides, and isolated forests. The model value of saprotroph occupancy (~9.2%) was a precise match for ground truth measurements. The ectomycorrhizal species recorded a moderate degree of occupancy (mean = 0.16) and were predominantly found under degraded stands of *Shorea robusta*. In contrast, higher termite mound distribution and corresponding termite-associated fungal species occupancy values were recorded at a mean of 0.08–0.15. This indicates the extent of the termite hill.

The Lateritic Plateau had the highest degree of functional differentiation. Ectomycorrhizal species had the highest spatial prevalence. *Laccaria laccata* (0.56), *Russula alatoreticula* (0.51), *Amanita vaginata* (0.48), *Lactarius zonarius* (0.45), and *Russula brunneoaurantiaca* (0.42) formed an uninterrupted ECM throughout sal forests. They quantitatively represented 38% of the fungal populations, though they comprise only 7% of the fungal species. There was strong spatial suppression among saprotrophic fungi (mean occupancy 0.041), which were mostly isolated in areas of wood debris. Gadgil's functional differentiation postulated that dominant ectomycorrhizal fungal communities are capable of retaining organic nutrients in place and preventing saprotrophic function in soils that lack adequate nutrients.

The Himalayan Hills had the most fragmented occupancy pattern. The overall mean occupancy was small (0.034), and singletons comprised a substantial percentage (41%) of species. Only a few cold-hardy saprobes, *Xylaria hypoxylon* (0.23) and *Trametes versicolor* (0.17),

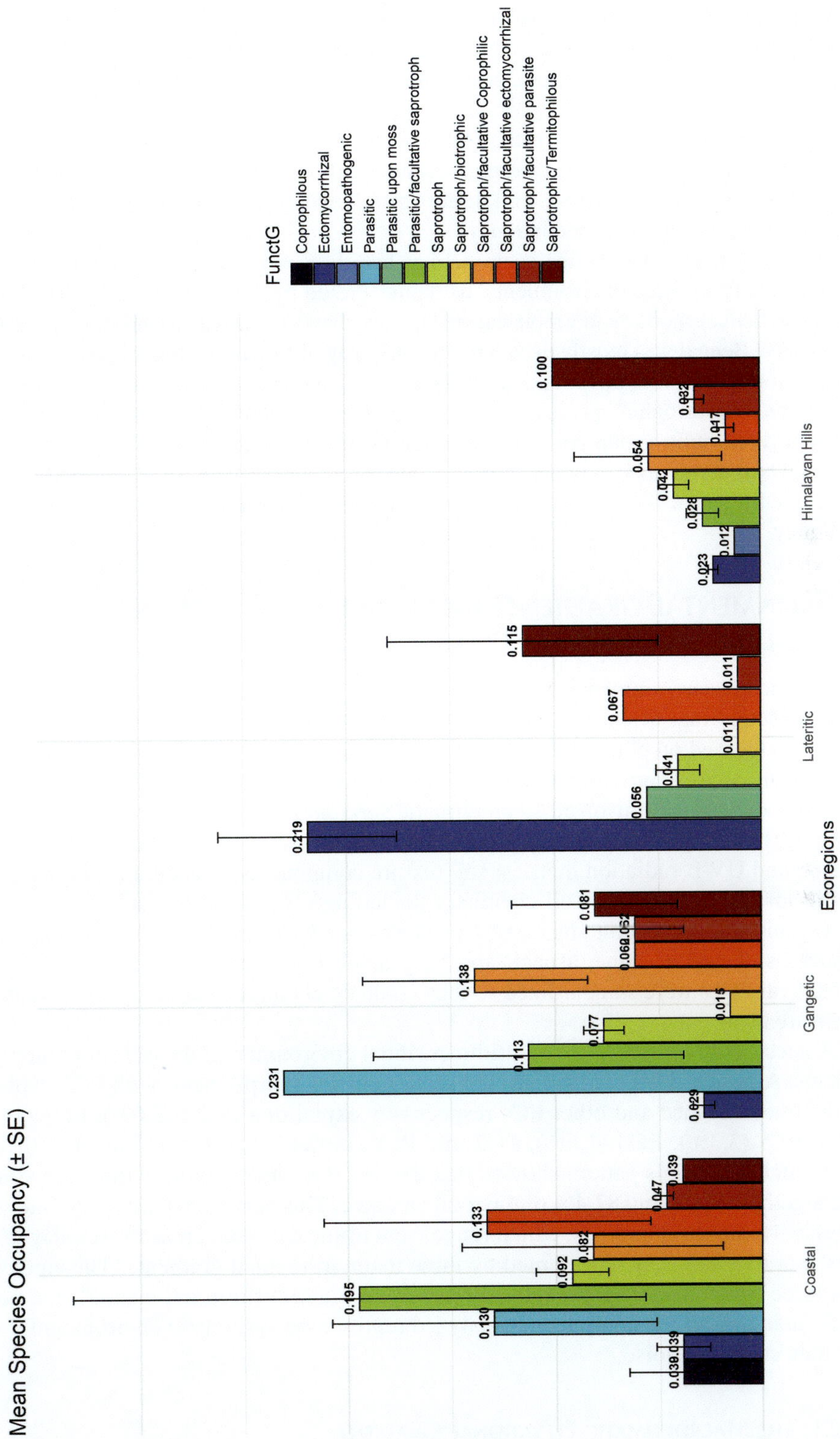

FIGURE 2.11 Mean species occupancy of different functional groups across ecoregions.

achieved a moderate level of occupancy. Ectomycorrhizal taxa, though represented, had a small mean occupancy (0.048), which is not surprising because trees occur patchily over a wide elevational range. The negative values in these interaction matrices represented a cumulative effect of climatic severity, terrain, and substrates, strengthening the hypothesis for niche fragmentation in these mountains.

For all ecoregions combined, 47% of the total species pool consisted of rare fungi with occupancy levels below 0.02. These species were associated with very fine habitats like dung, bamboo litter, advanced wood decay, moss mats, and insect hosts. Though those species were rare individually, they contributed to functional redundancy.

It illustrates just how valuable fine-scale habitat heterogeneity was for maintaining regional macrofungal diversity in these environments, as demonstrated by Pradhan et al. (2013). Based on the integrated statistical model and empirical occupancy matrices, it was found that species-level occupancy in West Bengal was mostly driven by the interplay of functional traits and environments, and not by environment alone, or functional traits and environments separately. For example, abiotic stress favored saprotrophic species that are physiologically tolerant in coastal areas; habitat heterogeneity favored multi-group coexistence in the Gangetic Plains; biotic components such as competition and host dominance in lateritic forests; and abiotic filtering for all species in Himalayan Hills ranges. Finally, species-level occupancy was made to be a useful metric for macrofungal community assembly.

2.6 ENVIRONMENTAL GRADIENTS INFLUENCING MACROFUNGAL ASSEMBLAGES: A PRINCIPAL COMPONENT ANALYSIS

We used PCA on 130 environmental variables measured at 150 sampling sites to find the main environmental factors that affect macrofungal distribution across West Bengal's diverse ecoregions. The study focused on climatic factors, soil properties, radiation, and topography factors to illustrate how a hierarchical environmental structure would work to assemble fungal assemblages. Such environmental variables involved 19 bioclimatic variables that were related to the annual and monthly averages of temperature, precipitation, water vapor pressure, and wind speed; monthly solar radiation and UV-B radiation metrics; soil texture components measured at six depth intervals from the surface to 2 meters; soil chemistry, including pH and cation exchange capacity at all depths; organic carbon content and density profiles; coarse fragment volume; and topographic variables such as elevation, slope, aspect, and geographic coordinates. Before the analysis, all environmental variables were standardized so they could be compared across different scales and units of measurement.

The PCA successfully reduced the multidimensional environmental dataset into comprehensible gradients. After the third component, the scree plot was sharply bent, with PC1 explaining 49.0% of the total variance and other PCs respectively explaining PC2 (23.6%), PC3 (14.8%), PC4 (2.7%), PC5 (2.3%), PC6 (1.5%), PC7 and PC8 (0.9%), PC9 (0.8%), and PC10 (0.5%) (Figures 2.12 and 2.13). This pattern showed that the first three axes captured the main structure of the data, together explaining 87.4% of the total variance. This very high percentage shows that West Bengal's diverse environments, which range from tropical coastal areas to Himalayan montane forests, can be accurately represented by three main orthogonal gradients. The eigenvalues for the first three components significantly surpassed the Kaiser criterion threshold, and parallel analysis validated that all ten components analyzed represented systematic environmental structure rather than sampling noise.

2.6.1 PC1: The Macroclimatic-Elevational Gradient

The first principal component captured the dominant environmental gradient across the study area, ranging from warm, humid, and thermally stable lowland landscapes to cooler, more seasonal, and topographically complex upland regions. This axis combines closely integrated gradients

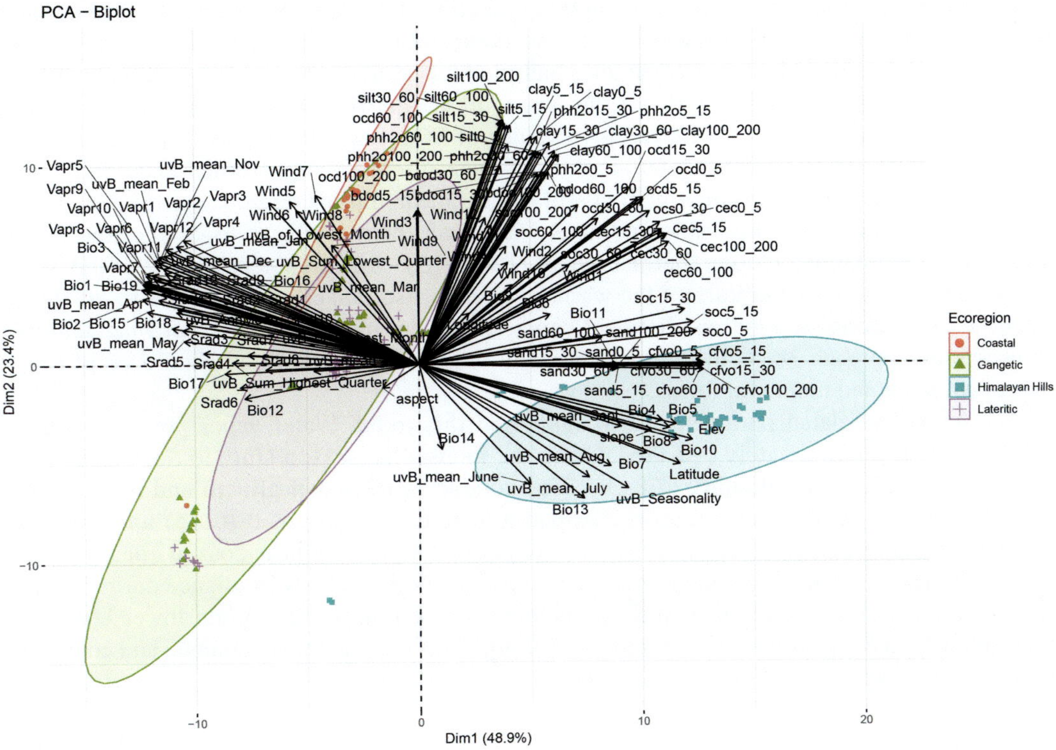

FIGURE 2.12 PCA biplot illustrating environmental variation across the four studied ecoregions.

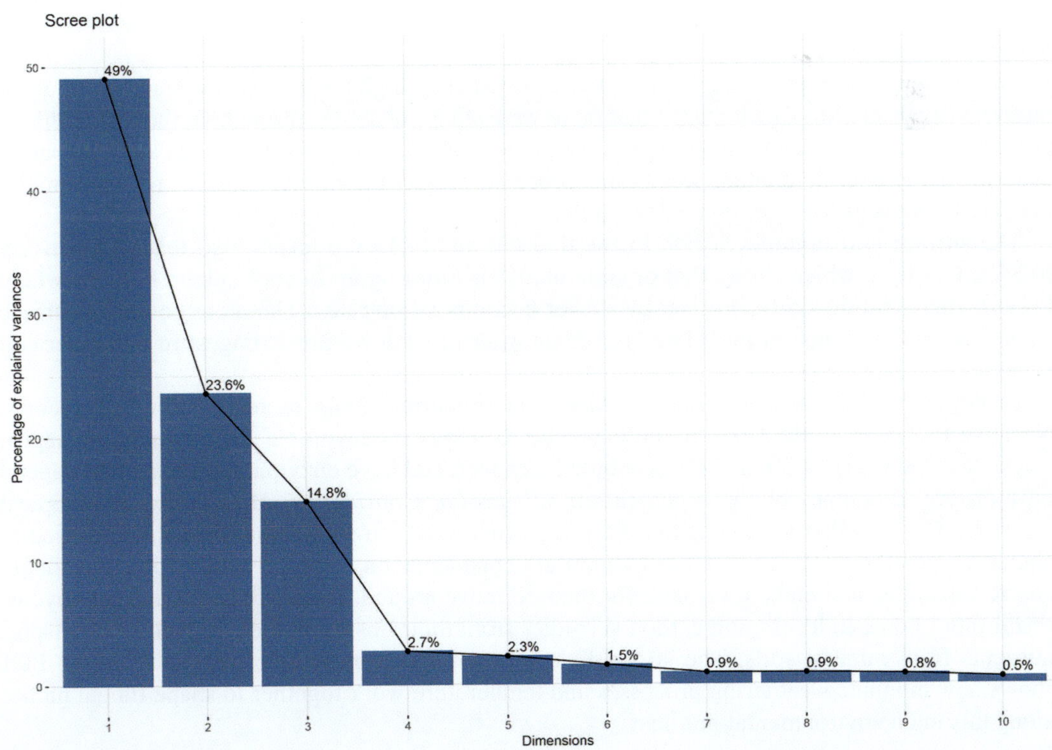

FIGURE 2.13 Scree plot of eigenvalues from the PCA.

of climate, topography, and soil development, reflecting the major transition of biogeographic structure. Variables of tropical lowland conditions ranked highest in loading on the negative side of PC1. Of these, annual mean temperature had the highest absolute value of loading (−0.116), suggesting that thermal conditions are the major dimension of this axis. For the Coastal and Gangetic ecoregions, stations with negative values of PC1 corresponded to areas having mean annual temperatures above 25 °C. Of similar importance in assessing thermal conditions, measured by negative values of isothermality (−0.116) and mean diurnal temperature range (−0.116), was the differentiation of lowland conditions with small thermal ranges (6–8 °C) from upland areas having substantially larger thermal ranges (10–14 °C). Maximum temperature of the warmest month and mean temperature of the warmest quarter also ranked high (−0.108 to −0.109) in importance, reflecting top thermal limits of conditions in pre-monsoon retrospect when temperatures readily approach 38 °C in summer, unlike in uplands having lowland conditions of low to absent cold-adapted mountain forms.

The humidity-related parameters also supported this trend. Water vapor pressure, a negative loading of approximately (−0.115) during the monsoonal season (July to September), represented high levels of humidity (2.8–3.4 kPa) conducive to fungal growth and reproduction. Precipitation during the driest season also had a negative loading of (−0.115) and suggested a continued supply of moisture in lowland ecosystems, especially near coastal zones. What is interesting is the negative loading of (−0.115) approximated by UV-B irradiance during the pre-monsoonal season of March and April, a counterintuitive finding suggesting lower UV irradiance in lowland ecosystems, probably a result of high humidity, aerosol contents, and convective cloud development, underlining the fact that processes, and not just latitudes, dominate lowland radiation patterns in the atmosphere.

In contrast, cooler, topographically more complex habitats with pronounced soil development showed positive load values for PC1. Elevation showed a strong positive loading value (0.115), supporting the fact that PC1 captures an elevation gradient from near sea level to well above 2400 meters in the Darjeeling Himalayas. Latitude showed a small but positive value (0.110), supporting the idea of transitioning from lowland habitats that are warm and wet to cooler, more seasonal conditions at higher elevations.

The content of coarse soil fragments showed consistently high positive loadings across all depth intervals (0.118–0.120), which means that there were stony substrates throughout the soil profile at positive PC1 sites. The signal's persistence with depth, peaking at 60–100 cm, indicates that parent material and long-term montane weathering processes, rather than merely surface rock accumulation, are responsible for upland soil differentiation.

The surface soil organic carbon in the 0–5 cm and 5–15 cm layers also loaded positively (0.117 and 0.115), which shows that organic matter is building up in cool upland systems where lower primary productivity is offset by slower decomposition rates. The more powerful surface signal shows that temperature-related limits on litter breakdown are strongest in the upper soil horizons.

Ecologically, PC1 acts as a master filter that organizes pools of regional fungal species. Negative PC1 sites in the Coastal and Gangetic lowlands are favorable for thermophilic saprotrophs that grow best at 30–35 °C, have quick turnover, and have enzymatic systems that depend on moisture. Examples of these genera are *Volvariella, Coprinellus, Marasmius*, and tropical *Agaricus*. On the other hand, positive PC1 sites choose for stress-tolerant taxa that are used to cooler temperatures (15–25 °C), have temporal dependence on resource availability, and rough, coarse substrates that make it difficult for them to move around. In these conditions, ectomycorrhizal fungi flourish, leveraging carbon subsidies from their hosts and their capacity to transport nutrients from mineral soils. The strong link between elevation and soil stoniness along PC1 shows how montane weathering processes and temperature work together to shape fungal niches along this main environmental gradient.

2.6.2 PC2: The Soil Texture-Chemistry Gradient

The second principal component was associated with orthogonal components of physical and chemical properties of soils, which grouped sites according to pedogenic development and parent material, irrespective of the climatic-elevational factor described by PC1. PC2 explained a cumulative percentage variance of 23.6%, which was mainly described by positive loadings associated with soil texture and fertility. The loadings for silt were highest on this component, which was 0.167 at depths ranging from 60 to 100 cm, 100 to 200 cm, and were slightly lower at other depths (0.163) in the top layers. The relative constancy of loadings for silt at various depths of this two-meter soil profile indicates that this factor describes relatively stable textural properties described by parent material rather than recent reworking at or near the surface.

The clay fraction components also had high positive loadings, with the largest in the surface horizons (0.155 in the 0–5 cm and 5–15 cm layers) and the smallest in the 30–60 cm range (0.147). This is an indication that the process of clay illuviation or in situ weathering was important in the delineation of PC2. High soil organic carbon density in the deeper soil horizons had positive loadings, with the largest in the 60–100 cm layer (0.150) and high in the deeper soil horizons (0.135 in the 100–200 cm layer), indicating the presence of significant amounts of mineral-associated carbon in high PC2 soils. This is an indicator that the organic matter has been relatively constant for a longer period of time in fine-proportioned matrices of minerals, which is different from the surface accumulation as demonstrated in PC1.

The soil pH in water was positive at all depths (≈0.144–0.145), indicating that PC2 is an acid and base gradient, where a positive factor represents a non-acidic to weakly basic (pH 6.5–7.5) area. Cation exchange capacity had moderately positive loadings (0.096 at the surface, decreasing to 0.083 at 30–60 cm), which indicate that PC2 not only considers nutrient retention capacity, but also considers texture and pH. Sand had a negative influence, which distinguished fine, non-acidic soils from coarse, possibly acidic substrates.

The Gangetic Plains have silty clay alluvial soils that are silty clay crumbs. These soils were formed due to the presence of Himalayan catchment sediments deposited for a long period of time. These soils have fixed moisture due to their ability to retain water due to capillary action and the presence of numerous nutrients. They also contain chemicals that can neutralize soil pH and provide optimal conditions for various saprotrophic organisms. There is a higher cation exchange capacity and soil base saturation that favors the presence of herbivores and termites. These organisms serve as nitrogen rich food sources, called "coprophilous fungi" and termitophilous species, like *Termitomyces*. The Gangetic Plains also have vertical variations in texture that allow the growth of species with varying hyphae exploration capacities.

Low scores in PC2 are associated with sandy shorelines and eroded lateritic top soils that are very harsh for plant and animal survival. Low cation exchange capacity indicates very poor plant nutrient retention, high drainage rates that stress plants by drying out too frequently, and very low pH (4.5–5.5) that reduces nutrient bioavailability and detoxifies enzyme systems, as well as very toxic aluminum. Reduced substrate chemistry and poor clay-organic interactions are characterized by poor niche partitioning and dominate stress-resistant functional groups such as melanized polypore wood decomposers, acid-loving ectomycorrhizal taxa (e.g., *Russula*), which are associated with nutrient-poor host trees, and pioneer fungal species adapted to temporary resource patches.

This orthogonality of PC2 to PC1 reveals that the factors of soil fertility and texture vary independently of temperature and elevation, causing the environment to be complex for the ecologists to interpret. When the complete range of the combined PCs of 1 and 2 is occupied, there emerge different environments such as warm/sandy, warm/silty, cool/sandy, and cool/silty, each supporting different species of fungi, yet they all occupy the same axis of environment. This observation is supported by the lateritic plateaus, which occupy the middle range of the PC1 axis while spread widely in the range of the second axis, depending on their erosion and clay composition, causing a notable variation in environment within their ecologic regions.

2.6.3 PC3: Seasonal Wind-Precipitation-Radiation Dynamics

The third main component explained 14.8% of the total variability, capturing temporal and spatial variability of dynamic climate processes. This component represents how changes in climate and other long-term changes occur without necessarily interfering with climate structure represented by PC1 and soil fertility represented by PC2.

Mean temperature of wettest quarter: PC3 loadings were most positive (0.186) and aided in the location of areas where hot conditions and high rainfall coincide. A wet and hot monsoon regime, which usually accompanies convective precipitation systems triggered by summer heating, is expected. Such conditions promote higher rates of decay and fungal metabolism. Transitional wind speeds included very high positive values, especially at the beginning and end of the monsoon (March: 0.166, September: 0.170, October: 0.163, and November: 0.159). This PC3 is able to detect variations in wind speeds at critical periods of the season. These correspond to times when fungal sporulation and spore discharge are at their peak, so atmospheric turbulence helps lead to the dissemination of spores. A positive load (0.151) indicated that the longitudinal component is more appreciably affected by the presence of the ocean, especially on the east and west sides of the Bay of Bengal, where monsoons, typhoons, and moist-wind patterns abound. Also positive (though less) was the load on the mean temperature of the coldest quarter (0.135), which aided in differentiating regions experiencing less cold conditions and thus longer periods of more or less continuous fungal growth.

Negative PC3 factors existed in areas experiencing much stress because of radiation and dry seasons. The rainfall in the driest month had a strong negative loading of −0.189, which shows that it was very dry that year, which prevents fungi from growing or puts them in dormancy until the monsoon. Rainfall in a year had a negative loading of −0.161, which shows that it was very dry that year, which prevents fungi from growing or puts them in dormancy until the monsoon. The negative loading indicates that PC3 distinguishes rainfall seasonality from total rainfall. This indicates that high PC3 values correspond to high rainfall, even when monsoon rainfall is low. From April to June, solar radiation in summer had a strong negative loading of −0.143 to −0.157, which indicates that it was forceful before the monsoon, which enhanced photo-inhibitory stress. The subsurface sand percentage in the 30–60 cm and 60–100 cm layers had strong negative loadings of −0.127 to −0.134, indicating that the sand is coarse and it can lose much moisture, especially during drought, thus increasing seasonal stress.

Ecologically speaking, PC3 is a "kinetic energy" gradient, which reflects the importance of winds, pulsed rains, and radiation-driven moisture derived from monsoons to the probability of range extension, the availability of moisture, and photobiological stress, respectively (Büntgen et al. 2012). In terms of seasonal change, when the warm, wet monsoon phases are associated with strong winds, which are conducive to the range extension of wind-borne fungi, the regions with high values of PC3, which are still found within the east Ganges River plains, are the favored habitats for the wind-borne fungi.

On the other hand, low PC3 areas in the Western Lateritic Interior and Rain Shadow regions are under greater stress due to the intensity of radiation preceding the monsoons as well as the aridity associated with the dry season. This provides an advantage to xerotolerant species, which exhibit strategies such as the presence of sclerotia, thick-spined or melanized spores, nocturnal fruiting, or survival via a well-developed mycorrhiza. The loading factors on the wind and rainfall variables suggest that there is an underlying trade-off in which the presence of strong seasonal wind is associated with episodic rather than regular rainfall. The periodicity that comes with this makes the environment less amenable to opportunistic reproduction opportunities, supporting fungi that live in highly synchronized biotas and perhaps accounting for the erratic yet highly productive fructification observed in the low PC3 areas.

2.6.4 Ecoregional Clustering and Its Biogeographic Implications

From the PC1-PC2 bi-plot, it was clear that the four ecoregions were dissimilar to each other based on environmental properties. The 95% confidence ellipses did not overlap; therefore, these units are more than randomly distributed areas since they belong to different ecological units. Coastal areas were distinguished jointly in the negative range of PC1 and PC2; hence, these areas were distinguished by warm and wet climates with sandy and nutrient-poor soils. Sites in the Gangetic Plains were distinguished in the negative range of PC1 and the positive range of PC2, moderate to high. This suggests that these sites were found in a lowland region that is thermally similar but with fertile soils that are fine and alluvial. Sites in the Himalayan Hills differentiated toward positive values of PC1 and negative to positive values of PC2, as expected in regions that are cooler and higher with a stony surface. Sites under Lateritic were predominantly differentiated based on high values of PC2 with varied values of PC1, as expected for regions with acidic and clay-rich soils and also at an intermediate elevation.

Pairwise ordinations were used for comparing the magnitude and direction of differences between environments. The Gangetic-Coastal environments were moderately differentiated, primarily because of PC2. The negative PC1 scores, however, jointly indicated a similarity in thermal conditions, yet a dissimilarity in edaphic factors. This contrasted with near-complete separation for Gangetic-Himalayan Hills environments, primarily because of PC1, encompassing the complete range from tropical lowlands and cool hill forests, respectively. A moderate degree of differentiation was recorded for Gangetic-Lateritic environments, where Lateritic environments were tending toward environments with higher scores for PC1 and indicating higher plateaus, respectively. Both regions were also extending into positive PC2, spanning regions with finer textures of soil.

Coastal and Himalayan Hills ecoregions had extreme comparisons. There is complete divergence between Coastal and Himalayan sites along both axes. Coastal sites are very similar, and Himalayan sites are distributed across the wide range of PC2 values, reflecting topographic and edaphic heterogeneity. Divergence between Coastal and Lateritic regions is more along PC1, which reflects variation in topographic elevation and temperature. In PC2, divergence is not so distinct, as both have similar soil types—sandy and poor in nutrients. Himalayan Hills-Lateritic sites had considerable overlap in the positive PC1 region; however, Himalayan sites had more extreme and greater dispersal, indicating greater heterogeneity within the group. No doubt this is responsible for the possibility of transitional fungal assemblages occurring at ecoregional boundaries, particularly at the ridge top forming the Kalimpong-Kurseong region, where the geological characteristics of the plains intersect and merge with the climatic characteristics typical of the Himalayas.

In general, the strong environment structuring confirms the hypothesis that ecoregions are suitable units for the estimation of fungal diversity. PC1 shows the main temperature-moisture gradient that causes a lot of functional turnovers. Saprotrophic groups are more common in lowland Coastal and Gangetic areas, while ectomycorrhizal groups are more common in upland lateritic and Himalayan areas. PC2 shows how changes in the chemistry and texture of the substrate affect how mycorrhizae break down and take up nutrients. PC3 adds seasonal changes that probably affect the timing of fruiting and dispersal.

The gradient of niche separation among ecoregion pairs, from sharp segregation between the Gangetic Plains and the Himalayan Hills to partial convergence between the Himalayan and Lateritic regions, suggests spatial variation in dispersal barriers and colonization constraints. Strong environmental discontinuities correspond with biogeographic boundaries, while environmentally analogous regions facilitate shared species pools and gradual transitions. In this context, the PCA offers a quantitative foundation for forecasting the reorganization of fungal functional groups due to warming, intensified monsoons, and modified precipitation seasonality, with boundary regions, especially the Himalayan-Lateritic transition zone, expected to undergo the most significant compositional changes in future climates.

2.7 SPECIES-ENVIRONMENT RELATIONSHIPS: VARIABLE IMPORTANCE ANALYSIS

Variable importance analysis showed that species vary in their response to different environments. For instance, some species are more affected by environmental factors compared to others

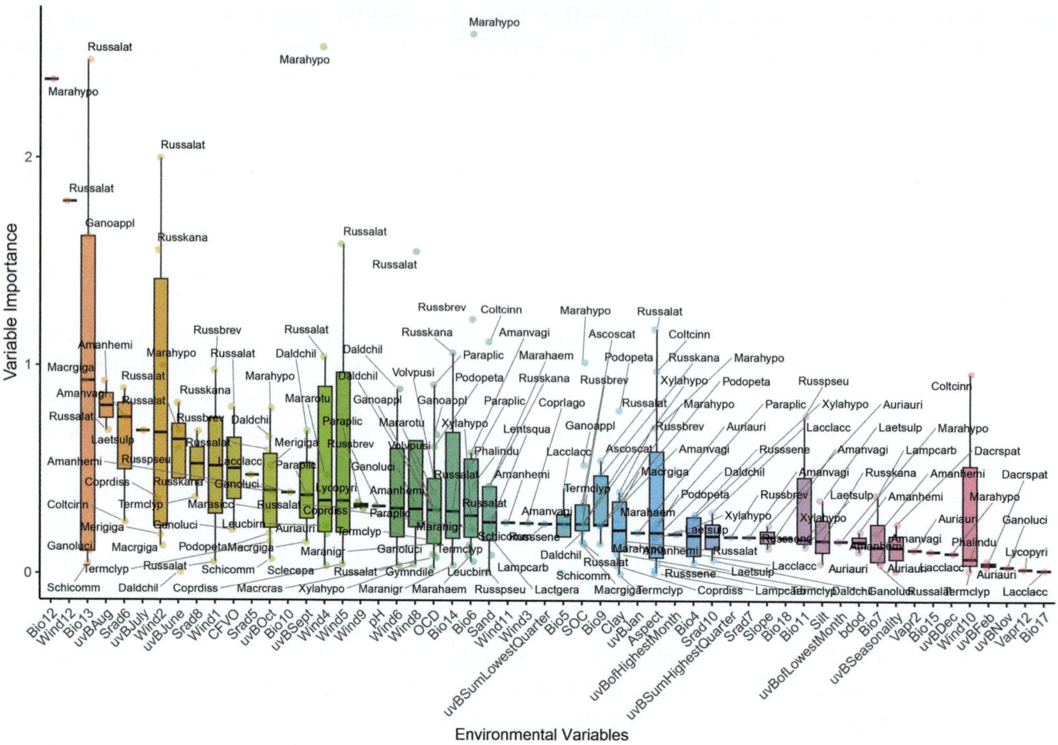

FIGURE 2.14 Plot of environmental variable importance of macrofungal species in West Bengal.

(Figure 2.14). Variable importance analysis showed that there exist 176 species-variable relationships of the 37 studied macrofungal species; this shows that the environment filters macrofungal communities through multi-dimensional filtering, as opposed to filtering through principal gradients. There existed five different types of variables: UV-B radiation conditions (monthly means and seasonality), wind conditions (monthly variability), bioclimatic data (temperature and precipitation variables), soil properties (texture, organic carbon, bulk density), and topographic variables (slope, orientation, solar radiation).

Factors related to climate had the greatest importance in species and environment interactions. Bioclimatic variables (primarily Bio6, Bio13, Bio5, and Bio9) and UV-B radiation had the highest predictive ability for 28 out of 37 species. *Marasmius hypochroides* (Marahypo) showed the strongest dependence, with an importance value of 2.60, on Bio6 (minimum temperature of the coldest month). The next most important factors for Marahypo were Wind4 (importance = 2.53) and Bio12 (annual precipitation; importance = 2.37). It shows that *M. hypochroides* is a temperature- and precipitation-sensitive species, as it favors transition habitats with low precipitation.

With an importance value of 2.47, *Russula alatoreticula* had high sensitivity for Bio13 (precipitation of the wettest month), followed by Wind2 (2.00) and Wind12 (1.79), indicating that this species relies largely on the monsoonal pattern of rainfall, probably aided by wind forces for spore distribution.

The ectomycorrhizal species had relatively high sensitivity for both temperature and soil chemical properties. *Russula alatoreticula* (Russalat), *Russula brevipes* (Russbrev), *Russula kanadii* (Russkana), and *Russula pseudocyanoxantha* (Russpseu) had important responses to Bio6 (minimum temperature of the coldest month), which had importance values ranging from 0.28 to 1.22, signifying the temperate limitations for ectomycorrhizal associations. The soil organic carbon (SOC) feature at various depths was also identified as a major predictor variable, especially for *Marasmius hypochroides* (Marahypo), which showed the highest importance value of 1.02 to SOC levels of 30–60 cm. The results illustrate the importance of fungal mutualisms being dependent on soil nutrient levels and distribution of carbon within the host root systems. The importance levels of wind variables of Wind1, Wind2, Wind4, and Wind8 signified the importance of wind variables for fungal spore dispersal and establishment of host-fungi mutualisms.

The saprotrophic species showed stronger correlations with the variables associated with substrates, especially organic carbon density (OCD) and soil texture. In Bio13, the importance value was 1.62 for *Ganoderma applanatum* (Ganoappl), and in OCD at a depth of 60–100 cm, it was 0.67, indicating its dependence on moisture-mediated decomposition and deep carbon accumulation. In *Parasola plicatilis* (Paraplic), importance was 0.91 for OCD at 100–200 cm, suggesting it depends on stable, long-lasting carbon stored in lateritic forests. Termitophilous fungi such as *Termitomyces clypeatus* (Termclyp) had moderate importance for solar radiation intensity, Srad8, at 0.37, and soil organic carbon, SOC, at 0.24, suggesting it depends on termite-induced breakdown and sunlight-exposed mounds.

There were distinct environment signatures for the parasitic and coprophilous assemblages. The species that responded most to the environment was *Coltricia cinnamomea* (Coltcinn), which was influenced by the presence of sand (0–5 cm; importance = 1.12), orientation (importance = 0.98), and Wind10 (importance = 0.97). This indicates an intolerance of *C. cinnamomea* to the ability of the surface soil to drain, the orientation of the microclimate, and the change in the wind over the year. The coprophilous species (including *Coprinellus disseminatus*, Coprdiss, *Coprinopsis lagopides*, Coprlago) were primarily associated with the wind (Wind2, Wind3, and Wind5) and Bio6, revealing the transient and spatially patchy nature of the environment.

Parameters of UV-B radiation, including monthly averages, turned out to be significant predictors for 18 species, with the highest importance values exceeding 0.92 for *Amanita hemibapha* (Amanhemi) (uvB_mean_Aug). The effect of monthly patterns of UV-B radiation (in June, August, September, and October) was large, indicating that UV radiation has consequently influenced fruiting, sporocarp growth, or perhaps even mycelial growth. *Daldinia childiae* (Daldchil) was most affected by uvB_mean_Sept, with an importance value of 0.69, and by uvB_mean_Oct, with an importance value of 0.66. This indicates that fruiting periods during the late monsoon seasons coincided with periods of decreasing UV stress but increasing soil water. *Lactiporus sulphureus* (Laetsulp) showed the highest importance for uvB_mean_July, with an importance value of 0.69, which was as expected, marking the onset of monsoons.

The importance of UV-B annual cycles (*Dacryopinax spathularia*, Dacrspat: importance = 0.24; *Auricularia auricula-judae*, Auriauri: importance = 0.02) suggests that fungi respond not only to the quantity of UV-B radiation but also to its annual cycles and possibly act as indicators for the timing of reproduction. These outcomes show the importance of the presence of fine-scale phenological constraints on the distributions of macrofungi rather than the broad climate zones with respect to predictions of the effects of climate change on fruiting.

Topographic variables, such as slope, aspect, and solar radiation (Srad), influenced species distribution due to variation in microclimate, moisture, and substrate. Auriauri has a very high slope importance (importance = 0.23), which is reasonable since this species grows in upland forests with well-drained soils. Aspect had a high influence on Coltcinn (importance = 0.98), Laetsulp (importance = 0.17), and Russalat (importance = 1.18). This is an indication that these species are very sensitive to temperature, moisture, and solar light.

The patterns of solar radiation varied between the months of the year for Srad5, Srad6, Srad8, and Srad10. For example, the species *Laccaria laccata* (Lacclacc) contributed the most to Srad5 with a measure of importance = 0.47 and Srad10 with a measure of importance = 0.30. This shows that the amount of solar radiation changes from season to season because it affects the quality of the substrate and the distribution of the photosynthates within the host tree.

In this assessment, the dependence of various species on the environment illustrates the underlying reason for the segregation of niches among ecoregions. Very important factors had high variability among the co-occurring species, indicating the presence of finer heterogeneity in the environment, which structures the macrofungal communities based on the niche. Marahypo, Russalat, and Ganoappl, some of the most frequent species, responded to various factors (thermal minima, wind-precipitation interaction, and environmental moisture versus carbon, respectively). This clearly reveals the presence of diverse species within a particular environment, without indicating the ecosystem to be similar among the various species.

These findings have practical implications for the work of conservation biologists. Those species that are restricted in the areas where they can exist (those of significant importance to individual factors) are particularly vulnerable to deforestation and global warming. Coltcinn, highly vulnerable to the presence of sand on the surface and the slope of the surface, is a case in point among those that are endangered due to soil disturbances and forest conversion. Those with significant importance to individual factors (Russalat and Marahypo) may perhaps prove more hardy as they are able to accommodate changes in their environment. They are nevertheless vulnerable to environmental changes that involve more than one factor.

Incorporation of variable importance measures into conservation planning will enhance identification of key thresholds of environmental importance. In order to maintain the safety of macrofungal diversity under a rapidly changing planet, it will become important to protect sites that are characterized by a set of important environmental variables, including soil organic carbon variation, temperature, humidity, and topographic variability. In addition, future predictive models will identify key changes within distribution and refugia under IPCC scenarios that are based on these environmental responses.

2.8 TROPICAL INDIAN MACROFUNGI WITHIN A GLOBAL FRAMEWORK

When gauged from a broader global perspective, tropical Indian macrofungal assemblages display similarities and dissimilarities concerning those observed from temperate and tropical forests across the planet. The extreme right-skewed distribution detected in this study closely aligns with that detected from European temperate forests, Neotropical rainforests, and Southeast Asian dipterocarp forests, dominated by a few generalist saprotrophs/symbionts that are most abundant spatially. This observation suggests that macrofungal assemblages, irrespective of their geographical distribution, are controlled by high-level limitations on dispersal, habitat, and mycelial survival (Peay et al. 2016). Yet again, while these underlying tenets are identical across tropical and temperate regions, unique dissimilarities generated by environmental constraints and biological evolution underlying macrofungal assemblages from tropical India must be explored further. As regards temperate forests, the richness of ECM macrofungi is generally characterized by a more stable climate and long exposure to host-symbiont co-evolutionary adaptations. In tropical India, ECM assemblages are dominated exclusively within select edaphic and vegetational types (e.g., Lateritic Sal forests) and are peripheral and fragmentary elsewhere.

However, the factor that makes these patterns in India unique is that the associated mechanisms for these patterns are unique to India. Temperate forests, where ECMs generally involve adaptability growing from a stable climate over long timescales of co-evolution between hosts and fungi, follow this trend. This is reflected in tropical India as well, but in a purely edaphic, or vegetation, niche, the Lateritic Sal forests. ECMs, by contrast, generally expand in relation to biotopes, or habitats, unlike tropical Southeast Asia, home to dipterocarp forests, where ECMs are now widespread as a fungal type. ECMs in India, therefore, display unique characteristics, as they are a habitat-conditioned symbiosis, where soil and not climate is the primary conditioning factor. Indian ECM fungi, therefore, most closely resemble those of African savannas and uprooted tropical forests of the Neotropics, where rapid turnover determines decomposer success factors, dependent on moisture seasonality (Steidinger et al. 2019). Simultaneously, Indian saprophytic fungi most closely respond to salinity, temperature, and drought, especially in coastal areas, making them more drought-weather-tolerant than their tropical counterparts. All of these make tropical Indian biomes especially enlightening sites in fungal evolution studies, especially as they adapt under definite stressful conditions, as noted by Akshaya et al. (2023).

In the lowland rivers of India, the prevalence of saprotrophic fungal groups is analogous to what is seen in African savannas and Neotropical disturbed forests, based on the rapid growth of decomposers, which is a result of the short turnover of the substrate and a periodic water flow of moisture. The saprotrophs of India are, however, very tolerant of high salinity, temperature, and aridity, particularly in the coastal rivers of India. This is the most stress-tolerant fungal community on earth. It is very significant to learn about the abrupt conditions under which fungal communities adapt in the tropical region of India (Niskanen et al. 2023).

There are a large number of endemic species in this region, as revealed by studies conducted in the Amazon as well as Southeast Asia. This corroborates the hypothesis that instead of large geographical dominance, fine-scale niche partitioning is necessary for the maintenance of fungal diversity. However, the unusually large proportion of singleton species in the Himalayan regions surpasses all previously documented global records, as it plays a pivotal role in the enhancement of fungal beta-diversity (Vandegrift et al. 2023).

From the international point of view, the macrofungi of the Indian subcontinent are a largely undervalued element of tropical biodiversity, a contribution that draws attention to their importance as a reflection of function and adaptation. Moreover, the macrofungi of tropical India are clearly of critical significance to the understanding of the impact of various degrees of edaphic stress, human-induced fragmentation, and climatic change, issues that are currently at the forefront of international concern. Tropical India must, therefore, be recognized not only as the fungal biodiversity hotspot but also as the key international natural laboratory.

2.9 MACROFUNGI AS BIOINDICATORS FOR ECOSYSTEMS

The functional composition of macrofungi is a sensitive and holistic measure integrating data on vegetation, substrates, soil chemistry, and microclimate conditions. Within the present data, marked inequalities in functional dominances, evenness, and spatial persistence of macrofungal communities within and between ecoregions demonstrate how macrofungal communities package information about disturbance and ecological maturity. Assemblages, which are dominantly saprotrophic in composition yet with lower functional evenness (as in the coastal and Gangetic Plains) of a region, display the effect of disturbance on substrates, while a balanced functional pattern indicates structurally complex and relatively undisturbed forests (Xing et al. 2024).

The efficacies of macrofungi as bioindicators are said to rely upon their reaction to changes in the environment, in addition to their key roles in the processing of organic matter in the ecosystem (Case et al. 2025). The saprotrophic cover was moderate in Coastal ecosystems (≈0.09–0.13), which was markedly lower for ectomycorrhizal fungi (~0.04), suggesting their limited host connectivity as well as disturbed sandy, nutrient-poor environments. The Gangetic ecosystems had similarly lower ectomycorrhizal coverage (~0.03), along with increased coverage of parasites (~0.11–>0.23) and facultative saprotrophs.

Measurements based on occupancy were very helpful in determining the health of the respective ecosystems. Lateritic forests had the maximum average occupancy by the ectomycorrhizal fungi network (~0.22), which was significantly greater compared to the Coastal and Gangetic areas. Since the maximum occupancy was in a small functional group, it depicts that the systems were filtered by the environment and that the host was well regulating the systems, typical of a well-developed forest. Conversely, the Himalayan Hills portrayed low but broad-spectrum occupancy measurements within the majority of the functional groups (~0.04–0.05). This depicts that there were a large number of niches and habitats and not just a few highly competitive species.

These patterns of occupancy demonstrate the early signs of either simplification or equilibrium for functions prior to patterns in species richness becoming more apparent. Much occupancy in a few functional groups indicates that things are becoming more stable or stressful, while occupancy that is more evenly distributed among taxa indicates that things are at a state of dynamic equilibrium in relatively unstable environments (Hazen et al. 2024).

Macrofungi are more valuable in the study of the dynamics of ecosystems with the changing climate since they link the temporal dynamics of the weather with the soil processes. A persistently lower occupancy rate in the Coastal and Gangetic ecosystems suggests lower host-symbiont connectivity and potential saturation of the nutrient flow routes. However, the non-fragmented high occupancy rate in the lateritic forests may indicate host functional stability but vulnerability to the loss of host species.

Saprotrophic fungi, especially lignicolous and litter-saprotrophic species, showed a marked response to disturbance regimes and substrate supply. High occupancy of generalist saprotrophic species in Coastal and Gangetic areas signified frequent disturbance, fast organic matter turnover, and high inputs of dead wood (Wang et al. 2021). However, their low occupancy in lateritic forests, approximately 0.04, signified species exclusion in stable, nutrient-scarce communities dominated by ectomycorrhizal species. Occupancy patterns of saprotrophic species could convey vast information about the strength of disturbance, rates of decomposition, and carbon flow in the environment.

Macrofungi belonging to specific functional groups enhance the effectiveness of macrofungi as bioindicators. Termitophilous fungi had fairly high occupancy values in the Himalayan (~0.10) and Lateritic Ecoregions (~0.11). This was due to stable soil conditions, retention of moisture, and active cycling of nutrients by the termites. However, their values were small in the Coastal ecoregion (≈ 0.04), indicating unstable soil conditions. The presence of markers of coprophilous fungi, grass-associated fungal communities, and entomopathogenic fungi even when their occupation values are generally small acts as an indicator of the presence of small-scale heterogeneity in the vegetation environment that is often ignored (Wang et al. 2023).

The presence of rare species that are endemic to a certain elevation range is one example that shows how macrofungi are very sensitive to differences in microclimate. As one element that should be observed, the reduction in occupancy for cold-tolerant saprotrophic and ectomycorrhizal species could indicate the initial sign for the range shift into higher elevations, which is caused by the microclimate. Macrofungi can be seen as a holistic measure of ecosystem integrity by merging climatic, soil, and biological information to improve upon the use of plant and animal indices.

Occupancy measures can then be used to quantify bioindicative processes rather than simply being inferred qualitatively. Mean, dispersion, or skewness changes can then be linked to factors such as the disturbance frequency, nutrient availability, or level of fragmentation. Using macrofungi as additional components of regular ecological studies can then be considered an inexpensive, sensitive, and mechanistically appropriate method of determining ecosystem health across the varied landscape of the Indian subcontinent.

2.10 SYNTHESIS, PRIMARY THREATS, AND CONSERVATION IMPERATIVES FOR MACROFUNGAL DIVERSITY

This section brings together the main ecological trends, assessments of threat, and climate-related problems that impact macrofungal groups in the regions of tropical and subtropical India. It emphasizes that ecoregional factors, determined by regional climate, soil, vegetation, and organic substrate content, are the key drivers of community composition, leading to significant spatial exclusivity and limited niche overlap among ecoregions such as the Himalayan Hills, Gangetic Plains, Lateritic forests, and Coastal ecosystems. Based on these findings, the following subsections deal with the associated risks of habitat loss, fragmentation, and climate change. All these suggest the need for conservation strategies that are distinctive to each ecoregion and focus on process-oriented strategies rather than uniform protected area plans.

2.10.1 Ecological Patterns and Conservation Implications

This study presents a very fine-grained, geographically explicit summary of macrofungal ecology for tropical and subtropical India, aggregated for species occupancy, functional group composition, niche overlap, and ecoregional filtering over a variety of landscape settings. Integrated species-level, niche overlap, and geographic analyses of these data verify that ecoregional affiliation is primarily driving macrofungal community composition for West Bengal. There is a very high degree of spatial exclusivity with very little niche overlap between ecoregions, and this clearly demonstrates that macrofungal diversity is tightly coupled with local climate conditions, soil types, vegetation types, and availability of organic substrate. The abundance of saprotrophs in southern parts of the country is clearly tied to high rates of organic matter turnover, while the limited distribution of biotrophs (ectomycorrhizal and parasitic fungi) suggests specialized ecological requirements and a dependence on host communities that are ecoregion-specific. Quantitative niche overlap analyses reveal very few niche dimensions shared between ecoregions, particularly between those with very disparate environments—for instance, between the niche of Himalayan Hills and coast, showing a complete niche partitioning.

Occupancy is a novel technique to detect the way in which macrofungal assemblages are assembled. It illustrates the importance of ecological strategy and biologically based conditions over and above taxonomic identity in this respect. The existence of strong right-skewed occupancy distributions in all regions, indicating that a few species contribute disproportionately largely to spatial occupancy, reveals a fundamental ecological asymmetry and is similar to the global patterns of macrofungal occupancy, while the factors causing it are entirely distinct in the Indian conditions of unique soil, climate, and vegetation properties.

Model-based inferential and empirical confirmation show that important conservation values cannot be established solely on the basis of species abundance. For example, lateritic forests contain lower numbers of saprotrophic dominant species, but the abundance of ectomycorrhizal fungi is remarkably high and spatially well-intertwined, thus implying the existence of a strong link between the symbionts and the host trees. Gangetic and Coastal forests contain an important level of functional diversity as a result of habitat complexity and disturbance-induced substrate turnover, despite the lower abundance of symbiotic categories. Himalayan forests contain an abundance of species but are moderately fragmented in space.

The observed frequency pattern has several implications regarding the functioning and protection of ecological systems. Biologically complex systems with a few high-frequency species, such as the Coastal and Lateritic Ecoregions, show reduced functional redundancy and intolerance to environmental disturbances. Other ecoregions that show less dominance, such as the Gangetic Plains, may potentially persist despite the loss of species but remain susceptible to human-induced changes that result in a less diverse substrate. In the Himalayan Hills regions, the occupancy level of the majority of species is extremely sensitive to climatic change; even a slight change in the temperature or moistness patterns may adversely affect geographically restricted fungi.

2.10.2 Threat Assessment

A number of threats, operating at varying spatial scales, are being experienced by the macrofungal assemblages in tropical India. Habitat destruction and fragmentation remain the chief causes, particularly in the Gangetic Plains and along the coasts. This is because the disruption of the habitat by agricultural expansion, urbanization, and plantation of forests has made it difficult for the hosts to locate food. This rate of loss of ectomycorrhizal host occurrence, particularly in lateritic forests, indicates that the disruption of the host community impacts the associated fungi rather than the other type of fungi.

Forest management and habitat damage: Intensive forestry operations, such as removing deadwood, damage saprotrophic diversity and nutrient cycles. Plantation forestry facilitates the presence of hosts, which simplifies ectomycorrhizal assemblages. Ectomycorrhizal fungi depend very selectively on some hosts, and hence, lateritic and Himalayan forests are particularly vulnerable to selective felling and substitution of plant species. Coastal ecosystems, which have less functional diversity, get affected further.

2.10.3 Consequences of Climate Change and Overharvesting

Some of the rising challenges posed by climate change include an increase in temperature and precipitation and intense events, particularly in Himalayan ranges. Climate change also affects the fruiting periods, as well as humidity within the atmosphere, particularly in coastal and mountain ranges. Some fungal species with narrow temperature ranges and elevational distributions show high susceptibility, as indicated by their low presence and high singletons. Climate change, particularly changing monsoon patterns, also affects coastal ranges through an influx of saltwater and a rise in ocean levels.

Although of local importance, the non-sustainable harvest of edible species poses secondary threats relative to habitat destruction. However, overharvesting could reduce sporocarp production and disrupt sustenance in species that are intensively harvested, such as termite associates and ectomycorrhizal species, which have a short fruiting period.

Large-scale host processes, host contiguity, or microhabitat contiguity can determine survival times for macrofungi. Conservation strategies ought thereby to promote more process-oriented protection, in addition to biota assessments. Host contiguity, termite hill ecosystems, microhabitat contiguity, elevation corridors in slightly undulating tops, or microhabitat contiguity could, or ought to, or must, respectively, guarantee host contiguity, termite hill ecosystems, coarse woody debris, elevation corridors, or microhabitat contiguity.

A uniform protected area plan is incapable of ensuring the conservation of macrofungal diversity in the various tropical landscapes within India. There is a need for plans that are applicable in each ecoregion, as there is niche partitioning. Lateritic systems require the conservation of host-dominated forests and non-replacing ectomycorrhizal communities that are predisposed to substitution. Coastal systems require the integrity of the substrate to remain intact despite the existence of abiotic stress, and they are also under threat from salinity intrusion and sea-level rise. There is a need for plans that conserve the opportunities for macrofungi along Himalayan systems, which are under threat from warming trends and contain a large number of evolutionarily unique species that occur in topographic corridors and microrefugia. Functional fungal diversity is constrained by environmental gradients that distinctly define ecoregional boundaries and show that habitat degradation or loss will uniquely impact declining macrofungal diversity.

The presence of a few species with high occupancy should not be mistaken for ecological redundancy, instead, it reflects a kind of specialization in terms of very fine-grained environmental filters. It is true that, individually, rare species do not occupy much space, but taken together, their presence increases ecosystem resilience to a certain extent as gaps are filled based on their functionalities.

The inclusion of occupancy data and fungal indicators in assessments of environmental impact, forest certification schemes, and biodiversity monitoring programs represents a mechanistic approach that can elevate the conservation of fungal organisms to the same plane as that of plant and animal conservation. Fungi are essential to the functioning of the environment, and as a result, they must be specifically included in national biodiversity conservation programs and inventories for biodiversity conservation. Failing to acknowledge their importance could result in conservation programs overlooking an integral component that would not be long functional with a rapidly changing environment. There are also some data gaps that require more sampling to be carried out to increase their resolution and representation in an ecoregion-structured fungal diversity in the tropics of India.

2.11 CONCLUSIONS AND PROSPECTIVE AVENUES

This study offered a quantitative, strategy-centered approach to the biogeography, community composition, and conservation value of tropical and subtropical macrofungal assemblages in the Indian tropics. This study showed that the organization of a macrofungal community occurs not only along the axis of species richness but also along a species occurrence persistence axis as a consequence of the combination of ecological strategy, species specialization, abiotic filtering, and species association. Species occurrence is a simple and useful ecological metric, as it indicates the elements of a particular species assemblage organization that are not apparent when either species richness or basidiocarp abundance is measured.

For every ecoregion, the occupancy distributions of macrofungi communities had strong right-skewed distributions. This indicates that there was a small number of species that could persist in one location for a long time, but most species had low population densities, which could only persist in particular locations. Such patterns reflect basic constraints of fungal dispersal, growth, or mycelial life span, since they follow patterns observed in global macrofungal networks. Meanwhile, both ecological processes, which shaped this distribution, were unique in every Indian ecoregion. Such data shows just how important the environment is in influencing movement patterns for fungus.

The Coastal ecoregion was a saprotrophic hotspot, where a large number of stress-tolerant fungi, known as wood decomposers and litter decomposers, occur. The presence of stable and constant supplies of organic matter, combined with support for tolerance to temperature, salinity, and moisture, favored generalist decomposing organisms that could maintain a large network of mycelia on various substrates. The high occupancy in this system reflects physiological tolerance and universality toward substrates under strong abiotic stress rather than a competitive position.

Functioning Gangetic Plains in this system functioned as a mixed-strategy transitional area and supported moderate levels of occupancy for saprotrophic, parasitic, termitophilous, and ectomycorrhizal fungal functional groups. This is how habitat heterogeneity at small spatial scales is fashioned through agro-habitat mosaics, forest patches, river corridors, and seasonal disturbance regimes. In this situation, neither functional group had dominant effects on spatial persistence, and functional group occupancy is governed by the shape and complexity of the landscape and not by strong abiotic and biotic filters. This is how functional fungal biodiversity is maintained through human-induced heterogeneity.

On the other hand, it was apparent that the Lateritic Plateau was a region of high dominance by ectomycorrhizal fungi. The dominance of ECM fungi was aided by the connectivity of the sal (*Shorea robusta*) hosts and the poor nutritional but high iron content of the soil. Also, the large reduction of saprotrophic dominance in the region shows spatially explicit data supporting the Gadgil effect, which states that a region of high dominance by ECM fungi will lock away organic resources and suppress the spatial spread of the free-living saprotrophs. Here, occupancy can be taken as a proxy for connectivity of hosts underground and the capacity to acquire nutrients through competition.

The Himalayan Hills supported the most punctuated fungal assemblages, with very low mean occupancy for nearly all functional groups and a high prevalence of singletons and range-restricted species. Topographic relief, climate niche partitioning, substrate disconnection, and host fragmentation all contributed to reducing mycelial connectivity and landscape occupancy. Despite high taxonomic diversity, occupancy analyses indicated a very selective environment with very limited niche dimensions. These findings further support the notions of mountain niche fragmentation, underscoring the importance of mountainous terrain for promoting beta-diversity while inhibiting landscape occupancy.

Taken together, these results indicate that macrofungal diversity in tropical India is impacted by ecological filters that are scale-dependent, ranging from the monsoon-driven climatic regimes demarcating ecoregions, through soil chemistry and host tree distribution, to microhabitat availabilities, which promote the maintenance of specialists. Macrofungi play an important role in forest ecosystems, not as passive elements but as actively interacting regulators of ecosystem processes of nutrient cycling, carbon sequestration, soil formation, and ecosystem functioning.

Species-level data on patterns of occupancy can be very helpful for planning conservation for fungi. First, high-occupancy saprotrophic genera recognize hotspots of ecological importance for carbon cycling processes, while a decrease in occupancy could be an indicator of carbon cycle disturbance in coastal and Gangetic ecosystems. Second, the dominance of ectomycorrhizal species in lateritic forests reveals the importance of maintaining continuity in trees and soil integrity. A reduction in ECM occupancy can be an indicator of disturbance affecting forests as a consequence of early signs of change in tree regrowth or soil chemistry processes. Third, conservation planning for the Himalayan Hills ranges should consider microhabitats such as ravines, dark slopes, debris, and bands of forests confined to specific ranges, as ranges where low-occupancy species exist reflect stable ecosystems due to their favorable ecological status.

Occupancy indexes are a robust method to determine the sensitivity of a species to the effects of climate change. Species with low occupancy values and small environmental niches, in particular those that are montane and host-specific fungi, are expected to be among the first to decline as a result of the effects of warmer temperatures and changes in rainfall patterns. The inclusion of the thresholds of occupancy in Red Listings would provide us a mechanism to gauge the effectiveness of the protection of fungi, which has been a significant lacuna in biodiversity politics for a long while.

Further studies must extend this paradigm in terms of both temporal and mechanistic approaches. To distinguish between the occurrence of true geographical rarity and temporal fluctuations in sporocarp production in monsoon-paced systems, long-term sampling is required. There needs to be integration of techniques that utilize the power of molecular tools such as eDNA and population genomics techniques in order to study cryptic diversity, mycelial continuity, and dispersal constraints. Functional ecology and traits of sporocarp morphology and enzymes could be combined to assess host specificity and stress tolerance.

REFERENCES

Akshaya, K.K., A. Karthikeyan & C. Kunhikannan. 2023. Status of macrofungal diversity in the wet evergreen forests of Agasthyamala Biosphere Reserve, Western Ghats, India. *Journal of Threatened Taxa* 15:23575–23586. https://doi.org/10.11609/jott.8469.15.7.23575-23586

Arnolds, E. 1992. The analysis and classification of fungal communities with special reference to macrofungi. In W. Winterhoff (ed.), *Fungi in Vegetation Science*. Handbook of Vegetation Science 19. Springer, Dordrecht. https://doi.org/10.1007/978-94-011-2414-0_2

Beckmann, M., T. Václavík, A.M. Manceur, L. Šprtová, H. von Wehrden, E. Welk & A.E. Cord. 2014. glUV: A global UV-B radiation dataset for macroecological studies. *Methods in Ecology and Evolution* 5:372–383. https://doi.org/10.1111/2041-210X.12168

Büntgen, U., H. Kauserud & S. Egli. 2012. Linking climate variability to mushroom productivity and phenology. *Frontiers in Ecology and the Environment* 10:14–19. https://doi.org/10.1890/110064

Case, N.T., S.J. Gurr, M.C. Fisher, et al. 2025. Fungal impacts on Earth's ecosystems. *Nature* 638:9–57. https://doi.org/10.1038/s41586-024-08419-4

Chatterjee, S., R. Datta, A. Dey, P. Pradhan & K. Acharya. 2012. In vivo hepatoprotective activity of ethanolic extract of *Russula albonigra* against carbon tetrachloride-induced hepatotoxicity in mice. *Research Journal of Pharmacy and Technology* 5:1034–1038.

Chattopadhyay, P., S. Biswas, S. Chandra & A.K. Dutta. 2025. *Marasmius pseudojasminodorus* (Marasmiaceae, Agaricales), a new jasmine-scented species from the lateritic region of West Bengal, India. *Brittonia*. https://doi.org/10.1007/s12228-025-09836-z

Chattopadhyay, P., M. Talukdar, J. Beypih, K. Tayung & A.K. Dutta. 2022. A new species of *Volvariella* (Agaricales, Basidiomycota) from West Bengal, India. *Phytotaxa* 567:36–48.

Dutta, A.K., P. Pradhan, S.K. Basu & K. Acharya. 2013. Macrofungal diversity and ecology of the mangrove ecosystem in the Indian part of Sundarbans. *Biodiversity* 14:196–206. https://doi.org/10.1080/14888386.2013.848824

Fick, S.E. & R.J. Hijmans. 2017. WorldClim 2: New 1 km spatial resolution climate surfaces for global land areas. *International Journal of Climatology* 37:4302–4315. https://doi.org/10.1002/joc.5086

Hawksworth, D.L. 2001. The magnitude of fungal diversity: The 1.5 million species estimate revisited. *Mycological Research* 105:1422–1432. https://doi.org/10.1017/S0953756201004725

Hazen, E.L., M.S. Savoca, T.J. Clark-Wolf, M. Czapanskiy, P.M. Rabinowitz & B. Abrahms. 2024. Ecosystem sentinels as early-warning indicators in the Anthropocene. *Annual Review of Environment and Resources* 49:573–598. https://doi.org/10.1146/annurev-environ-111522-102317

ISRIC. 2022. *SoilGrids Database*. The International Soil Reference and Information Centre (ISRIC). Accessed 10 December 2022. https://isric.org/explore/soilgrids

Mondal, S., P. Chattopadhyay, A. Banerjee & D. Palit. 2022a. Ecological assessment of tree communities and pedological characteristics of coal mine generated wastelands of Raniganj Coal Field Area, West Bengal, India. *Vegetos* 35:367–380. https://doi.org/10.1007/s42535-021-00305-y

Mondal, S., P. Chattopadhyay, A. Banerjee & D. Palit. 2022b. Floristic composition and inventorization of forest resources in some selected forest areas of Paschim Bardhaman District, West Bengal, India. In M. Sahana, G. Areendran & K. Raj (eds.), *Conservation, Management and Monitoring of Forest Resources in India*. Springer, Cham.

Mora, C., D.P. Tittensor, S. Adl, A.G.B. Simpson & B. Worm. 2011. How many species are there on Earth and in the ocean? *PLoS Biology* 9:e1001127.

Niskanen, T., R. Lücking, A. Dahlberg, et al. 2023. Pushing the frontiers of biodiversity research: Unveiling the global diversity, distribution, and conservation of fungi. *Annual Review of Environment and Resources* 48:149–176. https://doi.org/10.1146/annurev-environ-112621-090937

Peay, K.G., P.G. Kennedy & J.M. Talbot. 2016. Dimensions of biodiversity in the Earth mycobiome. *Nature Reviews Microbiology* 14:434–447. https://doi.org/10.1038/nrmicro.2016.59

Pradhan, P., A.K. Dutta, A. Roy, S.K. Basu & K. Acharya. 2012. Inventory and spatial ecology of macrofungi in the *Shorea robusta* forest ecosystem of lateritic region of West Bengal. *Biodiversity* 13:88–99. https://doi.org/10.1080/14888386.2012.690560

Pradhan, P., A.K. Dutta, A. Roy, S.K. Basu & K. Acharya. 2013. Macrofungal diversity and habitat specificity: A case study. *Biodiversity* 14:147–161. https://doi.org/10.1080/14888386.2013.805660

Pradhan, P., A.K. Dutta, S. Paloi, A. Roy & K. Acharya. 2016. Diversity and distribution of macrofungi in the Eastern Himalayan ecosystem. *EurAsian Journal of BioSciences* 10:1–12. https://doi.org/10.5053/ejobios.2016.10.0.1

Pradhan, P., S. Banerjee, A. Roy & K. Acharya. 2010. Role of wild edible mushrooms in the Santal livelihood in lateritic region of West Bengal. *Journal of the Botanical Society of Bengal* 64:61–65.

Roy, N., P. Chattopadhyay & A.K. Dutta. 2025. A new species of *Russula* subg. *Brevipes* (Russulaceae) from the lateritic regions of West Bengal, India. *New Zealand Journal of Botany* 63:331–341. https://doi.org/10.1080/0028825X.2024.2385663

Steidinger, B.S., T.W. Crowther & J. Liang. 2019. Climatic controls of decomposition drive the global biogeography of forest-tree symbioses. *Nature* 569:404–408. https://doi.org/10.1038/s41586-019-1128-0

Talukdar, M., P. Chattopadhyay, K. Acharya & A.K. Dutta. 2023. First report of *Lepiota bengalensis* (Agaricaceae) from India, validated by morphology and molecular phylogeny. *Biology Bulletin* 50:876–883. https://doi.org/10.1134/S106235902360201X

Vandegrift, R., D.S. Newman, B.T.M. Dentinger, et al. 2023. Richer than gold: The fungal biodiversity of Reserva Los Cedros, a threatened Andean cloud forest. *Botanical Studies* 64:17. https://doi.org/10.1186/s40529-023-00390-z

Wang, J.T., J.P. Shen, L.M. Zhang, B.K. Singh, M. Delgado-Baquerizo, H.W. Hu, L.L. Han, W.X. Wei, Y.T. Fang & J.Z. He. 2021. Generalist taxa shape fungal community structure in cropping ecosystems. *Frontiers in Microbiology* 12:678290. https://doi.org/10.3389/fmicb.2021.678290

Wang, M., C. Wang, Z. Yu, H. Wang, C. Wu, A. Masoudi & J. Li. 2023. Fungal diversities and community assembly processes show different biogeographical patterns in forest and grassland soil ecosystems. *Frontiers in Microbiology* 14:1036905. https://doi.org/10.3389/fmicb.2023.1036905

Watling, R. 1997. The business of fructification. *Nature* 385:299–300. https://doi.org/10.1038/385299a0

Xing, H., W. Chen, Y. Liu, et al. 2024. Local community assembly mechanisms and the size of species pool jointly explain the beta diversity of soil fungi. *Microbial Ecology* 87:58. https://doi.org/10.1007/s00248-024-02374-3

3 Biodiversity of Wood-Rotting Fungi in Tropical Ecosystems of China

Changlin Zhao

3.1 WOOD-ROTTING FUNGI

Fungi are the second most diverse group of organisms in the world, and more than 150,000 species of fungi have been reported (Hawksworth and Lücking 2017; Wang and Cai 2023). The wood-rotting (or wood decay) fungi are one of the most diverse groups of living organisms on earth. These fungi occur across a broad range of ecosystems, including extreme environments, in which wood-rotting fungi are primarily heterotrophs and occur as unicellular to multicellular organisms and are vital to ecosystem functioning (Dong et al. 2024; Deng et al. 2025, 2026; Hibbett et al. 2025). The wood-rotting fungi secrete a spectacular array of bioactive chemical compounds and enzymes, and these play crucial roles in the biosphere, from digesting organic matter and recycling nutrients from dead plant and animal tissues to mediating intimate and mutually beneficial associations with the roots of almost all land plants (Brundrett and Tedersoo 2018). The wood-rotting fungi are a key source of food and fermented products, including many beverages and industrial feedstocks, as well as dozens of important enzymes and drugs. Conversely, some species of fungi are pathogenic to plants. These pathogenic species jeopardize food security worldwide by infecting and contaminating forests and items of food. Consequently, they are of considerable economic importance (Bergman and Casadevall 2010).

The wood-rotting fungi of the class Agaricomycetes within the phylum Basidiomycota comprise more than 3,000 species with an essentially worldwide distribution (https://www.indexfungorum.org/). The wood-rotting fungi are a cosmopolitan group and have a rich diversity related to the high diversity of plants growing in boreal, temperate, subtropical, and tropical regions (Dai et al. 2014; Wang and Cai 2023; Zhao et al. 2023a, 2023b; Dong et al. 2024). Wood-rotting fungi have typically been classified as either causing white rot or brown rot, according to their abilities to degrade lignin, cellulose, and hemicellulose. White rot fungi can decompose all of the components of wood, while brown-rot fungi degrade only cellulose and hemicellulose (Riley et al. 2014; Chung et al. 2017; Hage et al. 2021; Yu et al. 2023; Luo et al. 2024). In addition to being saprotrophic, some polypores have been observed as being mycorrhizae (Alem et al. 2021). Some brown-rot fungi are forest pathogens, which means that they can infect different parts of the wood of trees, such as the branches, heartwood, trunk, and base (Yuan et al. 2023). Some brown-rotfungi are edible or have medicinal properties (Wu et al. 2019). Although the brown-rot fungi are dispersed in different orders of the Basidiomycota (He et al. 2019), the order Polyporales Gäum. still contains the majority of brown-rot fungi. The latter are divided into two clades—the antrodia clade and the leptoporus clade (Liu et al. 2023). In recent years, many mycologists have focused on the brown-rot fungi within the Polyporales and concluded that most of the brown-rot fungi belong to the antrodia clade, including many new taxa (Binder et al. 2013; Justo et al. 2017; Liu et al. 2023). The leptoporus clade is embedded in the family Irpicaceae Spirin & Zmitr. and includes only *Leptoporus mollis* (Pers.) Quél. and *L. submollis* B.K. Cui & Shun Liu (Liu et al. 2023). Liu et al. (2023) carried out a comprehensive systematic

DOI: 10.1201/9781003647140-3

classification and phylogenetic study of the brown-rot fungi within the Polyporales and revised the classification system of the brown-rot fungi within the Polyporales.

3.2 THE DIVERSITY RESEARCH OF WOOD-ROTTING FUNGI IN TROPICAL ECOSYSTEMS

Wood-rotting basidiomycetes have been extensively studied during the past 100 years, and almost all types of forest vegetation have been surveyed (Karsten 1879, 1881, 1892; Patouilliard 1900; Donk 1960, 1964; Jülich 1981; Ryvarden 1991; Zhao 1998; Kirk et al. 2008; Dai et al. 2014; Cui et al. 2019; Dong et al. 2024). Recently, mycologists from different countries are focusing on the wood-rotting fungi in tropical ecosystems (Figure 3.1), of which there are nearly 1,600 species of wood-rotting basidiomycetes have been found in China (Dai 2010, 2012; Cui et al. 2019; Sun et al. 2019, 2022; Li and Wei 2021; Dai et al. 2021a, 2021b, 2022; Liu et al. 2021, 2023; Ma et al. 2022; Wu et al. 2022; Dong et al. 2024; Yang et al. 2025). The majority of wood-rotting basidiomycetes are saprophytic and utilize dead wood as a substrate for growth and reproduction. However, some wood-rotting species attack living trees by either invading and killing the living sapwood or living in the heartwood of live host trees (Dai et al. 2007). The funga, species diversity, species composition, and nutritional modes of wood-rotting fungi in China, North America, and Europe have been compiled. Based on these data, significant regional differences exist in the number of species and genera and the proportions of nutritional modes in the three landmasses (Wu et al. 2022). China had the highest number of species/

FIGURE 3.1 Tropical forest in China with a very large liana in the foreground.

Source: Photo by Changlin Zhao

genera, followed by North America, and Europe had the least, with the highest wood-rotting fungal diversity among the three landmasses. White-rot is the most prevalent nutritional mode, followed by brown-rot, and the white-rot species as a proportion of all wood-rotting fungi in each of the three landmasses are highest in China, followed by North America and then Europe. The brown-rot polypore proportion is highest in Europe, followed by North America and then China (Wu et al. 2022). The latter has a diverse range of climates (including temperate monsoon, subtropical monsoon, tropical monsoon, temperate continental, plateau, and mountain) and numerous types of vegetation (including boreal, temperate, warm temperate, subtropical, and tropical forests) (Figure 3.2). China's diverse climate and vegetation provide favorable environments for wood-rotting fungi, especially as there is an abundance of gymnosperm and angiosperm trees that furnish suitable substrates for wood-rotting fungi. Gymnosperm trees, as do the brown-rot polypores, decrease from the temperate to subtropical and tropical zones because these fungi prefer coniferous substrates (Wu et al. 2021). In the temperate zones, wood-rotting fungi tend to occur in coniferous and broadleaf mixed forests compared to pure coniferous or broadleaf forests, and more wood-rotting fungal species are found in coniferous and broadleaf-mixed forests (Wang and Wei 2015; Gibertoni et al. 2016). The white rot wood-rotting fungal proportions decrease from China to North America and Europe (Wu et al. 2022).

Wood-inhabiting fungi, a large group within the Basidiomycota, exhibit diverse morphological features in their basidiomata, and this group includes fungi known as polypores, corticioids, and hydnoids within the aphyllophoroid fungi, in which the polypores own poroid hymenophores and those of the corticioids are smooth to tuberculate. In contrast, the hydnoids have a hydnoid or grandinoid or

FIGURE 3.2 Tropical broadleaf forest in Xishuangbanna, Yunnan Province, China.

Source: Photo by Changlin Zhao

odontioid hymenophore (Hjortstam et al. 1987). These fungi are cosmopolitan and rich in diversity, occurring in tropical, subtropical, and temperate regions (Gilbertson and Ryvarden 1987; Bernicchia and Gorjón 2010; Dai 2012; Dai et al. 2021; Zhao et al. 2023a, 2023b; Dong et al. 2025). Hence, wood-rotting fungi, as important strategic biological resources, represent a captivating facet of biodiversity, showcasing a remarkable tapestry of morphological, phylogenetic, and ecological diversity (Park et al. 2020; Liu et al. 2023). To enhance resource utilization and minimize the impact of detrimental species, it is essential to systematically identify and thoroughly sample these fungi. The accurate recognition of species diversity also holds a significant role in advancing global conservation initiatives for wood-rotting fungi (Krah et al. 2018). The phylogenetic diversity of this group has been extensively investigated recently (Justo et al. 2017; Guan et al. 2023; Zhao et al. 2023a, 2024). In China, aside from the nationwide records of wood-rotting fungi (Dai 2012, 2022), surveys have been conducted in several provinces (Dai et al. 2011) and renowned nature reserves (Dai et al. 2021b).

Yunnan Province is in the southwest of China on the Yunnan-Guizhou Plateau, and some southern regions are characterized by tropical ecosystems. The strategic position of Yunnan Province as an "Ecological Security Barrier in Southwest China" and "Biodiversity Treasure" underscores its crucial role in the overall national ecological safety and conservation of biodiversity (Yan et al. 2021). Northern Yunnan reaches elevations of more than 5,000 m.a.s.l and has a subalpine environment. Central Yunnan forms part of the Yunnan-Guizhou Plateau, and has an elevation of about 1,900 m.a.s.l and a temperate climate. The Southern Yunnan province lies in the lower elevations and has subtropical to tropical climatic zones (Figure 3.3). The climate in Yunnan offers environments conducive

FIGURE 3.3 Tropical mixed broadleaf-conifer forest on Gaoligong Mountain, Baoshan, Yunnan Province, China.

Source: Photo by Changlin Zhao

for the speciation and diversification of various life forms (Mi et al. 2021). The climatic diversity and varied landscapes in Yunnan with multiple habitats result in a high species diversity index, with over 6,000 recorded species of fungi (Liu et al. 2023). The forest coverage rate of the province has reached 65.04%, harboring 19,333 species of higher plant, which belong to 3,084 genera and 440 families, constituting 50.1% of the nation's total (Yan et al. 2021). Forty-three primary families (>100 species) account for more than 77% (11,662 species) of all species of angiosperm in Yunnan, and nearly half of all angiosperms in Yunnan are represented by three families. These are the Asteraceae Bercht. & J. Presl, Orchidaceae Juss., and Poaceae Barnhart (Qian et al. 2020). Thirteen main genera comprise over 100 species, including such core genera as *Rhododendron* L. (349 species, 54% of China) and *Pedicularis* L. (196 species, 59% of China, Qian et al. 2020). In addition, 2,273 species of vertebrates, accounting for 51.4% of the country's total, are found within a network of 166 conservation areas, including 21 designated as National Nature Reserves (Dong et al. 2024).

3.3 PATHOGENIC WOOD-ROTTING FUNGI IN TROPICAL ECOSYSTEMS

Pathogenic wood-rotting basidiomycetes have important implications in the forestry industry. Some species of *Heterobasidion* Bref. cause extensive root rot on coniferous trees in the Northern Hemisphere (Woodward et al. 1998). These pathogenic species are of interest not only to forest pathologists but also to taxonomists (Sinclair et al. 1987; Schulze and Bahnweg 1998; Sinclair and Lyon 2005; Linzer et al. 2008; Otrosina and Garbelotto 2010; Rajchenberg and Robledo 2013; Yuan et al. 2021). A special genus of wood-rotting basidiomycetes is *Phylloporia* Murrill, which exhibits high levels of host specificity for living trees or shrubs. The particular feature of *Phylloporia* is that some species in this genus do not decay the wood of the host immediately but rather coexist with the host for a period of time. The *Septobasidium* fungal disease is the primary disease of economic concern in forests (Figure 3.4).

3.4 MACROSCOPIC WOOD-ROTTING FUNGI IN TROPICAL ECOSYSTEMS

Macroscopic wood-rotting fungi have been used as an important type of food and for medicinal purposes in some ancient civilizations worldwide for thousands of years (Anonymous 1955). For example, the latest archaeological evidence has indicated the earliest use of *Ganoderma* spp. could date back 6,800 years ago in Neolithic China (Yuan et al. 2018).

In China, people began to collect wild mushrooms as food since ancient times (Zhang et al. 2015). Today, hundreds of wild species are still commonly collected and eaten around China, especially in the southwestern and northeastern portions. The cultivation of mushrooms dates back to the Tang Dynasty (618–907 AD). It has been recorded that the artificial cultivation of *Lentinula edodes* (Berk.) Pegler was invented about 800 years ago in the Song Dynasty. More recently, mushroom cultivation has become the fifth-largest agriculture industry in China. Around 200 species have been successfully cultivated in laboratories, nearly 100 edible mushroom species have been domesticated, and 60 have been commercially cultivated (Zhang et al. 2015). In 2010, 966 taxa (936 species, 23 varieties, three subspecies, and four forms) were summarized as edible macrofungi in China (Dai et al. 2010). The huge land area of China provides highly diverse habitats for macrofungi. Many of these macrofungi are directly related to people's daily lives and have been utilized by ancient Chinese for at least 6,800 years. In the study reported herein, we evaluated the currently known Chinese macrofungal resource diversity. This research yielded a total of 1,662 taxa, and all species names and their authorities were checked and corrected according to authentic mycological databases. Among the 1,662 taxa, 1,020, 692, and 480 are considered to be edible, medicinal, and poisonous mushrooms, respectively (Wu et al. 2019). A few edible macrofungi in China are commonly used for commercial production. All known medicinal functions are listed for medicinal species. The most common medicinal functions of Chinese macrofungi relate to their antitumor or anticancer, antioxidant, and antimicrobial properties. For example, *Vanderbylia cinnamomea* is an important medicinal fungus (Figure 3.5). A total of 277 Chinese macrofungi are edible and also have

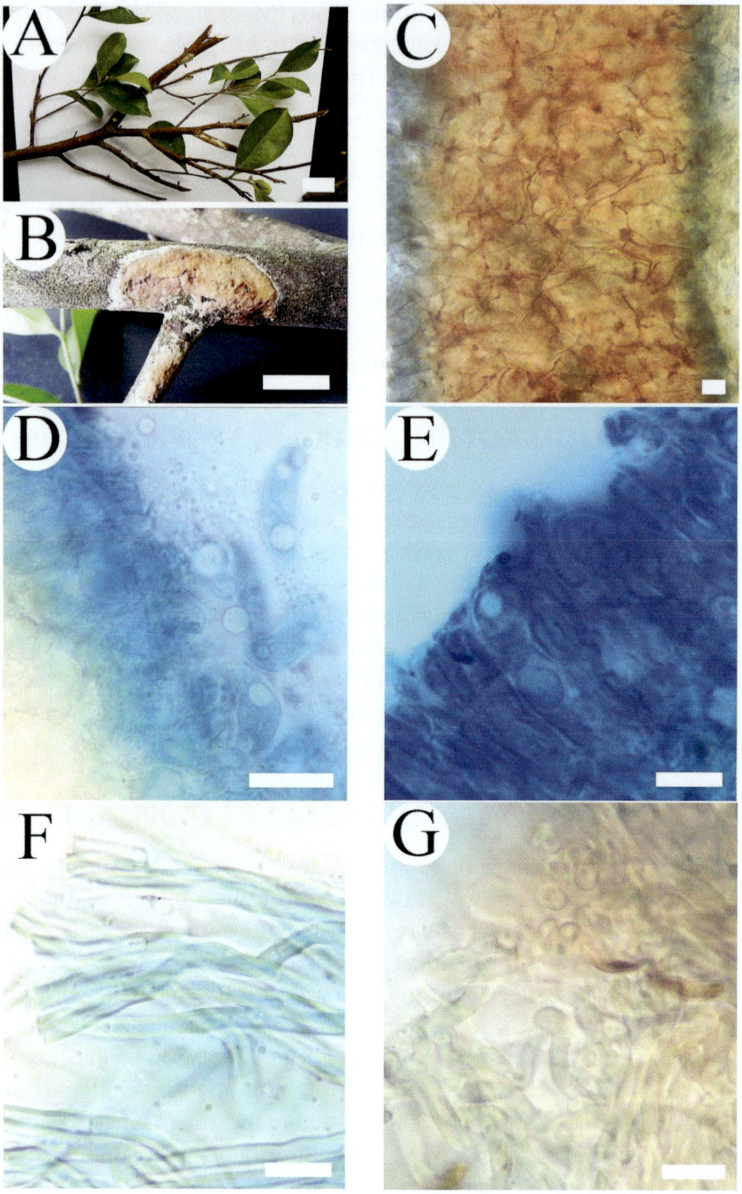

FIGURE 3.4 *Septobasidium aquilariae* (holotype, SWFC 006611). (A, B) Basidiomata on branch, (C) Sections of basidiomata, (D) Basidiospore, (E) Basidia, (F) Hyphae, (G) Haustoria. Bars: A = 5 cm; B = 2 cm; C–G = 10 μm.

Source: Photo by Changlin Zhao and Ruiling Ma

certain medicinal functions and no known toxicity (Figure 3.6). These species could be treated as "Gold Mushrooms." In contrast, 193 edible and/or medicinal species are also recognized as poisonous mushrooms (Wu et al. 2019).

Fungi belonging to the family Lycoperdaceae F. Berchtold & J. Presl are commonly known as puffballs. These fungi have globose to subglobose basidiomes, and the mature gleba releases a powdery mass of spores passively through the apical opening of the endoperidium. Puffballs are widely distributed in temperate, arid, and tropical climates, where they are found on the

FIGURE 3.5 Basidiomata of *Vanderbylia cinnamomea* (holotype, SWFC 008952). Bars A–B = 5 cm.

Source: Photo by Changlin Zhao

FIGURE 3.6 Basidiomata of *Artomyces yunnanensis* (holotype, SWFC 007118). Bars: a = 1 cm; b = 5 mm.

Source: Photo by Changlin Zhao

forest floor, rotting wood, or in open meadows, and some species within this family have a culinary and medicinal value. For example, *Calvatia craniiformis* (Schwein.) Fr. ex De Toni and *C. cyathiformis* (Bosc) Morgan are used as food in the United Kingdom, while *C. gigantea* (Batsch) Lloyd, *C. lilacina* (Mont. & Berk.) Henn are used in traditional medicine for their hemostatic properties and to treat various ailments (Van Wyk and Coetzee 2009). Terrestrial saprophytic species such as *Lycoperdon perlatum* Pers. and *Bovistella utriformis* (Bull.) Demoulin & Rebriev contribute to organic matter decomposition and heavy metal enrichment (Petkovšek and Pokorny 2013).

The order Thelephorales Corn ex. Oberw., a member of the Agaricomycetes (Basidiomycota), was established by Oberwinkler in 1976. Species within the Thelephorales play a vital role in the ecosystems in which they occur (Figure 3.7). Many species of the Thelephorales are ectomycorrhizal (EcM or ECM) fungi, which form an outer sheath around the smaller roots of the plants, and their hyphae penetrate the spaces between the root cortex cells and the epidermal cells, thus forming what is known as a Hartig network (Smith and Read 2008; Vasco-Palacios et al. 2018). In addition, most species within the Thelephorales, as EcM fungi, can absorb organic compounds from host plants (Smith and Read 2008) and thus increase the surface area of host plants to absorb nutrients, especially nitrogen and phosphorus (Smith and Read 2008). Therefore, species of the Thelephorales can participate in the energy flow and promote material circulation within an ecosystem. Some species of the Thelephorales are delicious edible fungi or biologically active fungi, which have substantial economic value in developing and utilizing edible and medicinal fungal resources. For example, *Sarcodon imbricatus* (L.) P. Karst. and *Thelephora ganbajun* M. Zang are staples at edible mushroom markets in southwest China and are very popular with the locals. Some species, such as *Hydnellum concrescens* (Pers.). Banker and *Phellodon melaleucus* (Sw. ex Fr.) P. Karst. have biological activities. The former species has an inhibitory action on syncytium formation and trafficking of glycoprotein and hemagglutinin-neuraminidase (HN) to the cell surface (Lee et al. 2012), while the latter has a unique antibiotic named phellodonic acid, first isolated by Stalpers (1993).

FIGURE 3.7 Basidiomata of *Thelephora yunnanensis* (holotype, SWFC 020929). Bars: A = 1 cm; B = 1 mm.

Source: Photo by Changlin Zhao

3.5 IMPORTANT REPRESENTATIVE POLYPORES IN TROPICAL ECOSYSTEMS

Polypores (Basidiomycota, Agaricomycetes) are wood-rotting macrofungi with a poroid hymenophore and occur on living trees, dead standing trees, fallen trunks, pieces of rotten wood, stumps, roots of trees, and even soil closely associated with the root system of trees (Gibertoni et al. 2016; Shen et al. 2019; Wu et al. 2022; Yang et al. 2025). They are the major group of wood-decomposing fungi and play a key role in the wood decomposition process, releasing carbon (as carbon dioxide) and nutrients from woody tissues (Palviainen et al. 2010; Berglund et al. 2011; Stokland and Larsson 2011; Miettinen et al. 2016, 2018; Huang et al. 2022).

In forest ecosystems, polypores are primarily saprotrophs; however, some are facultative parasites or ectomycorrhizae, causing either brown rot or white rot, and play a key role in energy recycling and forest regeneration. Polypores clearly play a crucial role in energy recycling and forest regeneration in forest ecosystems (Figure 3.8). The majority of these fungi are wood degraders, but some are forest pathogens and others are ectomycorrhizal symbionts (Karsten 1879, 1881, 1892; Patouilliard 1900; Donk 1960, 1964; Ryvarden 1991; Zhao 1998; Kirk et al. 2008; Cui et al. 2019; Dong et al. 2025). The basidiocarps provide food and shelter for many organisms, mostly invertebrates and also some vertebrates, as well as providing food and medicine for humans. The diversity patterns of polypores are probably closely related to tree species diversity and the distribution patterns of angiosperm and gymnosperm trees.

The concept of the Polyporaceae Fr. ex Cord was originally described by Fries (1838) as a family of the Aphyllophorales that included all fungi with poroid hymenophores. This artificial system emphasized the obvious poroid hymenophores, but the important microscopic features used for classification were largely ignored. Subsequently, many mycologists have revised the concept of Polyporaceae. The classifications of the Polyporaceae have been changing for a long time (Karsten 1879, 1881, 1892; Patouilliard 1900; Donk 1960, 1964; Ryvarden 1991; Zhao 1998; Kirk et al. 2008), and species in this family were usually regarded as polypores in previous studies. Among those studies, Leif Ryvarden alone or with collaborators made a major contribution to the taxonomy of polypores, and according to the 10th edition of the Dictionary of Fungi (Kirk et al. 2008), 92 genera and 636 species are accepted in the family Polyporaceae.

Polypores have been intensively studied in the past because these fungi can decompose cellulose, hemicellulose, and lignin in the plant cell walls and therefore play a key role in nutrient recycling in most forest ecosystems. Moreover, some polypores are economically important forest pathogens (Bader et al. 1995; Dai et al. 2007; Gilbert et al. 2008; Rajchenberg and Robledo 2013; Coetzee et al. 2015; Cui et al. 2014; Spirin et al. 2024) and medicinal mushrooms (Dai et al. 2009) and play an important role in industrial biotechnology (Moldes and Sanromán 2006; Marco-Urrea et al. 2009). In recent years, with the rapid development of molecular techniques, molecular data, mostly DNA sequence data, has been widely used in taxonomic studies of polypores and has largely

FIGURE 3.8 Basidiomata of *Perenniporia mopanshanensis* (holotype, SWFC 002404). Bars =3 cm.

Source: Photo by Changlin Zhao

contributed to a more natural classification system of the Polyporaceae. Justo et al. (2017) provided a revised family-level classification of the Polyporales based on phylogenetic analyses inferred from nrLSU, nrITS, and RPB1 genes. Studies in which the species diversity, geographic distribution, taxonomy, and phylogeny of the Polyporaceae have been investigated have resulted in the concept of the Polyporaceae being re-defined. Two hundred and seventeen species belonging to 42 genera of Polyporaceae are found in China, and species diversity and taxonomic classification of wood-rotting fungi in China are provided in Table 3.1. The phylogeny of accepted genera in the Polyporaceae found in China has been reconstructed based on DNA sequences of multiple loci, including the internal transcribed spacer (ITS) regions, the large subunit nuclear ribosomal RNA gene (nLSU), the small subunit nuclear ribosomal RNA gene (nSSU), the small subunit mitochondrial rRNA gene sequences (mtSSU), the translation elongation factor 1-a gene (TEF1), the b-tubulin gene (TBB1), the RNA polymerase II largest subunit (RPB1) and second-largest subunit (RPB2) genes (Cui et al. 2019). In recent years, systematic studies on the taxonomy and phylogeny of some genera of the Polyporaceae from China have been carried out, and molecular data have been widely used in these studies (Cui et al. 2019).

TABLE 3.1
Species Diversity and Taxonomic Classification of Wood-Rotting Fungi in China.

Order	Family	Genus	Species
Agaricales Underw.	*Cystostereaceae* Jülich	*Crustomyces Jülich*	*Crustomyces albidus* Yue Li, Nakasone, C.L. Zhao & S.H. He
			C. tephroleucus (J. Song, Y.C. Dai & B.K. Cui) Yue Li, Nakasone & S.H. He
	Fistulinaceae Lotsy	*Fistulina* Bull.	*Fistulina orientalis* Y.C. Dai, D.P. Bao & Y.W. Lim
			F. subhepatica B.K. Cui & J. Song
	Mycenaceae Overeem	*Favolaschia* (Pat.) Pat.	*Favolaschia bannaensis* Q.Y. Zhang & Y.C. Dai
			F. brevibasidiata Q.Y. Zhang & Y.C. Dai
			F. brevistipitata Q.Y. Zhang & Y.C. Dai
			F. calocera R. Heim
			F. crassipora Q.Y. Zhang & Y.C. Dai
			F. flabelliformis Q.Y. Zhang & Y.C. Dai
			F. imbricata Q.Y. Zhang, L.S. Bian, F. Wu & Y.C. Dai
			F. longistipitata Q.Y. Zhang & Y.C. Dai
			F. manipularis (Berk.) Teng
			F. minutissima Q.Y. Zhang & Y.C. Dai
			F. miscanthi Q.Y. Zhang, L.S. Bian, F. Wu & Y.C. Dai
			F. peziziformis (Berk. & M.A. Curtis) Kuntze
			F. pustulosa (Jungh.) Kuntze
			F. rigida Q.Y. Zhang & Y.C. Dai
			F. semicircularis Q.Y. Zhang & Y.C. Dai
			F. sinarundinariae Q.Y. Zhang, L.S. Bian, F. Wu & Y.C. Dai
			F. subpustulosa Q.Y. Zhang & Y.C. Dai
			F. tenuissima Q.Y. Zhang & Y.C. Dai
			F. tephroleuca Q.Y. Zhang & Y.C. Dai
			F. tonkinensis (Pat.) Kuntze
		Mycena (Pers.) Roussel	*Mycena indigotica* C.L. Wei & R. Kirschner
		Panellus P. Karst.	*Panellus alpinus* Q. Y. Zhang, F. Wu & Y. C. Dai

TABLE 3.1 ***(Continued)***
Species Diversity and Taxonomic Classification of Wood-Rotting Fungi in China.

Order	Family	Genus	Species
			P. bambusicola Q.Y. Zhang & Y.C. Dai
			P. crassiporus Q. Y. Zhang, F. Wu & Y. C. Dai
			P. longistipitatus Q. Y. Zhang, F. Wu & Y. C. Dai
			P. minutissimus Q. Y. Zhang, F. Wu & Y. C. Dai
			P. palmicola Q. Y. Zhang, F. Wu & Y. C. Dai
			P. pusillus (Pers. ex Lév.) Burds. & O.K. Mill.
			P. yunnanensis Q.Y. Zhang & Y.C. Dai
	Pleurotaceae Kühner	*Resupinatus* Nees ex Gray	*Resupinatus europaeus* Consiglio & Setti
			R. porosus M.P. Martín, Lodge & Thorn
			R. porrigens J.Z. Xu & Yu Li
			R. reviviscens Carpouron & Raspé
			R. yunnanensis Yang Yang & C.L. Zhao
	Porotheleaceae Murrill	*Porotheleum* Fr.	*Porotheleum fimbriatum* (Pers.) Fr
	Pterulaceae Corner	*Radulotubus* Y.C. Dai, S.H. He & C.L. Zhao	*Radulotubus resupinatus* Y.C. Dai, S.H. He & C.L. Zhao
	Radulomycetaceae	*Radulomyces* M.P. Christ.	*Radulomyces fissuratus* J.H. Dong & C.L. Zhao
			R. hydnoides J.H. Dong & C.L. Zhao
			R. sinensis J.H. Dong & C.L. Zhao
			R. yunnanensis J.H. Dong & C.L. Zhao
			R. zixishanensis J.H. Dong & C.L. Zhao
	Stephanosporaceae Oberw. & E. Horak	*Lindtneria* Pilát	*Lindtneria asiae-orientalis* S.L. Liu & S.H. He
	Incertae sedis	*Dendrothele* Höhn. & Litsch.	*Dendrothele papillosa* Höhn. & Litsch.
			Dendrothele yunnanensis J.H. Dong & C.L. Zhao
Amylocorticiales K.H. Larss., Manfr. Binder & Hibbett	*Amylocorticiaceae* Jülich	*Anomoloma* Niemelä & K.H. Larss.	*Anomoloma albolutescens* (Romell) Niemelä & K.H. Larss.
			A. denticulatum Y.C. Dai, Vlasák & Meng Zhou
			A. eurasiaticum Y.C. Dai, Vlasák & Meng Zhou
			A. flavissimum (Niemelä) Niemelä & K.H. Larss.
			A. luteoalbum J. Song & B.K. Cui
			A. rhizosum Y.C. Dai & Niemelä
			A. submyceliosum J. Song & B.K. Cui
		Anomoporia Pouzar	*Anomoporia vesiculosa* Y.C. Dai & Niemelä
		Ceraceomyces Jülich	*Ceraceomyces atlanticus* K.H. Larss. & Chikowski
			C. rhizomorphus C.L. Zhao & H.M. Zhou
			C. yunnanensis Qi Yuan & C.L. Zhao
Auriculariales Bromhead	*Auriculariaceae* Fr.	*Alloexidiopsis* L.W. Zhou & S.L. Liu	*Alloexidiopsis australiensis* S.L. Liu, Z.Q. Shen & L.W. Zhou
			A. grandinea J.H. Dong & C.L. Zhao
			A. nivea (J.J. Li & C.L. Zhao) L.W. Zhou & S.L. Liu
			A. sinensis J.H. Dong & C.L. Zhao
			A. xantha J.H. Dong & C.L. Zhao

(Continued)

TABLE 3.1 *(Continued)*
Species Diversity and Taxonomic Classification of Wood-Rotting Fungi in China.

Order	Family	Genus	Species
		Elmerina Bres.	*Elmerina caryae* (Schwein.) D.A. Reid
			E. cladophora (Berk.) Bres.
			E. foliacea Pat.
			E. fragilis F. Wu & Hai J. Li
			E. hispida (Imazeki) Y.C. Dai & L.W. Zhou
		Eichleriella Bres.	*Eichleriella aculeobasidiata* Hui Wang, Dong Qiong Wang, C.L. Zhao
			E. xinpingensis C.L. Zhao
			E. yunnanensis Y.L. Deng & C.L. Zhao
		Heteroradulum Lloyd ex Spirin & Malysheva	*Heteroradulum adnatum* Spirin & Malysheva
			H. niveum J.J. Li & C.L. Zhao
			H. yunnanense C.L. Zhao
	Incertae sedis	*Aporpium* Bondartsev & Singer	*Aporpium efibulatum* (Y.C. Dai & Y.L. Wei) Sotome & T. Hatt.
			A. miniporum F. Wu, Y.C. Dai & L.W. Zhou
			A. obtusisporum F.C. Huang & Bin Liu
		Adustochaete Alvarenga & K.H. Larss.	*Adustochaete albomarginata* J.H. Dong & C.L. Zhao
			A. punctata J.H. Dong & C.L. Zhao
			A. yunnanensis Y.F. Li & C.L. Zhao
		Basidiodendron Rick	*Basidiodendron yunnanense* C.L. Zhao
		Grammatus H.S. Yuan & C. Decock	*Grammatus labyrinthinus* H.S. Yuan & C. Decock
			G. semis (Spirin & Malysheva) H.S. Yuan & C. Decock
		Nodulochaete J.H. Dong & C.L. Zhao	*Nodulochaete fissurata* J.H. Dong & C.L. Zhao
			N. punctata J.H. Dong & C.L. Zhao
Boletales E.-J. Gilbert	*Coniophoraceae* Ulbr.	*Coniophora* DC.	*Coniophora sistotremoides* (Schwein.) Massee
			C. yunnanensis Yang Yang & C.L. Zhao
	Serpulaceae Jarosch & Bresinsky	*Serpula* (Pers.) Gray	*Serpula dendrocalami* C.L. Zhao
			S. himantioides (Fr.) P. Karst.
			S. lacrymans (Wulfen) J. Schröt
			S. similis (Berk. & Broome) Ginns
	Tapinellaceae C. Hahn	*Bondarcevomyces* Parmasto	*Bondarcevomyces taxi* (Bondartsev) Parmasto
Cantharellales Gäum	*Hydnaceae* Chevall.	*Sistotrema* Fr.	*Sistotrema confluens* Pers.
			S. muscicola (Pers.) S. Lundell
			S. subconfluens L.W. Zhou
			S. yunnanense L.Q. Cai & C.L. Zhao
Corticiales K.H. Larss.	*Corticiaceae* Herter	*Corticium* Pers.	*Corticium roseoalbum* J.H. Dong & C.L. Zhao
		Lyomyces P. Karst.	*Lyomyces allantosporus* Riebesehl, Yurch. & Langer
			L. albopulverulentus C.L. Zhao

TABLE 3.1 ***(Continued)***
Species Diversity and Taxonomic Classification of Wood-Rotting Fungi in China.

Order	Family	Genus	Species
			L. austro-occidentalis Xue W. Wang & L.W. Zhou
			L. bambusinus C.L. Zhao
			L. cremeus C.L. Zhao
			L. daweishanensis J.H. Dong & C.L. Zhao
			L. densiusculus Viner & Ryvarden
			L. elaeidicola Xue W. Wang & L.W. Zhou
			L. fimbriatus (Sheng H. Wu) Riebesehl & Yurchenko
			L. fissuratus C.L. Zhao
			L. fumosus C.L. Zhao
			L. gatesiae Xue W. Wang & L.W. Zhou
			L. granulosus Yurchenko & Langer
			L. guttulatus Xue W. Wang & L.W. Zhou
			L. hengduanensis Q. Yuan & C.L. Zhao
			L. incanus J.H. Dong & C.L. Zhao
			L. incrustatus (Kotir. & Saaren.) Hjortstam & Ryvarden
			L. leptocystidiatus Xue W. Wang & L.W. Zhou
			L. lincangensis J.H. Dong & C.L. Zhao
			L. luteoalbus J.H. Dong & C.L. Zhao
			L. macrescens (Banker) Riebesehl & Langer
			L. macrosporus C.L. Zhao
			L. mascarensis Riebesehl, Yurch. & Langer
			L. napoensis Yurchenko & Riebesehl
			L. neocrustosus Yurchenko & Riebesehl
			L. niveomarginatus Q. Yuan & C.L. Zhao
			L. niveus C.L. Zhao
			L. niveus C.L. Zhao ex L.W. Zhou & Xue W. Wang
			L. ochraceoalbus C.L. Zhao
			L. oleifer Yurchenko & Langer
			L. orarius Yurchenko & Riebesehl
			L. parvus Yurchenko & Langer
			L. pruni (Lasch) Riebesehl & Langer
			L. qujingensis J.H. Dong & C.L. Zhao
			L. sceptrifer Yurchenko & Langer
			L. sinensis J.H. Dong & C.L. Zhao
			L. stratosus (G. Cunn.) Riebesehl & Langer
			L. subcylindricus Yurchenko & Riebesehl
			L. tasmanicus Xue W. Wang & L.W. Zhou
			L. tenuissimus (Yurchenko & Sheng H. Wu) Riebesehl & Langer
			L. vietnamensis (Yurchenko & Sheng H. Wu) Riebesehl & Langer
			L. wuliangshanensis C.L. Zhao
			L. wumengshanensis Q. Yuan & C.L. Zhao
			L. yunnanensis C.L. Zhao
			L. zhaotongensis Q. Yuan & C.L. Zhao
Gloeophyllales Thorn	*Gloeophyllaceae* Jülich	*Gloeophyllum* P. Karst.	*Gloeophyllum abietinum* (Bull.) P. Karst.

(Continued)

TABLE 3.1 ***(Continued)***

Species Diversity and Taxonomic Classification of Wood-Rotting Fungi in China.

Order	Family	Genus	Species
			G. carbonarium (Berk. & M.A. Curtis) Ryvarden
			G. odoratum (Wulfen) Imazeki
			G. protractum (Fr.) Imazeki
			G. subferrugineum (Berk.) Bondartsev & Singer
			G. sepiarium (Wulfen) P. Karst.
			G. trabeum (Pers.) Murrill
		Griseoporia Ginns	*Griseoporia taiwanensis* Y.C. Dai & S.H. He
		Hispidaedalea Y.C. Dai & S.H. He	*Hispidaedalea imponens* (Ces.) Y.C. Dai & S.H. He
Gomphales Corner	*Gomphaceae* Donk	*Ramaria* Holmsk.	*Ramaria xizangensis* C.L. Zhao & H.M. Zhou
		Ramaricium J. Erikss.	*Ramaricium yunnanense* C.L. Zhao
Hymenochaetales Oberw.	*Hymenochaetaceae* Donk	*Coltricia* Gray	*Coltricia abieticola* Y.C. Dai
			C. austrosinensis L.S. Bian & Y.C. Dai
			C. baoshanensis (Y.C. Dai & B.K. Cui) Y.C. Dai & F. Wu
			C. cinnamomea (Jacq.) Murrill
			C. crassa Y.C. Dai
			C. dependens (Berk. & M.A. Curtis) Imazeki
			C. duportii (Pat.) Ryvarden
			C. fimbriata L.S. Bian, M. Zhou & J. Yu
			C. focicola (Berk. & M.A. Curtis) Murrill
			C. fragilissima (Mont.) Ryvarden
			C. globosa (L.S. Bian & Y.C. Dai) Y.C. Dai & F. Wu
			C. lateralis L.S. Bian & Y.C. Dai
			C. lenis L.S. Bian, M. Zhou & J. Yu
			C. macropora Y.C. Dai
			C. minima L.S. Bian & Y.C. Dai
			C. minor Y.C. Dai
			C. montagnei (Fr.) Murrill
			C. naviculiformis (Y.C. Dai & Niemelä) Y.C. Dai & F. Wu
			C. oblectabilis (Lloyd) Ryvarden
			C. perennis (L.) Murrill
			C. pseudodependens (L.S. Bian & Y.C. Dai) Y.C. Dai & F. Wu
			C. pusilla Imazeki & Kobayasi
			C. rigida L.S. Bian & Y.C. Dai
			C. strigosipes Corner
			C. sinoperennis Y.C. Dai & F. Wu
			C. subcinnamomea L.S. Bian & Y.C. Dai
			C. subglobosa (Y.C. Dai) Y.C. Dai & F. Wu
			C. subpicta (Lloyd) Imazeki & Kobayasi
			C. subverrucata L.S. Bian & Y.C. Dai
			C. tasmanica (Cleland & Rodway) Y.C. Dai & F. Wu
			C. tenuihypha L.S. Bian, M. Zhou & J. Yu
			C. tibetica Y.C. Dai & F. Wu
			C. tsugicola Y.C. Dai & B.K. Cui
			C. velutina Baltazar & Gibertoni

TABLE 3.1 *(Continued)*
Species Diversity and Taxonomic Classification of Wood-Rotting Fungi in China.

Order	Family	Genus	Species
			C. verrucata Aime, T.W. Henkel & Ryvarden
			C. weii Y.C. Dai
			C. wenshanensis L.S. Bian & Y.C. Dai
			C. zixishanensis Q. Zhou & C.L. Zhao
		Coltriciella Murrill	*Coltriciella yunnanensis* Q. Zhou & C.L. Zhao
		Coniferiporia L.W. Zhou & Y.C. Dai	*Coniferiporia qilianensis* (B.K. Cui, L.W. Zhou & Y.C. Dai) L.W. Zhou & Y.C. Dai
			C. sulphurascens (Pilát) L.W. Zhou & Y.C. Dai
			C. uzbekistanensis L.W. Zhou, Xue W. Wang & Gafforov
			C. weirii (Murrill) L.W. Zhou & Y.C. Dai
		Cylindrosporus L.W. Zhou	*Cylindrosporus flavidus* (Berk.) L.W. Zhou
		Fomitiporella Murrill	*Fomitiporella austroasiana* Y.C. Dai, X.H. Ji & J. Vlasák
			F. adhaerens (J.E. Wright & Blumenf.) Y.C. Dai & F. Wu
			F. caviphila L.W. Zhou
			F. chinensis (Pilát) Y.C. Dai, X.H. Ji & Vlasák
			F. crystallina X.H. Ji
			F. cyclobalanopsidis (T.T. Chang & W.N. Chou) Y.C. Dai & F. Wu
			F. shoushana (T.T. Chang & W.N. Chou) Y.C. Dai & F. Wu
			F. sinica Y.C. Dai, X.H. Ji & Vlasák
			F. subinermis Y.C. Dai, X.H. Ji & Vlasák
			F. tenuicula (B.K. Cui) Y.C. Dai & F. Wu
			F. tenuissima (H.Y. Yu, C.L. Zhao & Y.C. Dai) Y.C. Dai, X.H. Ji & Vlasák
			F. umbrinella (Bres.) Murrill
		Fomitiporia Murrill	*Fomitiporia alpina* B.K. Cui & Hong Chen
			F. bannaensis Y.C. Dai
			F. carpinea X.H. Ji, X.M. Tian & Y.C. Dai
			F. gaoligongensis B.K. Cui & Hong Chen
			F. hainaniana B.K. Cui & Hong Chen
			F. hartigii (Allesch. & Schnabl) Fiasson & Niemelä
			F. norbulingka B.K. Cui & Hong Chen
			F. pentaphylacis L.W. Zhou
			F. punctata (Fr.) Murrill
			F. punicata Y.C. Dai, B.K. Cui & Decock
			F. pusilla (Lloyd) Y.C. Dai
			F. rhamnoides T.Z. Liu & F. Wu
			F. subhippophaëicola B.K. Cui & H. Chen
			F. subrobusta B.K. Cui & Hong Chen
			F. subtropical B.K. Cui & Hong Chen
			F. tenuitubus L.W. Zhou
			F. torreyae Y.C. Dai & B.K. Cui
		Fulvifomes Murrill	*Fulvifomes allardii* (Bres.) Bondartseva & S. Herrera
			F. azonatus Y.C. Dai & X.H. Ji

(Continued)

TABLE 3.1 ***(Continued)***
Species Diversity and Taxonomic Classification of Wood-Rotting Fungi in China.

Order	Family	Genus	Species
			F. caligoporus Y.C. Dai & X.H. Ji
			F. collinus (Y.C. Dai & Niemelä) Y.C. Dai
			F. dracaenicola Z.B. Liu & Y.C. Dai
			F. durissimus (Lloyd) Bondartseva & S. Herrera
			F. fastuosus (Lév.) Bondartseva & S. Herrera
			F. fushanianus (T.T. Chang) Y.C. Dai & F. Wu
			F. hainanensis L.W. Zhou
			F. indicus (Massee) L.W. Zhou
			F. lloydii (Cleland) Y.C. Dai, X.H. Ji
			F. merrillii (Murrill) Baltazar & Gibertoni
			F. nonggangensis F.C. Huang, H.F. Zheng & Bin Liu
			F. rigidus (B.K. Cui & Y.C. Dai) X.H. Ji & Jia J Chen
			F. rimosus (Berk.) Fiasson & Niemelä
			F. subindicus Y.C. Dai & X.H. Ji
			F. submerrillii X.H. Ji & Jia J Chen
			F. tubogeneratus F.C. Huang, H.F. Zheng & Bin Liu
		Fulvoderma L.W. Zhou & Y.C. Dai	*Fulvoderma australe* L.W. Zhou & Y.C. Dai
			F. microporum Xue W. Wang & L.W. Zhou
			F. scaurum (Lloyd) L.W. Zhou & Y.C. Dai
			F. yunnanense C.L. Zhao
		Fuscoporia Murrill	*Fuscoporia acutimarginata* Y.C. Dai & Q. Chen
			F. ambigua P. Du, Q. Chen & Vlasák
			F. australasiatica Q. Chen, F. Wu & Y.C. Dai
			F. bambusicola (L.W. Zhou & B.S. Jia) Q. Chen, F. Wu & Y.C. Dai
			F. chinensis Q. Chen, F. Wu & Y.C. Dai
			F. chrysea (Lév.) Baltazar & Gibertoni
			F. contigua (Pers.) G. Cunn.
			F. discipes (Berk.) Y.C. Dai & Ghob.-Nejh
			F. ferrea (Pers.) G. Cunn., Bull. N. Z. Dept
			F. ferruginosa (Schrad.) Murrill
			F. gilva (Schwein.) T. Wagner & M. Fisch.
			F. formosana (T.T. Chang & W.N. Chou) T. Wagner & M. Fisch.
			F. karsteniana Q. Chen, F. Wu & Y.C. Dai
			F. montana Y.C. Dai & Niemelä
			F. monticola Y.C. Dai, Q. Chen & J. Vlasák
			F. ramulicola Y.C. Dai & Q. Chen
			F. rhabarbarina (Berk.) Groposo, Log.-Leite & Góes-Neto
			F. senex (Nees & Mont.) Ghob.-Nejh
			F. setifera (T. Hatt.) Y.C. Dai
			F. sinica Y.C. Dai, Q. Chen & J. Vlasák
			F. sinuosa Y.C. Dai & F. Wu
			F. subchrysea Q. Chen, F. Wu & Y.C. Dai
			F. subferrea Q. Chen & Yuan Yuan
			F. torulosa (Pers.) T. Wagner & M. Fisch.

TABLE 3.1 ***(Continued)***
Species Diversity and Taxonomic Classification of Wood-Rotting Fungi in China.

Order	Family	Genus	Species
			F. wahlbergii (Fr.) T. Wagner & M. Fisch.
			F. yunnanensis Y.C. Dai
		Hastodontia (Parmasto) Hjortstam & Ryvarden	*Hastodontia halonata* (J. Erikss. & Hjortstam) Hjortstam & Ryvarden
		Hymenochaete Lév.	*Hymenochaete campylopora* (Mont.) Spirin & Miettinen
			H. cinnamomea (Pers.) Bres.
			H. cyclolamellata T. Wagner & M. Fisch.
			H. microcycla (Zipp. ex Lév.) Spirin & Miettinen
			H. micropora L.W. Zhou & Y.C. Dai
			H. rubiginosa (Dicks.) Lév.
			H. setipora (Berk.) S.H. He & Y.C. Dai
			H. subporioides L.W. Zhou & Y.C. Dai
			H. sinensis Y.C. Li & C.L. Zhao
			H. xerantica (Berk.) S.H. He & Y.C. Dai
		Hydnoporia Murrill	*Hydnoporia lamellata* (Y.C. Dai & Niemelä) Miettinen & K.H. Larss.
			H. pinicola J.H. Dong & C.L. Zhao
			H. tabacina (Sowerby) Spirin, Miettinen & K.H. Larss.
			H. yunnanensis J.H. Dong & C.L. Zhao
		Inocutis Fiasson & Niemelä	*Inocutis levis* (P. Karst.) Y.C. Dai
			I. ludoviciana (Pat.) T. Wagner & M. Fisch.
			I. mikadoi (Lloyd) Y.C. Dai & F. Wu
			I. rheades (Pers.) Fiasson & Niemelä
			I. subdryophila Y.C. Dai & H.S. Yuan
			I. tamaricis (Pat.) Fiasson & Niemelä
			I. tenuicarnis (Pegler & D.A. Reid) Y.C. Dai & F. Wu
		Inonotopsis Parmasto	*Inonotopsis subiculosa* (Peck) Parmasto
		Inonotus P. Karst.	*Inonotus acutus* B.K. Cui & Y.C. Dai
			I. casuarinae L.S. Bian
			I. chihshanyenus T. T. Chang & W. N. Chou
			I. compositus Han C. Wang
			I. cuticularis (Bull.) P. Karst.
			I. diverticuloseta Pegler
			I. fushanianus T.T. Chang
			I. griseus L.W. Zhou
			I. hainanensis H.X. Xiong & Y.C. Dai
			I. henanensis Juan Li & Y.C. Dai
			I. *hispidus* (Bull.) P. Karst.
			I. indurescens Y.C. Dai
			I. krawtzewii (Pilát) Pilát
			I. latemarginatus Y.C. Dai
			I. magnisetus Y.C. Dai
			I. niveomarginatus H.Y. Yu, C.L. Zhao & Y.C. Dai
			I. obliquus (Fr.) Pilát
			I. ochroporus (Van der Byl) Pegler

(Continued)

TABLE 3.1 *(Continued)*
Species Diversity and Taxonomic Classification of Wood-Rotting Fungi in China.

Order	Family	Genus	Species
			I. patouillardii (Rick) Imazeki
			I. plorans (Pat.) Bondartsev & Singer
			I. rickii (Pat.) D.A. Reid
			I. rodwayi D.A. Reid
			I. sublevis Y.C. Dai & Niemelä
			I. subradiatus Y.C. Dai & F. Wu
		Mensularia Lázaro Ibiza	*Mensularia lithocarpi* L.W. Zhou
			M. radiata (Sowerby) Lázaro Ibiza
			M. rhododendri F. Wu, Y.C Dai & L.W. Zhou
		Neomensularia F. Wu, L.W. Zhou & Y.C. Dai	*Neomensularia castanopsidis* Y.C. Dai & F. Wu
			N. duplicata F. Wu, L.W. Zhou & Y.C. Dai
			N. kanehirae (Yasuda) F. Wu, L.W. Zhou & Y.C. Dai
			N. rectiseta X.H. Ji, L.W. Zhou & F. Wu
		Ochrosporellus (Bondartseva & S. Herrera) Bondartseva & S. Herrera	*Ochrosporellus chrysomarginatus* (B.K. Cui & Y.C. Dai) Y.C. Dai & F. Wu
			O. pachyphloeus (Pat.) Y.C. Dai & F. Wu
			O. puerensis (Hai J. Li & S.H. He) Y.C. Dai & F. Wu
			O. taiwanensis (Sheng H. Wu et al.) Y.C. Dai & F. Wu
			O. tricolor (Bres.) Y.C. Dai & F. Wu
		Onnia P. Karst.	*Onnia himalayana* Y.C. Dai, H. Zhao & Meng Zhou
			O. leporina (Fr.) H. Jahn
			O. microspora Y.C. Dai & L.W. Zhou
			O. tibetica Y.C. Dai & S.H. He
			O. tomentosa (Fr.) P. Karst.
			O. vallata (Berk.) Y.C. Dai & Niemelä
		Phellinidium (Kotl.) Fiasson & Niemelä	*Phellinidium aciferum* Y.C. Dai
			P. asiaticum Spirin, L.W. Zhou & Y.C. Dai
			P. ferrugineofuscum (P. Karst.) Fiasson & Niemelä
		Phellinopsis Y.C. Dai	*Phellinopsis asetosa* L.W. Zhou
			P. conchata (Pers.) Y.C. Dai
			P. helwingiae L.W. Zhou & W.M. Qin
			P. junipericola L.W. Zhou
			P. lonicericola L.W. Zhou
			P. resupinata L.W. Zhou
			P. tibetica L.W. Zhou
		Phellinus Quél.	*Phellinus alni* (Bondartsev) Parmasto
			P. austrosinensis L.S. Bian
			P. castanopsidis B.K. Cui, Y.C. Dai & Decock
			P. cuspidatus Y.C. Dai & F. Wu
			P. ellipsoideus (B.K. Cui & Y.C. Dai) B.K. Cui, Y.C. Dai & Decock
			P. igniarius (L.) Quél.
			P. lundellii Niemelä
			P. monticola L.W. Zhou & Y.C. Dai

TABLE 3.1 *(Continued)*
Species Diversity and Taxonomic Classification of Wood-Rotting Fungi in China.

Order	Family	Genus	Species
			P. mori Y.C. Dai & B.K. Cui
			P. nigricans (Fr.) P. Karst.
			P. orientoasiaticus L.W. Zhou & Y.C. Dai
			P. padicola L.W. Zhou & Y.C. Dai
			P. parmastoi L.W. Zhou & Y.C. Dai
			P. piceicola B.K. Cui & Y.C. Dai
			P. tremulae (Bondartsev) Bondartsev & P.N. Borisov
		Phellopilus Niemelä, T. Wagner & M. Fisch.	*Phellopilus nigrolimitatus* (Romell) Niemelä, T. Wagner & M. Fisch.
		Phylloporia Murrill	*Phylloporia alyxiae* Sheng H. Wu
			P. bibulosa (Lloyd) Ryvarden
			P. clausenae L.W. Zhou
			P. crataegi L.W. Zhou & Y.C. Dai
			P. cylindrispora L.W. Zhou
			P. cystidiolophora F. Wu, G.J. Ren & Y.C. Dai
			P. dependens Y.C. Dai
			P. flacourtiae L.W. Zhou
			P. fontanesiae L.W. Zhou & Y.C. Dai
			P. gutta L.W. Zhou & Y.C. Dai
			P. hainaniana Y.C. Dai & B.K. Cui
			P. homocarnica L.W. Zhou
			P. lespedezae G.J. Ren & F. Wu
			P. lonicerae W.M. Qin, Xue W. Wang, T. Sawahata & L.W. Zhou
			P. manglietiae Yuan Y. Chen & B.K. Cui
			P. minutipora L.W. Zhou
			P. minutissima Y.C. Dai & F. Wu
			P. mori Sheng H. Wu
			P. moricola Sheng H. Wu
			P. murrayae Sheng H.Wu
			P. oblongospora Y.C. Dai & H.S. Yuan
			P. oreophila L.W. Zhou & Y.C. Dai
			P. osmanthi L.W. Zhou
			P. pectinata (Klotzsch) Ryvarden
			P. pendula Yuan Y. Chen & B.K. Cui
			P. perangusta F. Wu, G.J. Ren & Y.C. Dai
			P. pseudopectinata Yuan Y. Chen & B.K. Cui
			P. pulla (Mont. & Berk.) Decock & Yombiyeni
			P. radiata L.W. Zhou
			P. rattanicola F. Wu, G.J. Ren & Y.C. Dai
			P. rubiacearum Sheng H. Wu
			P. splendida F. Wu, G.J. Ren & Y.C. Dai
			P. subpulla F.Wu, G.J. Ren & Y.C. Dai
			P. terrestris L.W. Zhou
			P. tiliae L.W. Zhou
		Porodaedalea Murrill	*Porodaedalea alpicola* S.J. Dai, F. Wu & Y.C. Dai
			P. chinensis S.J. Dai & F. Wu

(Continued)

TABLE 3.1 ***(Continued)***

Species Diversity and Taxonomic Classification of Wood-Rotting Fungi in China.

Order	Family	Genus	Species
			P. chrysoloma (Fr.) Fiasson & Niemelä
			P. himalayensis (Y.C. Dai) Y.C. Dai
			P. laricis (Jacz. ex Pilát) Niemelä
			P. microsperma S.J. Dai & Y.C. Dai
			P. mongolica Y.D. Wu & Y. Yuan
			P. pini (Brot.) Murrill
			P. qilianensis Y.C. Dai & F. Wu
			P. schrenkianae Y.C. Dai & F. Wu
			P. yunnanensis S.J. Dai, F. Wu & Y.C. Dai
		Pseudoinonotus T. Wagner & M. Fisch.	*Pseudoinonotus tibeticus* (Y.C. Dai & M. Zang) Y.C. Dai, B.K. Cui & Decock
		Pyrrhoderma Imazeki	*Pyrrhoderma adamantinum* (Berk.) Imazeki
			P. hainanense L.W. Zhou & Y.C. Dai
			P. lamaoense (Murrill) L.W. Zhou & Y.C.Dai
			P. nigra Meng Zhou, Yuan Yuan & Vlasák
			P. sublamaensis (Lloyd) Y.C. Dai & F. Wu
			P. yunnanense L.W. Zhou & Y.C. Dai
		Rigidonotus Y.C. Dai, F. Wu, L.W. Zhou, Vlasák & B.K. Cui	*Rigidonotus pruinosus* (Bondartsev) Y.C. Dai, F. Wu, L.W. Zhou, Vlasák & B.K. Cui
		Sanghuangporus Sheng H. Wu, L.W. Zhou & Y.C. Dai	*Sanghuangporus alpinus* (Y.C. Dai & X.M. Tian) L.W. Zhou & Y.C. Dai
			S. baumii (Pilát) L.W. Zhou & Y.C. Dai
			S. lonicericola (Parmasto) L.W. Zhou & Y.C. Dai
			S. mongolicus T. Bau
			S. quercicola Lin Zhu & B.K. Cui
			S. sanghuang (Sheng H. Wu, T. Hatt. & Y.C. Dai) Sheng H. Wu, L.W. Zhou & Y.C. Dai
			S. subbaumii Shan Shen, Y.C. Dai & L.W. Zhou
			S. toxicodendri Sheng H. Wu, B.K. Cui & Guo Z. Jiang
			S. vaninii (Ljub.) L.W. Zhou & Y.C. Dai
			S. vitexicola Sheng H. Wu
			S. weigelae (T. Hatt. & Sheng H. Wu) Sheng H. Wu, L.W. Zhou & Y.C. Dai
			S. zonatus (Y.C. Dai & X.M. Tian) L.W. Zhou & Y.C. Dai
		Tubulicrinis Donk	*Tubulicrinis pini* J.H. Dong & C.L. Zhao
			T. thermometrus (G. Cunn.) M.P. Christ.
			T. viticola (Schwein.) Z.S. Bi & G.Y. Zheng
			T. xantha C.L. Zhao
			T. yunnanensis C.L. Zhao
		Tropicoporus L.W. Zhou, Y.C. Dai & Sheng H. Wu	*Tropicoporus boehmeriae* L.W. Zhou & F. Wu
			T. cesatii (Bres.) Y.C. Dai & F. Wu
			T. detonsus (Fr.) Y.C. Dai & F. Wu

TABLE 3.1 *(Continued)*
Species Diversity and Taxonomic Classification of Wood-Rotting Fungi in China.

Order	Family	Genus	Species
			T. excentrodendri L.W. Zhou & Y.C. Dai
			T. glaucescens (Petch) Y.C. Dai & F. Wu
			T. hainanicus Y.C. Dai & F. Wu
			T. minisporus (B.K. Cui & Y.C. Dai) Y.C. Dai & F. Wu
			T. minus Y.C. Dai & F. Wu
			T. ravidus Y.C. Dai & F. Wu
			T. tenuis Y.C. Dai & F. Wu
			T. zuzaneae A.H. Zhu, Yuan Yuan & S.H. He
	Hyphodontiaceae Xue W. Wang & L.W. Zhou	*Kneiffiella* P. Karst.	*Kneiffiella abdita* Riebesehl & Langer
	Hirschioporaceae Y.C. Dai, Yuan Yuan & Meng Zhou	*Hirschioporus* Donk	*Hirschioporus abietinus* (Pers. ex J.F. Gmel.) Donk
			H. acontextus Y.C. Dai, Yuan Yuan & Meng Zhou
			H. beijingensis Y.C. Dai, Yuan Yuan & Meng Zhou
			H. chinensis Y.C. Dai, Yuan Yuan & Meng Zhou
			H. fuscoviolaceus (Ehrenb.) Donk
			H. montanus (T. Hatt.) Y.C. Dai, Yuan Yuan & Meng Zhou
			H. pubescens Y.C. Dai, Yuan Yuan & Meng Zhou
			H. tianschanicus Y.C. Dai, Yuan Yuan & Meng Zhou
		Nigrohirschioporus Y.C. Dai, Yuan Yuan & Meng Zhou	*Nigrohirschioporus durus* (Jungh.) Y.C. Dai, Yuan Yuan & Meng Zhou
		Pallidohirschioporus Y.C. Dai, Yuan Yuan & Meng Zhou	*Pallidohirschioporus biformis* (Fr.) Y.C. Dai, Yuan Yuan & Meng Zhou
			P. brastagii (Corner) Y.C. Dai, Yuan Yuan & Meng Zhou
			P. imbricatus (Y.C. Dai & B.K. Cui) Y.C. Dai, Yuan Yuan & Meng Zhou
			P. polycystidiatus (Pilát) Y.C. Dai, Yuan Yuan & Meng Zhou
			P. versicolor Y.C. Dai, Yuan Yuan & Meng Zhou
		Perennihirschioporus Y.C. Dai, Yuan Yuan & Meng Zhou	*Perennihirschioporus perennis* (Y.C. Dai & H.S. Yuan) Y.C. Dai, Yuan Yuan & Meng Zhou
	Neoantrodiellaceae Y.C. Dai, B.K. Cui, Jia J. Chen & H.S. Yuan	*Neoantrodiella* Y.C. Dai, B.K. Cui, Jia J. Chen & H.S. Yuan	*Neoantrodiella gypsea* (Yasuda) Y.C. Dai, B.K. Cui, Jia J. Chen & H.S. Yuan
			N. thujae (Y.C. Dai & H.S. Yuan) Y.C. Dai, B.K. Cui, Jia J. Chen & H.S. Yuan
	Nigrofomitaceae Jülich	*Nigrofomes* Murrill	*Nigrofomes sinomelanoporus* L.W. Zhou
			N. submelanoporus M. Zhou, F. Wu & X.M. Tian
	Oxyporaceae Zmitr. & V. Malysheva	*Leucophellinus* Bondartsev & Singer	*Leucophellinus hobsonii* (Berk. ex Cooke) Ryvarden

(Continued)

TABLE 3.1 ***(Continued)***
Species Diversity and Taxonomic Classification of Wood-Rotting Fungi in China.

Order	Family	Genus	Species
			L. irpicoides (Bondartsev ex Pilát) Bondartsev & Singer
		Oxyporus (Bourdot & Galzin) Donk	*Oxyporus bucholtzii* (Bondartsev & Ljub.) Y.C. Dai & Niemelä
			O. corticola (Fr.) Ryvarden
			O. cuneatus (Murrill) Aoshima
			O. ginkgonis Y.C. Dai
			O. latemarginatus (Durieu & Mont.) Donk
			O. macroporus Y.C. Dai & Y.L. Wei
			O. obducens (Pers.) Donk
			O. piceicola B.K. Cui & Y.C. Dai
			O. philadelphi (Parmasto) Ryvarden
			O. populinus (Schumach.) Donk
			O. sinensis X.L. Zeng
			O. subpopulinus B.K. Cui & Y.C. Dai
	Punctulariaceae Donk	*Punctulariopsis* Ghob.-Nejh.	*Punctulariopsis efibulata* (M.J. Larsen & Nakasone) Ghob.-Nejh.
			P. subglobispora (Hallenb. & Hjortstam) Ghob.-Nejh.
			P. yunnanensis A. Muhammad & C.L. Zhao
	Rickenellaceae Vizzini	*Peniophorella* P. Karst.	*Peniophorella cremea* C.L. Zhao
			P. daweishanensis J.H. Dong & C.L. Zhao
			P. fissurata C.L. Zhao
			P. guttulifera (P. Karst.) K.H. Larss.
			P. olivacea J.H. Dong & C.L. Zhao
			P. pubera (Fr.) P. Karst.
			P. sidera Xue W. Wang & L.W. Zhou
			P. subreticulata Xue W. Wang & L.W. Zhou
			P. yunnanensis C.L. Zhao
		Resinicium Parmasto	*Resinicium bicolor* (Alb. & Schwein.) Parmasto
			R. furfuraceum (Bres.) Parmasto
			R. yunnanense Jia Cai, Wei Zhang, H.G. Liu & C.L. Zhao
		Sidera Miettinen & K.H. Larss.	*Sidera borealis* Z.B. Liu & Yuan Yuan
			S. inflata Z.B. Liu & Y.C. Dai
			S. lenis (P. Karst.) Miettinen
			S. minutissima Y.C. Dai, F. Wu, G.M. Gates & Rui Du
			S. parallela Y.C. Dai, F. Wu, G.M. Gates & Rui Du
			S. punctata Z.B. Liu & Y.C. Dai
			S. roseobubalina Z.B. Liu & Y.C. Dai
			S. salmonea Z.B. Liu, Jian Yu & F. Wu
			S. tianshanensis B.K. Cui & T.M. Xu
			S. tibetica Z.B. Liu, Jian Yu & F. Wu
			S. vulgaris (Fr.) Miettinen
	Schizoporaceae Jülich	*Basidioradulum* Nobles	*Basidioradulum radula* (Fr.) Nobles
		Fasciodontia Yurchenko & Riebesehl	*Fasciodontia brasiliensis* Yurchenko & Riebesehl

TABLE 3.1 *(Continued)*
Species Diversity and Taxonomic Classification of Wood-Rotting Fungi in China.

Order	Family	Genus	Species
			F. bugellensis (Ces.) Yurchenko, Riebesehl & Langer
			Fasciodontia yunnanensis C.L. Zhao
		Poriodontia Parmasto	*Poriodontia subvinosa* Parmasto
		Xylodon (Pers.) Gray	*Xylodon acystidiatus* Xue W. Wang & L.W. Zhou
			X. asiaticus X.C. Zhang & C.L. Zhao
			X. angustisporus Viner & Ryvarden
			X. bamburesupinus J.H. Dong & C.L. Zhao
			X. bambusinus C.L. Zhao & X. Ma
			X. bubalinus (Min Wang, Yuan Y. Chen & B.K. Cui) Che C. Chen & Sheng H. Wu
			X. bugellensis (Ces.) Hjortstam & Ryvarden
			X. calongei (M. Dueñas, Tellería, Melo & M.P. Martín) Viner
			X. cremeoparinaceus Qi Yuan & C.L. Zhao
			X. cystidiatus (A. David & Rajchenb.) Riebesehl & Langer
			X. damansaraensis Xue W. Wang & L.W. Zhou
			X. daweishanensis C.L. Zhao
			X. dimiticus (Jia J. Chen & L.W. Zhou) Riebesehl & Langer
			X. fissilis J.H. Dong & C.L. Zhao
			X. fissuratus C.L. Zhao
			X. flaviporus (Berk. & M.A. Curtis ex Cooke) Riebesehl & Langer
			X. flocculosus C.L. Zhao
			X. gossypinus C.L. Zhao & K.Y. Luo
			X. gloeocystidiifer Yurchenko & Riebesehl
			X. grandineus K.Y. Luo & C.L. Zhao
			X. hallenbergii (Sheng H. Wu) Hjortstam & Ryvarden
			X. heterocystidiatus (H.X. Xiong, Y.C. Dai & Sheng H. Wu) Riebesehl, Yurch. & Langer
			X. hydnoides J.H. Dong & C.L. Zhao
			X. kunmingensis L.W. Zhou & C.L. Zhao
			X. laceratus C.L. Zhao
			X. lagenicystidiatus Xue W. Wang & L.W. Zhou
			X. laxiusculus Viner & Ryvarden
			X. luteodontioides Qi Yuan & C.L. Zhao
			X. membranaceus (Bull.) Chevall.
			X. macrosporus C.L. Zhao & K.Y. Luo
			X. montanus C.L. Zhao
			X. mollissimus (L.W. Zhou) C.C. Chen & Sheng H. Wu
			X. muchuanensis Xue W. Wang & L.W. Zhou
			X. mussooriensis Samita, Sanyal & Dhingra ex L.W. Zhou & T.W. May
			X. neotropicus Yurchenko, Motato-Vásq. & Viner
			X. niemelaei (Sheng H. Wu) Hjortstam & Ryvarden
			X. nongravis (Lloyd) C.C. Chen & Sheng H. Wu

(Continued)

TABLE 3.1 ***(Continued)***

Species Diversity and Taxonomic Classification of Wood-Rotting Fungi in China.

Order	Family	Genus	Species
			X. ovisporus (Corner) Riebesehl & E. Langer
			X. olivaceobubalinus J.H. Dong & C.L. Zhao
			X. paradoxus (Schrad.) Chevall
			X. pingbianensis J.H. Dong & C.L. Zhao
			X. poroides Qi Yuan & C.L. Zhao
			X. poroideoefibulatus (Sheng H. Wu) Hjortstam & Ryvarden
			X. pseudotropicus (C.L. Zhao, B.K. Cui & Y.C. Dai) Riebesehl, Yurchenko & Langer
			X. puerensis C.L. Zhao
			X. punctus K.Y. Luo & C.L. Zhao
			X. rhizomorphus (C.L. Zhao, B.K. Cui & Y.C. Dai) Riebesehl, Yurch. & Langer
			X. rhododendricola Xue W. Wang & L.W. Zhou
			X. subflaviporus C.C. Chen & Sheng H. Wu
			X. subtropicus (C.C. Chen & Sheng H. Wu) C.C. Chen & Sheng H. Wu
			X. syringae (Langer) Hjortstam & Ryvarden
			X. sinensis C.L. Zhao & K.Y. Luo
			X. subglobosus Samita, Sanyal & Dhingra ex L.W. Zhou & T.W. May
			X. subserpentiformis Xue W. Wang & L.W. Zhou
			X. subtropicus (Che C. Chen & Sheng H. Wu) Che C. Chen & Sheng H. Wu
			X. subscopinellus (G. Cunn.) Hjortstam & Ryvarden
			X. taiwanianus (Sheng H. Wu) Hjortstam & Ryvarden
			X. tropicus C.L. Zhao
			X. victoriensis Xue W. Wang & L.W. Zhou
			X. wenshanensis K.Y. Luo & C.L. Zhao
			X. wumengshanensis Qi Yuan & C.L. Zhao
			X. xinpingensis C.L. Zhao & X. Ma
			X. yarraensis Xue W. Wang & L.W. Zhou
			X. yunnanensis Xue W. Wang & L.W. Zhou
			X. zonatus (Berk.) Kuntze
	Trichaptaceae Y.C. Dai, Yuan Yuan & Meng Zhou	*Trichaptum* Murrill	*Trichaptum byssogenum* (Jungh.) Ryvarden
	Incertae sedis	*Podocarpioporus* Y.C. Dai, Yuan Yuan & Meng Zhou	*Podocarpioporus podocarpi* (Y.C. Dai) Y.C. Dai, Yuan Yuan & Meng Zhou
		Pseudotrichaptum Y.C. Dai, Yuan Yuan & Meng Zhou	*Pseudotrichaptum laricinum* (P. Karst.) Y.C. Dai, Yuan Yuan & Meng Zhou
		Skvortzovia Bononi & Hjortstam	*Skvortzovia dabieshanensis* Jia Yu, Xue W. Wang, S.L. Liu & L.W. Zhou
			S. incana J.H. Dong & C.L. Zhao
			S. yunnanensis C.L. Zhao
Polyporales Gäum	*Adustoporiaceae* Audet	*Adustoporia* Audet	*Adustoporia sinuosa* (Fr.) Audet

TABLE 3.1 *(Continued)*
Species Diversity and Taxonomic Classification of Wood-Rotting Fungi in China.

Order	Family	Genus	Species
		Amyloporia Singer	*Amyloporia subxantha* (Y.C. Dai & X.S. He) B.K. Cui & Y.C. Dai
			Amyloporia xantha (Fr.) Bondartsev & Singer
		Lentoporia Audet	*Lentoporia subcarbonica* B.K. Cui, Y.Y. Chen & Shun Liu
		Rhodonia Niemelä	*Rhodonia obliqua* (Y.L. Wei & W.M. Qin) B.K. Cui, L.L. Shen & Y.C. Dai
			R. placenta (Fr.) Niemelä, K.H. Larss. & Schigel
			R. subplacenta (B.K. Cui) B.K. Cui, L.L. Shen & Y.C. Dai
			R. tianshanensis Yuan Yuan & L.L. Shen
		Resinoporia Audet	*Resinoporia crassa* (P. Karst.) Audet
			R. pinea (B.K. Cui & Y.C. Dai) Audet
			R. sitchensis (D.V. Baxter) Audet
	Auriporiaceae B.K. Cui, Shun Liu & Y.C. Dai	*Auriporia* Ryvarden	*Auriporia aurea* (Peck) Ryvarden
			R. aurulenta A. David, Tortić & Jelić
			R. pileata Parmasto
	Cerrenaceae Miettinen, Justo & Hibbett	*Cerrena* Gray	*Cerrena albocinnamomea* (Y.C. Dai & Niemelä) H.S. Yuan
			C. unicolor (Bull.) Murrill
			C. zonata (Berk.) H.S. Yuan
		Pseudospongipellis Y.C. Dai & Chao G. Wang	*Pseudospongipellis litschaueri* (Lohwag) Y. C. Dai & Chao G. Wang
		Radulodon Ryvarde	*Radulodon aneirinus* (Sommerf.) Spirin
			R. yunnanensis Y.C. Dai & Chao G. Wang
	Climacocystaceae B.K. Cui, Shun Liu & Y.C. Dai	*Climacocystis* Kotl. & Pouzar	*Climacocystis borealis* (Fr.) Kotl. & Pouzar
			C. montana B.K. Cui & J. Song
		Diplomitoporus Domański	*Diplomitoporus crustulinus* (Bres.) Domański
			D. flavescens (Bres.) Domański
	Fibroporiaceae Audet	*Faerberia* Pouzar	*Fibroporia albicans* B.K. Cui & Yuan Y. Chen
			F. bambusae Yuan Y. Chen & B.K. Cui
			F. ceracea Yuan Y. Chen & B.K. Cui
			F. citrina (Bernicchia & Ryvarden) Bernicchia & Ryvarden
			F. gossypium (Speg.) Parmasto
			F. radiculosa (Peck) Parmasto
			F. vaillantii (DC.) Parmasto
		Pseudofibroporia Yuan Y. Chen, B.K. Cui & Y.C. Dai	*Pseudofibroporia citrinella* Yuan Y. Chen & B.K. Cui
	Fomitopsidaceae Jülich	*Anthoporia* Karasiński & Niemelä	*Anthoporia albobrunnea* (Romell) Karasiński & Niemelä
		Antrodia P. Karst.	*Antrodia aridula* Y.C. Dai, H.M. Zhou, Y.D. Wu & Shun Liu

(Continued)

TABLE 3.1 *(Continued)*
Species Diversity and Taxonomic Classification of Wood-Rotting Fungi in China.

Order	Family	Genus	Species
			A. bambusicola Y.C. Dai & B.K. Cui
			A. heteromorpha (Fr.) Donk
			A. macra (Sommerf.) Niemelä
			A. neotropica Kaipper-Figueiró, Robledo & Drechsler-Santos
			A. peregrina Spirin, Y.C. Dai & Vlasák
			A. subheteromorpha B.K. Cui, Y.Y. Chen & Shun Liu
			A. subserpens B.K. Cui & Yuan Y. Chen
			A. tanakae (Murrill) Spirin & Miettinen
			A. variispora Y.C. Dai, H.M. Zhou, Y.D. Wu & Shun Liu
		Brunneoporus Audet	*Brunneoporus malicola* (Berk. & M.A. Curtis) Audet
		Buglossoporus Kotl. & Pouzar	*Buglossoporus eucalypticola* M.L. Han, B.K. Cui & Y.C. Dai
		Cartilosoma Kotlába & Pouzar	*Cartilosoma ramentacea* (Berk. & Broome) Teixeira
		Daedalea Pers.	*Daedalea allantoidea* M.L. Han, B.K. Cui & Y.C. Dai
			D. circularis B.K. Cui & Hai J. Li
			D. dickinsii Yasuda
			D. hydnoides I. Lindblad & Ryvarden
			D. modesta (Kunze ex Fr.) Aoshima
			D. radiata B.K. Cui & Hai J. Li
		Flavidoporia Audet	*Flavidoporia pulverulenta* (B. Rivoire) Audet
			F. pulvinascens (Pilát) Audet
		Fomitopsis P. Karst.	*Fomitopsis abieticola* B.K. Cui, M.L. Han & Shun Liu
			F. bambusae Y.C. Dai, Meng Zhou & Yuan Yuan
			F. betulina (Bull.) B.K. Cui, M.L. Han & Y.C. Dai
			F. cana B.K. Cui, Hai J. Li & M.L. Han
			F. ginkgonis B.K. Cui & Shun Liu
			F. hengduanensis B.K. Cui & Shun Liu
			F. massoniana B.K. Cui, M.L. Han & Shun Liu
			F. nivosa (Berk.) Gilb. & Ryvarden
			F. palustris (Berk. & M.A. Curtis) Gilb. & Ryvarden
			F. resupinata B.K. Cui & Shun Liu
			F. submeliae B.K. Cui & Shun Liu
			F. subpinicola B.K. Cui, M.L. Han & Shun Liu
			F. subtropica B.K. Cui & Hai J. Li
			F. tianshanensis B.K. Cui & Shun Liu
			F. yimengensis B.K. Cui & Shun Liu
		Fragifomes B.K. Cui, M.L. Han & Y.C. Dai	*Fragifomes niveomarginatus* (L.W. Zhou & Y.L. Wei) B.K. Cui, M.L. Han & Y.C. Dai
		Melanoporia Murrill	*Melanoporia castanea* (Imazeki) T. Hatt. & Ryvarden
		Neoantrodia Audet	*Neoantrodia angusta* (Spirin & Vlasák) Audet
			N. leucaena (Y.C. Dai & Niemelä) Audet

TABLE 3.1 ***(Continued)***
Species Diversity and Taxonomic Classification of Wood-Rotting Fungi in China.

Order	Family	Genus	Species
			N. primaeva (Renvall & Niemelä) Audet
			N. serialis (Fr.) Audet
			N. serrata (Vlasák & Spirin) Audet
		Neolentiporus Rajchenb	*Neolentiporus tropicus* B.K. Cui & Shun Liu
		Niveoporofomes B.K. Cui, M.L. Han & Y.C. Dai	*Niveoporofomes orientalis* B.K. Cui & Shun Liu
			N. spraguei (Berk. & M.A. Curtis) B.K. Cui, M.L. Han & Y.C. Dai
		Pseudoantrodia B.K. Cui, Y.Y. Chen & Shun Liu	*Pseudoantrodia monomitica* B.K. Cui, Y.Y. Chen & Shun Liu
		Rhodoantrodia B.K. Cui, Y.Y. Chen & Shun Liu	*Rhodoantrodia purpurascens* L. Wang & C.L. Zhao
			R. tropica (B.K. Cui) B.K. Cui, Y.Y. Chen & Shun Liu
			R. subtropica B.K. Cui & Shun Liu
			R. yunnanensis (M.L. Han & Q. An) B.K. Cui & Shun Liu
		Rhodofomes Kotl. & Pouzar	*Rhodofomes cajanderi* (P. Karst.) B.K. Cui, M.L. Han & Y.C. Dai
			R. incarnatus (K.M. Kim, J.S. Lee & H.S. Jung) B.K. Cui, M.L. Han & Y.C. Dai
			R. roseus (Alb. & Schwein.) Kotl. & Pouzar
			R. subfeei (B.K. Cui & M.L. Han) B.K. Cui, M.L. Han & Y.C. Dai
		Rhodofomitopsis B.K. Cui, M.L. Han & Y.C. Dai	*Rhodofomitopsis monomitica* (Yuan Y. Chen) B.K. Cui, Yuan Y. Chen & Shun Liu
		Rubellofomes B.K. Cui, M.L. Han & Y.C. Dai	*Rubellofomes cystidiatus* (B.K. Cui & M.L. Han) B.K. Cui, M.L. Han & Y.C. Dai
		Ungulidaedalea B.K. Cui, M.L. Han & Y.C. Dai	*Ungulidaedalea fragilis* (B.K. Cui & M.L. Han) B.K. Cui, M.L. Han & Y.C. Dai
	Fragiliporiaceae Y.C. Dai, B.K. Cui & C.L. Zhao	*Fragiliporia* Y.C. Dai, B.K. Cui & C.L. Zhao	*Fragiliporia fragilis* Y.C. Dai, B.K. Cui & C.L. Zhao
	Ganodermataceae (Donk) Donk	*Foraminispora* Robledo, Costa-Rezende & Drechsler-Santos	*Foraminispora austrosinensis* (J.D. Zhao & L.W. Hsu) Y.F. Sun & B.K. Cui
			F. concentrica (J. Song, Xiao L. He & B.K. Cui) Y.F. Sun & B.K. Cui
			F. yinggelingensis Y.F. Sun & B.K. Cui
			F. yunnanensis (J.D. Zhao & X.Q. Zhang) Y.F. Sun & B.K. Cui
		Ganoderma P. Karst.	*Ganoderma ahmadii* Steyaert

(Continued)

TABLE 3.1 *(Continued)*
Species Diversity and Taxonomic Classification of Wood-Rotting Fungi in China.

Order	Family	Genus	Species
			G. alpinum B.K. Cui, J.H. Xing & Y.F. Sun
			G. angustisporum J.H. Xing, B.K. Cui & Y.C. Dai
			G. applanatum (Pers.) Pat.
			G. australe (Fr.) Pat.
			G. bambusicola Sheng H. Wu, C.L. Chern & T. Hatt.
			G. boninense Pat.
			G. bubalinomarginatum B.K. Cui, J.H. Xing & Y.F. Sun
			G. calidophilum J.D. Zhao, L.W. Hsu & X.Q. Zhang
			G. castaneum B.K. Cui, J.H. Xing & Y.F. Sun
			G. casuarinicola J.H. Xing, B.K. Cui & Y.C. Dai
			G. chuxiongense B.K. Cui, J.H. Xing & Y.F. Sun
			G. dianzhongense J. He, H.Y. Su & S.H. Li
			G. ellipsoideum Hapuar., T.C. Wen & K.D. Hyde
			G. esculentum J. He & S.H. Li
			G. flexipes Pat.
			G. gibbosum (Blume & T. Nees) Pat.
			G. guangxiense B.K. Cui, J.H. Xing & Y.F. Sun
			G. hoehnelianum Bres.
			G. leucocontextum T.H. Li, W.Q. Deng, Sheng H. Wu, Dong M. Wang & H.P. Hu
			G. lingzhi Sheng H. Wu, Y. Cao & Y.C. Dai
			G. lucidum (Curtis) P. Karst.
			G. magniporum J.D. Zhao & X.Q. Zhang
			G. multipileum Ding Hou
			G. mutabile Y. Cao & H.S. Yuan
			G. orbiforme (Fr.) Ryvarden
			G. ovisporum H.D. Yang, T.C. Wen
			G. philippii (Bres. & Henn. ex Sacc.) Bres.
			G. puerense B.K. Cui, J.H. Xing & Y.F. Sun
			G. sanduense Hapuar., T.C. Wen & K.D. Hyde
			G. shanxiense L. Fan & H. Liu
			G. sichuanense J.D. Zhao & X.Q. Zhang
			G. sinense J.D. Zhao, L.W. Hsu & X.Q. Zhang
			G. subangustisporum B.K. Cui, J.H. Xing & Y.F. Sun
			G. subflexipes B.K. Cui, J.H. Xing & Y.F. Sun
			G. tongshanense B.K. Cui, J.H. Xing & Y.F. Sun
			G. tropicum (Jungh.) Bres.
			G. tsugae Murrill
			G. weberianum (Bres. & Henn. ex Sacc.) Steyaert
			G. weixiense Karun. & J.C. Xu
			G. williamsianum Murrill
			G. yunlingense B.K. Cui, J.H. Xing & Y.F. Sun
		Magoderna Steyaert	*Magoderna subresinosum* (Murrill) Steyaert
		Sanguinoderma Y.F. Sun, D.H. Costa & B.K. Cui	*Sanguinoderma bataanense* (Murrill) Y.F. Sun & B.K. Cui
			S. elmerianum (Murrill) Y.F. Sun & B.K. Cui

TABLE 3.1 ***(Continued)***
Species Diversity and Taxonomic Classification of Wood-Rotting Fungi in China.

Order	Family	Genus	Species
			S. guangdongense B.K. Cui & Y.F. Sun
			S. infundibulare B.K. Cui & Y.F. Sun
			S. laceratum Y.F. Sun & B.K. Cui
			S. leucomarginatum B. K. Cui & Y. F. Sun
			S. longistipitum B.K. Cui & Y.F. Sun
			S. microporum Y.F. Sun & B.K. Cui
			S. perplexum (Corner) Y.F. Sun & B.K. Cui
			S. preussii (Henn.) B. K. Cui & Y. F. Sun
			S. rugosum (Blume & T. Nees) Y.F. Sun, D.H. Costa & B.K. Cui
		Sinoganoderma B.K. Cui, J.H. Xing & Y.F. Sun	*Sinoganoderma shandongense* (J.D. Zhao & L.W. Xu) B.K. Cui, J.H. Xing & Y.F. Sun
		Tomophagus Murrill	*Tomophagus colossus* (Fr.) Murrill
		Trachyderma Norman	*Trachyderma tsunodae* (Yasuda ex Lloyd) Imazeki
	Gelatoporiaceae Miettinen, Justo & Hibbett	*Cinereomyces* Jülich	*Cinereomyces fimbriatus* C.L. Zhao
			C. lindbladii (Berk.) Jülich
			C. wuliangshanensis C.L. Zhao & K.Y. Luo
		Sebipora Miettinen	*Sebipora aquosa* Miettinen
	Grifolaceae Jülich	*Grifola* Gray	*Grifola frondosa* (Dicks.) Gray
	Hyphodermataceae Jülich	*Hyphoderma* Wallr.	*Hyphoderma alienum* (Parmasto) Jülich
			H. anthracophilum (Bourdot) Jülich
			H. crystallinum C.L. Zhao & Q.X. Guan
			H. fissuratum C.L. Zhao & X. Ma
			H. flagellatum (G. Cunn.) Jülich
			H. floccosum C.L. Zhao & Q.X. Guan
			H. granuliferum P. Roberts
			H. griseoflavescens (Litsch.) Jülich
			H. guangdongense J.Q. Su & C.L. Zhao
			H. marginatum Z.Y. Duan & C.L. Zhao
			H. membranaceum C.L. Zhao & Q.X. Guan
			H. microporoides C.L. Zhao & Q.X. Guan
			H. mopanshanense C.L. Zhao
			H. naiophila Gilb. & Hemmes
			H. niveomarginatum Yang Yang & C.L. Zhao
			H. probatum (H.S. Jacks.) Jülich
			H. puerense C.L. Zhao & Q.X. Guan
			H. sinense C.L. Zhao & Q.X. Guan
			H. singularibasidium Dhingra, Avn.P. Singh & Singla
			H. sordidum Yang Yang & C.L. Zhao
			H. tenuissimum C.L. Zhao & Q.X. Guan
			H. tropicum Z.Y. Duan & C.L. Zhao
			H. tubulicystidium Hjortstam & Spooner
			H. weishanense Yang Yang & C.L. Zhao
			H. yunnanense Z.Y. Duan & C.L. Zhao

(Continued)

TABLE 3.1 ***(Continued)***
Species Diversity and Taxonomic Classification of Wood-Rotting Fungi in China.

Order	Family	Genus	Species
	Incrustoporiaceae Jülich	*Skeletocutis* Kotl. & Pouzar	*Skeletocutis albomarginata* (Zipp. ex Lév.) Rui Du & Y.C. Dai
			S. bambusicola L.W. Zhou & W.M. Qin
			S. cangshanensis B.K. Cui & Shun Liu
			S. carneogrisea A. David
			S. chrysella Niemelä
			S. cummata A. Korhonen & Miettinen
			S. fimbriata Juan Li & Y.C. Dai
			S. inflata B. K. Cui
			S. luteola B.K. Cui & Y.C. Dai
			S. mopanshanensis C.L. Zhao
			S. nivea (Jungh.) Jean Keller
			S. odora (Sacc.) Ginns
			S. pseudo-odora L.F. Fan & Jing Si
			S. rhizomorpha J.H. Dong & C.L. Zhao
			S. stellae (Pilát) Jean Keller
			S. subchrysella B.K. Cui & Shun Liu
			S. substellae Y.C. Dai
			S. subvulgaris Y.C. Dai
			S. yuchengii Miettinen & A. Korhonen
			S. yunnanensis L.S. Bian, C.L. Zhao & F. Wu
		Tyromyces P. Karst.	*Tyromyces chioneus* (Fr.) P. Karst.
			T. galactinus (Berk.) J. Lowe
			T. kmetii (Bres.) Bondartsev & Singe
			T. minutulus Y.C. Dai & C.L. Zhao
	Irpicaceae Spirin & Zmitr	*Ceriporia* Donk	*Ceriporia alachuana* (Murrill) Hallenb
			C. allantoidea Y.C. Dai, Chao G. Wang & Yuan Yuan
			C. arbuscula C.C. Chen & Sheng H. Wu
			C. aurantiocarnescens (Henn.) M. Pieri & B. Rivoire
			C. bubalinomarginata B.S. Jia & Y.C. Dai
			C. camaresiana (Bourdot & Galzin) Bondartsev & Singer
			C. crassa Y.C. Dai, Chao G. Wang & Yuan Yuan
			C. excelsa Parmasto Parmasto
			C. gossypinum Y.C. Dai, Chao G. Wang & Yuan Yuan
			C. hinnulea Y.C. Dai, Chao G. Wang & Yuan Yuan
			C. jiangxiensis B.S. Jia & B.K. Cui
			C. macrospora Y.C. Dai, Chao G. Wang & Yuan Yuan
			C. mellita (Bourdot & Galzin) Bondartsev & Singer
			C. orientalis Y.C. Dai, Chao G. Wang & Yuan Yuan
			C. pseudospissa Y.C. Dai, Chao G. Wang & Yuan Yuan
			C. punctata Y.C. Dai, Chao G. Wang & Yuan Yuan
			C. purpurea (Fr.) Donk

TABLE 3.1 *(Continued)*
Species Diversity and Taxonomic Classification of Wood-Rotting Fungi in China.

Order	Family	Genus	Species
			C. reticulata (Hoffm.) Domański
			C. sinospissa Y.C. Dai, Chao G. Wang & Yuan Yuan
			C. sinoviridans Y.C. Dai & Jia J. Chen
			C. spissa (Schwein. ex Fr.) Rajchenb
			C. subbadia Y.C. Dai, Chao G. Wang & Yuan Yuan
			C. subviridans Y.C. Dai, Chao G. Wang & Yuan Yuan
			C. sulphuricolor Bernicchia & Niemelä
			C. tarda (Berk.) Ginns
			C. totara (G. Cunn.) P.K. Buchanan & Ryvarden
			C. viridans (Berk. & Broome) Donk
		Efibula Sheng H. Wu	*Efibula daweishanensis* J.H. Dong & C.L. Zhao
			E. deflectens (P. Karst.) Sheng H. Wu
			E. yunnanensis C.L. Zhao
		Flavodon Ryvarden	*Flavodon flavus* (Klotzsch) Ryvarden
		Gloeoporus Mont.	*Gloeoporus citrinoalbus* Yuan Yuan & Jia J. Chen
			G. dichrous (Fr.) Bres.
			G. hainanensis Yuan Yuan & Jia J. Chen
			G. orientalis P.E. Jung & Y.W. Lim
			G. pannocinctus (Romell) J. Erikss
			G. septatus Y.C. Dai, Chao G. Wang & Yuan Yuan
			G. variiformis Y.C. Dai, Chao G. Wang & Yuan Yuan
		Irpex Fr.	*Irpex jinshaensis* Z.B. Liu, X.M. Tian & Y.C. Dai
			I. laceratus (N. Maek., Suhara & R. Kondo) C.C. Chen & Sheng H. Wu
			I. lacteus (Fr.) Fr.
			I. lenis C.C. Chen & Sheng H. Wu
			I. rosettiformis (Cooke) C.C. Chen & Sheng H. Wu
			I. subulatus (Ryvarden) Z.B. Liu & Y.C. Dai
		Leptoporus Quél.	*Leptoporus submollis* B.K. Cui & Shun Liu
		Meruliopsis Bondartsev	*Meruliopsis albomellea* (Yuan Yuan, Jia J. Chen & X.H. Ji) C.C. Chen & Sheng H. Wu
			M. bambusicola Y.C. Dai, Chao G. Wang & Yuan Yuan
			M. crassitunicata (Y.C. Dai & Sheng H. Wu) C.C. Chen & Sheng H. Wu
			M. leptocystidiata C.C. Chen & Sheng H. Wu
			M. marginata Y.C. Dai, Chao G. Wang & Yuan Yuan
			M. nanlingensis (B.K. Cui & B.S. Jia) C.C. Chen & Sheng H. Wu
			M. parvispora C.C. Chen & Sheng H. Wu
			M. pseudocystidiata (B.S. Jia & Y.C. Dai) C.C. Chen & Sheng H. Wu
			M. taxicola (Pers.) Bondartsev
			M. variegata (B.S. Jia & Y.C. Dai) Zmitr
		Trametopsis Tomšovský	*Trametopsis cervina* (Schwein.) Tomšovský
			T. abieticola B.K. Cui & Shun Liu
	Ischnodermataceae Jülich	*Ischnoderma* P. Karst.	*Ischnoderma benzoinum* (Wahlenb.) P. Karst.

(Continued)

TABLE 3.1 ***(Continued)***

Species Diversity and Taxonomic Classification of Wood-Rotting Fungi in China.

Order	Family	Genus	Species
			I. resinosum (Schrad.) P. Karst.
	Laetiporaceae Jülich	*Laetiporus* Murrill	*Laetiporus ailaoshanensis* B.K. Cui & J. Song
			L. cremeiporus Y. Ota & T. Hatt.
			L. medogensis J. Song & B.K. Cui
			L. montanus Cerný ex Tomsovský & Jankovský
			L. sulphureus (Bull.) Murrill
			L. versisporus (Lloyd) Imazeki
			L. xinjiangensis J. Song, Y.C. Dai & B.K. Cui
			L. zonatus B.K. Cui & J. Song
		Wolfiporiella B.K. Cui & Shun Liu	*Wolfiporiella cartilaginea* (Ryvarden) B.K. Cui & Shun Liu
			W. dilatohypha (Ryvarden & Gilb.) B.K. Cui & Shun Liu
			W. macrospora X.H. Ji, L.W. Zhou & S.L. Liu
		Wolfiporiopsis B.K. Cui & Shun Liu	*Wolfiporiopsis castanopsidis* (Y.C. Dai) B.K. Cui & Shun Liu
	Laricifomitaceae Jülich	*Laricifomes* Kotl. & Pouzar	*Laricifomes officinalis* (Vill.) Kotl. & Pouzar
	Meripilaceae Jülich	*Irpiciporus* Murrill	*Irpiciporus sinuosus* Y.C. Dai & Chao G. Wang
		Rigidoporus Murrill	*Rigidoporus crocatus* (Pat.) Ryvarden
			R. fibulatus H.S. Yuan & Y.C. Dai
			R. hypobrunneus (Petch) Corner
			R. microporus (Sw.) Overeem
			R. minutus B.K. Cui & Y.C. Dai
			R. submicroporus F. Wu, Jia J. Chen & Y.C. Dai
			R. ulmarius (Sowerby) Imazeki
			R. vinctus (Berk.) Ryvarden
		Spongipellis Pat.	*Spongipellis quercicola* Y.C. Dai & Chao G. Wang
			S. sibirica (Penzina & Ryvarden) Penzina & Kotir.
			S. spumea (Sowerby) Pat.
	Meruliaceae Rea	*Aurantiporus* Murrill	*Aurantiporus fissilis* (Berk. & M.A. Curtis) H. Jahn
		Crustodontia Hjortstam & Ryvarden	*Crustodontia chrysocreas* (Berk. & M.A. Curtis) Hjortstam & Ryvarden
			C. nigrodontea (C.L. Zhao & R.X. Huang) C.C. Chen & Sheng H. Wu
			C. rhododendri C.L. Zhao
			C. tongxiniana (C.L. Zhao) C.C. Chen & Sheng H. Wu
		Ceriporiopsis Domański	*Ceriporiopsis albonigrescens* Núñez, Parmasto & Ryvarden
			C. aurantitingens (Corner) T. Hatt.
			C. carnegieae (D.V. Baxter) Gilb. & Ryvarden
			C. egula C.J. Yu & Y.C. Dai
			C. tianshanensis B.K. Cui & T.M. Xu
			C. lavendula B. K. Cui
			C. mucida (Pers.) Gilb. & Ryvarden
			C. subrufa (Ellis & Dearn.) Ginns
			C. subvermispora (Pilát) Gilb. & Ryvarden

TABLE 3.1 *(Continued)*
Species Diversity and Taxonomic Classification of Wood-Rotting Fungi in China.

Order	Family	Genus	Species
		Ceriporiopsoides C.L. Zhao	*Ceriporiopsoides lagerheimii* (Læssøe & Ryvarden) C.L. Zhao
		Hydnophlebia Parmasto	*Hydnophlebia fimbriata* (C.L. Zhao & Y.C. Dai) C.L. Zhao
		Luteoporia F. Wu, Jia J. Chen & S.H. He	*Luteoporia albomarginata* F. Wu, Jia J. Chen & S.H. He
			L. lutea (G. Cunn.) C.C. Chen & Sheng H. Wu
			L. tenuissima K.Y. Luo, Yuan Yuan, Y.C. Dai & Ghobad-Nejhad
		Mycoacia Donk	*Mycoacia gilvescens* (Bres.) Zmitr.
			M. kunmingensis (C.L. Zhao) Zmitr.
			M. tuberculata (Berk. & M.A. Curtis) C.L. Zhao
		Phlebia Fr.	*Phlebia ailaoshanensis* C.L. Zhao
			P. fuscotuberculata C.L. Zhao
			P. nigrodontea C.L. Zhao & R.X. Huang
			P. niveomarginata C.L. Zhao
			P. poroides C.L. Zhao
			P. tomentopileata C.L. Zhao
			P. tongxiniana C.L. Zhao
			P. wuliangshanensis C.L. Zhao
		Phlebicolorata C.L. Zhao	*Phlebicolorata alboaurantia* (C.L. Zhao, B.K. Cui & Y.C. Dai) C.L. Zhao
			P. rosea (C.L. Zhao & Y.C. Dai) C.L. Zhao
		Pseudophlebia C.L. Zhao	*Pseudophlebia semisupina* (C.L. Zhao, B.K. Cui & Y.C. Dai) C.L. Zhao
		Phlebiporia Jia J. Chen, B.K. Cui & Y.C. Dai	*Phlebiporia bubalina* Jia J. Chen, B.K. Cui & Y.C. Dai
		Physisporinus P. Karst.	*Physisporinus castanopsidis* Jia J. Chen & Y.C. Dai
			P. cinereus (Núñez & Ryvarden) F. Wu, Jia J. Chen & Y.C. Dai
			P. crataegi F. Wu, Jia J. Chen & Y.C. Dai
			P. eminens (Y.C. Dai) F. Wu, Jia J. Chen & Y.C. Dai
			P. furcatus (Núñez & Ryvarden) F. Wu, Jia J. Chen & Y.C. Dai
			P. lavendulus F. Wu, Jia J. Chen & Y.C. Dai
			P. lineatus (Pers.) F. Wu, Jia J. Chen & Y. C. Dai
			P. pouzarii (Vampola & Vlasák) F. Wu, Jia J. Chen & Y.C. Dai
			P. roseus Jia J. Chen & Y.C. Dai
			P. subcrocatus F. Wu, Jia J. Chen & Y.C. Dai
			P. tibeticus F. Wu, Jia J. Chen & Y.C. Dai
			P. vitreus (Pers.) P. Karst.
			P. yunnanensis C.L. Zhao
		Resiniporus Zmitr.	*Resiniporus pseudogilvescens* (Pilát) Zmitr.
			R. resinascens (Romell) Zmitr.
		Scopuloides (Massee) Höhn. & Litsch.	*Scopuloides ellipsoidea* S.H. He, Yue Li & Nakasone

(Continued)

TABLE 3.1 ***(Continued)***
Species Diversity and Taxonomic Classification of Wood-Rotting Fungi in China.

Order	Family	Genus	Species
			S. yunnanensis Z.R. Gu & C.L. Zhao
	Phanerochaetaceae Jülich	*Bjerkandera* P. Karst.	*Bjerkandera adusta* (Willd.) P. Karst.
			B. atroalba (Rick) Westphalen & Tomšovský
			B. fulgida Y.C. Dai & Chao G. Wang
			B. fumosa (Pers.) P. Karst.
			B. minispora Y.C. Dai & Chao G. Wang
			B. resupinata Y.C. Dai & Chao G. Wang
		Crepatura C.L. Zhao	*Crepatura ellipsospora* C.L. Zhao
		Donkiella J.H. Dong & C.L. Zhao	*Donkiella yunnanensis* J.H. Dong & C.L. Zhao
		Geliporus Yuan Yuan, Jia J. Chen & S.H. He	*Geliporus exilisporus* (Y.C. Dai & Niemelä) Yuan Yuan, Jia J. Chen & S.H. He
		Hapalopilus P. Karst.	*Hapalopilus flavus* B.K. Cui & Y.C. Dai
			H. priscus (Niemelä, Miettinen & Manninen) Melo & Ryvarden
			H. rutilans (Pers.) P. Karst.
		Hyphodermella J. Erikss. & Ryvarden	*Hyphodermella poroides* Y.C. Dai & C.L. Zhao
			H. sinensis L. Wang & C.L. Zhao
		Oxychaete Miettinen	*Oxychaete cervinogilva* (Jungh.) Miettinen
		Phanerina Miettinen	*Phanerina mellea* (Berk. & Broome) Miettinen
		Phanerochaete P. Karst.	*Phanerochaete capitata* Sheng H. Wu
			P. daliensis J. Yu & C.L. Zhao
			P. inflata (B.S. Jia & B.K. Cui) Miettinen
			P. mopanshanensis J.H. Dong & C.L. Zhao
			P. pruinosa C.L. Zhao & D.Q. Wang
			P. rhizomorpha C.L. Zhao & D.Q. Wang
			P. subtropica J. Yu & C.L. Zhao
			P. subtuberculata J. Yu & C.L. Zhao
			P. tongbiguanensis Y.L. Deng & C.L. Zhao
			P. velutina (DC.) P. Karst.
		Phlebiopsis Jülich	*Phlebiopsis castanea* (Lloyd) Miettinen & Spirin
			P. daweishanensis J.H. Dong & C.L. Zhao
			P. fissurata Y.L. Deng & C.L. Zhao
			P. gigantea (Fr.) Jülich
			P. lacerata C.L. Zhao
			P. yunnanensis C.L. Zhao
		Riopa D. A. Reid	*Riopa pudens* Miettinen
	Phaeolaceae Jülich	*Phaeolus* (Pat.) Pat.	*Phaeolus asiae-orientalis* Y.C. Dai & Yuan Yuan
			P. schweinitzii (Fr.) Pat.
			P. yunnanensis Y.C. Dai & Yuan Yuan
		Wolfiporia Ryvarden & Gilb.	*Wolfiporia hoelen* (Fr.) Y.C. Dai & V. Papp
			W. pseudococos F. Wu, J. Song & Y.C. Dai
	Piptoporellaceae B.K. Cui, Shun Liu & Y.C. Dai	*Piptoporellus* B.K. Cui, M.L. Han & Y.C. Dai	*Piptoporellus hainanensis* M.L. Han, B.K. Cui & Y.C. Dai

TABLE 3.1 *(Continued)*
Species Diversity and Taxonomic Classification of Wood-Rotting Fungi in China.

Order	Family	Genus	Species
			P. soloniensis (Dubois) B.K. Cui, M.L. Han & Y.C. Dai
			P. triqueter M.L. Han, B.K. Cui & Y.C. Dai
	Podoscyphaceae D.A. Reid	*Abortiporus* Murrill	*Abortiporus biennis* (Bull.) Singer
	Polyporaceae Fr. ex Corda	*Abundisporus* Ryvarden	*Abundisporus fuscopurpureus* (Pers.) Ryvarden
			A. mollissimus B.K. Cui & C.L. Zhao
			A. pubertatis (Lloyd) Parmasto
			A. quercicola Y.C. Dai
			A. roseoalbus (Jungh.) Ryvarden
		Amylosporia B.K. Cui, C.L. Zhao & Y.C. Dai	*Amylosporia hattorii* (Y.C. Dai & B.K. Cui) B.K. Cui, C.L. Zhao & Y.C. Dai
		Aurantioporia B.K. Cui & Xing Ji	*Aurantioporia bambusicola* (Choeyklin, T. Hatt. & E.B.G. Jones) B.K. Cui & Xing Ji
		Bresadolia Speg.	*Bresadolia cuticulata* (Y.C. Dai, Jing Si & Schigel) Audet
		Cerarioporia F. Wu, L.W. Zhou & Jing Si	*Cerarioporia cystidiata* F. Wu, L.W. Zhou & Jing Si
		Cerioporus Quél.	*Cerioporus austrosinensis* (B.K. Cui, Xing Ji & J.L. Zhou) B.K. Cui & Xing Ji
			C. cucullatus (Mont.) B.K. Cui & Xing Ji
			C. gaoligongensis (B.K. Cui, Hai J. Li & Y.C. Dai) B.K. Cui & Xing Ji
			C. lamelliporus (B.K. Cui, Xing Ji & J.L. Zhou) B.K. Cui & Xing Ji
			C. melanocarpus (B.K. Cui, Hai J. Li & Y.C. Dai) Zmitr
			C. mollis (Sommerf.) Zmitr. & Kovalenko
			C. parvovarius (H. Lee, N.K. Kim & Y.W. Lim) B.K. Cui & Xing Ji
			C. scutellatus (Schwein.) Zmitr
			C. sinensis (B.K. Cui, Hai J. Li &Y.C. Dai) B.K. Cui & Xing Ji
			C. squamosus (Huds.) Quél.
			C. subtropicus (B.K. Cui, Hai J. Li & Y.C. Dai) Zmitr.
			C. tibeticus (B.K. Cui, Hai J. Li & Y.C. Dai) Zmitr.
			C. tropicus (B.K. Cui, Hai J. Li & Y.C. Dai) Zmitr
			C. varius (Pers.) Zmitr. & Kovalenko
			C. yunnanensis Y. Yang & C.L. Zhao
		Citrinoporia B.K. Cui & Xing Ji	*Citrinoporia citrinoalba* (B.K. Cui, C.L. Zhao & Y.C. Dai) B.K. Cui & Xing Ji
		Coriolopsis Murrill	*Coriolopsis brunneoleuca* (Berk.) Ryvarden
			C. dendriformis Hai J. Li, Y.C. Dai & B.K. Cui
			C. glabrorigens (Lloyd) Núñez & Ryvarden
			C. hainanensis Hai J. Li, Y.C. Dai & B.K. Cui
			C. retropicta (Lloyd) Teng
			C. strumosa (Fr.) Ryvarden

(Continued)

TABLE 3.1 *(Continued)*
Species Diversity and Taxonomic Classification of Wood-Rotting Fungi in China.

Order	Family	Genus	Species
		Crassisporus B.K. Cui & Xing Ji	*Crassisporus imbricatus* B.K. Cui & Xing Ji
			C. macroporus B.K. Cui & Xing Ji
			C. microsporus B.K. Cui & Xing Ji
		Cryptoporus (Peck) Shear	*Cryptoporus sinensis* Sheng H. Wu & M. Zang
			C. volvatus (Peck) Shear
		Cyanoporus Y.C. Dai, W.L. Mao & Yuan Yuan	*Cyanoporus camptogrammus* (Pat.) Y.C. Dai, W.L. Mao & Yuan Yuan
			C. fuligo (Berk. & Broome) Y.C. Dai, W.L. Mao & Yuan Yuan
		Cystidioporia B.K. Cui & Xing Ji	*Cystidioporia piceicola* (Y.C. Dai) B.K. Cui & Xing Ji
		Daedaleopsis J. Schröt	*Daedaleopsis confragosa* (Bolton) J. Schröt
			D. hainanensis Hai J. Li & S.H. He
			D. purpurea (Cooke) Imazeki & Aoshima
			D. sinensis (Lloyd) Y.C. Dai
			D. tricolor (Bull.) Bondartsev & Singer
		Dendroporia B.K. Cui & Xing Ji	*Dendroporia cinereofusca* (B.K. Cui & C.L. Zhao) B.K. Cui & Xing Ji
		Dextrinoporus H.S. Yuan	*Dextrinoporus aquaticus* H.S. Yuan
		Dichomitus D.A. Reid	*Dichomitus campestris* (Quél.) Domański & Orlicz
			D. sinuolatus H.S. Yuan
			D. squalens (P. Karst.) D.A. Reid
		Donkioporia Kotl. & Pouzar	*Donkioporia expansa* (Desm.) Kotl. & Pouzar
		Donkioporiella L.W. Zhou	*Donkioporiella mellea* L.W. Zhou
		Earliella Murrill	*Earliella scabrosa* (Pers.) Gilb. & Ryvarden
		Echinochaete D.A. Reid	*Echinochaete brachypora* (Mont.) Ryvarden
			E. ruficeps (Berk. & Broome) Ryvarden
			E. russiceps (Berk. & Broome) D.A. Reid
		Favolus Fr.	*Favolus acervatus* (Lloyd) Sotome & T. Hatt.
			F. emerici (Berk. ex Cooke) Imazeki
			F. philippinensis (Berk.) Sacc.
			F. niveus J.L. Zhou & B.K. Cui
			F. pseudoemerici J.L. Zhou & B.K. Cui
			F. septatus J.L. Zhou & B.K. Cui
			F. spathulatus (Jungh.) Lév.
			F. subtropicus J.L. Zhou & B.K. Cui
		Flammeopellis Y.C. Dai, B.K. Cui & C.L. Zhao	*Flammeopellis bambusicola* Y.C. Dai, B.K. Cui & C.L. Zhao
		Fomes (Fr.) Fr.	*Fomes fomentarius* (L.) Fr.
		Funalia Pat.	*Funalia cystidiata* Hai J. Li, Y.C. Dai & B.K. Cui
			F. gallica (Fr.) Bondartsev & Singer
			F. sanguinaria (Klotzsch) Zmitr. & Malysheva

TABLE 3.1 ***(Continued)***
Species Diversity and Taxonomic Classification of Wood-Rotting Fungi in China.

Order	Family	Genus	Species
			F. subgallica Hai J. Li & S.H. He
			F. trogii (Berk.) Bondartsev & Singer
		Grammothele Berk. & M.A. Curtis	*Grammothele denticulata* Y.C. Dai & L.W. Zhou
			G. duportii (Pat.) Y.C. Dai, W.L. Mao & Yuan Yuan
			G. hainanensis F Wu & L.W. Zhou
			G. lineata Berk. & M.A. Curtis
			G. quercina (Y.C. Dai) B.K. Cui & Hai J. Li
			G. taiwanensis C.C. Chen
		Grammothelopsis Jülich	*Grammothelopsis asiatica* Y.C. Dai & B.K. Cui
			G. subtropica B.K. Cui & C.L. Zhao
		Haploporus Bondartsev & Singer	*Haploporus alabamae* (Berk. & Cooke) Y.C. Dai & Niemelä
			H. angustisporus Meng Zhou & Y.C.Dai
			H. bicolor Y.C. Dai, Meng Zhou & Yuan Yuan
			H. crassus Meng Zhou & Y.C.Dai
			H. cylindrosporus L.L. Shen, Y.C. Dai & B.K. Cui
			H. latisporus Juan Li & Y.C. Dai
			H. microsporus L.L. Shen, Y.C. Dai & B.K. Cui
			H. monomitica Y.C. Dai
			H. nepalensis (T. Hatt.) Piątek
			H. odorus (Sommerf.) Bondartsev & Singer
			H. papyraceus (Cooke) Y.C. Dai & Niemelä
			H. septatus L.L. Shen, Y.C. Dai & B.K. Cui
			H. subpapyraceus L.L. Shen, Y.C. Dai & B.K. Cui
			H. subtrameteus (Pilát) Y.C. Dai & Niemelä
			H. thindii (Natarajan & Koland.) Y.C. Dai
		Hexagonia Fr.	*Hexagonia apiaria* (Pers.) Fr.
			H. aspera (Jungh.) Imazeki
			H. glabra (P. Beauv.) Ryvarden
			H. hirta (P. Beauv.) Fr.
			H. tenuis (Fr.) Fr.
		Hirticrusta Matozaki, T. Hatt. & Sotome	*Hirticrusta subradiata* (Lloyd) Matozaki, T. Hatt. & Sotome
		Hornodermoporus Teixeira	*Hornodermoporus latissimus* (Bres.) B.K. Cui & Y.C. Dai
			H. martius (Berk.) Teixeira
		Jorgewrightia Gibertoni & C.R.S. Lira	*Jorgewrightia bambusae* (Y.C. Dai, Yuan Yuan & Ya.R. Wang) Gibertoni
			J. cystidiolophora (B. K. Cui & Y. C. Dai) Gibertoni & C.R.S. Lira
			J. ellipsoidea (B.K. Cui & P. Du) C.R.S. Lira & Gibertoni
			J. irregularis Y. C. Dai, Yuan Yuan, Ya R. Wang & Y. D. Wu
			J. guangdongensis (B.K. Cui & Hai J. Li) Gibertoni & C.R.S. Lira

(Continued)

TABLE 3.1 *(Continued)*
Species Diversity and Taxonomic Classification of Wood-Rotting Fungi in China.

Order	Family	Genus	Species
			J. hengduanensis (B.K. Cui & Hai J. Li) Gibertoni & C.R.S. Lira
			J. major (G.Y. Zheng & Z.S. Bi) C.R.S. Lira & Gibertoni
			J. rimosa (Y. Yuan, X.H. Ji & Y.C. Dai) C.R.S. Lira & Gibertoni
			J. tenuis Y. C. Dai, Yuan Yuan, Ya R. Wang & Y. D. Wu
			J. tropica (Y. Yuan, X.H. Ji & Y.C. Dai) Gibertoni & C.R.S. Lira
			J. violacea (B.K. Cui & P. Du) Gibertoni & C.R.S. Lira
			J. yunnanensis (Y. Yuan, X.H. Ji & Y.C. Dai) C.R.S. Lira & Gibertoni
		Leifiporia Y.C. Dai, F. Wu & C.L. Zhao	*Leifiporia rhizomorpha* Y.C. Dai, F. Wu & C.L. Zhao
		Lentinus Fr.	*Lentinus arcularius* (Batsch) Zmitr
			L. brumalis (Pers.) Zmitr
			L. substrictus (Bolton) Zmitr. & Kovalenko
			L. thailandensis (Sotome) B.K. Cui, Xing Ji & J.L. Zhou
		Lignosus Lloyd ex Torrend	*Lignosus hainanensis* B.K. Cui
			L. rhinocerotis (Cooke) Ryvarden
		Luteoperenniporia B.K. Cui & Xing Ji	*Luteoperenniporia bannaensis* (B.K. Cui & C.L. Zhao) B.K. Cui & Xing Ji
			L. mopanshanensis (C.L. Zhao) B.K. Cui & Xing Ji
			L. yinggelingensis (B.K. Cui & Y.C. Dai) B.K. Cui & Xing Ji
		Macroporia B.K. Cui & Xing Ji	*Macroporia lacerata* (B.K. Cui & C.L. Zhao) B.K. Cui & Xing Ji
			M. macropora (B.K. Cui & C.L. Zhao) B.K. Cui & Xing Ji
			M. subrhizomorpha (Xue W. Wang, L.W. Zhou & X.M. Tian) B.K. Cui & Xing Ji
			M. tibetica (B.K. Cui & C.L. Zhao) B.K. Cui & Xing Ji
		Macrosporia B.K. Cui & Xing Ji	*Macrosporia nanlingensis* (B.K. Cui & C.L. Zhao) B.K. Cui & Xing Ji
		Mariorajchenbergia Gibertoni & C.R.S. Lira	*Mariorajchenbergia hubeiensis* (Hai J. Li & B.K. Cui) Gibertoni & C.R.S. Lira
			M. pseudocavernulosa (B.K. Cui & Hai J. Li) Gibertoni & C.R.S. Lira
			M. rhododendri (Y.C. Dai & Y.L. Wei) Gibertoni & C.R.S. Lira
			M. subcavernulosa (Y.C. Dai & Sheng H. Wu) Gibertoni & C.R.S. Lira

TABLE 3.1 *(Continued)*
Species Diversity and Taxonomic Classification of Wood-Rotting Fungi in China.

Order	Family	Genus	Species
		Megasporia B.K. Cui, Y.C. Dai & Hai J. Li	*Megasporia hexagonoides* (Speg.) B.K. Cui, Y.C. Dai & Hai J. Li
			M. olivacea Y. C. Dai, Yuan Yuan, Ya R. Wang & Y. D. Wu
			M. sinuosa Y. C. Dai, Yuan Yuan, Ya R. Wang & Y. D. Wu
		Megasporoporia Ryvarden & J.E. Wright	*Megasporoporia bannaensis* B.K. Cui & Hai J. Li
			M. minor B.K. Cui & Hai J. Li
			M. minuta Y.C. Dai & X.S. Zhou
			M. setulosa (Henn.) Rajchenb
		Melanoderma B.K. Cui & Y.C. Dai	*Melanoderma disciforme* H.S. Yuan
			M. microcarpum B.K. Cui & Y.C. Dai
		Microporellus Murrill	*Microporellus obovatus* (Jungh.) Ryvarden
			M. celtis (T.T. Chang & W.N. Chou) Decock
			M. subadustus (Z.S. Bi & G.Y. Zheng) B.K. Cui & Xing Ji
		Microporus P. Beauv.	*Microporus affinis* (Blume & T. Nees) Kuntze
			M. subaffinis (Lloyd) Imazeki
			M. vernicipes (Berk.) Kuntze
			M. xanthopus (Fr.) Kuntze
		Minoporus B.K. Cui & Xing Ji	*Minoporus minor* (Y.C. Dai & H.X. Xiong) B.K. Cui & Xing Ji
		Neofavolus Sotome & T. Hatt.	*Neofavolus alveolaris* (DC.) Sotome & T. Hatt.
			N. cremeoalbidus Sotome & T. Hatt.
			N. mikawae (Lloyd) Sotome & T. Hatt
			N. squamatus J.H. Xing, J.L. Zhou & B.K. Cui
			N. yunnanensis C.L. Zhao
		Neofomitella Y.C. Dai, Hai J. Li & Vlasák	*Neofomitella fumosipora* (Corner) Y.C. Dai, Hai J. Li & Vlasák
			N. guangxiensis B.K. Cui & Xing Ji
		Neoporia B.K. Cui & Xing Ji	*Neoporia rhizomorpha* (B.K. Cui, Y.C. Dai & Decock) B.K. Cui & Xing Ji
		Niveoporia B.K. Cui & Xing Ji	*Niveoporia decurrata* (Corner) B.K. Cui & Xing Ji
			N. russeimarginata (B.K. Cui & C.L. Zhao) B.K. Cui & Xing Ji
			N. subrusseimarginata B.K. Cui & Xing Ji
		Perenniporia Murrill	*Perenniporia aridula* B.K. Cui & C.L. Zhao
			P. bannaensis B.K. Cui & C.L. Zhao
			P. bostonensis C.L. Zhao
			P. cinereofusca B.K. Cui & C.L. Zhao
			P. citrinoalba B.K. Cui, C.L. Zhao & Y.C. Dai
			P. cystidiata Y.C. Dai, W.N. Chou & Sheng H. Wu
			P. contraria (Berk. & M.A. Curtis) Ryvarden

(Continued)

TABLE 3.1 ***(Continued)***
Species Diversity and Taxonomic Classification of Wood-Rotting Fungi in China.

Order	Family	Genus	Species
			P. dendrohyphidia Ryvarden
			P. fergusii Gilb. & Ryvarden
			P. gomezii Rajchenb. & J.E. Wright
			P. hainaniana B.K. Cui & C.L. Zhao
			P. inflexibilis (Berk.) Ryvarden
			P. isabellina (Pat.) Ryvarden
			P. luteola B.K. Cui & C.L. Zhao
			P. medulla-panis (Jacq.) Donk
			P. macropora B.K. Cui & C.L. Zhao
			P. mopanshanensis C.L. Zhao
			P. nanlingensis B.K. Cui & C.L. Zhao
			P. nonggangensis F.C. Huang & Bin Liu
			P. puerensis C.L. Zhao
			P. russeimarginata B.K. Cui & C.L. Zhao
			P. substraminea B.K. Cui & C.L. Zhao
			P. subtephropora B.K. Cui & C.L. Zhao
			P. tianmuensis B.K. Cui & C.L. Zhao
			P. tibetica B.K. Cui & C.L. Zhao
			P. tephropora (Mont.) Ryvarden
			P. xantha Decock & Ryvarden
		Perenniporiopsis C.L. Zhao	*Perenniporiopsis minutissima* (Yasuda) C.L. Zhao
			P. sinensis Y. Yang & C.L. Zhao
		Picipes Zmitr. & Kovalenko	*Picipes admirabilis* (Peck) J.L. Zhou & B.K. Cui
			P. ailaoshanensis B.K. Cui, Xing Ji & J.L. Zhou
			P. annularius B.K. Cui, Xing Ji & J.L. Zhou
			P. atratus B.K. Cui, Xing Ji & J.L. Zhou
			P. auriculatus B.K. Cui, Xing Ji & J.L. Zhou
			P. badius (Pers.) Zmitr. & Kovalenko
			P. baishanzuensis J.L. Zhou & B.K. Cui
			P. brevistipitatus B.K. Cui, Xing Ji & J.L. Zhou
			P. conifericola (H.J. Xue & L.W. Zhou) J.L. Zhou & B.K. Cui
			P. dictyopus (Mont.) B.K. Cui, Xing Ji & J.L. Zhou
			P. fraxinicola (L.W. Zhou & Y.C. Dai) J.L. Zhou & B.K. Cui
			P. hainanensis J.L. Zhou & B.K. Cui
			P. jiajinensis J.L. Zhou & B.K. Cui
			P. nigromarginatus B.K. Cui, Xing Ji & J.L. Zhou
			P. pseudovarius J.L. Zhou & B.K. Cui
			P. pumilus (Y.C. Dai & Niemelä) J.L. Zhou & B.K. Cui
			P. rhizophilus (Pat.) J.L. Zhou & B.K. Cui
			P. subdictyopus (H. Lee, N.K. Kim & Y.W. Lim) B.K. Cui, Xing Ji & J.L. Zhou
			P. submelanopus (H.J. Xue & L.W. Zhou) J.L. Zhou & B.K. Cui
			P. subtropicus J.L. Zhou & B.K. Cui

TABLE 3.1 ***(Continued)***
Species Diversity and Taxonomic Classification of Wood-Rotting Fungi in China.

Order	Family	Genus	Species
			P. subtubaeformis J.L. Zhou & B.K. Cui
			P. taibaiensis (Y.C. Dai) J.L. Zhou & B.K. Cui
			P. tibeticus J.L. Zhou & B.K. Cui
			P. ulleungus (H. Lee, N.K. Kim & Y.W. Lim) B.K. Cui, Xing Ji & J.L. Zhou
			P. wuyishanensis B.K. Cui, Xing Ji & J.L. Zhou
		Polyporus P. Micheli ex Adans.	*Polyporus auratus* B.K. Cui, Xing Ji & J.L. Zhou
			P. hapalopus H.J. Xue & L.W. Zhou
			P. hemicapnodes Berk. & Broome
			P. mangshanensis B.K. Cui, J.L. Zhou & Y.C. Dai
			P. megasporoporus Y.C. Dai, Yuan Yuan & Ya.R. Wang
			P. subvarius C.J. Yu & Y.C. Dai
			P. tuberaster (Jacq. ex Pers.) Fr.
			P. umbellatus (Pers.) Fr.
		Poriella C.L. Zhao	*Poriella africana* (Ipulet & Ryvarden) B.K. Cui & Xing Ji
			P. ellipsospora (Ryvarden & Gilb.) B.K. Cui & Xing Ji
			P. subacida (Peck) C.L. Zhao
		Porogramme (Pat.) Pat.	*Porogramme bubalina* (H.S. Yuan) Y.C. Dai, W.L. Mao & Yuan Yuan
			P. cylindrica Y.C. Dai, W.L. Mao & Yuan
			P. epimiltina (Berk. & Broome) Y.C. Dai, W.L. Mao & Yuan Yuan
			P. hinnulea (H.S. Yuan) Y.C. Dai, W.L. Mao & Yuan Yuan
			P. yunnanensis Y.C. Dai, W.L. Mao & Yuan Yuan
		Pseudogrammothele Y.C. Dai, W.L. Mao & Yuan Yuan	*Pseudogrammothele separabillima* (H.S. Yuan) Y.C. Dai, W.L. Mao & Yuan Yuan
		Pyrofomes Kotl. & Pouzar	*Pyrofomes castanopsidis* B.K. Cui & Y.C. Dai
			P. demidoffii (Lév.) Kotl. & Pouzar
		Rhizoperenniporia B.K. Cui & Xing Ji	*Rhizoperenniporia japonica* (Yasuda) B.K. Cui & Xing Ji
		Sparsitubus L.W. Hsu & J.D. Zhao	*Sparsitubus nelumbiformis* L.W. Xu & J.D. Zhao
		Szczepkamyces Zmitr.	*Szczepkamyces quercicola* Y.C. Dai, Yuan Yuan & Meng Zhou
		Theleporus Fr.	*Theleporus calcicolor* (Sacc. & P. Syd.) Ryvarden
			T. membranaceus Y.C. Dai & L.W. Zhou
			T. minisporus Y.C. Dai & L.W. Zhou
			T. rimosus H.S. Yuan
		Trametes Fr.	*Trametes acuta* (Berk.) Imazeki
			T. betulina (L.) Pilát
			T. cinnabarina (Jacq.) Fr.
			T. coccinea (Fr.) Hai J. Li & S.H. He

(Continued)

TABLE 3.1 ***(Continued)***

Species Diversity and Taxonomic Classification of Wood-Rotting Fungi in China.

Order	Family	Genus	Species
			T. conchifer (Schwein.) Pilát
			T. cystidiolophora B.K. Cui & H.J. Li
			T. duplexa Hai J. Li, Y.C. Dai & B.K. Cui
			T. ectypa (Berk. & M.A. Curtis) Gilb. & Ryvarden
			T. elegans (Spreng.) Fr.
			T. ellipsoidea Hai J. Li, Y.C. Dai & B.K. Cui
			T. ellipsospora Ryvarden
			T. gibbosa (Pers.) Fr.
			T. hirsuta (Wulfen) Lloyd
			T. lactinea (Berk.) Sacc.
			T. ljubarskyi Pilát
			T. manilaensis (Lloyd) Teng
			T. maxima (Mont.) A. David & Rajchenb
			T. menziesii (Berk.) Ryvarden
			T. mimetes (Wakef.) Ryvarden
			T. ochracea (Pers.) Gilb. & Ryvarden
			T. orientalis (Yasuda) Imazeki
			T. pavonia (Hook.) Ryvarden
			T. pocas (Berk.) Ryvarden
			T. polyzona (Pers.) Justo
			T. pubescens (Schumach.) Pilát
			T. sanguinea (L.) Lloyd
			T. stipitata Hai J. Li, Y.C. Dai & B.K. Cui
			T. suaveolens (L.) Fr.
			T. subsuaveolens B.K. Cui & Y.C. Dai
			T. tephroleuca Berk.
			T. thujae J.D. Zhao
			T. versicolor (L.) Lloyd
			T. vespacea (Pers.) Zmitr., Wasser & Ezhov
		Tropicoporia B.K. Cui & Xing Ji	*Tropicoporia aridula* (B.K. Cui & C.L. Zhao) B.K. Cui & Xing Ji
		Truncatoporia B.K. Cui & Xing Ji	*Truncatoporia pyricola* (Y.C. Dai & B.K. Cui) B.K. Cui & Xing Ji
			T. truncatospora (Lloyd) B.K. Cui & Xing Ji
		Truncospora Pilát	*Truncospora macrospora* B.K. Cui & C.L. Zhao
			T. ochroleuca (Berk.) Pilát
			T. ornata Spirin & Bukharova
		Vanderbylia D.A. Reid	*Vanderbylia cinnamomea* C.L. Zhao
			V. delavayi (Pat.) B.K. Cui & Y.C. Dai
			V. fraxinea (Bull.) D.A. Reid
			V. robiniophila (Murrill) B.K. Cui & Y.C. Dai
			V. vicina (Lloyd) D.A. Reid
		Vanderbyliella B.K. Cui & Xing Ji	*Vanderbyliella tianmuensis* (B.K. Cui & C.L. Zhao) B.K. Cui & Xing Ji
		Whitfordia Murrill	*Whitfordia scopulosa* (Berk.) Núñez & Ryvarden
		Xanthoperenniporia B.K. Cui & Xing Ji	*Xanthoperenniporia maackiae* (Bondartsev & Ljub.) B.K. Cui & Xing Ji
			X. punctata (Hai J. Li & Jing Si) B.K. Cui & Xing Ji

TABLE 3.1 *(Continued)*
Species Diversity and Taxonomic Classification of Wood-Rotting Fungi in China.

Order	Family	Genus	Species
			X. subcorticola (Chao G. Wang & F. Wu) B.K. Cui & Xing Ji
			X. tenuis (Schwein.) B.K. Cui & Xing Ji
		Yuchengia B.K. Cui & K.T. Steffen	*Yuchengia narymica* (Pilát) B.K. Cui, C.L. Zhao & K.T. Steffen
			Y. kilemariensis (Spirin & Shirokov) B.K. Cui & Xing Ji
	Postiaceae B.K. Cui, Shun Liu & Y.C. Dai	*Amaropostia* B.K. Cui, L.L. Shen & Y.C. Dai	*Amaropostia hainanensis* B.K. Cui, L.L. Shen & Y.C. Dai
			A. stiptica (Pers.) B.K. Cui, L.L. Shen & Y.C. Dai
		Amylocystis Bondartsev & Singer ex Singer	*Amylocystis lapponicus* (Romell) Bondartsev & Singer
		Calcipostia B.K. Cui, L.L. Shen & Y.C. Dai	*Calcipostia guttulata* (Sacc.) B.K. Cui, L.L. Shen & Y.C. Dai
		Cyanosporus McGinty	*Cyanosporus alni* (Niemelä & Vampola) B.K. Cui, L.L. Shen & Y.C. Dai
			C. auricomus (Spirin & Niemelä) B.K. Cui & Shun Liu
			C. bifaria (Spirin) B.K. Cui & Shun Liu
			C. bubalinus B.K. Cui & Shun Liu
			C. coeruleivirens (Corner) B.K. Cui, Shun Liu & Y.C. Dai
			C. comatus (Miettinen) B.K. Cui & Shun Liu
			C. flavus B.K. Cui & Shun Liu
			C. fusiformis B.K. Cui, L.L. Shen & Y.C. Dai
			C. glauca (Spirin & Miettinen) B.K. Cui & Shun Liu
			C. hirsutus B.K. Cui & Shun Liu
			C. magnus (Miettinen) B.K. Cui & Shun Liu
			C. microporus B.K. Cui, L.L. Shen & Y.C. Dai
			C. piceicola B.K. Cui, L.L. Shen & Y.C. Dai
			C. populi (Miettinen) B.K. Cui & Shun Liu
			C. rigidus B.K. Cui & Shun Liu
			C. subhirsutus B.K. Cui, L.L. Shen & Y.C. Dai
			C. submicroporus B.K. Cui & Shun Liu
			C. subungulatus B.K. Cui & Shun Liu
			C. tenuicontextus B.K. Cui & Shun Liu
			C. tenuis B.K. Cui, Shun Liu & Y.C. Dai
			C. tricolor B.K. Cui, L.L. Shen & Y.C. Dai
			C. ungulatus B.K. Cui, L.L. Shen & Y.C. Dai
		Cystidiopostia B.K. Cui, L.L. Shen & Y.C. Dai	*Cystidiopostia hibernica* (Berk. & Broome) B.K. Cui, L.L. Shen & Y.C. Dai
			C. inocybe (A. David & Malençon) B.K. Cui, L.L. Shen & Y.C. Dai
			C. pileata (Parmasto) B.K. Cui, L.L. Shen & Y.C. Dai
			C. simanii (Pilát) B.K. Cui, Shun Liu & L.L. Shen
			C. subhibernica B.K. Cui & Shun Liu
		Fuscopostia B.K. Cui, L.L. Shen & Y.C. Dai	*Fuscopostia avellanea* B.K. Cui & Shun Liu

(Continued)

TABLE 3.1 *(Continued)*
Species Diversity and Taxonomic Classification of Wood-Rotting Fungi in China.

Order	Family	Genus	Species
			F. duplicata (L.L. Shen, B.K. Cui & Y.C. Dai) B.K. Cui, L.L. Shen & Y.C. Dai
			F. fragilis (Fr.) B.K. Cui, L.L. Shen & Y.C. Dai
			F. leucomallella (Murrill) B.K. Cui, L.L. Shen & Y.C. Dai
			F. persicina B.K. Cui & Shun Liu
			F. subfragilis B.K. Cui & Shun Liu
			F. tomentosa B.K. Cui & Shun Liu
		Jahnoporus Nuss	*Jahnoporus brachiatus* Spirin, Vlasák & Miettinen
			J. hirtus (Cooke) Nuss
			J. pekingensis (J.D. Zhao & L.W. Xu) Y.C. Dai
		Oligoporus Bref.	*Oligoporus podocarpi* Y.C. Dai, Chao G. Wang & Yuan Yuan
			O. rennyi (Berk. & Broome) Donk
			O. romellii (M. Pieri & B. Rivoire) Niemelä
			O. sericeomollis (Romell) Bondartseva
		Osteina Donk	*Osteina obducta* (Berk.) Donk
			O. subundosa (Peck) B.K. Cui, Shun Liu & L.L. Shen
			O. undosa (Peck) Zmitr
		Postia Fr.	*Postia amurensis* Y.C. Dai & Penttilä
			P. calcarea Y.L.Wei & Y.C. Dai
			P. cana H.S. Yuan & Y.C. Dai
			P. cylindrica H.S. Yuan
			P. gloeocystidiata Y.L. Wei & Y.C. Dai
			P. hirsuta L.L. Shen & B.K. Cui
			P. lactea (Fr.) P. Karst.
			P. lowei (Pilát) Jülich
			P. ochraceoalba L.L. Shen, B.K. Cui & Y.C. Dai
			P. qinensis Y.C. Dai & Y.L. Wei
			P. sublowei B.K. Cui, L.L. Shen & Y.C. Dai
			P. tephroleuca (Fr.) Jülich
			P. ptychogaster (F. Ludw.) Vesterh.
		Resupinopostia B.K. Cui & Shun Liu	*Resupinopostia lateritia* (Renvall) B.K. Cui & Shun Liu
			R. sublateritia B.K. Cui & Shun Liu
		Spongiporus Murrill	*Spongiporus balsameus* (Peck) A. David
			S. cerifluus (Berk. & M.A. Curtis) A. David
			S. floriformis (Quél.) Zmitr
			S. gloeoporus (L.L. Shen, B.K. Cui & Y.C. Dai) B.K. Cui, L.L. Shen & Y.C. Dai
			S. japonica (Y.C. Dai & T. Hatt.) B.K. Cui, Shun Liu & L.L. Shen
			S. persicinus (Niemelä & Y.C. Dai) B.K. Cui, Shun Liu & L.L. Shen
			S. zebra (Y.L. Wei & W.M. Qin) B.K. Cui, L.L. Shen & Y.C. Dai
	Pycnoporellaceae Audet	*Pycnoporellus* Murrill	*Pycnoporellus fulgens* (Fr.) Donk

TABLE 3.1 *(Continued)*
Species Diversity and Taxonomic Classification of Wood-Rotting Fungi in China.

Order	Family	Genus	Species
	Sarcoporiaceae Audet	*Sarcoporia* P. Karst.	*Sarcoporia polyspora* P. Karst.
			S. yunnanensis Y. Yang & C.L. Zhao
	Steccherinaceae Parmasto	*Antella* Miettinen	*Antella chinensis* (H.S. Yuan) Miettinen
		Antrodiella Ryvarden & I. Johans	*Antrodiella citripileata* H.S.Yuan
			A. mentschulensis (Pilát) Melo & Ryvarden
			A. micra Y.C. Dai
			A. nanospora H.S. Yuan
			A. onychoides (Egeland) Niemelä
			A. pendulina H.S.Yuan
			A. perennis B.K. Cui & Y.C. Dai
			A. romellii (Donk) Niemelä
			A. semisupina (Berk. & M.A. Curtis) Ryvarden s. l
			A. stipitata H.S. Yuan & Y.C. Dai
			A. ussurii Y.C. Dai & Niemelä
		Atraporiella Ryvarden	*Atraporiella yunnanensis* C.L. Zhao
		Butyrea Miettinen	*Butyrea japonica* (Nuñez & Ryvarden) Miettinen & Ryvarden
			B. luteoalba (P. Karst.) Miettinen
		Citripora Miettinen	*Citripora bannaensis* Miettinen
		Elaphroporia Z.Q. Wu & C.L. Zhao	*Elaphroporia ailaoshanensis* Z.Q. Wu & C.L. Zhao
		Etheirodon Banker	*Etheirodon fimbriatum* (Pers.) Banker
			E. purpureum Westphalen
			E. roseoalbum J.H. Dong & C.L. Zhao
		Flaviporus Murrill	*Flaviporus citrinellus* (Niemelä & Ryvarden) Ginns
			F. liebmannii (Fr.) Ginns
			F. stramineus (Bres.) Ginns
		Frantisekia Spirin & Zmitr	*Frantisekia abieticola* H.S. Yuan
		Junghuhnia Corda	*Junghuhnia crustacea* (Jungh.) Ryvarden
			J. flabellata H.S. Yuan & Y.C. Dai
			J. minor H.S. Yuan
			J. nitida (Pers.) Ryvarden
			J. pseudocrustacea H.S. Yuan
			J. pseudominuta H.S. Yuan & Y.C. Dai
			J. rhizomorpha H.S. Yuan & Y.C. Dai
			J. taiwaniana H.S. Yuan, Sheng H. Wu & Y.C. Dai
			J. tropica H.S. Yuan, Sheng H. Wu & Y.C. Dai
		Mycorrhaphium Maas Geest	*Mycorrhaphium subadustum* T. Cao & H.S. Yuan
		Niemelaea Zmitr., Ezhov & Khimich	*Niemelaea balaenae* (Niemelä) V. Papp
			N. consobrina (Bres.) Zmitr., Ezhov & Khimich
			N. cremea (Parmasto) Zmitr., Ezhov & Khimich
		Nigroporus Murrill	*Nigroporus ussuriensis* (Bondartsev & Ljub.) Y.C. Dai & Niemelä

(Continued)

TABLE 3.1 ***(Continued)***

Species Diversity and Taxonomic Classification of Wood-Rotting Fungi in China.

Order	Family	Genus	Species
			N. vinosus (Berk.) Murrill
		Rhomboidia C.L. Zhao	*Rhomboidia wuliangshanensis* C.L. Zhao
		Steccherinum Gray	*Steccherinum aurantilaetum* (Corner) Bernicchia & Gorjón
			S. austrosinense (F. Wu, P. Du & X.M. Tian) Z.B. Liu, Y.C. Dai & Jing Si
			S. fissuratum J.H. Dong & C.L. Zhao
			S. fragile Z.B. Liu & Y.C. Dai
			S. formosanum (T.T. Chang & W.N. Chou) Miettinen
			S. hirsutum Y.X. Wu & C.L. Zhao
			S. incrustans Z.B. Liu, Y.C. Dai & Jing Si
			S. juniperi Z.B. Liu, Y.C. Dai & Jing Si
			S. lincangense J.H. Dong & C.L. Zhao
			S. longiaculeiferum J.H. Dong & C.L. Zhao
			S. nandinae (F. Wu, P. Du & X.M. Tian) Z.B. Liu, Y.C. Dai & Jing Si
			S. puerense Y.X. Wu, J.H. Dong & C.L. Zhao
			S. punctatum J.H. Dong & C.L. Zhao
			S. rubigimaculatum Y.X. Wu, J.H. Dong & C.L. Zhao
			S. semisupiniforme (Murrill) Miettinen
			S. subcollabens (F. Wu, P. Du & X.M. Tian) Z.B. Liu & Y.C. Dai
			S. subnitidum (H.S. Yuan & Y.C. Dai) Miettinen
			S. subglobosum H.S. Yuan & Y.C. Dai
			S. subtropicum J.H. Dong & C.L. Zhao
			S. subulatum H.S. Yuan & Y.C. Dai
			S. tenuissimum C.L. Zhao & Y.X. Wu
			S. weishanense J.H. Dong & C.L. Zhao
			S. wumengshanense Lu Wang & C.L. Zhao
			S. xanthum C.L. Zhao & Y.X. Wu
			S. yunnanense Y.X. Wu & C.L. Zhao
			S. zeylanicum Maas Geest.
		Trullella Zmitr	*Trullella conifericola* T. Cao & H.S. Yuan
			T. yunnanensis C.L.Zhao
		Xanthoporus Audet	*Xanthoporus peckianus* (Cooke) Audet
			X. syringae (Parmasto) Audet
	Taiwanofungaceae B.K. Cui, Shun Liu & Y.C. Dai	*Taiwanofungus Sheng* H. Wu, Z.H. Yu, Y.C. Dai & C.H. Su	*Taiwanofungus camphoratus* (M. Zang & C.H. Su) Sheng H. Wu, Z.H. Yu, Y.C. Dai & C.H. Su
			T. salmoneus (T.T. Chang & W.N. Chou) Sheng H. Wu, Z.H. Yu, Y.C. Dai & C.H. Su
	Incertae sedis	*Candelabrochaete* Boidin	*Candelabrochaete africana* Boidin
			C. dispar Hjortstam & Ryvarden
			C. yunnanensis Lu Wang & C.L. Zhao
		Neohypochnicium N. Maek. & R. Sugaw.	*Neohypochnicium daweishanense* J.H. Dong & C.L. Zhao
			N. farinaceum J.H. Dong & C.L. Zhao

TABLE 3.1 *(Continued)*
Species Diversity and Taxonomic Classification of Wood-Rotting Fungi in China.

Order	Family	Genus	Species
			N. murinum J.H. Dong & C.L. Zhao
			N. subrigescens (Boidin) N. Maek.
			N. velutinum J.H. Dong & C.L. Zhao
			N. wakefieldiae (Bres.) N. Maek.
			N. yunnanense J.H. Dong & C.L. Zhao
			N. zixishanense J.H. Dong & C.L. Zhao
Russulales Kreisel ex P.M. Kirk, P.F. Cannon & J.C. David	*Albatrellaceae* Nuss	*Albatrellopsis* Teixeira	*Albatrellopsis confluens* (Alb. & Schwein.) Teixeira
			A. flettioides J. Kahn
		Albatrellus Gray	*Albatrellus alpinus* H.M. Zhou & Y.C. Dai
			A. avellaneus Pouzar
			A. cantharellus (Lloyd) Pouzar
			A. citrinus Ryman
			A. fumosus H.D. Zheng & P.G. Liu
			A. ovinus (Schaeff.) Kotl. & Pouzar
			A. piceiphilus B.K. Cui & Y.C. Dai
			A. subrubescens (Murrill) Pouzar
			A. tianschanicus (Bondartsev) Pouzar
			A. zhuangii Y.C. Dai & Juan Li
			A. tibetanus H.D. Zheng & P.G. Liu
		Neoalbatrellus Audet	*Neoalbatrellus caeruleoporus* (Peck) Audet
			N. odorus Yuan Y. Chen & B.K. Cui
			N. yasudae (Lloyd) Audet
		Scutiger Paulet	*Scutiger pes-caprae* (Pers.) Bondartsev & Singer
	Auriscalpiaceae Maas Geest.	*Artomyces* Jülich	*Artomyces candelabrus* (Massee) Jülich
			A. niveus J.H. Dong & C.L. Zhao
			A. yunnanensis J.H. Dong & C.L. Zhao
	Bondarzewiaceae Kotl. & Pouzar	*Amylonotus* Ryvarden	*Amylonotus labyrinthinus* (T. Hatt.) Y.C. Dai, Jia J. Chen & B.K. Cui
		Amylosporus Ryvarden	*Amylosporus casuarinicola* (Y.C. Dai & B.K. Cui) Y.C. Dai, Jia J. Chen & B.K. Cui
			A. rubellus (Y.C. Dai) Y.C. Dai, Jia J. Chen & B.K. Cui
			A. succulentus Jia J. Chen & L.L. Shen
			A. sulcatus F.C. Huang & Bin Liu
		Bondarzewia Singer	*Bondarzewia dickinsii* (Berk.) Jia J. Chen, B.K. Cui & Y.C. Dai
			B. podocarpi Y.C. Dai & B.K. Cui
			B. submesenterica Jia J. Chen, B.K. Cui & Y.C. Dai
			B. tibetica B.K. Cui, J. Song & Jia J. Chen
		Heterobasidion Bref.	*Heterobasidion amyloideum* Y.C. Dai, Jia J. Chen & Korhonen
			H. amyloideopsis Saba, C.L. Zhao, Khalid & Pfister
			H. armandii Y. C. Dai, Jia J. Chen & Yuan Yuan
			H. australe Y.C. Dai & Korhonen
			H. ecrustosum Tokuda, T. Hatt. & Y.C. Dai

(Continued)

TABLE 3.1 *(Continued)*
Species Diversity and Taxonomic Classification of Wood-Rotting Fungi in China.

Order	Family	Genus	Species
			H. insulare (Murrill) Ryvarden
			H. linzhiense Y.C. Dai & Korhonen
			H. orientale Tokuda, T. Hatt. & Y.C. Dai
			H. parviporum Niemelä & Korhonen
			H. subinsulare Y. C. Dai, Jia J. Chen & Yuan Yuan
			H. subparviporum Y. C. Dai, Jia J. Chen & Yuan Yuan
			H. tibeticum Y.C. Dai, Jia J. Chen & Korhonen
		Rigidoporopsis I. Johans. & Ryvarden	*Rigidoporopsis tegularis* Juan Li & Y.C. Dai
		Wrightoporia Pouzar	*Wrightoporia austrosinensis* Y.C. Dai
			W. borealis Y.C. Dai
			W. cinnamomea Ryvarden
			W. lenta (Oveh. & J. Lowe) Pouzar
			W. nigrolimitata Jia J. Chen
			W. ochrocrocea (Henn. & E. Nyman) A. David & Rajchenb
			W. subavellanea Jia J. Chen & B.K. Cui
			W. unguliformis Y.C. Dai & B.K. Cui
	Echinodontiaceae Donk	*Larssoniporia* Y.C. Dai, Jia J. Chen & B.K. Cui	*Larssoniporia incrustatocystidiata* Y.C. Dai, Jia J. Chen & B.K. Cui
	Hericiaceae Donk	*Dentipellicula* Y.C. Dai & L.W. Zhou	*Dentipellicula austroafricana* Jia J. Chen, L.L. Shen & Y.C. Dai
			D. leptodon (Mont.) Y.C. Dai & L.W. Zhou
			D. taiwaniana (Sheng H. Wu) Y.C. Dai & L.W. Zhou
		Pseudowrightoporia Y.C. Dai, Jia J. Chen & B.K. Cui	*Pseudowrightoporia crassihypha* Y.C. Dai, Jia J. Chen & B.K. Cui
			P. gillesii (A. David & Rajchenb.) Y.C. Dai, Jia J. Chen & B.K. Cui
			P. hamata Y.C. Dai, Jia J. Chen & B.K. Cui
			P. japonica (Núñez & Ryvarden) Y.C. Dai, Jia J. Chen & B.K. Cui
			P. oblongispora Y.C. Dai, Jia J. Chen & B.K. Cui
		Wrightoporiopsis Y.C. Dai, Jia J. Chen & B.K. Cui	*Wrightoporiopsis amylohypha* Y.C. Dai, Jia J. Chen & B.K. Cui
			W. biennis (Jia J. Chen & B.K. Cui) Y.C. Dai, Jia J. Chen & B.K. Cui
	Peniophoraceae Lotsy	*Asterostroma* Massee	*Asterostroma muscicola* (Berk. & M.A. Curtis) Massee
			A. paramuscicola J.H Dong & C.L. Zhao
			A. rhizomorpharum H.M. Zhou & C.L. Zhao
			A. roseoalbum J.H Dong & C.L. Zhao
			A. vararioides S.L. Liu & S.H. He
			A. yunnanense Y.L. Deng & C.L. Zhao
		Peniophora Cooke	*Peniophora roseoalba* L. Zou & C.L. Zhao
			P. yunnanensis L. Zou & C.L. Zhao
		Scytinostroma Donk	*Scytinostroma daweishanense* J.H. Dong & C.L. Zhao

TABLE 3.1 *(Continued)*
Species Diversity and Taxonomic Classification of Wood-Rotting Fungi in China.

Order	Family	Genus	Species
			S. duriusculum (Berk. & Broome) Donk
			S. galactinum (Fr.) Donk
			S. yunnanense C.L. Zhao
		Vararia P. Karst.	*Vararia bispora* S.L. Liu & S.H. He
			V. daweishanensis L. Zou & C.L. Zhao
			V. fissurata Y.L. Deng & C.L. Zhao
			V. fragilis L. Zou & C.L. Zhao
			V. isabellina Y.L. Deng & C.L. Zhao
			V. lincangensis Y.L. Deng & C.L. Zhao
			V. punctata Y.L. Deng & C.L. Zhao
			V. sinensis Y.L. Deng & C.L. Zhao
			V. tenuata Ghobad-Nejhad
			V. yaoshanensis Y.L. Deng & C.L. Zhao
			V. yingjiangensis Y.L. Deng & C.L. Zhao
			V. yunnanensis Y.L. Deng & C.L. Zhao
	Stereaceae Pilát	*Aleurodiscus* Rabenh.	*Aleurodiscus isabellinus* S.H. He & Y.C. Dai
			A. mirabilis (Berk. & M.A. Curtis) Höhn.
		Gloeocystidiellum Donk	*Gloeocystidiellum yunnanense* Y.L. Zhao & C.L. Zhao
	Xenasmataceae Oberw.	*Xenasmatella* Oberw.	*Xenasmatella ailaoshanensis* C.L. Zhao ex C.L. Zhao & T.K. Zong
			X. athelioidea (N. Maek.) N. Maek.
			X. bambusicola Qi Yuan & C.L. Zhao
			X. gossypina (C.L. Zhao) G. Gruhn & Trichies
			X. hjortstamii S.L. Liu & L.W. Zhou
			X. jilongensis S.L. Liu & L.W. Zhou
			X. nigroidea K.Y. Luo & C.L. Zhao
			X. rhizomorpha C.L. Zhao
			X. roseobubalina Z.B. Liu & Yuan Yuan
			X. tenuis C.L. Zhao
			X. wuliangshanensis (C.L. Zhao) G. Gruhn & Trichies
			X. xinpingensis C.L. Zhao
	Incertae sedis	*Polypus* Audet	*Polypus dispansus* (Lloyd) Audet
		Xeroceps Audet	*Xeroceps skamania* (Murrill) Audet
			X. yunnanensis (H.D. Zheng & P.G. Liu) Audet
Septobasidiales Couch ex Donk	*Septobasidiaceae* Racib.	*Septobasidium* Pat.	*Septobasidium abnorme* (Henn.) Höhn. & Litsch.
			S. albiziae S.Z. Chen & L. Guo
			S. aquilariae C.L. Zhao
			S. brunneum Wei Li bis & L. Guo
			S. cremeum Couch ex L.D. Gómez & Henk
			S. macrobasidium Q.Q. Jiang & C.L. Zhao
			S. puerense Q.Q. Jiang & C.L. Zhao
			S. wuliangshanense Q.Q. Jiang & C.L. Zhao
			S. yunnanense S.Z. Chen & L. Guo
Stereopsidales Sjökvist, E. Larss., B.E. Pfeil & K.H. Larss.	*Stereopsidaceae* Sjökvist, E. Larss., B.E. Pfeil & K.H. Larss.	*Stereopsis* D.A. Reid	*Stereopsis africana* Ryvarden

(Continued)

TABLE 3.1 ***(Continued)***
Species Diversity and Taxonomic Classification of Wood-Rotting Fungi in China.

Order	Family	Genus	Species
			S. burtiana (Peck) D.A. Reid
			S. yunnanensis Qi Yuan & C.L. Zhao
Thelephorales Corner ex Oberw	*Bankeraceae* Donk	*Boletopsis* Fayod	*Boletopsis grisea* (Peck) Bondartsev & Singer
			B. macrocarpa Y.C. Dai, F. Wu & H.M. Zhou
			B. tibetana Y.C. Dai, F. Wu & H.M. Zhou
	Thelephoraceae Chevall	*Lenzitopsis* Malenc¸on & Bertault	*Lenzitopsis daii* L.W. Zhou & Kõljalg
		Tomentella Pat.	*Tomentella ferruginea* (Pers.) Pat.
			T. olivaceobasidiosa X.J. Zhang & C.L. Zhao
			T. tenuifarinacea X.J. Zhang & C.L. Zhao
			T. velutina X.J. Zhang & C.L. Zhao
			T. wumenshanensis X.J. Zhang & C.L. Zhao
			T. yunnanensis X.J. Zhang & C.L. Zhao
			T.strum umbrinum (Fr.) Svrček
		Thelephora Ehrh. ex Willd.	*Thelephora grandinioides* C.L. Zhao & X.F. Liu
			T. resupinata X.J. Zhang & C.L. Zhao
			T. reticulata Fr.
			T. reticulata Pat.
			T. subtropica X.J. Zhang & C.L. Zhao
			T. wuliangshanensis C.L. Zhao & X.F. Liu
			T. yunnanensis X.J. Zhang & C.L. Zhao
Trechisporales K.H. Larss.	*Hydnodontaceae* Jülich	*Porpomyces* Jülich	*Porpomyces submucidus* F. Wu & C.L. Zhao
		Subulicystidium Parmasto	*Subulicystidium yunnanense* J.H. Dong & C.L. Zhao
		Trechispora P. Karst.	*Trechispora alba* S.L. Liu, G. He & L.W. Zhou
			T. bambusicola C.L. Zhao
			T. candidissima (Schwein.) Bondartsev & Singer
			T. dentata Z.B. Liu & Yuan Yuan
			T. dimitiella Z.B. Liu & Yuan Yuan
			T. damansaraensis S.L. Liu & L.W. Zhou
			T. daweishanensis C.L. Zhao
			T. dealbata (Berk.) L.W. Zhou & S.L. Liu
			T. fimbriata C.L. Zhao
			T. fissurata C.L. Zhao
			T. murina K.Y. Luo & C.L. Zhao
			T. mollusca (Pers.) Liberta
			T. odontioidea K.Y. Luo & C.L. Zhao
			T. olivacea K.Y. Luo & C.L. Zhao
			T. suberosa H.S. Yuan & Y.C. Dai
			T. subhymenocystis S.L. Liu, H.S. Yuan & L.W. Zhou
			T. tropica S.L. Liu & L.W. Zhou
			T. tuberculata S.L. Liu & L.W. Zhou
			T. xantha C.L. Zhao
			T. yunnanensis C.L. Zhao

Polypores constitute a very well-known group among the wood-inhabiting basidiomycetes. Before the advent of modern DNA methods, the taxonomy of wood-inhabiting basidiomycetes in general and polypores in particular relied on two main cornerstones—the type of wood decay and anatomical traits of basidiocarps. Depending on wood-decay characteristics, these fungi were traditionally divided into two groups—brown-rot- and white-rot-producing species. The type of rot was recognized as a taxonomically significant character sufficient for rearranging morphologically similar species into separate genera or supra-generic units after the studies by Nobles (1958, 1971). It is particularly likely in well-studied conspicuous fungi such as the polypores that many potentially new species already have an older name. The only way to diminish the risk of introducing superfluous names is to investigate the type material of all taxa, both accepted and considered synonyms, which appear in phylogenetic analyses as close relatives of a newly detected species (Spirin et al. 2024). The next challenge facing taxonomists arises from the binomial nature of a species name. According to the Code (Art. 35.1), the valid publication of a new species implies the species is going to be described as a member of an already existing or simultaneously introduced genus. Traditionally, a single feature that, in the opinion of the author, distinguishes a newly introduced taxon from other ones is enough for introducing a new species or a new genus (Code Art. 38.2). This somewhat ambiguous provision in the Code opens a door for describing new genera based on arbitrarily selected traits, without a critical evaluation of their constancy and taxonomic significance in a given group. As a consequence, many of these genera turn out to be monotypic (Parmasto 1994). Currently, it is not an exceptional situation that DNA data show a newly found species is equally distant from already described and sequenced members of other genera, and thus seemingly requiring its own genus. The next step often taken is using any features differentiating the new species from its phylogenetic neighbors as justification for recognizing a new genus for it (Spirin et al. 2024).

3.6 THE IMPORTANT WOOD-INHABITING FUNGAL FAMILY HYMENOCHAETACEAE DONK

The important wood-inhabiting fungal family Hymenochaetaceae Donk (Hymenochaetales in the Agaricomycetes) is characterized by its annual to perennial, resupinate, effused-reflexed, pileate to stipitate, brownish basidiomata with a xanthochroic reaction in KOH, a mono-, di- or trimitic hyphal system with simple septate generative hyphae, the presence or absence of setal elements, and smooth, thin- to thick-walled, hyaline or brownish, indextrinoid or dextrinoid, cyanophilous or acyanophilous basidiospores, and causing a white rot (Liu et al. 2025).

The Hymenochaetaceae includes both poroid and nonporoid species (Figure 3.9). The species diversity and phylogeny of poroid species were studied by Wu et al. (2022), and 32 poroid genera with 672 species were recognized. The results of a parallel study are presented herein, in which the remaining genera of the Hymenochaetaceae, Hymenochaete Lév., and Hydnoporia Murrill, the species of which have mostly a corticioid or hydnoid hymenophore, were considered.

Many members of *Hymenochaete* and *Hydnoporia* are medicinal mushrooms with a significant economic value, while some other species are forest pathogens. Some species are pathogenic, affecting many different living angiosperm trees in native forests and plantations (Yuan et al. 2023). Another species, *Hydnoporia corrugata* (Fr.) K.H. Larss. & Spirin., functions as a facultative parasite and is responsible for the death of several species of angiosperm trees and shrubs (Graves 1914); it also is known for its medicinal properties (Watling and Harper 1998).

Hymenochaete is the type genus of the family Hymenochaetaceae. The genus, typified by *H. rubiginosa*, was established by Léveillé (1846) based primarily on the presence of setae in the hymenium. Burt (1918) was the first to recognize the three distinct layers within the basidiomata of *Hymenochaete*. These are a setal layer (containing the setae), a context/subiculum (hyphal layer without setae), and a cortex (consisting of agglutinated hyphae with, in some instances, a tomentum covering the cortex also developing). Based on the specific occurrence of these three layers within

FIGURE 3.9 Basidiomata of *Hymenochaete sinensis* (holotype, SWFC 026040). Bars: A = 1 cm; B = 1 mm.

Source: Photo by Changlin Zhao

taxa, Léger (1998) divided *Hymenochaete* into four sections in his monograph. A total of 264 species were listed, but he accepted only 116 species. *Hymenochaete*, *Phellinus sensu lato*, and *Inonotus* sensu lato are the three largest genera in the Hymenochaetaceae (Dai 2010). However, compared to *Phellinus sensu lato* and *Inonotus sensu lato*, the diversity of *Hymenochaete* (and *Hydnoporia*) is poorly known. Many species remain undiscovered, especially in China (Liu et al. 2025).

REFERENCES

Alem, D., T. Dejene, J.A. Oria-de-Rueda & P. Martín-Pinto. 2021. Survey of macrofungal diversity and analysis of edaphic factors influencing the fungal community of church forests in dry Afromontane areas of Northern Ethiopia. *Forest Ecology and Management* 496:119391.

Anonymous. 1955. *Shen Nong Materia Medica, 102–200 AD* (E. Reprinted). People's Hygiene Press, Beijing, China.

Bader, P., S. Jansson & B.G. Jonsson. 1995. Wood-rotting fungi and substratum decline in selectively logged boreal spruce forests. *Biological Conservation* 72:355–362.

Berglund, H., M.T. Jonsson, R. Penttilä & I. Vanha-Majamaa. 2011. The effects of burning and dead-wood creation on the diversity of pioneer wood-rotting fungi in managed boreal spruce forests. *Forest Ecology and Management* 261:1293–1305.

Bergman, A. & A. Casadevall. 2010. Mammalian endothermy optimally restricts fungi and metabolic costs. *mBio* 1(5):e00212–10.

Bernicchia, A. & S.P. Gorjón. 2010. *Fungi Europaei 12: Corticiaceae s.l.* Edizioni Candusso, Alassio, Italy. pp. 1–1008.

Binder, M., A. Justo, R. Riley, A. Salamov, F. Lopez-Giraldez, E. Jökvist, A. Copeland, B. Foster, H. Sun, E. Larsson, K.H. Larsson, J. Townsend, I.V. Grigoriev & D.S. Hibbet. 2013. Phylogenetic and phylogenomic overview of the polyporales. *Mycologia* 105(6):1350–1373.

Brundrett, M.C. & L. Tedersoo. 2018. Evolutionary history of mycorrhizal symbioses and global host plant diversity. *New Phytologist* 220:1108–1115.

Burt, E.A. 1918. The Thelephoraceae of North America 10. *Hymenochaete. Annals of the Missouri Botanical Garden* 5:301–372.

Chung, C.L., T.J. Lee, M. Akiba, H.H. Lee, T.H. Kuo, D. Liu, H.M. Ke, T. Yokoi, M.B. Roa, M.Y.J. Lu, Y.Y. Chang, P.J. Ann, J.N. Tsai, C.Y. Chen, S.S. Tzean, Y. Ota, T. Hattori, N. Sahashi, R.F. Liou, T. Kikuchi & I.J. Tsai. 2017. Comparative and population genomic landscape of *Phellinus noxious*: A hypervariable fungus causing root rot in trees. *Molecular Ecology* 26:6301–6316.

Coetzee, M.P.A., S. Marincowitz, V.G. Muthelo & M.J. Wingfield. 2015. *Ganoderma* species, including new taxa associated with root rot of the iconic *Jacaranda mimosifolia* in Pretoria, South Africa. *IMA Fungus* 6:249–256.

Cui, B.K., H.J. Li, X. Ji, J.L. Zhou, J. Song, J. Si, Z.L. Yang & Y.C. Dai. 2019. Species diversity, taxonomy and phylogeny of Polyporaceae (Basidiomycota) in China. *Fungal Diversity* 97:137–392.

Cui, B.K., C.L. Zhao, J. Vlasák & Y.C. Dai. 2014. A preliminary report on decay and canker of *Acacia richii* caused by *Inonotus rickii* in China. *Forest Pathology* 44:82–84.

Dai, Y.C. 2010. Hymenochaetaceae (Basidiomycota) in China. *Fungal Diversity* 45:131–343.

Dai, Y.C. 2012. Polypore diversity in China with an annotated checklist of Chinese polypores. *Mycoscience* 53:49–80.

Dai, Y.C. 2022. *Resources and Diversity of Wood-Decaying Fungi in Yunnan*. Science Press, Beijing, China.

Dai, Y.C., B.K. Cui, H.S. Yuan & B.D. Li. 2007. Pathogenic wood decaying fungi in China. *Forest Pathology* 37:105–120.

Dai, Y.C., L.F. Fan, J.J. Chen, C.P. Wu, Y.D. Wu & Y. Yuan. 2021a. Diversity of *Heterobasidion* causing root and butt rotting of conifers and proposal for entry wood quarantine aiming at *Heterobasidion* species. *Mycosystema* 40:1958–1964.

Dai, Y.C., H.J. Xue, J. Vlasák, M. Rajchenberg, B. Wang & L.W. Zhou. 2014. Phylogeny and global diversity of *Polyporus* group *Melanopus* (Polyporales, Basidiomycota). *Fungal Diversity* 64:133–144.

Dai, Y.C., Z.L. Yang, B.K. Cui, G. Wu, H.S. Yuan, L.W. Zhou, S.H. He, Z.W. Ge, F. Wu, Y.L. Wei, Y. Yuan & J. Si. 2021b. Diversity and systematics of the important macrofungi in Chinese forests. *Mycosystema* 40:770–805.

Dai, Y.C., Z.L. Yang, B.K. Cui, C.J. Yu & L.W. Zhou. 2009. Species diversity and utilization of medicinal mushrooms and fungi in China. *International Journal of Medicinal Mushrooms* 11:287–302.

Dai, Y.C., L.W. Zhou & K. Steffen. 2011. Wood-decaying fungi in eastern Himalayas 1. Polypores from Zixishan Nature Reserve, Yunnan Province. *Mycosystema* 30:674–679.

Dai, Y.C., Zhou, L.W., Yang, Z.L., Wen, H.A., Bau, Tolgor, Li, T.H. 2010. A revised checklist of edible fungi in China. *Mycosystema* 29:1–21.

Deng, Y.L., M. Chen, K.S.W. Liu, D. Chen, S. Yang, W. Li, H. Zhou & C.L. Zhao. 2025. Morphological and phylogenetic analyses reveal one new genus and six new species in Irpicaceae and Steccherinaceae (Polyporales, Basidiomycota) from the Yunnan–Guizhou Plateau, Asia. *IMA Fungus* 16:e172367.

Deng, Y.L., M. Chen, S. Zhang, K. Wang, W. Liu, Y. Qiu, Y. Dou, X. Liu, A.S. Wijesinghe, H. Zhou, S. Jabeen & C.L. Zhao. 2026. Notes, taxonomy, and phylogeny of wood-inhabiting fungi in Russulales. *Mycosphere* 17:e003.

Dong, J.H., M.L. Chen, M. Chen, Q. Li, Y. Zhu, X. Zhang, C. Zhou, W. Li, A. Muhammad, C. Zhou, S. Jabeen & C.L. Zhao. 2025. Notes, outline, taxonomy and phylogeny of wood-inhabiting Agaricales. *Mycosphere* 16:2599–2711.

Dong, J.H., Q. Li, Q. Yuan, Y.X. Luo, X.C. Zhang, Y.F. Dai, Q. Zhou, X.F. Liu, Y.L. Deng, H.M. Zhou, A. Muhammad & C.L. Zhao. 2024. Species diversity, taxonomy, molecular systematics and divergence time of wood-rotting fungi in Yunnan-Guizhou Plateau, Asia. *Mycosphere* 15:1110–1293.

Donk, M.A. 1960. The generic names proposed for Polyporaceae. *Persoonia* 1:173–302.

Donk, M.A. 1964. The conspectus of the families of the Aphyllophorales. *Persoonia* 3:199–324.

Fries, E. 1838. *Epicrisis Systematis Mycologici*. Typographia Academica, Upsaliae, Sweden.

Gibertoni, T.B., P.S. Medeiros, A.M.S. Soares, A.C. Gomes-Silva, P.J.P. Santos, H.M.P. Sotao, L.V. Ferreira & E. Savino. 2016. The distribution of polypore fungi in endemism centres in Brazilian Amazonia. *Fungal Ecology* 20:1–6.

Gilbert, G.S., J. Gorospe & L. Ryvarden. 2008. Host and habitat preferences of polypore fungi in Micronesian tropical flooded forests. *Mycological Research* 112:674–680.

Gilbertson, R.L. & L. Ryvarden. 1987. *North American Polypores*, Vols. 1–2. Fungiflora, Oslo: Lubrecht and Cramer Ltd, Norway.

Graves, A.H. 1914. Parasitism in *Hymenochaete agglutinans*. *Mycologia* 6:279–284.

Guan, Q.X., J. Huang, J. Huang & C.L. Zhao. 2023. Five new species of Schizoporaceae (Basidiomycota, Hymenochaetales) from East Asia. *MycoKeys* 96:25–56.

Hage, H., S. Miyauchi, M. Viragh, E. Drula, B. Min, D. Chaduli, D. Navarro, A. Favel, M. Norest, L. Lesage-Meessen, B. Balint, Z. Merenyi, L. Eugenio, E. Morin, A.T. Martínez, P. Baldrian, M. Stursova, M.J. Martínez, C. Novotny, J.K. Magnuson, J.W. Spatafora, S. Maurice, J. Pangilinan, W. Andreopoulos, K. LaButti, H. Hundley, H. Na, A. Kuo, K. Barry, A. Lipzen, B. Henrissat, R. Riley, S. Ahrendt, L.G. Nagy, I.V. Grigoriev, F. Martin & M.N. Rosso. 2021. Gene family expansions and transcriptome signatures uncover fungal adaptations to wood decay. *Environmental Microbiology* 23:5716–5732.

Hawksworth, D.L. & R. Lücking. 2017. Fungal diversity revisited: 2.2 to 3.8 million species. *Microbiology Spectrum* 5:10.

He, M.Q., R.L. Zhao, K.D. Hyde & D. Begerow. 2019. Notes, outline and divergence times of Basidiomycota. *Fungal Diversity* 99:105–367.

Hibbett, D., L.G. Nagy & R.H. Nilsson. 2025. Fungal diversity, evolution, and classification. *Current Biology* 35(11):R463–R469.

Hjortstam, K., K.H. Larsson & L. Ryvarden. 1987. *The Corticiaceae of North Europe*, Vol. 1. Fungiflora, Oslo, Norway.

Huang, C.J., X.Q. Wu, X.Y. Liu, Y.T. Fang, L. Liu & C.S. Wu. 2022. Functional fungal communities dominate wood decomposition and are modified by wood traits in a subtropical forest. *The Science of The Total Environment* 806:1–11.

Jülich, W. 1981. Higher taxa of basidiomycetes. *Bibliotheca Mycologica* 85:1–485.

Justo, A., O. Miettinen, D. Floudas, B. Ortiz-Santana, E. Sjökvist, D. Lindner, K. Nakason, T. Niemelä, K.H. Larsson, L. Ryvarden & D.S. Hibbett. 2017. A revised family-level classification of the Polyporales (Basidiomycota). *Fungal Biology* 121:798–824.

Karsten, P.A. 1879. Symbolae ad mycologiam Fennicam. 6. *Acta Societatis pro Fauna et Flora Fennica* 5:15–51.

Karsten, P.A. 1881. Symbolae ad mycologiam Fennicam. 8. *Acta Societatis pro Fauna et Flora Fennica* 6:7–13.

Karsten, P.A. 1892. Mycetes aliquot in Mongolia et China boreali a clarissimo C.N. Potanin lecti. *Hedwigia* 31:38–40.

Kirk, P.M., P.F. Cannon, D.W. Minter & J.A. Stalpers. 2008. *Dictionary of the Fungi*, 10th ed. CAB International, Oxfordshire, Britain.

Krah, F.S., S. Seibold, R. Brandl, P. Baldrian, J. Müller & C. Bässler. 2018. Independent effects of host and environment on the diversity of wood-inhabiting fungi. *Journal of Ecology* 106:1428–1442.

Lee, D., K.H. Boo, J.M. Lee, C.D. Viet, N. Quyen, T. Unno, M. Cho, K.Z. Riu & D.S. Lee. 2012. Anti-viral activity of *Hydnellum concrescens*, a medicinal mushroom. *African Journal of Biotechnology* 11:15241–15245.

Léger, J.C. 1998. Le genre *Hymenochaete* Léveillé. *Bibliotheca Mycologica* 171:1–319.

Léveillé, J.H. 1846. Descriptions de champignons de l'herbier du Muséum de Paris. *Annales des Sciences Naturelles, Botanique III* 5:111–167.

Li, Q.S. & Y.L. Wei. 2021. Species composition and distribution characteristics of wood-decaying fungi in tropical areas of China. *Mycosystema* 40:2653–2672.

Linzer, R.E., W.J. Otrosina, P. Gonthier, J. Bruhn, G. Laflamme, G. Bussières & M. Garbelotto. 2008. Inferences on the phylogeography of the fungal pathogen *Heterobasidion annosum*, including evidence of interspecific horizontal genetic transfer and of human-mediated, long-range dispersal. *Molecular Phylogenetics and Evolution* 46:844–862.

Liu, S., Y.Y. Chen, Y.F. Sun, X.L. He, C.G. Song, J. Si, D.M. Liu, G. Gates & B.K. Cui. 2023. Systematic classification and phylogenetic relationships of the brown-rot fungi within the Polyporales. *Fungal Diversity* 118:1–94.

Liu, S., M.L. Han, T.M. Xu, Y. Wang, D.M. Wu & B.K. Cui. 2021. Taxonomy and phylogeny of the *Fomitopsis pinicola* complex with descriptions of six new species from East Asia. *Frontiers in Microbiology* 12:1–19.

Liu, Z.B., Y. Yuan, Y.C. Dai, H.G. Liu, J. Vlasák, G.Y. Zeng, S.H. He & F. Wu. 2025. Global diversity and systematics of Hymenochaetaceae with non-poroid hymenophore. *Fungal Diversity* 131:1–97.

Luo, K.Y., J.Q. Su & C.L. Zhao. 2024. Morphological and molecular identification for four new wood inhabiting species of *Trechispora* (Basidiomycota) from China. *MycoKeys* 105:155–178.

Ma, H.X., J. Si, Y.C. Dai, A.H. Zhu, B.K. Cui, Y.G. Fan, Y. Yuan & S.H. He. 2022. Diversity of wood-rotting macrofungi in Hainan Province, South China. *Mycosystema* 41:695–712.

Marco-Urrea, E., M. Pérez-Trujillo, T. Vicent & G. Caminal. 2009. Ability of white-rot fungi to remove selected pharmaceuticals and identification of degradation products of ibuprofen by *Trametes versicolor*. *Chemosphere* 74:765–772.

Mi, X.C., G. Feng, Y.B. Hu, J. Zhang, L. Chen, R.T. Corlett, A.C. Hughes, S. Pimm, B. Schmid, S.J. Shi, J.C. Svenning & K.P. Ma. 2021. The global significance of biodiversity science in China: An overview. *National Science Review* 8:7.

Miettinen, O., R. Riley, K. Barry, D. Cullen, R.P. de Vries, M. Hainaut, A. Hatakka, B. Henrissat, K. Hilden, R. Kuo, K. LaButti, A. Lipzen, M.R. Makela, L. Sandor, J.W. Spatafora, I.V. Grigoriev & D.S. Hibbett. 2016. Draft genome sequence of the white-rot fungus *Obba rivulosa* 3A-2. *Genome Announcements* 4:e00976–16.

Miettinen, O., J. Vlasak, B. Rivoire & V. Spirin. 2018. *Postia caesia* complex (Polyporales, Basidiomycota) in temperate northern hemisphere. *Fungal Systematics and Evolution* 1:101–129.

Moldes, D. & M.Á. Sanromán. 2006. Amelioration of the ability to decolorize dyes by laccase: Relationship between redox mediators and laccase isoenzymes in *Trametes versicolor*. *World Journal of Microbiology and Biotechnology* 22:1197–1204.

Nobles, M.K. 1958. Cultural characters as a guide to the taxonomy and the phylogeny of the Polyporaceae. *Canadian Journal of Botany* 36:883–926.

Nobles, M.K. 1971. Cultural characters as a guide to the taxonomy of the Polyporaceae. In R. Petersen (ed.), *Evolution in the Higher Basidiomycetes*, pp. 169–196. University of Tennessee Press, Knoxville.

Otrosina, W.J. & M. Garbelotto. 2010. *Heterobasidion occidentale* sp. nov. and *Heterobasidion irregulare* nom. nov.: A disposition of North American *Heterobasidion* biological species. *Fungal Biology* 114:16–25.

Palviainen, M., L. Finer, R. Laiho, E. Shorohova, E. Kapitsa & I. Vanha-Majamaa. 2010. Carbon and nitrogen release from decomposing Scots pine, Norway spruce and silver birch stumps. *Forest Ecology and Management* 256:390–398.

Park, J.H., I.N. Pavlov, M.J. Kim, M.S. Park, S.Y. Oh, K.H. Park, J.J. Fong & Y.W. Lim. 2020. Investigating wood decaying fungi diversity in Central Siberia, Russia using ITS sequence analysis and interaction with host trees. *Sustainability* 12:2535.

Parmasto, E. 1994. Limits of splitting (on schizotaxia). *Mycologia Helvetica* 6:8–34.

Patouilliard, N.T. 1900. *Essai taxonomique sur les familles et les genres des Hyme´nomycetes*. Lucien Declume, Lons-le-Saunier, France.

Petkovšek, S.A.S. & B. Pokorny. 2013. Lead and cadmium in mushrooms from the vicinity of two large emission sources in Slovenia. *The Science of The Total Environment* 443:944–954.

Qian, L.S., J.H. Chen, T. Deng & H. Sun. 2020. Plant diversity in Yunnan: Current status and future directions. *Plant Diversity* 42:281–291.

Rajchenberg, M. & G. Robledo. 2013. Pathogenic polypores in Argentina. *Forest Pathology* 43:171–184.

Riley, R., A.A. Salamov, D.W. Brown, L.G. Nagy, D. Floudas, B.W. Held, A. Levasseur, V. Lombard, E. Morin, R. Otillar, E.A. Lindquist, H. Sun, K.M. LaButti, J. Schmutz, D. Jabbour, H. Luo, S.E. Baker, A.G. Pisabarro, J.D. Walton, R.A. Blanchette, B. Henrissat, F. Martin, D. Cullen, D.S. Hibbett & I.V. Grigoriev. 2014. Extensive sampling of basidiomycete genomes demonstrates inadequacy of the white-rot/brown-rot paradigm for wood decay fungi. *Proceedings of the National Academy of Sciences* 111:9923–9928.

Ryvarden, L. 1991. *Genera of Polypores. Nomenclature and Taxonomy.* Synopsis Fungorum 5. Fungiflora, Oslo, Norway.

Schulze, S. & G. Bahnweg. 1998. Critical review of identification techniques for *Armillaria* spp. and *Heterobasidion annosum* root and butt rot diseases. *Journal of Phytopathology* 146:61–72.

Shen, L.L., M. Wang, J.L. Zhou, J.H. Xing, B.K. Cui & Y.C. Dai. 2019. Taxonomy and phylogeny of *Postia*. Multi-gene phylogeny and taxonomy of the brown-rot fungi: *Postia* and its related genera. *Persoonia* 42:101–126.

Sinclair, W. & H.H. Lyon. 2005. *Diseases of Trees and Shrubs*, 2nd ed. Cornell University Press, Ithaca, New York.

Sinclair, W., H.H. Lyon & W.T. Johnson. 1987. *Diseases of Trees and Shrubs*. Cornell University Press, Ithaca, New York.

Smith, S.E. & D.J. Read. 2008. *Mycorrhizal Symbiosis*, 3rd ed. Academic Press, London, Britain.

Spirin, V., K. Runnel, J. Vlasák, I. Viner, M.D. Barrett, L. Ryvarden, A. Bernicchia, B. Rivoire, A.M. Ainsworth, T. Grebenc, M. Cartabia, T. Niemelä, K.H. Larsson & O. Miettinen. 2024. The genus Fomitopsis (Polyporales, Basidiomycota) reconsidered. *Studies in Mycology* 107:149–249.

Stalpers, J.A. 1993. The Aphyllophoraceous Fungi: Keys to the Species of the Thelephorales. I. Centraalbureau voor Schimmelcultures. *Studies in Mycology* 35:1–168.

Stokland, J.N. & K.H. Larsson. 2011. Legacies from natural forest dynamics: Different effects of forest management on wood-rotting fungi in pine and spruce forests. *Forest Ecology and Management* 161:1707–1721.

Sun, Y.F., D.H. Costa-Rezende, J.H. Xing, J.L. Zhou, B. Zhang, T.B. Gibertoni, G. Gates, M. Glen, Y.C. Dai & B.K. Cui. 2019. Multigene phylogeny and taxonomy of *Amauroderma* s. lat. (Ganodermataceae). *Persoonia* 44:206–239.

Sun, Y.F., J.H. Xing, X.L. He, D.M. Wu, C.G. Song, S. Liu, J. Vlasák, G. Gates, T.B. Gibertoni & B.K. Cui. 2022. Species diversity, systematic revision, and molecular phylogeny of Ganodermataceae (Polyporales, Basidiomycota) with an emphasis on Chinese collections. *Studies in Mycology* 101:287–415.

Van Wyk, A.E. & J.C. Coetzee. 2009. The genus *Calvatia* ("Gasteromycetes", Lycoperdaceae): A review of its ethnomycology and biotechnological potential. *African Journal of Biotechnology* 8:6007–6015.

Vasco-Palacios, A.M., J. Hernandez, M.C. Peñuela-Mora, A.E. Franco Molano & T. Boekhout. 2018. Ectomycorrhizal fungi diversity in a white sand forest in western Amazonia. *Fungal Ecology* 31:9–18.

Wang, K. & L. Cai. 2023. Overview of the historical and current status of fungal taxonomy and diversity in China. *Mycosystema* 42:50–62.

Wang, X.Y. & Y.L. Wei. 2015. Population difference of polypores in northwest and southeast of China. *The Journal of Applied Ecology* 26:3160–3166.

Watling, R. & D.B. Harper. 1998. Chloromethane production by wood-rotting fungi and an estimate of the global flux to the atmosphere. *Mycological Research* 102:769–787.

Woodward, S., J. Stenlid, R. Karjalainen & A. Huttermann. 1998. *Heterobasidion annosum: Biology, Ecology, Impact and Control*. CAB International, United States.

Wu, F., S.J. Dai, J. Vlasák, V. Spirin & Y.C. Dai. 2019. Phylogeny and global diversity of *Porodaedalea*, a genus of gymnosperm pathogens in the Hymenochaetales. *Mycologia* 111:40–53.

Wu, F., X.W. Man, A. Tohtirjap & Y.C. Dai. 2022. A comparison of polypore funga and species composition in forest ecosystems of China, North America, and Europe. *Forest Ecosystems* 9:100051.

Wu, Y.D., W.L. Mao & Y. Yuan. 2021. Comparison of polypore florae and diversity from temperate to subtropical forest zones in China. *Biodiversity Science* 29:1369–1376.

Yan, S., M. Zhong, B.Q. Luo & W. Li. 2021. Innovation-driven high-level conservation of biodiversity in Yunnan. *Journal of West China Forestry Science* 50:5–8.

Yang, Y., Y. Xu, L. Wang, Q.Q. Jiang, J.Q. Su, R. Li, H.M. Zhou & C.L. Zhao. 2025. Multigene phylogeny of seven wood-inhabiting fungal orders in Basidiomycota, and proposal of a new genus and thirteen new species. *Mycosphere* 16:245–295.

Yu, J., J. Cai, H.G. Liu & C.L. Zhao. 2023. Molecular systematics and taxonomy reveal three new taxa in *Phanerochaete* (Phanerochaetaceae, Polyporales) from the Yunnan-Guizhou plateau, East Asia. *Mycological Progress* 22:59.

Yuan, Y., L.S. Bian, Y.D. Wu, J.J. Chen, F. Wu, H.G. Liu, G.Y. Zeng & Y.C. Dai. 2023. Species diversity of pathogenic wood-rotting fungi (Agaricomycetes, Basidiomycota) in China. *Mycology* 14:204–226.

Yuan, Y., J.J. Chen, K. Korhonen, F. Martin & Y.C. Dai. 2021. An updated global species diversity and phylogeny in the forest pathogenic genus *Heterobasidion* (Basidiomycota, Russulales). *Frontiers in Microbiology* 11:596393.

Yuan, Y., Y.J. Wang, G.P. Sun, Y.R. Wang, L.J. Cao, Y.M. Shen, B. Yuan, D. Han & L.Q. Huang. 2018. Archaeological evidence suggests earlier use of *Ganoderma* in Neolithic China. *Chinese Science Bulletin* 63:1180–1188.

Zhang, J.X., Q. Chen, C.Y. Huang, W. Gao & J.B. Qu. 2015. History, current situation, and trend of edible mushroom industry development. *Mycosystema* 34:524–540.

Zhao, C.L., M.H. Qu, R.X. Huang & S.C. Karunarathna. 2023a. Multi-gene phylogeny and taxonomy of the wood-rotting fungal genus *Phlebia sensu lato* (Polyporales, Basidiomycota). *Journal of Fungi* 9:1–41.

Zhao, H., Y.D. Nie, T.K. Zong, K. Wang, M.L. Lv, Y.J. Cui, A. Tohtirjap, J.J. Chen, C.L. Zhao, F. Wu, B.K. Cui, Y. Yuan, Y.C. Dai & X.Y. Liu. 2023b. Species diversity, updated classification, and divergence times of the phylum Mucoromycota. *Fungal Diversity* 123:49–157.

Zhao, H., Y.D. Wu, Z.R. Yang, H.G. Liu, F. Wu, Y.C. Dai & Y. Yuan. 2024. Polypore funga and species diversity in tropical forest ecosystems of Africa, America, and Asia, and a comparison with temperate and boreal regions of the Northern Hemisphere. *Forest Ecosystem* 11:100200.

Zhao, J.D. 1998. *Flora Fungorum Sinicorum 3, Polyporaceae*. Science Press, Beijing, China.

4 Diversity and Ecology of Macrofungi in Sri Lanka

Buddhika P. Dharmasena, Kasun M. Thambugala, Bhagya M. Premarathne, Isuru Sakbo Uyangoda, Samantha Chandranath Karunarathna, and A. K. Hasith Priyashantha

4.1 INTRODUCTION

Sri Lanka is known as the 'Emerald Isle' due to its lush greenery and the rich biodiversity owing to its unique topography, climatic conditions, and strong coastal influences. The island's biodiversity is spread over several levels, including its ecosystem, species, and genetic diversity. Its ecosystems include marine and coastal ecosystems, natural forest ecosystems, semi-natural and natural grasslands, and inland wetlands. Along with the Western Ghats, Sri Lanka is one of the 34 globally identified hotspots facing sustainability challenges and under immense threat, as it possesses extraordinary levels of endemic species, which are unique and found nowhere else in the world. In addition to these natural ecosystems, Sri Lanka also has artificial ecosystems such as agroecosystems and plantations, which add to its biodiversity. Furthermore, Sri Lanka is renowned for having a diverse range of species, both endemic to the island and found abundantly throughout numerous taxonomic groups, showcasing the genetic variation within Sri Lankan species populations (Gunatilleke et al. 2017; Sarathchandra et al. 2021).

Along with all other groups of living organisms, fungal diversity in Sri Lanka is very high, with an estimated number of more than 33,000 species arising from a unique combination of geographical, climatic, and ecological factors. As an island separated from the Indian subcontinent for millions of years, its isolation has created ideal conditions for speciation and endemism, not just in its famous flora and fauna but equally in its fungal populations (Galappaththi et al. 2023). The country's dramatic climatic variation plays a pivotal role. The southwestern wet zone, with its year-round rainfall and humidity exceeding 80%, creates perfect conditions for moisture-loving species like wood-decaying fungi, species forming intricate mycorrhizal networks, and diverse lichen communities. In contrast, the northern and eastern dry zones, with their prolonged droughts and seasonal rainfall patterns, are selected for resilient fungal species capable of surviving extreme conditions, including specialized soil fungi and those forming symbiotic relationships with drought-adapted plants. This ecological diversity is further amplified by Sri Lanka's elevational range, from coastal mangroves to 2,500-meter-high montane forests. Each elevation supports distinct fungal communities: lowland rainforests such as Sinharaja host complex fungal associations with dipterocarp trees, while higher elevation ecosystems such as the Horton Plains nurture cold-tolerant fungi rarely found in tropical regions. The wide diversity of habitats provided by the tropical evergreen forests of the country contributes to much of the fungal diversity, and the highest number of species have been recorded in these forests (Figure 4.1). The availability of constant moisture and numerous different substrates to be colonized, including living plants, plant debris, decaying logs, and ground litter, in these forests, is one reason for this. Although a large number of species of fungi inhabit Sri Lanka, only about 2,000 species have been taxonomically recorded, while some additional species have been identified only recently. Among the studied fungi, macrofungi (Figure 4.2)

DOI: 10.1201/9781003647140-4

are the most widely studied group. They are primarily members of the phyla Ascomycota and Basidiomycota. Of the species of fungi recorded from Sri Lanka, more than 513 species belong to the subphylum Agaricomycotina (Pegler 1986), the group to which most macrofungi belong. According to Karunarathna et al. (2023), 725 species of basidiomycetes belonging to 138 genera have been reported, with nearly half of these from the genera *Agaricus*, *Marasmius*, *Polyporus*, and *Uredo*. Macrofungi of the classes Agaricomycetes, Dacrymycetes, and Tremellomycetes (Basidiomycota) have been extensively recorded in Sri Lanka (Karunarathna 2012; Konara et al. 2022; Dasanthi et al. 2024). Also recorded are macrofungi of the subphyla Pezizomycotina and Taphrinomycotina (Ascomycota) (Coomaraswamy and Kumarasingham 1988; CAB International UK 2008; Ediriweera et al. 2014a; Adikaram et al. 2020).

The recorded macrofungi in Sri Lankan ecosystems occupy a wide range of ecological niches, portraying the significance of their life cycles in the ecosystem balance of Sri Lanka. Most of the macrofungi are saprobic, meaning they nourish themselves by decomposing dead organic matter, which includes fallen leaves, wood, and other types of plant debris. They are fundamentally involved in the decomposition of complex organic compounds into lesser forms, thus enabling nutrient recycling in the soil, a process that not only aids in breaking down organic debris but also facilitates the release of locked nutrients essential to plant growth, such as nitrogen, phosphorus, carbon, and other vital elements (Fukasawa et al. 2009).

Combined with saprobes, numerous filmmakers with parasitic lifestyles infect living hosts along with other organisms and, at times, initiate diseases. Athough parasitic fungi can be detrimental to the host organism, they play an integral part in population control and tend to promote greater biodiversity because no single species can hold dominance (Lisiewska 1992; Leake 2005). Together with other macrofungi, these fungi form a combination with plant roots and mycorrhizal associations. In stressed or nutrient-poor environments, these mutually beneficial fungi assist in improving the health and resiliency of host plants by increasing their water and nutrient absorption. The extensive hyphal networks formed by mycorrhizal fungi aid in binding particulate soil, enabling greater soil stabilization and maintenance, and thus preventing erosion. Furthermore, certain species of macrofungi assist in natural pest control by producing bioactive substances that suppress phytopathogens or repel plant-feeding insects. Their ability to transform harmful pollutants through bioremediation underscores their ecological relevance, as it they metabolize toxic materials, clean contaminated soils and waters (Bonfante and Anca 2009; Willis et al. 2013; Madhushan et al. 2023). In general, these fungi have both direct and indirect impacts on Sri Lanka's environment, with the most

FIGURE 4.1 Tropical forest ecosystems in Sri Lanka.

Source: Photo by Thivanka Peiris

FIGURE 4.2 Some of the species of macrofungi that occur in Sri Lanka. (1) *Dacrymyces spathularia*, (2) A myxomycete, (3) *Pterula* sp., (4) *Geastrum* sp., (5) *Xylaria* sp., (6) *Galiella* sp., (7) *Cyathus* sp., (8) *Earliella* sp., (9) *Hexagonia* sp., (10) *Rigidoporus* sp., (11) *Trametes coccinea*, (12) *Microporus xanthopus*, (13) *Ganoderma* sp., (14) *Earliella* sp., (15) *Sarcoscypha* sp., (16) *Auricularia* sp., (17) *Phallus* sp., (18) *Psathyrella* sp., (19) *Crinipellis* sp., (20) *Mycena* sp., (21) *Coprinellus dissaminatus*, (22) *Marasmiellus* sp., (23) *Marasmius* sp., (24) *Schizophyllum* sp., (25) *Pleurotus* sp., (26) *Russula* sp., (27) Asexual morph of *Xylaria*, (28) *Calvatia* sp., (29) *Termitomyces microcarpus*, (30) *Volvariella* sp.

important being decomposition, nutrient recycling, symbiotic associations, plant growth promotion, maintaining soil stability, pest control, and bioremediation (Weerasinghe et al. 2021; Subasinghe and Hettiarachchige 2022; Niego et al. 2023; Sridhar and Deshmukh 2024).

Some species of macrofungi are distributed throughout the island, but others are restricted to certain eco-climatic zones. Although macrofungal diversity is high throughout the country, knowledge of their traditional uses and possible applications is scarce, while scientific studies of fungal phylogeny, ecology, physiology, and biochemistry have rarely been carried out. While many species of macrofungal, mainly agaricoid mushrooms occurring in Sri Lanka are edible (Coomaraswamy 1979b), only a handful are consumed by locals (Ranawake 2021). Species of *Termitomyces* such as *T. microcarpus*, *T. eurrhizus*, *T. heimii*, and *T. cartilaginous* are the most preferred. Some other mushrooms consumed are *Auricularia* spp., *Tremella* spp., *Pleurotus* spp., *Volvariella* spp., *Schizophyllum commune*, *Coprinus* spp., *Pleurotus gigantius*, *Macrocybe gigantea*, *Lentinus squarrosulus*, *Flammulaster fulvoalbus*, and *Macrolepiota* spp. The medicinal use of the species of macrofungi in Sri Lanka remains very limited.

Sri Lanka is considered one of the countries in Asia where the earliest mycological studies in Asia were carried out, and these date back to the nineteenth century (Hyde et al. 2020). The earliest recorded macrofungus in Sri Lanka was *Lentinus tuberregium* from an illustration labelled *Peziza Ceylonche* by Houttuyn (1783). During the period of 1844–1849, Miles Joseph Berkeley and Christopher Edmund Broome described various macrofungi collected by George Gardner and H. K. Thwaites. In the late nineteenth century, William De Alwis Seneviratne (1842–1916) made watercolor paintings of macrofungi collected in Sri Lanka. Thomas Petch (1870–1948) was appointed mycologist to the government of Ceylon from 1905, and until 1925, he collected and described many species of macrofungi occurring in Sri Lanka (Pegler 1986). Later, Coomaraswamy and Kumarasingham (1988) mentioned macrofungi in the orders Tuberales, Pezizales, Sphaeriales, Tulasnellales, Agaricales, Aphyllophorales, Lycoperdales, Phallales, Hymenogastrales, Nidulariales, and Sclerodermatales found in Sri Lanka.

4.2 DIFFERENT GROUPS OF MACROFUNGI RECORDED IN SRI LANKA AND THEIR IMPORTANCE

Most terrestrial macrofungi function as saprobes or mycorrhizal symbionts, but some are parasites of various plants, animals, and other fungi. The Agaricales are among the most well-known fungal groups in scientific research. Taxa in just this group have been documented as saprotrophs, symbionts, and parasites. These fungi inhabit a wide array of ecosystems, ranging from arctic regions to the tropics (A.N. Ediriweera et al. 2023). Mycorrhizal macrofungi have symbiotic relationships with plant roots. Notably, many symbiotic types, such as most ectomycorrhizae, also depend on their host plants for dispersal and reproduction, as they cannot reproduce independently (Horton 2017; Hoeksema et al. 2018). Some fungi act as parasites, infecting various plant tissues and frequently leading to plant diseases and sometimes the death of the infected plant. Certain species seriously impact forest and shade trees, causing root rot and heartwood decay. Examples include *Armillaria mellea* (Vahl) P. Kumm., *Phellinus* sp., *Fomes* sp., *Rossellinia* sp., and *Corticium* sp. (Coomaraswamy and Kumarasingham 1988).

Sri Lanka is home to a remarkable variety of macrofungi encompassing several important taxonomic groups that greatly enhance the ecosystems of the country. Macrofungi, primarily represented by mushrooms and other spore-bearing fungi, are crucial for nutrient cycling, decomposition, and symbiotic relationships with plants. Their diverse habitats closely align with local species of trees and substrates, underscoring their ecological significance in preserving regional biodiversity. The two predominant groups of macrofungi identified in Sri Lanka are the Ascomycota and Basidiomycota, with the former the larger group but the latter representing the majority of described taxa (Coomaraswamy and Kumarasingham 1988; Ediriweera et al. 2014a). According to Ediriweera et al. (2014a), saprobic genera such as *Schizophyllum commune* Fr. and *Auricularia* sp. are acknowledged for their culinary benefits, while *Daldinia concentrica* (Bolton) Ces. and De Not. and several species in the genus *Ganoderma* P. Karst. are well known for their medicinal value (Coomaraswamy and Kumarasingham 1988; Ediriweera et al. 2014a).

In addition, mushrooms have been valued for their nutritional and culinary value and their historical use as tonics and medicinal remedies. In a contemporary context, they are increasingly perceived as functional foods, providing health benefits beyond their vital nutritional value. Numerous reports in the literature have reported the medicinal and therapeutic attributes of various traditional macrofungi such as *Ganoderma*, shiitake, and *Cordyceps*, making this knowledge readily available to the public (Cheung 2008). Specific macrofungi indicate environmental conditions, rendering them vital for assessing the effects of human activities, such as pollution and deforestation, on forest ecosystems (Ediriweera et al. 2022).

Species of *Ganoderma* are recognized for causing root and butt rot diseases in various woody plants by breaking down lignin, cellulose, and other related polysaccharides to meet their nutritional needs. However, the cultivation and industrial applications of these fungi are significantly impeded by the threat of fungicolous fungi. Among these, species of *Trichoderma* are particularly prominent, as they are responsible for what is known as the green mold disease, recognized as the most severe disease impacting species of *Ganoderma*. This disease poses a considerable barrier to the growth and sustainability of the *Ganoderma*-based industry (Konara et al. 2024).

4.3 SAPROBIC FUNGI

Most fungi are saprobic organisms that serve as decomposers, utilizing dead wood and woody debris along with colonizing leaf litter, twigs, trapped debris, and lianas. Hence, they serve many important ecological roles in an ecosystem. Wild edible saprotrophic fungi contribute minimally to local sales, but some species now serve as the cornerstone of the global cultivated mushroom industry, which is valued at around $23 billion annually (Boa 2004). Ediriweera et al. (2014a) identified saprobic species from various orders, including the Agaricales, Auriculariales, Boletales, Dacrymycetales, Hymenochaetales, Hypocreales, Pezizales, Polyporales, Poriales, Tremellales, and Xylariales. These species are associated with decaying wood, leaf litter, dead branches, decaying logs, and soil. Species such as *Trametes versicolor* (L.) Lloyd and *Schizophyllum commune* Fr. have been identified as particularly prevalent, highlighting their pivotal role as essential decomposers within the forest ecosystem.

Numerous studies have investigated the ecological preferences of these saprobic macrofungi and their optimal growth conditions. For example, a notable comparative analysis of substrates indicated that specific local materials such as coconut leaves and coir dust yielded optimal results for the growth of *Lentinus squarrosulus* Mont., *Schizophyllum commune*, and *Trametes versicolor*. The latter is a species recognized for both its edibility and medicinal properties (S.S. Ediriweera et al. 2015). Ediriweera et al. (2023b) discovered a novel taxon, *Candolleomyces ruhunensis* A.N. Ediriweera, Voto, Karun. and Kumla that belongs to the genus *Candolleomyces*, along with the first record of *Crepidotus striatus* T. Bau and Y.P. Ge from the wet zone of Sri Lanka. Both of these were identified as saprotrophic fungi (Ediriweera et al. 2023b).

4.4 ANIMAL PARASITES

Fungal infections in insects are common, frequently resulting in significant epizootics that severely diminish insect populations. Nearly all insect orders are vulnerable to these infections. Fungi invade insects by penetrating the cuticle of the host and represent the primary pathogens affecting sucking insects, as their hosts are unable to ingest other pathogens that enter through the digestive tract. In addition, fungi are crucial in regulating beetle populations, since viral and bacterial infections are relatively scarce in this order of insects. Fungal insect infections often lead to the demise of the host in situ; however, this does not always happen. Certain parasitic fungi, such as the host-specific pathogen *Ophiocordyceps unilateralis sensu lato*, possess the ability to influence the brain functions of insects and modify their behavior, steering it toward locations that are optimal for spore dispersal (Hajek and St. Leger 1994; de Bekker et al. 2014).

Species of *Cordyceps* are macrofungi recognized for their unique parasitic behavior affecting insect larvae, pupae, and adults. The earliest recorded reference to this taxon dates back to 620 AD in China, where it was described as an organism undergoing a yearly transformation from animal to plant during the summer and returning from plant to animal in the winter (Rathi et al. 2021). *Cordyceps sinensis* (Berk) Sacc., commonly referred to as the caterpillar fungus, is recognized as the most extensively studied species within the genus *Cordyceps* (Yan et al. 2014). In Chinese and Tibetan medicine, *C. sinensis* is esteemed as a potent tonic frequently recommended for lung, liver, and kidney ailments. Furthermore, it is noted for its potential antitumor, antiviral, anticancer effects, immunomodulatory benefits, antioxidant properties, and cholesterol-lowering abilities (Chen et al. 2013; Tuli et al. 2014; Sen et al. 2023).

Sri Lanka is home to a notable variety of animal-parasitic fungi, particularly those species within the genus *Torrubiella* Boud. that target insects and other arthropods. Species such as *T. tomentosa* Pat., *T. rubra* Pat. and Lagerh., *T. luteorostrata* Zimm., *T. brunnea* Keissl., *T. lecanii* J.R. Johnst., *T. tenuis* Petch, *T. sublintea* Petch, and *T. barda* Petch are known to be parasites of coccids, significantly impacting their populations. Several species of *Cordyceps* have also been documented in Sri Lanka, including *C. barnesi* Thwaites, which affects cockchafer larvae, while *C. dipterigena* Berk. and Broome targets flies from the genus *Mydaea* Robineau-Desvoidy. *Cordyceps unilateralis* var. *javanica* Höhn. It is a fungal parasite on ants. *Ophiocordyceps blattae* (Petch) Petch, which was discovered in Sri Lanka, infects *Blattella germanica* L. but remains inadequately studied, as its taxonomy is reliant solely on initial descriptions and illustrations provided by Petch, with no accompanying molecular data available (Petch 1923).

Members of the genus *Isaria* Pers. (e.g., *I. sinclairi* [Berk.] Lloyd), target cicada pupae and one species of *Isaria* affects *Orthoptera* Latreille, cocoons of *Thosea recta* Hampson, and larvae of *Homona coffearia* Nietner represent another important group. Species of *Hirsutella* Pat., such as *H. saussurei* (Cooke ex Sacc.) Speare, which impacts hornets, *H. citriformis* Speare, which affects pentatomids, and *H. floccosa* Speare, which impacts leafhoppers, are commonly found in tropical regions. *Spicaria prasina* (Maubl.) Sawada was initially identified in Japan on the larvae of *Phlyctaenia* (*Pionea*) *forficalis* Li. However, a researcher also documented its presence on the larvae of *Spodoptera mauritia* Boisduval in Sri Lanka (Petch 1923; Kepler et al. 2017).

4.5 PLANT PARASITES

Fungal plant parasites have a large scope that impacts plant health, populations, and even biodiversity if they function in an ecosystem. Some fungi act as mutualists, but most of them are pathogenic and attack crops, trees used in forestry, and even ornamental plants. Such infections have profound economic impacts globally and endanger food safety along with ecological balance. These pathogens adapt to a large diversity of environmental conditions, hence persistently posing danger in both temperate and tropical regions (Doehlemann et al. 2017; Adikaram and Yakandawala 2020). Plant parasitic macrofungi are fungi that have prominently visible parts like mushrooms or brackets and allow them to gain nutrients from healthy plants, usually resulting in disease and decay. This group includes well-known fungi like *Armillaria* (honey fungus) and *Ganoderma* species, which inhabit plants, causing significant damage to the plant or structures on the plant, which can lead to the weakening of the plant and sometimes death. They gain support through root wounds and spore routes, via mycelial growth. In addition, these forms of fungi can remain long after killing their host. They are crucial in nutrient recycling as well as forming habitats but are highly phytopathogenic in natural habitats, for agriculture, forestry, and city trees (Loyd et al. 2018; Devkota and Hammerschmidt 2020).

Since many aspects of the Sri Lankan economy depend upon agriculture, phytopathogenic fungi are considered immensely important. Although most of these fungi are microfungi, several can be considered macrofungi. According to Coomaraswamy (1979a), 235 genera of phytopathogenic fungi have been recorded in Sri Lanka. Adikaram and Yakandawala (2020) have recorded 404 species of plant pathogenic fungi and members of the Oomycota, belonging to 110 genera. Among these phytopathogens, macrofungi are observed in several orders. *Tremella ferruginea* Schumach., *T. fuciformis* Berk., *T. guttiformis* (Berk. and Broome) Bandoni, *T. indurata* Berk. and Broome, *T.*

scutelliformis Berk. and Broome, and *T. versicolor* Berk. and Broome have been recorded from the order *Tremellales* and are considered phytopathogenic (Coomaraswamy and Kumarasingham 1988; Karunarathna et al. 2023). Species of *Tremella* are also considered mycoparasitic and are sustained on different species of ascomycete and basidiomycete fungi, including lichens (Gams et al. 2004). Four species of *Poria* (Polyporales) were originally described from Sri Lanka. These are *Poria albobrunnea* (Romell) Karasiński and Niemelä, *P. endoxantha* Petch, *P. hypobrunnea* Petch, and *P. rubrochorda* Petch (Karunarathna et al. 2023). These species cause red root disease in many species of plants, and their basidiocarps are entirely resupinate with extended patches (Coomaraswamy 1979a). *Fomes* is an important plant pathogenic fungal genus of the order Polyporales. Many species of *Fomes* are wood-colonizing saprobic fungi and are also tree pathogens that cause wood degradation as brown rots (Gründemann et al. 2020). According to Karunarathna et al. (2023), *Fomes imitator* Petch can be found in Sri Lanka. *Fomitiporella caryophylli* (Racib.) T. Wagner and M. Fisch. (formerly *Fomes caryophylli*) has been recorded as parasitic on clove, *Navisporus floccosus* (Bres.) Ryvarden (formerly *Fomes floccosus*) on *Delonix regia*, and *Bombax ceiba* and *Pyrrhoderma lamaoense* (Murrill) L.W. Zhou and Y.C. Dai (formerly *Fomes lamaoensis*) on rubber and tea (Coomaraswamy 1979a). *Phellinus noxius* (Corner) G. Cunn. occurs on rubber plants in Sri Lanka as a pathogen and causes devastating effects on the plants (Silva et al. 2016; M K R Silva et al. 2017). Several species of the genus *Irpex* can be found in Sri Lanka, including *Irpex colliculosu* Berk. and Broome, *I. destruens* Petch, *I. lacteus* (Fr.) Fr., and *I. vellereus* Berk. and Broome (Karunarathna et al. 2023). These species grow as a crust on the dead hardwood. Some of these species are pathogenic on various plant species, where they cause white root rot (Novotný et al. 2009; Cisneros-Zambrano et al. 2024). *Pseudolagarobasidium subvinosum* (Berk. and Broome) Sheng H. Wu (formerly *Irpex subvinosus*) has been recorded as pathogenic on plants such as *Acacia decurrens*, *Camellia sinensis*, and *Tephrosia candida* (Coomaraswamy 1979a). Other than these, various species of *Ganoderma* have been recorded in Sri Lanka, and some of these show plant pathogenicity and cause white rot in rotting stumps, roots, and living trunks by decomposing lignin, cellulose, and related polysaccharides for nutrient mobilization (Konara et al. 2022; Konara et al. 2024).

Ganoderma lucidum has been identified as causing general wood rot in *Camellia sinensis* (tea) (Balasuriya 2008), *Ganoderma* root and butt rot in *Cassia javanica* sub sp. *nodosa*, *Cassia fistula*, and *Delonix regia* (Fernando 2008). *Ganoderma orbiforme* (Fr.) Ryvarden also has been recorded in Sri Lanka and causes basal stem rot and root and bole rot of *Cocos nucifera* (coconut) (Peries 1974; Wijesekara et al. 1996). *Rigidoporus microporus* (Sw.) Overeem is also a macrofungus in the order Polyporales, and it is pathogenic primarily on *Hevea brasiliensis* (rubber), causing a heavy economic impact. This species has also been identified as infecting *Artocarpus nobilis* (Jayasuriya and Thennakoon 2007; Madushani et al. 2014; Silva et al. 2017). *Perenniporia mesoleuca* (Petch) Ryvarden is also considered phytopathogenic and has been identified in Sri Lanka (Decock and Ryvarden 1999). *Pyrrhoderma noxium* (Corner) L.W. Zhou and Y.C. Dai, which is pathogenic on various plants (Bragard et al. 2024), has been recorded in Sri Lanka as a pathogen on *Cinnamomum verum* (cinnamon) (Jayasinghe et al. 2017), and *Hevea brasiliensis* (M K R Silva et al. 2017) is known to cause brown root disease. *Pyrrhoderma nigrum* Meng Zhou, Yuan Yuan and Vlasák, has been recorded on *Artocarpus heterophyllus* in Sri Lanka and can be considered a phytopathogenic macrofungus (Thambugala et al. 2023). The species *Cyphella grisea* Petch, *C. parasitica* Berk. and Broome, *Corticium hypophyllum* Petch, *Ceratobasidium cornigerum* (Bourdot) D.P. Rogers (formerly *Corticium invisum*), *Rhizoctonia cornigera* (Bourdot) Zmitr. (formerly *Corticium pervagum*), *Necator salmonicolor* (Berk. and Broome) K.H. Larss., Redhead and T.W. May (formerly *Corticium salmonicolor*), and *Agroathelia rolfsii* (Sacc.) Redhead and Mullineux (formerly *Corticium rolfsii*) are other plant parasitic and pathogenic macrofungi recorded in Sri Lanka (Coomaraswamy 1979a).

Most macrofungal members of the order Agaricales are saprobic, and very few species are plant parasites. Most of the species of *Armillaria* are considered plant parasites, causing white rot in plants (Worrall 2004). In Sri Lanka, *Armillaria fuscipes* Petch has been recorded infecting tea plants, causing Armillaria root disease (Pegler 1986). In addition, *A. mellea*, occurring on *Acacia decurrens*, has been documented (Coomaraswamy 1979a). Several species of the genus *Marasmius* found in Sri Lanka are considered plant parasites (Coomaraswamy 1979a), while *Marasmius crinis-equi* F. Muell. ex Kalchbr. causes horsehair blight in tea plants in Sri Lanka (Arulpragasam 1989; Dassanayake et al. 2010). *Cyphella pulchra* Berk. and Broome (formerly *Marasmius pulcher*), *Marasmius rigidichorda* Petch, and *Collybiopsis stenophylla* (Mont.) R.H. Petersen (formerly *Marasmius stenophyllus*) has also been recorded in Sri Lanka as parasitic on various plants (Coomaraswamy 1979a).

4.6 ECTOMYCORRHIZAL MACROFUNGI

Ectomycorrhizal macrofungi are plant root symbionts of perennial woody plants in the phyla Basidiomycota and Ascomycota that form macroscopic fruiting bodies (Tedersoo et al. 2010; Lehmann et al. 2017). This group comprises more than 5,000 species belonging to more than 250 genera (Niego et al. 2023). Most ectomycorrhizal macrofungi belong to the Agaricales, Boletales, Pezizales, Helotiales, Cantharellales, Russulales, Sebacinales, Thelephorales, and Trechisporales (Sharma 2017; Santolamazza-Carbone et al. 2021).

Truffles belonging to the genus *Tuber* are among the fungi with the greatest economic importance. These ectomycorrhizal macrofungi produce an underground ascocarp surrounded by a distinctive sheath, which can vary in form and texture, and is irregularly shaped, globose, and fleshy in some species while woody in others. In Sri Lanka, a single species (*Tuber zeylanicum* Berk. and Broome) of the genus has been recorded (Park 1932; Coomaraswamy and Kumarasingham 1988). *Amanita*, *Lactarius,* and *Russula* are some of the main clades of ectomycorrhizal basidiomycetes (Watkinson 2016), and all these genera have been recorded in Sri Lanka (Karunarathna et al. 2023). Most of these macrofungi (mushrooms) are edible, and some are considered culinary delicacies (Boa 2004; The Russulales News Team 2010; Voces et al. 2012). *Amanita hemibapha* (Berk. and Broome) Sacc. and *A. vaginata* (Bull.) Lam. are two of the ectomycorrhizal species of *Amanita* occurring in Sri Lanka (Coomaraswamy 1979b). Although most species of *Boletus* are ectomycorrhizal, none of the species recorded in Sri Lanka are ectomycorrhizal. *Suillus luteus* (L.) Roussel (formerly *Boletus luteus*) occurs in Sri Lanka and is recognized as an ectomycorrhizal edible mushroom that is associated with the plants in the family Pinaceae (Coomaraswamy 1979b; Murata et al. 2015). Several species of *Lactarius* have been recorded in Sri Lanka. These are *Lactarius echinellus* Verbeken and Stubbe, *L. echinus* Stubbe and Verbeken, *L. pomiolens* Verbeken and Stubbe, *L. shoreae* Stubbe and Verbeken, and *L. saturnisporus* Verbeken and Stubbe (Karunarathna et al. 2023). These species form ectomycorrhizal relationships with both deciduous and coniferous trees (Lee et al. 2019). It is believed that the diversity of species in this genus in Sri Lanka is higher than what has been estimated (Verbeken et al. 2014). *Russula* is another genus that includes many ectomycorrhizal species (Horton and Bruns 2001; Wang et al. 2015). Although more than 750 species of *Russula* have been reported to have a cosmopolitan distribution (Kirk et al. 2008), only three species have been recorded in Sri Lanka. These are *Russula fuscogrisea* Petch, *R. periglypta* Berk. and Broome, and *R. purpureonigra* Petch, all of which are edible and only rarely occur (Coomaraswamy 1979b).

4.7 THE PROGRESSION OF STUDIES OF MACROFUNGI AND THE NOVEL SPECIES RECORDED

As already mentioned, Sri Lanka is a tropical country with a high diversity of fungi. It is worth noting that rather than fungi-specific features (spore production rate and dissemination), local species composition and spatial distribution are generally associated with local environmental factors such as topography, soil, and level of light (Chen et al. 2018). With a favorable tropical climate, and consideration of these factors, the diversity of fungi in Sri Lanka should be greater than what is currently known; however, relevant studies are lagging.

Most of the early mycological studies, mainly directed toward macrofungi in Sri Lanka, were conducted by Miles Joseph Berkeley, Christopher Edmund Broome, and Thomas Petch in the late nineteenth and early twentieth centuries. In the late twentieth century, David Norman Pegler validated most of the fungal taxa previously reported. Currently, the mycological studies in Sri Lanka are increasing, and novel species of fungi are being discovered. *Ganoderma* can be considered one of the most intensively studied taxa in recent studies. For example, *G. philippi* (Bres. and Henn. ex Sacc.) Bres., *G. lucidum*, and *G. applanatum* (Pers.) Pat. were discovered to be associated with a variety of hosts. At the same time, *G. applanatum* and *G. lucidum* were also identified as the most frequently occurring Sri Lankan polypores found throughout the island (Konara et al. 2022). Tibpromma et al. (2017) described *Boletellus emodensis* (Berk.) Singer, *Favolaschia auriscalpium* (Mont.) Henn., *F. manipularis* (Berk.) Teng, *Lentinus sajor-caju* (Fr.) Fr., *Leucocoprinus cretaceus* (Bull.) Locq., *Russula* cf. *virescens* and *Tremella fuciformis* Berk. for the first time in Sri Lanka. *Calvatia candida* (Rostk.) Hollós and *Schizophyllum radiatum* Fr. have been newly reported by Dasanthi et al. (2024), and both have potential edibility. The rare *Helvella crispa* (Scop.) Fr. has been recorded for the first time in Sri Lanka (Adikaram et al. 2020). Ediriweera et al. (2014b) conducted a study of macrofungi in a forest surrounding Sigiriya Rock in the Central Province of Sri Lanka. They collected 49 species that were identified based on macromorphological and microscopic characteristics. Of the total collection, 39% were from the order Polyporales, while 31% of the others were representatives of the Agaricales (*Anthracophyllum* sp., *Marasmius* sp., *Mycena* sp., *Xerula* sp., *Psathyrella* sp., and *Schizophyllum commune*). These researchers also identified (making up 8% of the total collection) the jelly fungi *Tremella* sp., *Calocera* sp., and *Auricularia* sp. Aside from these primary categories, their collection encompassed several different types of macrofungi, including *Ramaria* sp., also known as a coral fungi. (A. Ediriweera et al. 2023) introduced a new species of *Termitomyces* in Sri Lanka, *T. srilankensis*. Meanwhile, another new species, *Candolleomyces ruhunensis* (Psathyrellaceae), was described, along with *Crepidotus striatus* T. Bau and Y.P. Ge (Crepidotaceae), which is newly recorded in Sri Lanka (Ediriweera et al. 2023b). Furthermore, Ediriweera et al. (2014c) identified two species of wood-inhabiting fungi, *Coriolopsis byrsina* (Mont.) Ryvarden and *Flavodon flavus* (Klotzsch) Ryvarden, from dry zone forest reserves (Minneriya and Kaudulla, North Central Province, Sri Lanka). The study was a benchmark in Sri Lankan macrofungal research since morphological and molecular-based methods have been used to identify the species of fungi collected. Since then, many studies have focused on molecular-level phylogeny for a more reliable understanding of species concepts and taxonomic relationships. *Ganoderma angustisporum* J.H. Xing, B.K. Cui and Y.C. Dai, *G. ellipsoideum* Hapuar., T.C. Wen and K.D. Hyde, and *G. orbiforme* (Fr.) Ryvarden (Konara et al. 2024), *Fuscoporia senex* (Nees and Mont.) Ghobad-Nejhad, *Rhodofomitopsis feei* (Fr.) B.K. Cui, M.L. Han and Y.C. Dai, *Phanerodontia chrysosporium* (Burds.) Hjortstam and Ryvarden, *Earliella scabrosa* (Pers.) Gilb. and Ryvarden, *Panus conchatus* (Bull.) Fr., *Lentinus similis* Berk. and Broome, *Cubamyces cubensis* (Mont.) Murrill, *Trametes elegans* (Spreng.) Fr., and *Trichaptum byssogenum* (Jungh.) Ryvarden (Ediriweera et al. 2021) have also been recorded in Sri Lanka. In one study, Ediriweera et al. (2014a) recorded the presence of *Fulvifomes fastuosus* (Lév.) Bondartseva and S. Herrera in Dambulla, Central Province, Sri Lanka. In their next study, Ediriweera et al. (2015b) found *Panaeolus sphinctrinus* (Fr.) Quél. and *Panaeolus foenisecii* (Pers.) J. Schröt. in Sri Lanka, associated with elephant dung collected from the Sigiriya Wilderness area and Kaudulla National Park. Ediriweera et al. (2015a) further recorded *Schizophyllum commune* and *Auricularia polytricha* (Mont.) Sacc. from Dambulla (recorded from an arboretum) and *Lentinus squarrosulus* from Matara, Southern Province.

Due to Sri Lanka being an island, the endemism of the assemblage of fungi present is remarkable. Most macrofungi originally described in Sri Lanka have not been identified elsewhere. Most Sri Lankan species of macrofungi have not been extensively studied for their endemism, ecology, and phylogeny. As a result, a study gap exists that needs to be reevaluated regarding the macrofungal diversity in the country.

4.8 ETHNOMYCOLOGY AND MISLEADING FOLKLORE

Unlike such countries as China and Thailand, traditional ethnomycological knowledge and the use of macrofungi are not well recognized in Sri Lanka (Wijesundera 1986; Karunarathna 2012). Interestingly, in Sri Lanka, most traditional healers use plant-based materials for treatment rather than fungi-based materials, and this is greatly appreciated even today (Attanayake et al. 2018). However, a handful of studies show some limited practices in which fungi are used. In Sri Lanka, *Macrocybe gigantea* (Massee) Pegler and Lodge, locally known as 'pol hathu' in Sinhalese, is a delicacy primarily found in the wet zone. Locals collect these mushrooms in the season as culinary delicacies. Some locals embed discarded mushroom fragments into decaying organic matter, such as leaf litter, to induce regrowth (Peiris et al. 2024). Hewage (2015) detailed how villagers in 'Kudava' village near the Sinharaja Forest harvest and consume wild mushrooms. During the mushroom season, villagers go into the forest to pick mushrooms, alone or in groups. The collectors focus on several types of mushrooms such as *Flammulaster fulvoalbus* (Berk. and Broome) Pegler, *Pleurotus giganteus* (Berk.) S.C. Karunarathna and K.D. Hyde, *Schizophyllum commune,* and *Termitomyces* spp. They do not try to collect other peculiar, unknown, or lesser-known varieties. In general, when the villagers enter the forest for their other activities and find a mushroom-rich area, they mark the place and then come back to collect from those areas when the mushroom season comes. Hewage (2015) further recognized that the mushroom locally identified as 'behet hathu' was the only type of mushroom used for medicinal purposes. People use this mushroom in the local ayurvedic system to treat broken or fractured bones. The study of Hewage (2015) is also a good example of getting to know about the poor local knowledge of the use of macrofungi in medical systems. Researchers also failed to find very much more details about the traditional knowledge of mushrooms other than the use mentioned above. Furthermore, some medicinal species in Sri Lanka, especially *Phallus* sp., have been rejected by people because of misleading folklore on their poisonous nature (Gamage and Ohga 2018). There are several misleading folklore stories in Sri Lanka such as that people believe mushrooms with rings, bright colors, and not eaten by animals or insects are poisonous. Fungal or mushroom poisoning seems to be rarely reported in the country. A handful of such news articles have become headlines in newspapers, but little public attention has been paid to them. In one study, Fernando and Fernando (1990) reported that based on the local hospital reports, there were about 2.5% (out of a total of 4,556 cases) instances of plant and mushroom poisoning. Unfortunately, there have been no more recent studies in this regard on the island, to the best of our knowledge.

4.9 THREATS TO MACROFUNGAL DIVERSITY IN SRI LANKA AND CONSERVATION

Despite their ecological significance, macrofungi in Sri Lanka are understudied and also face growing threats. Habitat loss and fragmentation, climate change, environmental pollution, the spread of pathogens, and environmental pollution are the main threats to macrofungi in the country (Karunarathna 2012; Pethiyagoda 2012; Liu et al. 2018; Surasinghe et al. 2019). In Sri Lanka, loss of habitat decreases the forests critical for macrofungi, eliminating their substrates and plant associates. Human activities that cause fragmentation in an area, like creating microclimates with lower humidity, greatly impact moisture levels and slow down the process of fungal growth and which threatens specialized and endemic fungi (Perera et al. 2017; Pullaiah 2023). Climate change disrupts the equilibrium of temperature with moisture required for the growth and fruiting bodies of fungi, which shifts or reduces the population of species. Environmental pollution, such as chemicals, airborne contaminants, and others, harms the soil, compromising the health of soil fungi. The influx of invasive pathogens causes the native fungi to succumb to new diseases, posing a significant decline in their population. All the issues mentioned above worsen the decline in the abundance and diversity of Sri Lankan macrofungi and their ecological roles. Managing these factors is, therefore, crucial (Dahlberg et al. 2010; Wimalasena et al. 2024).

The lack of awareness and limited research intensifies these threats. Furthermore, the exacerbation of wild mushroom harvesting is unsustainable and poses a risk to pre-existing vulnerable species, particularly those with little range or slow growth. The forest cover of Sri Lanka has been reduced from 40% to approximately 30% during the period from 1940 to 2017 (Sri Lanka's forest reference level submission to the UNFCCC 2017). Reduction of forest cover reduces the habitats for macrofungi, including the organic matter for saprobes and symbiotic plants for plant symbiotic macrofungi. Changes in the monsoon rain, annual precipitation, lightning patterns, and temperature fluctuations attributed to global climatic change also have impacted the reproduction, geographical distribution, physiology, and metabolism of macrofungi (Bidartondo et al. 2018), affecting their production of fruiting bodies. Plant-fungal interactions have also been altered due to climate change, having devastating effects on fungal ecology and diversity (Waheed et al. 2023). Different conservation strategies can be implemented to enhance macrofungal diversity, including habitat protection and restoration, research and documentation, awareness programs, and integration of the appropriate policies.

4.10 CURRENT KNOWLEDGE AND RESEARCH GAPS ON SRI LANKAN MACROFUNGI

Based on reported literature, some local people have some ethnobotanical and ethnomycological knowledge regarding the value of macrofungi. For instance, some rural folk have regarded some of these species as food, medicines, and for other cultural purposes, but there is no systematic documentation of such information. Thus, the regionally available macrofungi are overlooked for their potential usefulness in the fields of nutrition, pharmaceuticals, and biotechnological research (Azeem et al. 2020; Premarathne et al. 2025). Considering the current knowledge, the research performed on Sri Lankan macrofungi is limited, and most of the studies emphasize certain specifics like fungal phylogeny, ecology, physiology, and biochemistry. Work on the phylogeny of fungi is the systematic study of fungi which should concern the evolutionary relationships and classification of species, as it is crucial in the effort of describing new species that have attributes or uses unique to them. Ecological work is useful in establishing what role the macrofungi play in the ecosystem in relation to other organisms, their contribution to nutrient cycling, and their response to changes in the environment. Physiological and biochemical work is devoted to describing the functions of living organisms, in this case, fungi, how they withstand the stress of their environment, and essential active principles that can be used as raw materials for pharmaceuticals and industrial purposes (Tóth and Barta 2010). Agaricoid mushrooms, which are popular with the public because of their classic mushroom shapes, were pioneered in Sri Lanka. Many of these species are considered edible and foraged for food. Edible mushrooms are especially harvested from the forests in rainy seasons when there is a bounty of mushrooms. For most of the rural population, these mushrooms serve as an additional source of nutrition because they are rich in protein, vitamins, and minerals. While it is evident that macrofungi have a large potential to be cultivated in Sri Lanka, the country has yet to realize its economic value. Sustainable methods that can be embraced for these resources cause minimal damage to the wild macrofungi population while creating an opportunity to cultivate mushrooms and economically benefit the people in rural areas (Karunarathna 2012; Gamage and Ohga 2018).

From the ecological perspective, macrofungi extend their significance beyond utility as a food. Some fungi make associations with the roots of plants, forming mycorrhizal symbiotic relationships, and they are crucial to Sri Lanka's forest ecosystems, particularly concerning the lack of nutrients in the soil. For sustainable conservation and restoration of forests, the taxonomy and ecology of mycorrhizal fungi must be understood. In addition to aiding in the decomposition of leaves and wood, macrofungi de ale also transform complex organic substances into simpler products that can be utilized by lower life forms and plants. Decomposition of these nutrients increases soil fertility and adds to the carbon cycle, which makes ecosystems better and more productive (Arnolds 1992; Zotti et al. 2013). Incorporation of education about fungi has to occur simultaneously with the conservation of macrofungi as part of biodiversity plans. Such boundaries to habitats rich in fungal diversity provide additional benefits for restoring health to ecosystems and other life forms.

There is much scope in pursuing the study and conservation of macrofungi in Sri Lanka using various approaches. The preparation of systematic documentation and databases that include species distribution, their ecological requirements, and current conservation status is necessary. The participation of local communities and policymakers in documenting traditional knowledge integrated with scientific research will complete mycological and ecological work through collaborative efforts. Furthermore, Sri Lanka could use research on the use of macrofungi in biotechnology, medicine, and agriculture. Some species of fungi produce commercially useful enzymes and antimicrobial compounds with therapeutic potential (Karunarathna et al. 2017; Semwal et al. 2023). Addressing these challenges will increase the economic and social well-being of the country while conserving biological resources.

4.11 FUTURE RESEARCH DIRECTIONS

Due to limited studies, extensive and immediate research is needed to effectively determine species diversity, phylogeny, and distribution of macrofungi in Sri Lanka. There are many tropical evergreen forests with a very high diversity of macrofungi (e.g., the Sinharaja Forest Reserve), and surveys could be conducted to better understand the diversity within those ecologically sensitive forest reserves. The edible potential of most macrofungi has not been determined, and research should focus on this aspect of these organisms. This will lead to identifying potential species for domestication and, eventually, industrial cultivation. Wild-picked edible species such as *Flammulaster fulvoalbus* can be studied and cultivated in artificial conditions. With the high percentage of endemic and native species, many macrofungi from Sri Lanka are still undiscovered. They may be good sources of secondary metabolites, and many have unknown biochemical and pharmacological properties. Studies can be conducted to identify their bioactive properties and nutraceutical potential. However, these advantages cannot be attained due to the absence of detailed chemical profiling and pharmacological work.

The ecological scenarios of most macrofungi have not been revealed. The cultivation of *Pinus* trees in the barren tea estates in the hill country of Sri Lanka has been successful due to the plant-fungal interactions, and studies on plant-fungal interactions in Sri Lanka can be extended beyond this. Habitat-specific studies will further the studies relating to the conservation of these species. The fungi of this particular region contain different species of macrofungi that have either been undocumented or inadequately researched. The endemic species of macrofungi present in Sri Lanka add to the country's biodiversity. The unique biology of these macrofungi adds to Sri Lanka's biodiversity. Investing more in scientific research can lead to the exploration of new macrofungi. Moreover, improvement in cultivation methods will be critical in tapping these therapeutic potentials. Enhanced methods of cultivation and chemical profiling would facilitate the sterile mass production of medicinal and edible fungi with guaranteed consistent quality and availability adequate for nutritional and pharmaceutical needs.

4.12 CONCLUDING REMARKS

The diversity of macrofungi in Sri Lanka exceeds that of most other countries. The ecological diversity of macrofungi beyond the species level is also remarkable. Based on ecological relationships, saprobic, plant-parasitic, animal-parasitic, and mycorrhizal macrofungi occur in Sri Lanka. Although a considerable number of species of macrofungi have been described from Sri Lanka, most of those studies are decades old and based solely on morphological characteristics. Currently, fungal taxonomy mainly depends on molecular-level relationships. Consequently, Sri Lankan macrofungi should be revalidated with molecular characterization. The available information on Sri Lankan is dispersed through numerous sources, many of which are currently inaccessible or unavailable. Furthermore, the synonymy of the fungi has made the studies more complicated. In addition to molecular characterization, phylogenomics can be applied to obtain a solid and whole picture of fungal phylogeny. Also, establishing a centralized fungarium in Sri Lanka would address these issues and facilitate mycological studies. Apart from the macrofungi proper, lichens, which represent a fungal association, also exhibit rich diversity in Sri Lanka. Some species of fungi that cause smuts in plants, producing macroscopic masses of fungal mycelia and spores, can be observed in Sri Lanka. However, these two groups are seriously understudied in the country.

REFERENCES

Adikaram, N.K.B. & D.M.D. Yakandawala. 2020. A checklist of plant pathogenic fungi and Oomycota in Sri Lanka. *Ceylon Journal of Science* 49:93.

Adikaram, N.K.B., D.M.D. Yakandawala & L. Jayasinghe. 2020. The first report of *Helvella crispa* (Ascomycota, Pezizales), a rare fungal species in Sri Lanka. *Ceylon Journal of Science* 49:485.

Arnolds, E. 1992. The role of macrofungi in environmental conservation. *Giornale Botanico Italiano* 126:779–795.

Arulpragasam, P.V. 1989. *Studies on the Low Country Stem Canker Disease of Tea in Sri Lanka*. University of Kelaniya, Kelaniya.

Attanayake, A.M.H.S., U.M.G.D. De Silva, J.A.A.S. Jayaweera & D.L. Perera. 2018. Use of Ayurveda and Sri Lankan traditional medicine for healing shaft of humerus fracture following nonunion. *Journal of Ayurveda and Integrative Medicine* 9:217–220.

Azeem, U., K.R. Hakeem & M. Ali. 2020. Ethnomycology. In U. Azeem, K.R. Hakeem & M. Ali (eds.), *Fungi for Human Health*, pp. 23–39. Springer International Publishing.

Balasuriya, A. 2008. Common diseases of tea and their management. In A.K.N. Zoysa (ed.), pp. 173–z209. *Handbook on Tea*. Tea Research Institute of Sri Lanka, Talawakelle.

Bidartondo, M.I., C. Ellis, H. Kauserud, P.G. Kennedy, E. Lilleskov, L. Suz & C. Andrew. 2018. Climate change: Fungal responses and effects. In K.J. Willis (ed.), *State of the World's Fungi 2018*, pp. 62–69. Royal Botanical Gardens, Kew.

Boa, E.R. 2004. *Wild Edible Fungi: A Global Overview of Their Use and Importance to People (Non-Wood Forest Products)*. Food and Agriculture Organization of the United Nations, New York.

Bonfante, P. & I.-A. Anca. 2009. Plants, mycorrhizal fungi, and bacteria: A network of interactions. *Annual Review of Microbiology* 63:363–383.

Bragard, C., P. Baptista, E. Chatzivassiliou, F. Di Serio, P. Gonthier, J.A. Jaques Miret, A.F. Justesen, A. MacLeod, C.S. Magnusson, P. Milonas, J.A. Navas-Cortes, S. Parnell, R. Potting, E. Stefani, H. Thulke, W. Van der Werf, A. Vicent Civera, J. Yuen, L. Zappalà & P.L. Reignault. 2024. Pest categorisation of *Pyrrhoderma noxium*. *EFSA Journal* 22:e8667.

CAB International UK. 2008. *Taphrina deformans* [Distribution map]. *Distribution Maps of Plant Diseases*. CABI International, United Kingdom.

Chen, P.X., S. Wang, S. Nie & M. Marcone. 2013. Properties of *Cordyceps sinensis*: A review. *Journal of Functional Foods* 5:550–569.

Chen, Y., Z. Yuan, S. Bi, X. Wang, Y. Ye & J.-C. Svenning. 2018. Macrofungal species distributions depend on habitat partitioning of topography, light, and vegetation in a temperate mountain forest. *Scientific Reports* 8:13589.

Cheung, P.C.K. 2008. Nutritional value and health benefits of mushrooms. In P.C.K. Cheung (ed.), *Mushrooms as Functional Foods*, pp. 71–109. John Wiley & Sons, Hoboken, New Jersey.

Cisneros-Zambrano, A., J. Mendoza-Churape, H.A. Contreras-Cornejo, Y.A. Raya Montaño, C.R. Martínez-González, T. Raymundo, R. Valenzuela, M. Vargas-Sandoval, J.A. Ruiz-Valencia & Ma.B.N. Lara-Chávez. 2024. First report of *Irpex rosettiformis* causing white root rot in avocado trees in Michoacán, México. *Plant Disease* 108:805.

Coomaraswamy, U. 1979a. *A Hand book to the Fungi Parasitic on the Plants of Sri Lanka*, Vol. 4. National Science Council of Sri Lanka, Colombo.

Coomaraswamy, U. 1979b. *Hand Book to the Agarics of Sri Lanka*, Vol. 5. National Science Council of Sri Lanka, Colombo.

Coomaraswamy, U. & S. Kumarasingham. 1988. *Hand Book to the Macrofungi of Sri Lanka*, Vol. 17. Natural Resources, Energy & Science Authority of Sri Lanka, Colombo.

Dahlberg, A., D.R. Genney & J. Heilmann-Clausen. 2010. Developing a comprehensive strategy for fungal conservation in Europe: Current status and future needs. *Fungal Ecology* 3:50–64.

Dasanthi, N., K.M. Thambugala, S.C. Karunarathna, A. Ediriweera, H. Munasinghe, A.M. Elgorban, H. Harischandra & B.P. Dharmasena. 2024. Morphological and molecular insights into the hidden edible mushroom diversity in Sri Lanka. *Phytotaxa* 645:192–213.

Dassanayake, N., W. Wanigasundara, A. Balasuriya & M. Amaratunge. 2010. A field assessment of the factors affecting horse hair blight (*Marasmius equicrinis*) in tea in the Ratnapura District. *Journal of Agricultural Sciences–Sri Lanka* 4:59–66.

de Bekker, C., L.E. Quevillon, P.B. Smith, K.R. Fleming, D. Ghosh, A.D. Patterson & D.P. Hughes. 2014. Species-specific ant brain manipulation by a specialized fungal parasite. *BMC Evolutionary Biology* 14:166.

Decock, C. & L. Ryvarden. 1999. Studies in Perenniporia: *Perenniporia detrita* and its taxonomic synonyms. *Mycologia* 91:386.

Devkota, P. & R. Hammerschmidt. 2020. The infection process of *Armillaria mellea* and *Armillaria solidipes*. *Physiological and Molecular Plant Pathology* 112:101543.

Doehlemann, G., B. Ökmen, W. Zhu & A. Sharon. 2017. Plant pathogenic fungi. In J. Heitman & B.J. Howlett (eds.), *The Fungal Kingdom*, pp. 701–726. ASM Press, Washington, D.C.

Ediriweera, A.N., S.C. Karunarathna, P.N. Yapa, D.A. Schaefer, A.K. Ranasinghe, N. Suwannarach & J. Xu. 2022. Ectomycorrhizal mushrooms as a natural bio-indicator for assessment of heavy metal pollution. *Agronomy* 12:1041.

Ediriweera, A.N., V. Pietro, S.C. Karunarathna & C. Dilshan. 2023a. *Termitomyces srilankensis* sp. nov. (Lyophyllaceae, Agaricales), a new species from Sri Lanka. *Mycological Observations* 6:47–56.

Ediriweera, A.N., P. Voto, S.C. Karunarathna, J. Kumla, V. Thiyagaraja, M.K. Wijesooriya, K.S. Wadduwage, W. Lu, S.L. Stephenson & X. Jianchu. 2023b. A new species and a new record in the Agaricales from Sri Lanka. *Mycological Observations* 6:92–107.

Ediriweera, S.S., C.M. Nanayakkara, O. Vithanage, D. Sisira, J. Weerasena, S. Karunarathna, R. Lakshman, C. Wijesundera, M. Arachchige & S. Piyatissa. 2021. Morphology and phylogeny reveal nine new records of polypores from dry zone of Sri Lanka. *Chiang Mai Journal of Science* 48:893–908.

Ediriweera, S.S., R.L.C. Wijesundera, C.M. Nanayakkara & J. Weerasena. 2014a. A new record of *Fulvifomes fastuosus* from Sri Lanka. *Journal of the National Science Foundation of Sri Lanka* 42:375–377.

Ediriweera, S.S., R.L.C. Wijesundera, C.M. Nanayakkara & J. Weerasena. 2014b. Macrofungi from the Sigiriya wilderness in Sri Lanka. *Journal of Mycopathological Research* 52:47–51.

Ediriweera, S.S., R.L.C. Wijesundera, C.M. Nanayakkara & J. Weerasena. 2014c. Characterization of two wood inhabiting macrofungal species (*Ciriolopsis byrsina* and *Flavodon flavus*) from dry zone forest reserves in Sri Lanka. In *Annual Proceedings of the Sri Lanka Association for the Advancement of Science*, 1 December 2024. Colombo.

Ediriweera, S.S., R.L.C. Wijesundera, C.M. Nanayakkara & J. Weerasena. 2015a. First report of *Panaeolus sphinctrinus* and *Panaeolus foenisecii* (Psathyrellaceae, Agaricales) on elephant dung from Sri Lanka. *Applied and Environmental Microbiology* 1:19–23.

Ediriweera, S.S., R.L.C. Wijesundera, C.M. Nanayakkara & O. Weerasena. 2015b. Comparative study of growth and yield of edible mushrooms, *Schizophyllum commune* Fr., *Auricularia polytricha* (Mont.) Sacc. and *Lentinus squarrosulus* Mont. on lignocellulosic substrates. *Mycosphere* 6:760–765.

Fernando, K.M.E.P. 2008. The host preference of a *Ganoderma lucidum* strain for three tree species of Fabaceae family; *Cassia nodosa*, *Cassia fistula* and *Delonix regia*. *Journal of the National Science Foundation of Sri Lanka* 36:323–326.

Fernando, R. & D.N. Fernando. 1990. Poisoning with plants and mushrooms in Sri Lanka: A retrospective hospital based study. *Veterinary and Human Toxicology* 32:579–581.

Fukasawa, Y., T. Osono & H. Takeda. 2009. Effects of attack of saprobic fungi on twig litter decomposition by endophytic fungi. *Ecological Research* 24:1067–1073.

Galappaththi, M.C.A., S.C. Karunarathna, C.M. Nanayakkara, S.L. Stephenson, L. Dauner, N. Wijayawardene & U. Jayalal. 2023. Overview of Sri Lankan fungi and lichen research. In T. Pullaiah (ed.), *Biodiversity Hotspot of the Western Ghats and Sri Lanka*, pp. 447–456. Apple Academic Press, New York.

Gamage, S. & S. Ohga. 2018. A comparative study of technological impact on mushroom industry in Sri Lanka: A review. *Advances in Microbiology* 8:665–686.

Gams, W., P. Diederich & K. Poldmaa. 2004. Fungicolous fungi. In G.M. Mueller, G.F. Bills & M.S. Foster (eds.), *Biodiversity of Fungi: Inventory and Monitoring Methods*, pp. 343–392. Elsevier Science & Technology.

Gründemann, C., J.K. Reinhardt & U. Lindequist. 2020. European medicinal mushrooms: Do they have potential for modern medicine? An update. *Phytomedicine* 66:153131.

Gunatilleke, N., R. Pethiyagoda & S. Gunatilleke. 2017. Biodiversity of Sri Lanka. *Journal of the National Science Foundation of Sri Lanka* 36:25.

Hajek, A.E. & R.J. St. Leger. 1994. Interactions between fungal pathogens and insect hosts. *Annual Review of Entomology* 39:293–322.

Hewage, D. 2015. Traditional knowledge of edible wild mushrooms in a village adjacent to the Sinharaja forest. *Journal of the Royal Asiatic Society of Sri Lanka* 60:77–95.

Hoeksema, J.D., J.D. Bever, S. Chakraborty, V.B. Chaudhary, M. Gardes, C.A. Gehring, M.M. Hart, E.A. Housworth, W. Kaonongbua, J.N. Klironomos, M.J. Lajeunesse, J. Meadow, B.G. Milligan, B.J. Piculell, A. Pringle, M.A. Rúa, J. Umbanhowar, W. Viechtbauer, Y.-W. Wang & P.C. Zee. 2018. Evolutionary history of plant hosts and fungal symbionts predicts the strength of mycorrhizal mutualism. *Communications Biology* 1:116.

Horton, T.R. 2017. *Spore Dispersal in Ectomycorrhizal Fungi at Fine and Regional Scales*, pp. 61–78. Springer, Cham. https://link.springer.com/chapter/10.1007/978-3-319-56363-3_3

Horton, T.R. & T.D. Bruns. 2001. The molecular revolution in ectomycorrhizal ecology: Peeking into the black-box. *Molecular Ecology* 10:1855–1871.

Houttuyn, M. 1783. *Handleiding tot de plant–en kruidkunde, benevens eene uitvoerige beschrijving der boomen, planten, heesters, kruiden, varens, mossens, bollen grasplanten, volgens het zamenstel van C. Linnaeus* (Vol. 14). Bij Lodewyk van Es, Amsterdam.

Hyde, K.D., K.W.T. Chethana, R.S. Jayawardena, T. Luangharn, M.S. Calabon, E.B.G. Jones, S. Hongsanani & S. Lumyong. 2020. The rise of mycology in Asia. *ScienceAsia* 46S:1.

Jayasinghe, G.G., W.K. Liyanage, M. Wijayawardhana, K.M.M. Priyangika, D.N. Samaraweera & K.G.G. Wijesinghe. 2017. A study of rough bark disease on cinnamon (*Cinnamomum zeylanicum* Blume); disease symptoms, development and the causal agent with special reference to its morphology, histopathology and nutritional statutes of affected plants. In B. Marambe (ed.), *Proceedings of the Symposium on Minor Export Crops, 16–17 March 2017*, pp. 63–72. Peradeniya, Sri Lanka.

Jayasuriya, K.E. & B.I. Thennakoon. 2007. Biological control of *Rigidoporus microporus*, the cause of white root disease in rubber. *Ceylon Journal of Science (Biological Sciences)* 36:9–16.

Karunarathna, S.C. 2012. Current status of knowledge of Sri Lankan mycota. *Current Research in Environmental & Applied Mycology* 2:18–29.

Karunarathna, S.C., P.E. Mortimer, J. Xu & K.D. Hyde. 2017. Overview of research of mushrooms in Sri Lanka. *Revista Fitotecnia Mexicana* 40:399–403.

Karunarathna, S.C., A.K.H. Priyashantha, S. Tibpromma, M.C.A. Galappaththi, B.M. Premarathne, N.N. Wijayawardene, M.K. Wimalasena, R.G.U. Jayalal, K.D. Wickramanayake & H. Dangalla. 2023. Checklist, typification details, and nomenclature status of Basidiomycota, originally described from Sri Lanka. *Phytotaxa* 621:1–86.

Kepler, R.M., J.J. Luangsa-ard, N.L. Hywel-Jones, C.A. Quandt, G.-H. Sung, S.A. Rehner, M.C. Aime, T.W. Henkel, T. Sanjuan, R. Zare, M. Chen, Z. Li, A.Y. Rossman, J.W. Spatafora & B. Shrestha. 2017. A phylogenetically-based nomenclature for Cordycipitaceae (Hypocreales). *IMA Fungus* 8:335–353.

Kirk, P., P. Cannon, D.W. Minter & J. Stalpers. 2008. *Dictionary of the Fungi*, 10th ed. CABI Publishing, Wallingford, England.

Konara, U.A., K.M. Thambugala & K.K. Hapuarachchi. 2022. *Ganoderma* (Ganodermataceae, Polyporales): Historical perspectives, recent advances, and future research in Sri Lanka. *Studies in Fungi* 7:1–7.

Konara, U.A., K.M. Thambugala, S.C. Karunarathna, A. Ediriweera & K.K. Hapuarachchi. 2024. Unveiling the hidden diversity of *Ganoderma* (Ganodermataceae, Polyporales) in Sri Lanka: The first report of *G. angustisporum*, *G. ellipsoideum* and *G. orbiforme*. *New Zealand Journal of Botany* 63:1–25.

Leake, J.R. 2005. Plants parasitic on fungi: Unearthing the fungi in myco-heterotrophs and debunking the plant myth. *Mycologist* 19:113.

Lee, H., K. Wissitrassameewong, M.S. Park, A. Verbeken, J. Eimes & Y.W. Lim. 2019. Taxonomic revision of the genus *Lactarius* (Russulales, Basidiomycota) in Korea. *Fungal Diversity* 95:275–335.

Lehmann, A., E.F. Leifheit & M.C. Rillig. 2017. Mycorrhizas and soil aggregation. In N.C. Johnson, C.A. Gehring & J. Jansa (eds.), *Mycorrhizal Mediation of Soil*, pp. 241–262. Elsevier, Amsterdam.

Lisiewska, M. 1992. Macrofungi on special substrates. In W. Winterhoff (ed.), *Fungi in Vegetation Science*, pp. 151–182. Springer Publishing, Netherlands.

Liu, D., L. Cai, K. Wang, J. Li, T. Wei & Y. Yao. 2018. Threat assessments, problems and countermeasures of China's macrofungi. *Biodiversity Science* 26:1236–1242.

Loyd, A.L., E.R. Linder, N.A. Anger, B.S. Richter, R.A. Blanchette & J.A. Smith. 2018. Pathogenicity of *Ganoderma* species on landscape trees in the southeastern United States. *Plant Disease* 102:1944–1949.

Madhushan, K.W.A., S.C. Karunarathna, D.M.D. Dissanayake, T.D.C. Priyadarshani, S.L. Stephenson, A.M. Elgorban, T.M. Dawoud, A.K.H. Priyashantha, D. Dai, P.N. Yapa & X. Wang. 2023. Effect of arbuscular mycorrhizal fungi on lowland rice growth and yield (*Oryza sativa* L.) under different farming practices. *Agronomy* 13:2803.

Madushani, H., T. Fernando, R. Wijesundara & D. Siriwardane. 2014. First Report of white root disease of *Artocarpus nobilis* in Sri Lanka caused by *Rigidoporus microporus*. *Journal of the National Science Foundation of Sri Lanka* 42:197.

Murata, H., A. Yamada, S. Yokota, T. Maruyama, T. Shimokawa & H. Neda. 2015. Innate traits of Pinaceae-specific ectomycorrhizal symbiont *Suillus luteus* that differentially associates with arbuscular mycorrhizal broad-leaved trees *in vitro*. *Mycoscience* 56:606–611.

Niego, A.G.T., S. Rapior, N. Thongklang, O. Raspé, K.D. Hyde & P. Mortimer. 2023. Reviewing the contributions of macrofungi to forest ecosystem processes and services. *Fungal Biology Reviews* 44:100294.

Novotný, Č., T. Cajthaml, K. Svobodová, M. Šušla & V. Šašek. 2009. *Irpex lacteus*, a white-rot fungus with biotechnological potential—a review. *Folia Microbiologica* 54:375–390.

Park, M. 1932. *Tuber zeylanicum* B. & Br. and *Sclerotium rolfsii* Sacc. *Transactions of the British Mycological Society* 17:179–181.

Pegler, D.N. 1986. *Agaric Flora of Sri Lanka (Kew Bulletin Additional Series XII)*. Her Majesty's Stationery Office (H.M.S.O.), Royal Botanic Gardens, Kew.

Peiris, T.M., M. Perera, H.H. Munasinghe, K.M. Thambugala, B.P. Dharmasena, P. Suttiprapan & R. Cheewangkoon. 2024. The treasured giants: A current overview on agricultural, nutritional, bioactive, and economic potential of *Macrocybe* species (Agaricales, Basidiomycota). *Frontiers in Cellular and Infection Microbiology* 14:1493532.

Perera, P., S. Wijesinghe, N. Dayawansa, S. Marasinghe & C. Wickramarachchi. 2017. Response of tropical birds to habitat modifications in fragmented forest patches: A case from a tropical lowland rainforest in south-west Sri Lanka. *Community Ecology* 182:175–183.

Peries, O.S. 1974. *Ganoderma* basal stem rot of coconut: A new record of the disease in Sri Lanka. *The Plant Disease Reporter* 58:293.

Petch, T. 1923. Parasites of scale-insect fungi. *Transactions of the British Mycological Society* 8:206–212.

Pethiyagoda, R. 2012. Biodiversity conservation in Sri Lanka's novel ecosystems. *Ceylon Journal of Science (Biological Sciences)* 41:1–10.

Premarathne, B.M., M.C.A. Galappaththi, N.M. Patabendige, S.C. Karunarathna, N.N. Wijayawardene, Y.A.P.K. Dayasena, K.L.W. Kumara, D.S.A. Wijesundara, A. Ediriweera & S. Madawala. 2025. A review of wild edible and medicinal mushrooms in Sri Lanka: Systematic exploration and industrial applications. *MycoAsia* 1:2582–7278.

Pullaiah, T. 2023. *Biodiversity Hotspot of the Western Ghats and Sri Lanka*. Apple Academic Press, Florida.

Ranawake, A.L. 2021. The underutilized resources in the lowland wet zone forests of Sri Lanka and untapped Indigenous knowledge of peripheral households. *Journal of the University of Ruhuna* 9:47–71.

Rathi, A., C. Singh, P. Pathak, N. Chaudhary & D. Vyas. 2021. *A thematic approach on cordyceps*. Global Books Organisation, Delhi.

The Russulales News Team. 2010. *Edibility and Toxicity of Russulales*. Museo Tridentino di Scienze Naturali.

Santolamazza-Carbone, S., L. Iglesias-Bernabé, E. Sinde-Stompel & P.P. Gallego. 2021. Ectomycorrhizal fungal community structure in a young orchard of grafted and ungrafted hybrid chestnut saplings. *Mycorrhiza* 31:189–201.

Sarathchandra, C., Y.A. Abebe, I.L. Wijerathne, S.T. Aluthwattha, S. Wickramasinghe & Z. Ouyang. 2021. An overview of ecosystem service studies in a tropical biodiversity hotspot, Sri Lanka: Key perspectives for future research. *Forests* 12:540.

Semwal, K., S.L. Stephenson & A. Husen. 2023. *Wild Mushrooms and Health*. CRC Press. Boca Raton, Florida, United States.

Sen, S., D. Karati, R. Priyadarshini, T.K. Dua, P. Paul, R. Sahu & G. Nandi. 2023. *Cordyceps sinensis* (yar-sagumba): Pharmacological properties of a mushroom. *Pharmacological Research—Modern Chinese Medicine* 8:100294.

Sharma, R. 2017. Ectomycorrhizal mushrooms: Their diversity, ecology and practical applications. In A. Varma, R. Prasad & N. Tuteja (eds.), *Mycorrhiza—Function, Diversity, State of the Art*, pp. 99–131. Springer International Publishing, Cham, Switzerland.

Silva, M.K.R., H.P.S. Fernando & B.I. Tennakoon. 2016. Incidence of brown root disease in rubber plantations of Sri Lanka. *Bulletin of the Rubber Research Institute of Sri Lanka* 53:29–34.

Silva, M.K.R., T. Fernando, R.L.C. Wijesundara, C.M. Nanayakkara & B.I. Tennakoon. 2017. Development of brown root disease in Sri Lankan rubber plantations: Possible involvement of other tree species. *Journal of the Rubber Research Institute of Sri Lanka* 97:47–57.

Silva, M.K.R., C.K. Jayasinghe & B.I. Tennakoon. 2017. Evaluation of the antagonistic effect. *Journal of the Rubber Research Institute of Sri Lanka* 94:25–32.

Sri Lanka's Forest Reference Level Submission to the UNFCCC. 2017. UN-REDD Colombo, Sri Lanka.

Sridhar, K.R. & S.K. Deshmukh. 2024. Ecology of macrofungi. In K.R. Sridhar & S.K. Deshmukh (eds.), *Ecology of Macrofungi*, pp. 1–14. CRC Press, Boca Raton, Florida, United States.

Subasinghe, S.M.C.U.P. & R.P. Hettiarachchige. 2022. Endophytic fungal species in tropical trees: A review. *Journal of Tropical Forestry and Environment* 11:1–9.

Surasinghe, T., R. Kariyawasam, H. Sudasinghe & S. Karunarathna. 2019. Challenges in biodiversity conservation in a highly modified tropical river basin in Sri Lanka. *Water* 12:26.

Tedersoo, L., T.W. May & M.E. Smith. 2010. Ectomycorrhizal lifestyle in fungi: Global diversity, distribution, and evolution of phylogenetic lineages. *Mycorrhiza* 20:217–263.

Thambugala, K., S. Karunarathna, C. Zhao, A. Elgorban, D. Tennakoon & U. Konara. 2023. *Perenniporiopsis srijayewardenepurana* sp. nov. and validation of *Pyrrhoderma nigrum* in Sri Lanka. *Phytotaxa* 619:123–136.

Tibpromma, S., K.D. Hyde, R. Jeewon, S.S.N. Maharachchikumbura, J.-K. Liu, D.J. Bhat, E.B.G. Jones, E.H.C. McKenzie, E. Camporesi, T.S. Bulgakov, M. Doilom, A.L.C.M. de Azevedo Santiago, K. Das, P. Manimohan, T.B. Gibertoni, Y.W. Lim, A.H. Ekanayaka, B. Thongbai, H.B. Lee & S.C. Karunarathna. 2017. *Fungal Diversity Notes* 491–602: Taxonomic and phylogenetic contributions to fungal taxa. *Fungal Diversity* 83:1–261.

Tóth, B.B. & Z. Barta. 2010. Ecological studies of ectomycorrhizal fungi: An analysis of survey methods. *Fungal Diversity* 45:3–19.

Tuli, H.S., S.S. Sandhu & A.K. Sharma. 2014. Pharmacological and therapeutic potential of *Cordyceps* with special reference to Cordycepin. *3 Biotech* 4:1–12.

Verbeken, A., D. Stubbe, K. van de Putte, U. Eberhardt & J. Nuytinck. 2014. Tales of the unexpected: Angiocarpous representatives of the *Russulaceae* in tropical South East Asia. *Persoonia: Molecular Phylogeny and Evolution of Fungi* 32:13–24.

Voces, R., L. Diaz-Balteiro & Ó. Alfranca. 2012. Demand for wild edible mushrooms. The case of *Lactarius deliciosus* in Barcelona (Spain). *Journal of Forest Economics* 18:47–60.

Waheed, A., Y. Haxim, W. Islam, M. Ahmad, M. Muhammad, F.M. Alqahtani, M. Hashem, H. Salih & D. Zhang. 2023. Climate change reshaping plant-fungal interaction. *Environmental Research* 238:117282.

Wang, P., Y. Zhang, F. Mi, X. Tang, X. He, Y. Cao, C. Liu, D. Yang, J. Dong, K. Zhang & J. Xu. 2015. Recent advances in population genetics of ectomycorrhizal mushrooms *Russula* spp. *Mycology* 6:110–120.

Watkinson, S.C. 2016. Mutualistic symbiosis between fungi and autotrophs. In S.C. Watkinson, L. Boddy & N.P. Money (eds.), *The Fungi*, pp. 205–243. Elsevier, Amsterdam.

Weerasinghe, R.H., K. Maduranga, R.N. Attanayake, C. Shevkar, A.S. Kate, G. Weerakoon, K. Kalia & P.A. Paranagama. 2021. Bioactive properties and metabolite profiles of endolichenic fungi in mangrove ecosystem of Negombo lagoon, Sri Lanka. *Natural Product Communications* 16(10).

Wijesekara, H.T.R., R.L.C. Wijesundera & C.N.K. Rajapakse. 1996. Hyphal interactions between *Trichoderma viridae* and *Ganoderma boninense* Pat., the cause of coconut root and bole rot. *Journal of the National Science Council Sri Lanka* 24:217–219.

Wijesundera, R.L.C. 1986. Floristic studies and distribution of fungi in Sri Lanka. *Proceedings of the Indian Academy of Sciences* 96:375–377.

Willis, A., B.F. Rodrigues & P.J.C. Harris. 2013. The ecology of arbuscular mycorrhizal fungi. *Critical Reviews in Plant Sciences* 32:1–20.

Wimalasena, M.K., N.N. Wijayawardene, T.C. Bamunuarachchige, R.G.U. Jayalal, D.J. Bhat, T.M. Dawoud, R.M.K. Rathnayaka & D.-Q. Dai. 2024. Ongoing study reveals high aquatic fungal diversity: An untouched component in Sri Lankan biodiversity. *Chiang Mai Journal of Science* 51:1–31.

Worrall, J. 2004. Armillaria root disease. *The Plant Health Instructor*. https://doi.org/10.1094/PHI-I-2004-0706-01

Yan, J.-K., W.-Q. Wang & J.-Y. Wu. 2014. Recent advances in *Cordyceps sinensis* polysaccharides: Mycelial fermentation, isolation, structure, and bioactivities: A review. *Journal of Functional Foods* 6:33–47.

Zotti, M., A.M. Persiani, E. Ambrosio, A. Vizzini, G. Venturella, D. Donnini, P. Angelini, S. Di Piazza, M. Pavarino, D. Lunghini, R. Venanzoni, E. Polemis, V.M. Granito, O. Maggi, M.L. Gargano & G.I. Zervakis. 2013. Macrofungi as ecosystem resources: Conservation versus exploitation. *Plant Biosystems—An International Journal Dealing with All Aspects of Plant Biology* 147:219–225.

5 Costa Rican Fungi
Present and Future

Melissa Mardones Hidalgo, Julieta Carranza Velázquez, and Carlos Rojas Alvarado

5.1 INTRODUCTION

Costa Rica, a nation celebrated for its extraordinary concentration of biodiversity, stands as a testament to the intricate and vibrant tapestry of life in the Neotropics. Occupying a mere 0.03% of the Earth's landmass, this small Central American country is home to an estimated 5% of the world's species, a statistic that has rightfully positioned it as a global icon for conservation and ecological study (Obando 2007). While the verdant rainforests, iconic vertebrates, and dazzling insects often capture the popular imagination, a far more vast and fundamentally crucial kingdom of organisms underpins the health and function of these ecosystems: the Fungi. The mycobiota of Costa Rica is a frontier of biological discovery, a realm of immense, largely unexplored diversity that is integral to the nation's ecological wealth (Kapelle 2016).

Geographically, Costa Rica is a land bridge, a biological corridor connecting the great landmasses of North and South America. This unique position has allowed for the mingling of flora and fauna from both continents, creating a rich and complex biological heritage. Coupled with its dramatic topography—rising from sea level to the paramo ecosystems of the high Talamanca mountains at over 3 800 meters—the country encompasses an astonishing array of microclimates and life zones (Solano Quintero and Villalobos Flores 2001). From the humid Caribbean lowlands and the seasonally dry forests of the Pacific Northwest to the montane cloud forests and the subalpine paramos, each habitat hosts a distinct and specialized community of fungi. This environmental heterogeneity seems to be the primary driver of the nation's fungal megadiversity (Mardones et al. 2026).

Quantifying this diversity remains a monumental task. A landmark effort, the "First annotated checklist of fungi of Costa Rica," represents the culmination of over 170 years of mycological research and collaboration, documenting approximately 5 676 species and subspecific taxa (Mardones et al. 2024). While this figure is impressive for a country of its size, it represents only the tip of the iceberg. Globally, estimates for the ratio of fungal to plant species have ranged from 6:1 to as high as 10:1 (Hawksworth 2001; Cheek et al. 2020), and a recent estimation (Niskanen et al. 2023) considered that a 3.5:1 ratio for tropical areas is the most accurate. With over 12,000 known vascular plant species in Costa Rica, a conservative application of these ratios suggests a potential fungal diversity of 42,000 to 72,000 species. Other estimates (e.g., Hawksworth and Lücking 2017) have placed the potential figure even higher, suggesting that the true number of fungal species in Costa Rica could exceed 200,000. In all cases, the vast majority of Costa Rica's fungi remain unknown to science, awaiting discovery and description.

5.2 THE SOCIOCULTURAL CONTEXT OF THE HUMAN-FUNGUS INTERACTION IN COSTA RICA

The utilization of mushrooms in Costa Rica has been shaped by a unique demographic history, a blend of pre-Hispanic, European, Caribbean, and Asian influences (Arroyo and Rojas 2025). While other regions of Mesoamerica show extensive documentation of pre-Hispanic mushroom use, direct

DOI: 10.1201/9781003647140-5

evidence in Costa Rica is scarce. Isolated archaeological findings, such as an engraving of *Amanita muscaria* and a mushroom-shaped sculpture suggesting hallucinogenic use, hint at ancient interactions, likely influenced by Mesoamerican trade networks (Jones 1991). More recent evidence from archaeological excavations in the Diquís area suggests the use of *Psilocybe* species for recreational purposes (Arce-Cerdas and Rodríguez-Arce 2019). Among contemporary indigenous groups, the use of fungi remains limited and is not a significant part of their cultural traditions, a point highlighted by the observation from an indigenous leader that "young people do not know this form of food" (Rojas et al. 2023).

While some groups, like the Bribri and Cabécar, have documented uses of species like *Pleurotus, Cookeina* or *Auricularia* (Prieto de Lima et al. 2024), the practice of consuming wild mushrooms is not a large-scale endeavor. The Boruca people, while having a more developed vocabulary for fungi and various preparation methods, have also seen a decline in this tradition (Arroyo and Rojas 2025). The scattered nature of ethnomycological documentation underscores the need for more focused research to identify the precise species consumed by these groups.

In the general, non-indigenous population, particularly in the densely populated Central Valley, there is a greater awareness and appreciation for fungi, primarily as a food resource. A recent study (Molina-Murillo et al. 2015) has shown that in urban areas with high education and purchasing power, over 50% of people consume mushrooms monthly, and nearly a quarter consume them weekly. This contrasts with nationwide data, which shows lower but still significant consumption rates (Rojas and Molina-Murillo 2018). This emerging appreciation, concentrated in specific, highly developed regions, has been described as "mycoindifference" rather than the typical "mycophilia-mycophobia" dialectic found in ethnomycological literature (Arroyo and Rojas 2025). This new perception is not rooted in traditional cultural backgrounds but is a recent construct influenced by social factors like internet-mediated globalization, the rising trend of vegan diets, and the increasing availability of cultivated mushrooms in supermarkets and farmers' markets.

The public's view of fungi is also evolving beyond just food. A significant portion of the population (26%) sees the highest value of mushrooms in the medical industry (Molina-Murillo et al. 2015). This social receptivity for mycological products, whether for culinary or medicinal purposes, provides a powerful impetus for the development of new mycological applications. Efforts by institutions like the National Museum of Costa Rica or the University of Costa Rica to raise public awareness on alternative uses of fungi (i.e., medicinal or engineering applications) further support this trend.

5.2.1 Mycological Research in Costa Rica

The last three decades of the twentieth century and the beginning of the twenty-first have been a transformative period for mycology in Costa Rica, marking a shift from sporadic collections by foreign visitors to a systematic, country-led effort to inventory and understand the national mycobiota. This "golden age" was catalyzed by a unique confluence of institutional vision, international collaboration, and the dedication of a core group of researchers.

The pivotal moment in this history was the establishment of the Instituto Nacional de Biodiversidad (INBio) in 1989. This institute was a revolutionary non-governmental organization created with the ambitious mission to know what biological resources Costa Rica had. Its mandate was to conduct a comprehensive inventory of the country's biodiversity, a task of immense proportions, which from the beginning included the kingdom Fungi. This institutional commitment provided the framework and resources for the first large-scale, systematic fungal inventory in a tropical country (Figure 5.1).

To achieve this, INBio forged powerful collaborations with international experts and foreign institutions. This initiative generated a vast collection of specimens and data, mainly through a series of parataxonomists—local people instructed to collect and identify specimens—who were responsible for the enormous efforts (see Mardones and Rojas 2024). As such, even though INBio never engaged actively in the training of mycologists, a task mostly carried out by the University of

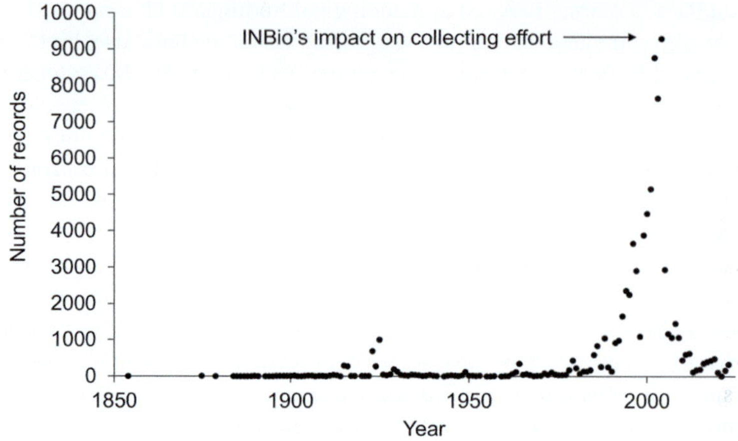

FIGURE 5.1 Number of fungal records in Costa Rica collected between 1854 and 2023 highlighting the impact of INBio's collecting efforts peaking in 2002.

Costa Rica, these efforts brought the attention of non-specialists to the world of fungi and ultimately provided strong contextual elements for strengthening formal mycology in Costa Rica. As part of these efforts, for instance, the publication of both the first illustrated field guides of Costa Rican macrofungi (Mata 1999; Mata et al. 2003) and lichens (Umaña and Sipman 2002) took place.

While INBio was leading the national inventory, the Universidad de Costa Rica was steadily building its capacity and emerging as the central academic hub for mycology in the country, a role it keeps today. This task involved a series of foreign experts collaborating with the training of Costa Rica's next generation of mycologists and culminated in the publication of both a special issue of the *Revista de Biología Tropical* called "Fungi of Costa Rica: Selected Studies on Biodiversity and Ecology" and the book *Common Mushrooms of the Talamanca Mountains, Costa Rica* (Halling and Mueller 2005), which made the country's macrofungal diversity accessible to a broader audience for the first time.

During that time, the Luis Fournier Origgi Herbarium (USJ) began to expand its fungal collection, and a number of key faculty members dedicated their careers to the study of fungi. Foremost among them is the second coauthor of this chapter, Dr. Julieta Carranza, who is widely regarded as a foundational figure in modern Costa Rican mycology and a role model for Latin American women in mycology (Carranza Velázquez et al. 2025). Over her long career at UCR, she mentored numerous students (including both other coauthors of this chapter!) and conducted pioneering research, particularly on plant pathogenic fungi and wood-decaying polypores. Her work and her dedication to teaching inspired the current generation of mycologists in Costa Rica, and more broadly in the Central American context.

The Herbario Nacional de Costa Rica (CR), part of Costa Rica's National Museum, has also been a key player. Curators have been responsible of contributing to the national inventory efforts and for managing important fungal collections, including taking care of INBio's complete inventory after its closure in 2015 (all fungal cultures are now maintained by the Center for Research on Biodiversity and Tropical Ecology—CIBET—in UCR). The work of INBio (during its operational years), UCR, and the National Museum has created a robust national network for mycological research, that albeit focusing on the basic inventory at first, very quickly diversified national research to encompass ecological, agricultural and engineering emphases as well (Rojas et al. 2023).

5.3 DIVERSITY PATTERNS

The fungal diversity of Costa Rica has been a central topic of discussion among researchers during the last decades. Only recently, however, an effort led by the first two coauthors of this chapter culminated with the construction of a database containing the metadata of all fungal collections in Costa Rican fungaria as well as information from a series of international institutions and scientific literature. Advances in the publication of results have been recently made (Mardones et al. 2024; Mardones et al. 2026), and they show that biological collections are essential to closing the gaps in taxonomic and geographical knowledge both in Costa Rica as well as in other neotropical territories.

Nevertheless, as revealed through the examination of over 78 000 digitized records, taxonomically, the collections of fungi from Costa Rica are dominated by the Phyla Basidiomycota and Ascomycota, which together account for nearly all of the records (54.4% and 45.4%, respectively). This is an expected pattern for fungal collections (see Piepenbring et al. 2020), but it highlights a contrasting taxonomic profile between the two main national collections: the Herbarium at the Museo Nacional de Costa Rica (CR) and the Luis Fournier Origgi Herbarium at the Universidad de Costa Rica (USJ). The CR collection has a higher proportion of ascomycetes, which may reflect a focus on lichens and microfungi, while the USJ collection's dominance of basidiomycetes could be linked to research projects centered on macromycetes (Table 5.1).

At the order level, a few groups are notably well represented, accounting for approximately 88% of the records among a total of 142 orders. The most abundant orders are Agaricales, Polyporales, Ostropales, and Lecanorales. These groups include a mix of macrofungi, such as wood-decaying fungi (Polyporales, Hymenochaetales), ectomycorrhizal fungi (Agaricales, Boletales, Russulales), lichens (Lecanorales, Ostropales, Peltigerales), and specific microfungi (Chaetosphaeriales, Sordariales, Xylariales). Certain plant-pathogenic groups, including rusts, smuts and tropical tar spot fungi, are also well-documented in Costa Rica (Piepenbring 1996).

TABLE 5.1
Records of Fungi from Costa Rica in the 18 Fungaria (Only Acronyms Provided) with the Most Collections, Arranged by Phylum.

	Phylum*						
Herbarium	**ASC**	**BAS**	**CHY**	**ENT**	**GLO**	**MUC**	**Total number of records**
CR	30603	23547					54150
USJ	1390	9546			2		10938
F	194	4688					4882
BPI	689	856	2				1547
S	938	380					1318
NY	35	1166					1201
IMI	587	183		1		2	773
TENN-F	22	688					710
PUR		602					602
ILL	517	80	2				599
LSUM	115	252					367
CUP	183	63	2				248
DUKE	2	185					187
CBS	140	41					181
USAM		158					158
KANU						117	117
WSP	25	73					98
ILLS	23	13					36
Other herbaria	126	92	3				221

*ASC: Ascomycota, BAS: Basidiomycota, CHY: Chytidriomycota, ENT: Entomophthoromycota, GLO: Glomeromycota, MUC: Mucoromycota.

A striking pattern of diversity is the high percentage of rare taxa. Approximately 90% of the species in fungarium datasets are known from five or fewer records, and 65% are represented by a single collection. This phenomenon is common in biodiversity inventories and can reflect a combination of true rarity and under-sampling (Cazabonne et al. 2024). It also poses significant challenges for conservation and species delimitation (Longton and Hedderson 2000). Despite this, the fungal data analysis from Costa Rica identified 68 taxa with more than 30 records, primarily wood-decaying and ectomycorrhizal fungi, which are considered sufficient for comparative studies.

Based on available data, the distribution of fungi is not even across the country (Figure 5.2). This pattern reflects the fact that efforts to study and document biological species are affected by external factors such as road conditions, accessibility, research infrastructure in place, climate, and research biases (Isaac and Pocock 2015). In Costa Rica, for instance, there are eight areas–equivalent to counties or districts—that despite encompassing only 3.5% of the territory are associated with one third of all fungal collections (they are the municipalities of Monteverde, Grecia, Santo Domingo, Montes de Oca, La Unión, El Guarco, Dota y Coto Brus). When the coefficient of variation is calculated for all administrative units in the country, data revealed a value of 118, which not only is extremely high, but supports quantitatively the unevenness of collecting efforts.

5.3.1 Reliability of Available Fungal Data

While fungarium data offer invaluable insights, their reliability for ecological and diversity studies is subject to significant biases (Isaac and Pocock 2015). As such, every country should perform a critical assessment of these limitations, which are essential for any mycologist to consider when using such data for research or conservation applications.

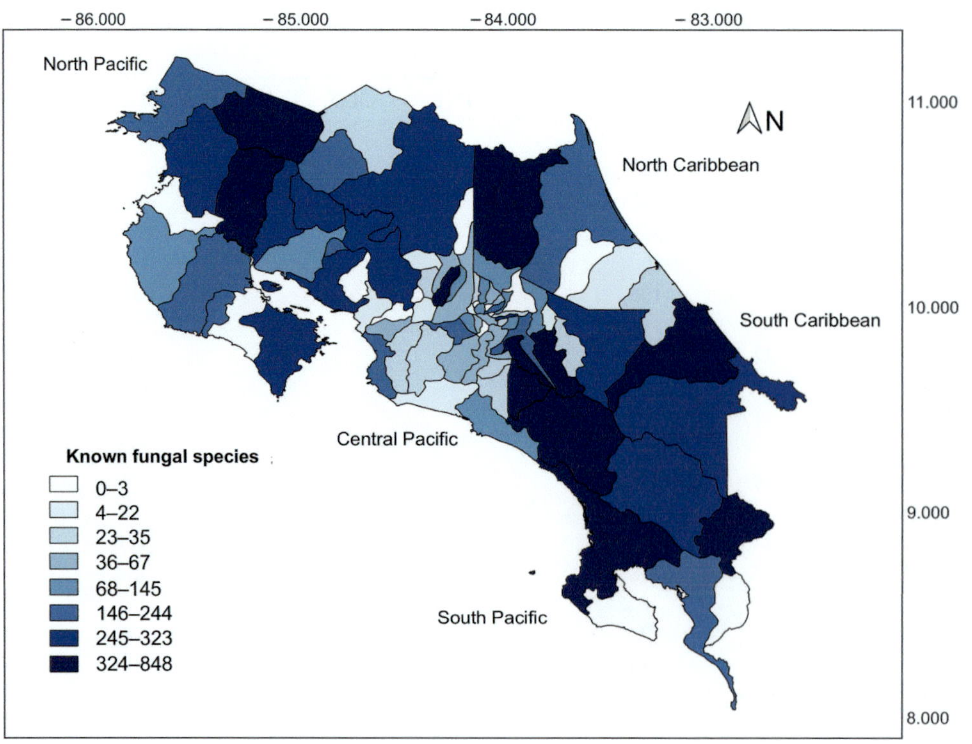

FIGURE 5.2 Map of Costa Rica showing the uneven distribution of recognized species of fungi for different administrative units within the country.

5.3.1.1 Costa Rican Inherent Biases and Limitations

The most pervasive issue in fungal data from Costa Rica is the presence of sampling and observation biases. As mentioned before, the collections reflect an uneven sampling effort across space and time, often driven by the research interests and access of specific collectors rather than a systematic sampling design. For example, about 50% of the records were collected in just three life zones: the lower montane wet forest, tropical wet forest, and premontane pluvial forest (forest types per Holdridge 1967). Just like the spatial unevenness of effort, this observation also indicates a localized effort in particular forest or climate conditions, with most other options (e.g., forest types) remaining under-sampled.

Taxonomic bias is also a major concern. The high proportion of macrofungi in the collections is attributed to their visibility and ease of collection, a phenomenon known as detectability bias (Hutto 2016). This leads to a significant underrepresentation of microfungi and basal fungal groups, which are more challenging to collect and are often preserved in culture collections. The absence of a national culture collection in Costa Rica exacerbates this issue, as it limits the ability to store and access these important fungal groups.

A critical finding is the high number of specimens identified only to the genus level or higher, particularly in the national collections (Figure 5.3). This suggests that local efforts have historically prioritized specimen collection over detailed curation and identification. The ratio of genus-level to species-level records is 18 times higher in Costa Rican collections compared to international ones. This "curation gap" means that a wealth of potential new species or first records are likely hidden within these uncurated collections (Mardones et al. 2026).

With these data, clear legal and institutional biases were also documented. About 72% of type specimens associated with collections from Costa Rica are stored in international collections (Table 5.2). Despite a 1998 law requiring the deposit of duplicates in national collections, this regulation is not necessarily followed. This practice restricts national access to reference material and hinders the development of local research capacity.

5.3.1.2 Costa Rican Unique Strengths and Valuable Insights

Despite these biases, fungal data, when carefully curated and analyzed, can provide invaluable insights. The long temporal span of the systematically collected data from Costa Rica (1854–2023)

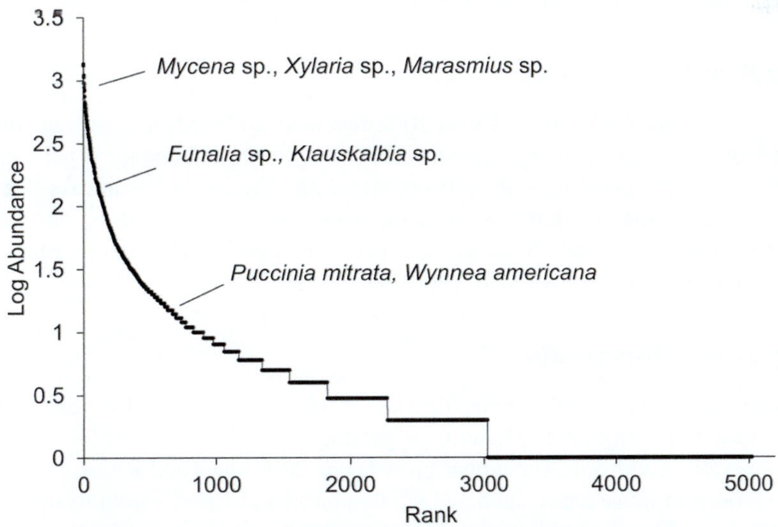

FIGURE 5.3 Rank abundance curve of Costa Rican fungi showing a normal trend of decreasing abundance in higher ranks and highlighting the fact that most of the identifications in the most abundant categories correspond to collections only identified to genus level.

TABLE 5.2
Distribution of Type Specimens of Fungi from Costa Rica Based on Location of Storage and Phylum.

	Location	
Phylum/group	**National**	**International**
Ascomycota	416	1100
Basidiomycota	183	477
Blastocladiomycota		1
Chytridiomycota	1	1
Glomeromycota		1
Mucoromycota		14
Other		8

is a unique strength, offering a historical baseline for understanding long-term changes in fungal populations and phenology. These collections can serve as a critical reference for future resampling efforts across regions, habitats, and specific groups.

The data analysis using the Costa Rican case also demonstrates that such information can be used to reveal meaningful patterns, especially for well-represented taxa. The use of subdatasets and the application of statistical methods, such as rarefaction, can mitigate some of the most significant biases, although these techniques have to be used carefully considering the characteristics of the organisms and the conditions used to acquire the data (Willis 2019). The analysis of specific, well-documented orders—such as Agaricales, Boletales, Polyporales, and Chaetosphaeriales—is a testament to the utility of these collections for ecological studies (see Section 5.4.2).

The high number of rare species and singletons, while challenging, also represents a significant strength. These specimens, often unique and irreplaceable, are crucial for understanding biodiversity patterns and can attract the attention of policymakers for conservation efforts (Cadotte and Davis 2010). The existence of these specimens underscores the importance of selective fieldwork and the digitization of small herbaria in biodiverse countries like Costa Rica.

5.4 ECOLOGICAL PATTERNS

The analysis of the fungal data from Costa Rica revealed a clear link between fungal diversity, collection intensity, and environmental factors. However, from the ecological perspective, as mentioned before, this is only possible with well-curated data. The Costa Rican case is a rare example for tropical regions, because it counts on more intensive studies across all regions within the territory and has had systematic research carried out over a longer period than other nations. Based on these facts, extra information is available from Costa Rica.

5.4.1 Geographic Distribution

The areas where most of the fungal records from Costa Rica were collected are characterized by mild to warm temperatures and high precipitation, suggesting that collectors have historically focused on regions with favorable conditions for fungal growth (see Mardones and Rojas 2024). The presence of biological stations in those areas, such as La Selva and Las Cruces—both from the Organization for Tropical Studies—have been important in the regional exploration of biological groups in this country. Research efforts in these lands have include fungi and therefore is not surprising that most fungal collections in Costa Rica come from warmer and wetter areas.

An analysis of taxonomic groups across different life zones revealed particular patterns for different groups. For example, the ectomycorrhizal macrofungi orders Russulales and Boletales are more abundant in the lower montane wet forests (Figure 5.4), which are home to oak trees. In contrast, the saprophytic orders Polyporales and Xylariales are most abundant in lowland forests (see Rojas and Calvo 2014; Carranza et al. 2018). The order Agaricales is abundant across all forest types in the country. This differential distribution of fungal groups across distinct ecosystems provides valuable ecological insights into their habitat preferences and ecological roles.

5.4.2 Phenological Patterns and the Role of Precipitation

Costa Rican fungal data can provide a rare, long-term analysis of fungal phenology in a Neotropical region, using a temporal dataset spanning over 100 years. Information shows a clear link between fungal fructification and precipitation that, albeit being somehow intuitive, has rarely been documented with data. In Costa Rica, the median number of records of fungi during rainy days is about 85% higher than the value for any day during the dry season. This pattern can be observed in Figure 5.5 where the probability of finding fungi from four different orders is linked with the different times of the year. In this example, all orders reacted negatively (less fungi in ecosystems) to the reduced precipitation brought by the dry season, and positively to the rainy season. Transitional periods between dry and rainy seasons had different effects.

This finding aligns with previous studies in other Neotropical regions and contrasts with research in temperate zones, where temperature and solar radiation are often the primary drivers of fungal fructification (e.g., Krah et al. 2023). The pronounced dry season in Costa Rica from November to April makes rainfall a more crucial trigger for sporocarp production. The data suggest that while some groups may be less sensitive to seasonal changes (see Carranza et al. 2018), precipitation remains a key ecological driver for most fungi in this tropical environment.

Costa Rican data also highlights the importance of analyzing different ecological groups separately. The contrasting patterns between the groups, such as the wood-decaying Polyporales versus the ectomycorrhizal Boletales, indicate that a single analysis for all fungi would obscure important ecological distinctions. This nuanced approach is essential for understanding how different fungal lifestyles and associations (e.g., mycorrhizal, saprophytic) respond to environmental cues. The fact

FIGURE 5.4 Examples of the orders Polyporales (A), Xylariales (B) and Pezizales (C) commonly found in Costa Rican lowland forests, as well as Russulales (D), Boletales (E), and Agaricales (F) from high-elevation ecosystems.

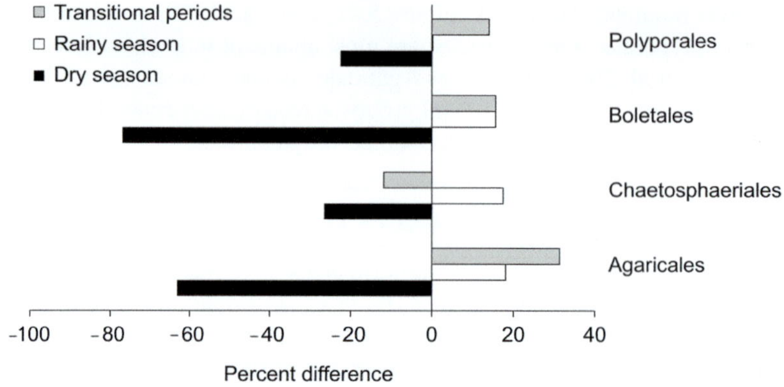

FIGURE 5.5 Difference in the observed number of records of fungi in comparison with a homogeneous distribution across the year (neutralizing the effect of seasons) for four orders with different ecological characteristics.

that precipitation is a key factor influencing fungal fructification in tropical regions represents a significant contribution to a field that has historically focused on temperate climates.

5.4.3 Functional and Structural Relationships With Forests

Macrofungi possess structural elements, such as the size of their sporocarps (fruiting bodies), biomass, and reproductive area, which are considered functional characters that express their ecological strategies. Recent fungal studies in Costa Rica (Rojas and Calvo 2014; Rojas et al. 2017) highlight that two specific functional variables—pileus diameter and the ratio of hymenial area to biomass—can be particularly useful for understanding the relationship between fungi and their environment.

A significant finding of these studies is the strong correlation between fungal characteristics and forest attributes. Pileus diameter, for example, showed a high positive correlation with light intensity and canopy openness, and a negative correlation with tree density, tree biomass, and the diversity of seedlings and saplings. Similar observations showing the link between forest structure and fungal characteristics have been made in other studies (e.g., Dove and Keeton 2015). This suggests that larger fungi tend to be found in more open, less dense forest patches. The ratio of hymenial area to biomass, which is indicative of a fungus's reproductive strategy, also showed strong correlations with environmental factors, being positively linked to temperature and canopy openness (also in Schreiber et al. 2024), and negatively correlated with leaf area index and canopy height. This implies that fungi in warmer, more open environments allocate more resources to reproductive structures relative to their overall body mass.

The distribution of fungal functional groups also seems to vary across forest types. Data from Costa Rica identified that a planted *Cupressus lusitanica* forest (an introduced species), for instance, had the highest tree density but the lowest number of fungal records in a particular region (Rojas and Calvo 2014). In contrast, a native and close by *Quercus seemannii* forest, which is ectomycorrhizal-dominant, yielded the largest fungal biomass and hymenial area. This finding is particularly important as it underscores the specialized relationships between certain fungi and host trees, such as the ectomycorrhizal species that are vital for the health of tropical forests (Corrales et al. 2018).

5.4.4 The Role of Fungi in Carbon Sequestration

One of the most compelling ecological aspects of fungi is their direct and indirect role in carbon cycling. The study by Rojas and Calvo (2014) in Costa Rica explicitly linked fungal dynamics to forest biomass and carbon stocks. The study found that while above-ground carbon stocks and tree biomass were highest in forests dominated by an introduced species, fungal biomass was highest in native forests. Similar observations have been made in Europe (Ferreira et al. 2006) and suggest a complex relationship where tree biomass and fungal biomass do not necessarily correlate directly, especially given the different carbon-uptake strategies of saprobes versus mycorrhizal fungi.

Furthermore, the data from Costa Rica showed that fungal productivity was higher in forest patches with greater tree density but was independent of the specific life zone (Rojas and Calvo 2014). This highlights the importance of microclimatic conditions and forest structure, rather than just regional climate, in determining fungal productivity and aligns with the findings of Lodge and Cantrell (1995). Costa Rican data shows that a focus on above-ground biomass and carbon is a critical first step to fully understand how fungal communities contribute to overall ecosystem carbon. The role of fungi in nutrient cycling, decomposition, and symbiotic relationships with plants imply that they are integral to a forest's ability to sequester and store carbon. Important work on the relationship between tropical fungi and the carbon cycle has been carried out in the last decade (e.g., Nottingham et al. 2022), after being referred to as "a neglected research imperative" (Alexander and Selosse 2009), but the ecological dynamics of tropical fungi in the context of carbon cycles are still an incipient topic of research.

5.5 CLIMATE EFFECT ON FUNGI

Costa Rican fungal data explicitly shows the relationships of fungi and aspects such as climate change, habitat degradation, and forest fragmentation, recognizing fungi as potential indicators of environmental stress (Rojas et al. 2017; Carranza et al. 2018). Costa Rican macrofungal populations are indeed sensitive to changes in climate and forest structure.

5.5.1 Sensitivity to Macroclimatic Events

Functional variables of macrofungal populations, such as biomass and hymenial areas, have been documented to display high variability over a short period (Rojas et al. 2017). This variability seems to be partially induced by macroclimatic events such as the ENSO phenomenon, suggesting a direct link between large-scale weather patterns and fungal fruiting. Data propose that ecosystems with a higher percentage of mycorrhizal fungi, such as oak forests, may be more sensitive to these events. This is likely due to the intimate, obligate relationship between mycorrhizal fungi and their host trees, which makes them highly dependent on the stability of the host's ecosystem (Fernandez et al. 2023). Any stress on the host tree from climatic events will therefore affect the associated fungal community.

Similarly, fluctuations in macroclimatic events seem to favor the formation of fungal structures, independent of the taxonomic group represented, and are associated with higher levels of fungal diversity (Carranza et al. 2018). In a high elevation oak forest study in the Talamanca Mountains, data revealed that acclimation processes are indeed being displayed by fungi and could be partially responsible for some of the patterns observed. Future hypothesis-driven investigations might be very relevant to understand the responses of fungi to climate in tropical areas, but the data from Costa Rica seem to indicate that high-elevation fungal communities are sensitive to these variables.

5.5.2 The Role of Forest Type and Structure

Costa Rican data also reveal that different forest types exhibit varying levels of vulnerability to external pressures. Moister oak forests have been linked to significantly larger hymenial areas for their fungi compared to the dryer forests (Rojas et al. 2017). This suggests that moisture availability is a critical factor for reproductive success. Plantations showing only introduced tree species, despite being a successful reforestation effort in terms of above-ground biomass, have been observed to have the lowest number of fungal records, indicating that monoculture plantations may not support the same diversity or density of fungal life as more complex native forests (Rojas and Calvo 2014). However, a study in a lowland forest was unable to determine differences in the community structure of endomycorrhizal fungi among monocultures and polycultures (Lovelock and Ewel 2005). As such, these results highlight the idea that forest management practices could have a direct impact on the resilience of fungal populations, but due to the scarcity of this type of research in Costa Rica, they must be studied with more detail in the future.

5.6 LEGAL FRAMEWORK AND CONSERVATION

One of the most significant factors distinguishing Costa Rica's mycological landscape from its regional counterparts is the nation's proactive legal framework. In 1998, the Costa Rican government enacted the Biodiversity Law (Ley 7788), which explicitly defines fungi as part of the nation's biodiversity. This simple yet powerful legislative act has had a profound impact, fully regulating fungal resources and promoting both research and conservation efforts.

This legal foundation has enabled an attractive model of integrated biological conservation, connecting fungal resources directly to government policy. The National Institute of Biodiversity (INBio), active from 1989 to 2015, played a pivotal role in this process. Researchers characterized approximately 50,000 fungal specimens, which are now legally protected as the nation's natural heritage and are deposited at the National Herbarium of Costa Rica (CR). This institutional effort, along with collections at the University of Costa Rica herbarium, has established the country as having the largest scientific collection of fungi in Central America, with about 1.4 specimens per square kilometer of land.

This legal and institutional support is crucial for the country's bioprospection efforts. The Biodiversity Law provides the legal basis for protecting the biochemical and genetic components of fungal biodiversity, paving the way for applied studies in fields like medicine and industry. This is in stark contrast to many other developing regions where such a clear legal framework is absent, making fungal conservation and management far more challenging.

5.7 WHERE DOES COSTA RICAN MYCOLOGY STAND NOW?

Costa Rican mycologists were among the first to analyze phenological data on a national scale in a Neotropical country using a temporal dataset spanning over a century. Since knowledge of fungi is strongly biased toward non-tropical regions (Paton et al. 2020), this fact serves as a crucial benchmark for the entire Neotropical region. It provides a model for how other biodiverse nations in the Global South can leverage their own biological collections to understand fungal diversity as well as to incorporate fungal information into socioeconomic development (Niego et al. 2023). In this sense, a recent evaluation in Latin America highlights the cost-benefit relationship between public investment and fungal research in Costa Rica (Parra López et al. 2024).

Fungal data from Costa Rica has exposed the dual nature of fungaria. They are indispensable tools for biodiversity documentation, but they are also repositories of inherent biases as well (Paton et al. 2020). This pattern is also amplified by digital information in the era of Artificial Intelligence (Parra López et al. 2024). By openly discussing issues like uneven spatial and temporal sampling, taxonomic biases, and the high proportion of records without full identification, mycologists can provide a transparent framework for future researchers. Fungarium data, for instance, are still invaluable for studying well-represented groups and for discovering long-term trends (see Mardones and Rojas 2024).

In Costa Rica, the need to prioritize the curation and identification of existing specimens is crucial. The high number of unidentified records and the fact that most type specimens are housed internationally reveal significant gaps in local research capacity and institutional support. Those issues affect much more than just mycological research in this country and represent one of the most important limitations to fostering a true innovation ecosystem in Costa Rica (Monge-González 2016). This constraint could be equilibrated by public investment in national culture collections and by enforcing regulations on specimen repatriation that are practical and essential for strengthening Costa Rica's role as a leader in regional biodiversity documentation. Fungal field work should continue but must be designed to fill geographic and taxonomic gaps (Mardones et al. 2026). Also, the development of standardized sampling methodologies, the implementation of tools as DNA barcoding and metagenomics, and a greater emphasis on data publication and quality control should be priorities soon.

Fungal data from Costa Rica provide concrete evidence that precipitation is a key driver of fungal fructification in the tropics, a finding that is particularly relevant in a region with an unfavorable climate change forecast (Quesada-Chacón et al. 2020). The demonstration that different fungal groups exhibit unique phenological patterns and that fungarium data can reveal these trends is a significant contribution to fungal ecology. This has important implications for conservation and ecological monitoring in the face of global climate phenomena. However, more ecologically centered studies could help us understand the dynamics of tropical fungi in the future. Recent citizen-based initiatives of fungal conservation in Costa Rica could use ecological data for the design of activities and to promote public policies.

The future of Costa Rican fungal research could also develop in the direction of bioprospection, a much-needed area if tangible benefits for society are sought. This endeavor should be accompanied by support from public agencies or universities, with the idea of creating public patents. Exploring the use of Costa Rican fungi in medicine and engineering as well as in other industrial tasks is an area ripe for new discoveries. Recently, interdisciplinary efforts in Costa Rica have been carried out in this direction, with studies intended to develop biological materials with fungi (Coto-López et al. 2023; Jiménez-Rey 2024).

Finally, future efforts could also concentrate on areas with high social impact such as food security. In this regard, for instance, the growing market for cultivated mushrooms and the public's positive reception to fungi as a food source (Rojas and Molina-Murillo 2018) suggests huge potential for developing new edible species for cultivation. This task requires standardizing analysis methods and creating an evidence-based classification system for edibility to prevent accidental poisonings. Those two efforts already started in Costa Rica are will likely generate tangible products in the years to come.

REFERENCES

Alexander, I. & M.A. Selosse. 2009. Mycorrhizas in tropical forests: A neglected research imperative. *New Phytologist* 182:1–3.

Arce-Cerdas, M.A. & J.M. Rodríguez-Arce. 2019. Ritual consumption of psychoactive fungi and plants in ancestral Costa Rica. *Journal of Psychedelic Studies* 3:179–197.

Arroyo, I. & C. Rojas. 2025. Breve reseña del enfoque sociocultural y potencial de los recursos micológicos en Costa Rica. *Lilloa* 62 (suplemento 1):79–101.

Cadotte, M.W. & T.J. Davis. 2010. Rarest of the rare: Advances in combining evolutionary distinctiveness and scarcity to inform conservation at biogeographical scales. *Diversity and Distributions* 16:376–385.

Carranza Velázquez, J., J.F. Di Stefano, W. Marín & M. Mata. 2018. Estudio comparativo de los macrohongos presentes en troncos de roble en dos bosques montanos neotropicales de Costa Rica. *Polibotánica* 45:35–56.

Carranza Velázquez, J., M. Mardones, W. Arce, R. Batallas Molina, T. Baptista Gibertoni, A. Betancourt, A.C. Bolaños Rojas, R. Mara Borges da Silveira, O.M. Correira Magalhães, L. Costa Maia, L.R. Dávila Giraldo, J. Duchicela, A.E. Franco Molano, N. Gómez Montoya, M.M. Granados Montero, A. de Mello Gugliotta, L. Guzmán Dávalos, T. Antje Hofmann, L. Lüdke Falcão, T. Iturriaga, D. Jaikel Víquez, R. Medel-Ortiz, I. Oliveira Lima, R. Maziero, E. Melgarejo-Estrada, L. Mesquita Paiva, V. Motato-Vásquez, G. Navarro Velasco, M.A. Neves, M. Pavlich, D.I. Brayer Pereira, R. Pereira Neves, D. Rocabado, E.I. Rojas, V. Salazar, N. Sanabria, C.M. de Souza-Motta, P. Vieira Tiago, L. Trierveiler-Pereira & P.A. Zapata Ocampo. 2025. Aportes de las mujeres a la micología de Latinoamérica. *Funga Latina* 2:V3A2.

Cazabonne, J., A.K. Walker, J. Lesven & D. Haelewaters. 2024. Singleton-based species names and fungal rarity: Does the number really matter? *IMA Fungus* 15:7.

Cheek, M., E. Nic Lughadha, P. Kirk, H. Lindon, J. Carretero, B. Looney, B. Douglas, D. Haelewaters, E. Gaya, T. Llewellyn, A.M. Ainsworth, Y. Gafforov, K. Hyde, P. Crous, M. Hughes, B.E. Walker, R. Campostrini Forzza, K. Meng Wong & T. Niskanen. 2020. New scientific discoveries: Plants and fungi. *Plants, People, Planet* 2:371–388.

Corrales, A., T. Henkel & M. Smith. 2018. Ectomycorrhizal associations in the tropics—biogeography, diversity patterns and ecosystem roles. *New Phytologist* 220:1076–1091.

Coto-López, C., M. Mardones-Hidalgo, R. Aguilar-Álvarez & C. Rojas. 2023. ¿Podría un biomaterial hecho con *Pleurotus ostreatus* funcionar como filtro de agua? *Funga Latina* 1:V1A2.
Dove, N.C. & W.S. Keeton. 2015. Structural complexity enhancement increases fungal species richness in northern hardwood forests. *Fungal Ecology* 13:181–192.
Fernandez, C.W., L. Mielke, A. Stefanski & P.G. Kennedy. 2023. Climate change–induced stress disrupts ectomycorrhizal interaction networks at the boreal–temperate ecotone. *PNAS* 120(34):e2221619120.
Ferreira, V., A. Elosegi, V. Gulis, J. Pozo & M.A.S. Graça. 2006. *Eucalyptus* plantations affect fungal communities associated with leaf-litter decomposition in Iberian streams. *Archiv für Hydrobiologie* 166:467–490.
Halling, R.E. & G.M. Mueller. 2005. *Common Mushrooms of the Talamanca Mountains, Costa Rica*. The New York Botanical Garden, USA.
Hawksworth, D.L. 2001. The magnitude of fungal diversity: The 1.5 million species estimate revisited. *Mycological Research* 105:1422–1432.
Hawksworth, D.L. & R. Lücking. 2017. Fungal diversity revisited: 2.2 to 3.8 million species. *Microbiology Spectrum* 5:FUNK-0052–2016.
Holdridge, L.R. 1967. *Life Zone Ecology*. Tropical Science Center, Costa Rica.
Hutto, R.L. 2016. Should scientists be required to use a model-based solution to adjust for possible distance-based detectability bias? *Ecological Applications* 26:1287–1294.
Isaac, N.J.B. & M.J.O. Pocock. 2015. Bias and information in biological records. *Biological Journal of the Linnean Society* 115:522–531
Jiménez-Rey, F. 2024. Densidad y flexibilidad de dos materiales biológicos elaborados con *Pleurotus ostreatus*. *Funga Latina* 2:V2A2.
Jones, U. 1991. Metates and hallucinogens in Costa Rica. *Papers from the Institute of Archaeology* 2:29–34.
Kapelle, M. 2016. *Costa Rican Ecosystems*. The University of Chicago Press, USA.
Krah, F.S., U. Büntgen & C. Bässler. 2023. Temperature affects the timing and duration of fungal fruiting patterns across major terrestrial biomes. *Ecology Letters* 26:1572–1583.
Lodge, D.J. & S. Cantrell, S. 1995. Fungal communities in wet tropical forests: Variation in time and space. *Canadian Journal of Botany* 73:1391–1398.
Longton, R.E. & T.A. Hedderson. 2000. What are rare species and why conserve them? *Lindbergia* 25:53–61.
Lovelock, C.E. & J.J. Ewel. 2005. Links between tree species, symbiotic fungal diversity and ecosystem functioning in simplified tropical ecosystems. *New Phytologist* 167:219–228.
Mardones, M., J. Carranza Velázquez & C. Rojas. 2026. *Exploring Costa Rica's Fungal Trends: Insights from Digitized Specimens*. Plants, People, Planet. https://doi.org/10.1002/ppp3.70203
Mardones, M. & C. Rojas. 2024. El curioso caso de los hongos de Coto Brus y Osa. *Funga Latina* 2:V2A6.
Mardones, M., L. Umaña, M.M. Granados-Montero, M. Mata-Hidalgo, A. Ruiz-Boyer, M. Piepenbring, D. Minter, C. Coto-López & J. Carranza. 2024. First annotated checklist of Costa Rican fungi. *Funga Latina* 2:V2A7.
Mata, M. 1999. *Macrohongos de Costa Rica = Costa Rica Mushrooms Vol 1*. Instituto Nacional de Biodiversidad, Costa Rica.
Mata, M., R. Halling & G.M. Mueller. 2003. *Macrohongos de Costa Rica Vol. 2 = Costa Rica Macrofungi Vol. 2*. Instituto Nacional de Biodiversidad, Costa Rica.
Molina-Murillo, S., C. Rojas, S. Somerville, R.G. Doss & B. Lemus. 2015. What do you think of fungi? Perceptions by "educated" groups in Honduras and Costa Rica. *Brenesia* 83–84:30–36.
Monge-González, R. 2016. *Innovation, Productivity, and Growth in Costa Rica: Challenges and Opportunities*. Inter-American Development Bank, New York.
Niego, A.G.T., C. Lambert, P. Mortimer, N. Thongklang, S. Rapior, M. Grosse, H. Schrey, E. Charria-Girón, A. Walker, K.D. Hyde & M. Stadler. 2023. The contribution of fungi to the global economy. *Fungal Diversity* 121:95–137.
Niskanen, T., R. Lücking, A. Dahlberg, E. Gaya, L.M. Suz, V. Mikryukov, K. Liimatainen, I. Druzhinina, J.R.S. Westrip, G.M. Mueller, K. Martins-Cunha, P. Kirk, L. Tedersoo & A. Antonelli. 2023. Pushing the frontiers of biodiversity research: Unveiling the global diversity, distribution, and conservation of fungi. *Annual Review of Environment and Resources* 48:149–176.
Nottingham, A.T., E. Gloor, E. Bååth & P. Meir. 2022. Soil carbon and microbes in the warming tropics. *Functional Ecology* 36:1338–1354.
Obando, V. 2007. *Biodiversidad de Costa Rica: Un tesoro por descubrir*. Editorial INBio.
Parra López, C., M. Mardones & C. Rojas. 2024. Normalized results of Latin America's mycological visibility: Are they a good proxy of potential impact? *Funga Latina* 2:V2A8.
Paton, A., A. Antonelli, M. Carine, R. Campostrini Forzza, N. Davies, S. Demissew, G. Dröge, T. Fulcher, A. Grall, N. Holstein, M. Jones, U. Liu, J. Miller, J. Moat, N. Nicolson, M. Ryan, S. Sharrock, D. Smith, B. Thiers, J. Victor, T. Wilkinson & J. Dickie. 2020. Plant and fungal collections: Current status, future perspectives. *Plants, People, Planet* 2:499–514.

Piepenbring, M. 1996. *Smut Fungi (Ustilaginales and Tilletiales) in Costa Rica*. J. Cramer, Germany.
Piepenbring, M., J.G. Maciá-Vicente, J.E.I. Codjia, C. Glatthorn, P. Kirk, J. Meswaet, D. Minter, B.M. Olou, K. Reschke, M. Schmidt & N.S. Yorou. 2020. Mapping mycological ignorance—checklists and diversity patterns of fungi known for West Africa. *IMA Fungus* 11:13.
Prieto de Lima, T.S., A. Orellana Mondol, I. Solano Rodríguez, L.F. Ledezma Vázquez, A. Quirós Castegnaro, A. Taylor Rojas, R. Segura Segura & D. Martínez Martínez. 2024. *Ski' Kuö Pàke: hablemos de hongos*. Universidad de Costa Rica, Costa Rica.
Quesada-Chacón, D., K. Barfus & C. Bernhofer. 2020. Climate change projections and extremes for Costa Rica using tailored predictors from CORDEX model output through statistical downscaling with artificial neural networks. *International Journal of Climatology* 41:211–232.
Rojas, C., I. Arroyo & R.G. Doss. 2023. Ethnomycology, bioprospection and uses of mushrooms in Costa Rica. In K.C. Semwal, S.L. Stephenson & A. Husen (eds.), *Wild Mushrooms and Health: Diversity, Phytochemistry, Medicinal Benefits and Cultivation*, pp. 51–64. CRC Press, Florida.
Rojas, C. & E. Calvo. 2014. Forest biomass, carbon stocks and macrofungal dynamics: A study case in Costa Rica. *International Journal of Forestry Research* 2014:07372.
Rojas, C. & S. Molina-Murillo. 2018. Percepción ciudadana de los recursos fúngicos y potencial de micoalfabetización en Costa Rica. *Scientia Fungorum* 48:23–31.
Rojas, C., R. Valverde & R. Morales. 2017. Functional variability of macrofungal populations in four different forest types of Costa Rica. *Mycosphere* 8:1288–1296.
Schreiber, J., P. Baldrian, V. Brabcová, R. Brandl, H. Kellner, J. Müller, F. Roy, C. Bässler & F.S. Krah. 2024. Effects of experimental canopy openness on wood-inhabiting fungal fruiting diversity across succession. *Scientific Reports* 14:16135.
Solano Quintero, J. & R. Villalobos Flores. 2001. Aspectos fisiográficos aplicados a un bosquejo de regionalización geográfico climático de Costa Rica. *Tópicos Meteorológicos y Oceanográficos* 8:26–39.
Umaña, L. & H.J.M. Sipman. 2002. *Líquenes de Costa Rica*. Editorial INBio, Costa Rica.
Willis, A.D. 2019. Rarefaction, alpha diversity, and statistics. *Frontiers in Microbiology* 10:2407.

6 Macrofungal Diversity in Colombia's Tropical Forests

Insight From the Agaricomycetes (Basidiomycota)

Nataly Gómez-Montoya, Viviana Motato-Vásquez, Ana Cristina Bolaños Rojas, Carolina Ríos Sarmiento, Hannya A. Chamorro-Martínez, Yulena S. Osorio-Navarro, Cristian Zambrano-Forero, Laura Vélez Jaramillo, Lina Rocío Dávila Giraldo, Paola Andrea Zapata Ocampo, Salomé Gómez-Gómez, Alvaro Alexander Dávila Giraldo, Ana Esperanza Franco-Molano, and Denis Cristina Benjumea-Aristizábal

6.1 INTRODUCTION

Fungi represent one of the most diverse clades of eukaryotes, with recent estimates suggesting between 2 and 3 million species worldwide (Niskanen et al. 2023), although only about 150,000 species have been formally described to date (Lücking et al. 2021). Fungal diversity is fundamental to ecosystem functioning and health, mediating interactions among organisms and ecosystems functioning and stability, mediating interactions among organisms and across trophic levels, and fulfilling essential ecological roles that sustain life and drive the dynamics of natural systems (Bahram and Netherway 2021).

Among their most crucial contributions is the decomposition of organic matter, as fungi are uniquely capable of breaking down cellulose and lignin, positioning them as key agents in carbon and nutrient cycling across terrestrial and aquatic ecosystems (Begon et al. 2006; Runnel et al. 2025). Symbiotic root-associated fungi, such as arbuscular and ectomycorrhizal taxa, enhance plant growth and productivity by improving mineral nutrition and water uptake (Tedersoo et al. 2022). Conversely, numerous fungal species act as pathogens of plants, animals, other fungi, and other microorganisms, profoundly influencing ecosystem characteristics such as community structure, diversity, and productivity (Runnel et al. 2025).

From a human perspective, fungi hold enormous biotechnological, cultural, and economic significance, offering compounds and processes used in food production, medicine, agriculture, and industry (Hyde et al. 2019). Their multifaceted contributions align with 10 of the 17 United Nations Sustainable Development Goals (UN General Assembly 2015; Meyer et al. 2020) and 12 targets of the Kunming-Montréal Global Biodiversity Framework (Fungal Conservation Network 2024), underscoring the idea that global biodiversity conservation cannot be achieved without integrating fungi (Cannon et al. 2018; Bahram and Netherway 2021).

Fungi belonging to various taxonomic groups that produce conspicuous sporomes are collectively known as macrofungi. This group includes the so-called gilled fungi, jelly fungi, coral fungi,

DOI: 10.1201/9781003647140-6

stink fungi, bracket fungi, puffballs, truffles, and bird's-nest fungi. Macrofungal diversity represents an important component of global diversity, particularly within fungal communities, and constitutes a key dimension of fungal diversity at large.

Macrofungi are classified within the phyla Ascomycota and Basidiomycota, understanding their diversity and taxonomy is essential for unravelling their ecological functions, evolutionary relationships, and potential applications. Particularly, the phylum Basidiomycota stands out as one of the most ecologically and economically significant fungal lineages (Lu et al. 2023). Members of this phylum are characterized by the production of sexual spores, known as basidiospores, developed on specialized structures called "basidia" (Webster and Weber 2007). This phylum encompasses not only a wide variety of macrofungi but also several important plant pathogens, such as rusts (*Puccinia* spp.) and smuts (*Ustilago* spp.) (Kirk et al. 2008).

Despite their global importance, knowledge of macrofungal diversity in tropical regions remains fragmentary. This is particularly evident in Colombia, one of the world's megadiverse countries, hosting close to 10% of the planet's biodiversity (Chaves et al. 2021). The country harbors a wide array of tropical forests, including rainforests, montane cloud forests, and tropical dry forests, that provide ideal habitats for macrofungi. Yet, estimates suggest that while Colombia could host between 27,400 and 380,000 fungal species, only about 7,650 have been formally recorded (Gaya et al. 2022; Gómez-Montoya et al. 2022; Charria-Girón et al. 2023). This striking gap reflects both the challenges of studying fungi in biodiverse but underexplored ecosystems and the historical underrepresentation of fungi in conservation and biodiversity policies. The limited state of knowledge about Colombia macrofungi results from multiple factors including a scarcity of trained mycologists, restricted access to remote or conflict-affected regions, and a historical bias toward flora and fauna in conservation efforts (Simijaca et al. 2022).

This chapter seeks to address this existing knowledge gap by providing an integrative overview of the diversity of macrofungi (Agaricomycetes, Basidiomycota) in Colombia's tropical forests. By combining ecological, morphological, and molecular perspectives, we aim to highlight both the advances and the existing knowledge shortfalls in the study of Colombian macrofungi. Furthermore, we emphasize the urgency of documenting fungal diversity in a country undergoing rapid environmental transformation, where many ecosystems are critically endangered (Etter et al. 2008).

In the following sections, we provide a comprehensive synthesis of the diversity, ecology, and conservation of macrofungi in Colombia's tropical forests. We examine the main taxonomic groups reported to date, highlight their key ecological interactions, and discuss the current challenges facing fungal research and conservation in the country. Finally, we outline future perspectives for integrating mycological knowledge into biodiversity management and national policy frameworks.

6.2 FUNGA OF COLOMBIA: THE HISTORICAL CONTEXT

A remarkable milestone in the history of tropical mycology is that Colombia was among the first tropical countries to undertake systematic efforts to document its regional fungi. In the early twentieth century, the first expeditions that gave notable attention to fungi in Colombia emerged within the field of phytopathology. One of the pioneering efforts was the *Helvetic Expedition* conducted by Fuhrmann and Mayor in 1914, who collected specimens from several regions, including the Andean forests of Cundinamarca (Fuhrmann and Mayor 1914; Quiroga 2011; Buriticá-Céspedes et al. 2014). The German lichenologist and mycologist Gustav Lindau, also based in Berlin, studied the lichens collected during this expedition, reporting 64 taxa, including the then-new species *Lecidea mayorii* Lindau. Unfortunately, the holotype of this species was destroyed during the 1943 bombing of the Berlin herbarium (Botanischer Garten und Botanisches Museum Berlin, B; Vasco-Palacios and Moncada 2022).

The *Helvetic expedition* inspired the Puerto Rican mycologist and plant pathologists Carlos Chardón and Rafael Toro to conduct the *Mycological Explorations of Colombia,* a project supported by the Colombian government (Chardón and Toro 1930). Their work expanded these early efforts through a national expedition that explored the mountain ranges of Cundinamarca, Antioquia, and Tolima, as well as the Pacific coast, the Magdalena Valley, and the Valle del Cauca (Dumont and Umaña 1978).

In the second half of the twentieth century, the focus of mycological research in Colombia shifted toward taxonomy and ecology. A key figure from this period was Kent P. Dumont (New York Botanical Garden NYBG), who during the 1960s and 1970s conducted comprehensive inventories of macrofungi. The *Mycological Flora of Colombia* project was established in 1974 as a collaboration between the New York Botanical Garden and the Instituto de Ciencias Naturales (ICN) at the Universidad Nacional de Colombia. Between 1978 and 1983, this initiative amassed about 4,000 collections from the departments of Antioquia, Boyacá, Cauca, Cundinamarca, Caquetá, the Pacific coast and Valle del Cauca (Vasco-Palacios and Moncada 2022). As a result of this project, ten publications were produced under the series *Los Hongos de Colombia I–X*, which included descriptions and records of macrofungi in the Ascomycota and Basidiomycota, as well as phytopathogenic fungi, molds, lichens, and myxomycetes (e.g., Dumont et al. 1978; Guzmán and Varela 1978; Pulido 1983). Additionally, the national mycological collections were reorganized and many of them are now housed in the Herbario Nacional Colombiano (COL) of the Universidad Nacional de Colombia (UN), Bogotá, one of the oldest and most extensive collections in the country.

During the same period, the renowned mycologist Rolf Singer described numerous new species, particularly from oak forests in Boyacá and Cundinamarca. Other foreign researchers, including Guzmán, Lowy, Farr, Dennis, Halling, Horak, and Mueller, also made fundamental contributions to the understanding of fungal diversity in Colombia (Vasco-Palacios and Moncada 2022).

Since then, the development of mycological programs at universities in the Andean region of Colombia has fostered the emergence of a national academic community. At the Universidad de Antioquia (UdeA), home of one of the most comprehensive mycological collections in the country, researchers such as L.F. Velásquez, Y. Saldarriaga, L.G. Enao-M. and L.E. Tobón has advanced the study of the Polyporales and Hymenochaetales, while A.E. Franco-Molano has consolidated research on the Agaricales and Boletales (Franco-Molano et al. 2010; Higuita et al. 2019).

In parallel, the establishment of the *Museo Micológico* of the Universidad Nacional de Colombia in Medellín in 1983, together with the integration of the *Colección Fúngica del Herbario Nacional*, greatly enhanced the country's taxonomic and conservation capabilities. Key figures such as Pablo Buriticá made significant contributions to the study of rust fungi (Uredinales), and his legacy continues through institutional mycological collections (Buriticá-Céspedes et al. 2014).

More recently, the field of ethnomycology has been advanced by researchers such as Rocío Peña-Cañón, who has documented the traditional use of edible mushrooms by peasant communities in the highlands of Boyacá (Peña-Cañón and Henao-Mejía 2014; Peña-Cañón et al. 2023). The publication of the *Catalogue of the Fungi of Colombia*, in collaboration with the Instituto Alexander von Humboldt and the Royal Botanic Kew Gardens, marked another milestone in the systematization of national fungal knowledge (De Almeida et al. 2022).

On the other hand, the Andean forest stands out as the region with the longest research tradition and greatest coverage in Colombian mycology. Although less studied, several initiatives have sought to characterize the mycobiota of Colombia's lowland regions. In the Amazonian forests, for instance, projects initiated more than two decades ago have aimed to document fungal diversity and associated ecological roles. In Caquetá, diversity studies have also incorporated ethnomycological knowledge in collaborations with local indigenous communities (Vasco-Palacios et al. 2008).

More recently, additional lowland ecosystems such as dry forests and mangroves have begun to be surveyed and characterized in terms of their fungal communities, largely driven by the strengthening and consolidation of mycological research groups in regions such as the Caribbean coast and Valle del Cauca (García-Lemos and Bolaños-Rojas 2010; Lombana et al. 2016; Luna-Fontalvo et al. 2021, 2023; Motato-Vásquez et al. 2025).

Today, Colombia is home to a growing number of researchers exploring not only species diversity, but also the ecological functions, medicinal properties, and biotechnological potential of fungi

(Vasco-Palacios et al. 2025). This research is no longer limited to the Andean region but is steadily expanding into other areas of the country.

6.3 ECOSYSTEMS OF COLOMBIA

Colombian ecosystems are distributed across six natural regions: The Amazon, Orinoco, Andes, Caribbean, Insular, and Pacific region. Despite its relatively small size, Colombia is the second most biologically diverse country on Earth, harboring about 10% of the world´s species. This exceptional biodiversity stems from the country's wide range of ecosystems, from humid tropical rainforest and coastal cloud forest to open savannas and dry forests. Currently, Colombia hosts approximately 31,141,338 biological species records across all kingdoms (SiB Colombia 2025), of which around 7,650 correspond to fungi (Gaya et al. 2021; Vasco-Palacios et al. 2025).

For the purposes of this chapter, vegetation cover is considered following the categorization of Colombia's six natural regions, distinguishing between lowland and highland forests. Lowland forests, extending from 0 to 1,200 m elevation, are characterized by temperatures ranging from 25 to 35 °C. These areas experience seasonal droughts, with precipitation between 500 and 2,000 mm per year, and host a great diversity of plants, animals, and fungi adapted to withstand water scarcity during the dry season. The structure and physiognomy of dry dense forest vary from closed canopies 25–30 m tall to more open, low, thorny thickets with cacti (Dryflor et al. 2016).

And from 1201 to 3400 m, referred to herein as highland forests regarding their plant composition, they stand out for their high diversity, hosting approximately 25,000 plant species, of which about 10% are endemic. Particularly, it has been identified that in the ecosystems between these altitudes, diversity is divided into a few dominant species and many rare species. Among the dominant species, the oak (*Quercus humboldtii*) is especially notable in high Andean forests. This species is of particular relevance in the context of fungal diversity, since most studies on macrofungi in Colombia have focused on oak forests (Franco-Molano and Uribe-Calle 2000).

6.4 DIVERSITY OF MACROFUNGI (AGARICOMYCETES, BASIDIOMYCOTA)

The Funga of Colombia's tropical lowlands is represented by more than 300 species, distributed across 11 departments: Amazonas, Antioquia, Boyacá, Caquetá, César, Chocó, Guajira, Magdalena, Tolima, Valle del Cauca, and Sucre (Saldarriaga et al. 1988; Franco-Molano et al. 2005; Bolaños-Rojas and Soto-Medina 2011; García-Lemos and Bolaños-Rojas 2010; Palacio et al. 2015; Bolaños-Rojas et al. 2016, 2024; Lombana et al. 2016; Villalobos et al. 2017; Robledo et al. 2020; Luna-Fontalvo et al. 2021, 2023; Zambrano-Forero et al. 2021, 2023). Most of the records originate from the Andean Region and are represented by species with a pantropical distribution or even cosmopolitan distribution, such as *Schizophyllum commune* Fr. (Boiko 2021) (see some representative species in Figure 6.1).

Eighteen orders have been recorded in the Colombian tropical lowlands. The orders with the highest numbers of records are Agaricales and Polyporales, with 85 and 64 species, respectively (Franco-Molano and Uribe-Calle 2000; Gómez-Montoya et al. 2022) (Figure 6.3). In contrast, the orders Dacrymycetales, Geastrales, and Tremellales are represented by only one to three species each, likely reflecting the limited number of studies on these fungal groups in tropical lowland forests. More than 40 fungal families are known from the Colombian tropical lowlands. The most diverse families include Polyporaceae (46 species), Hymenochaetaceae (29 species), and Agaricaceae (18 species). Notably, Hymenochaetaceae and Polyporaceae dominate the polyporoid fungi, which tend to be particularly abundant and diverse in lowland forests characterized by high tree diversity and favorable environmental conditions throughout the year (Mueller et al. 2006). Approximately 90% of the fungal records from these tropical lowlands correspond to saprobic species (thriving on wood, soil, and litter) (Figure 6.3), while a smaller proportion are plant pathogens that play crucial ecological roles by interacting with forest plant communities.

FIGURE 6.1 Fresh basidiomes of macrofungi from tropical lowlands in Colombia. A. *Phellinotus teixerae* on *Phithecelobium dulce* (Benth), Fabaceae, B. *Fabisporus sanguineus* C. *Cymatoderma dendritica* D. *Pleurotus* aff. *ostreatus*. E–F. *Oudemansiella platensis*. Scale bar = 2.5 cm.

Source: Photos by Motato-Vásquez & A.C. Bolaños-Rojas

An ecosystem of particular importance at sea level is represented by mangrove forests, which in Colombia occur along both the Pacific and Caribbean coasts, covering approximately 70% and 30% of the total mangrove area, respectively (Bernal et al. 2017). These forests are composed of plants adapted to salinity, tidal flooding, strong winds, high solar radiation, and elevated

FIGURE 6.2 Fresh basidiomes of macrofungi from tropical highlands in Colombia. A. *Lentinus crinitus*. B. *Auricularia delicata*. C. *Fuscoporia gilva*. D. *Myrmecopterula aff. moniliformis*. E. *Cyptotrama asprata*. F. *Amanita flavoconia*. G. *Stereum aff. sanguinolentum*. Scale bar = 2.5 cm.

Source: Photos by G. V. Motato-Vásquez; E–F. C. Franco; L. Velez-Jaramillo.

temperatures (Tomlinson 1986; Ball 1988; Krauss et al. 2008), as well as muddy, anaerobic soils (Guo et al. 2017). A recent exploratory study on the composition of macrofungi in a mangrove forest on the Colombian Pacific coast revealed remarkable findings. These included the discovery of three new species (*Neohypochnicium manglarense* Motato-Vásq. & Bolaños-Rojas, *Phlebiopsis colombiana* Motato-Vásq. & Bolaños-Rojas, and *Porogramme bononiae* Bolaños-Rojas & Motato-Vásq.) and eight new records for both Colombia and mangrove ecosystems—*Microporus affinis* (Blume & T. Nees) Kuntze, *Paramarasmius palmivorus* (Sharples) Antonín & Kolaric, *Phlebiopsis flavidoalba* (Cooke) Hjortstam, *Porogramme brasiliensis* (Ryvarden) Y.C. Dai, W. Li Mao & Yuan

Yuan, *Resinicium grandisporum* G. Gruhn, Dumez & Schimann, *Trametes ellipsospora* Ryvarden, *T. menziesii* (Berk.) Ryvarden, and *T. polyzona* (Pers.) Justo. In addition, *Lentinus scleropus* (Pers.) Fr. and *Oudemansiella platensis* (Speg.) Speg. were reported for the first time worldwide from mangrove habitats, although they had previously been recorded in Colombian terrestrial ecosystems. Furthermore, *Fomitopsis nivosella* (Murill) Spirin & Vlasák and *Punctularia strigosozonata* (Schwein.) P.H.B. Talbot were recorded for the first time from Colombia (Motato-Vásquez et al. 2025). These findings highlight mangrove ecosystems as a key setting for advancing the study and documentation of fungal diversity of Colombia.

On the other hand, fungal diversity in the Andean forests of Colombia is represented by more than 400 species distributed across 11 departments: Antioquia, Boyacá, Caldas, Cauca, Cundinamarca, Huila, Nariño, Quindío, Risaralda, Tolima and Valle del Cauca. The fungal assemblage comprises 15 orders, with Agaricales and Polyporales being the most common, accounting for 27 and 9 families, respectively (Gómez-Montoya et al. 2022; Pinzón-Osorio et al. 2022; Zambrano-Forero et al. 2023; Benjumea-Aristizábal et al. 2023; Franco-Molano et al. 2024) (see some representative species in Figure 6.2). Other relevant orders include the Russulales and Boletales (Figure 6.3), characterized by ectomycorrhizal species that form associations with *Quercus humboldtii*, thereby contributing to the maintenance of fertility in acidic and nutrient-poor soils (Corrales et al. 2018).

In terms of families, the fungal assemblage includes more than 60, with *Polyporaceae* (35 species), *Marasmiaceae* (35 species), *Hymenogastraceae* (30 species), and *Boletaceae* (28 species) being the most prominent (Gómez-Montoya et al. 2022; Pinzón-Osorio et al. 2022; Zambrano-Forero et al. 2023; Benjumea-Aristizábal et al. 2023; Franco-Molano et al. 2024), all of them belong to the phylum Basidiomycota. The dominance of saprobic families underscores the importance of these fungi in the decomposition of wood and leaf litter, processes fundamental to the biogeochemical dynamics of high Andean forests (Gibertoni et al. 2016). Approximately 85% of the known records correspond to saprobic species (occurring on wood, soil, and leaf litter), whereas 10% are ectomycorrhizal and mainly associated with oak (*Q. humboldtii*) (Figure 6.3).

6.5 TRADITIONAL KNOWLEDGE AND BIOTECHNOLOGICAL OPPORTUNITIES OF WILD BASIDIOMYCOTA

Wild basidiomycota in Colombia represent a valuable component of traditional knowledge and an emerging resource for biotechnology. Historically, Indigenous and rural communities have used various macrofungal species for food, medicinal, cosmetic, and cultural purposes. Recent studies have addressed the domestication, cultivation, and nutritional characterization of native edible fungi, as well as their potential for the production of enzymes, exopolysaccharides, and other functional bioproducts. These findings highlight their relevance for the development of bioeconomy initiatives and novel biotechnological applications.

6.5.1 Ethnomycological Knowledge in Colombia

Regarding the use of species of the Basidiomycota species in Colombia's tropical lowlands, *Podaxis pistillaris* (L.) Fr., *Inonotus rickii* (Pat.) D.A. Reid, and *Phellorinia herculeana* (Pers.) Kreisel are recognized by the Wayuu people, an indigenous community inhabiting the La Guajira Peninsula, which lies along the Caribbean coast of northeastern Colombia and extends into Venezuela (ANLA 2020). The Wayuu prepare a black, pigment-like powder from the spores of these fungi, which is applied in their face as sun protection and as cosmetic. In religious rituals, the same powder is used to cover the faces of the deceased, preparing them to endure the sun on their long journey to the afterlife (Villalobos et al. 2017; Ocando-Rincón 2014; Vergara-González 1986; Candelier 1893).

In the Andean region, particularly in the department of Boyacá, several species of wild macrofungi are recognized and collected by local communities for consumption (Henao and Ruíz 2006; Peña-Cañón and Henao-Mejía 2014; Pérez and Piragauta 2006; Ruiz-Roa et al. 2008). Elders play a key

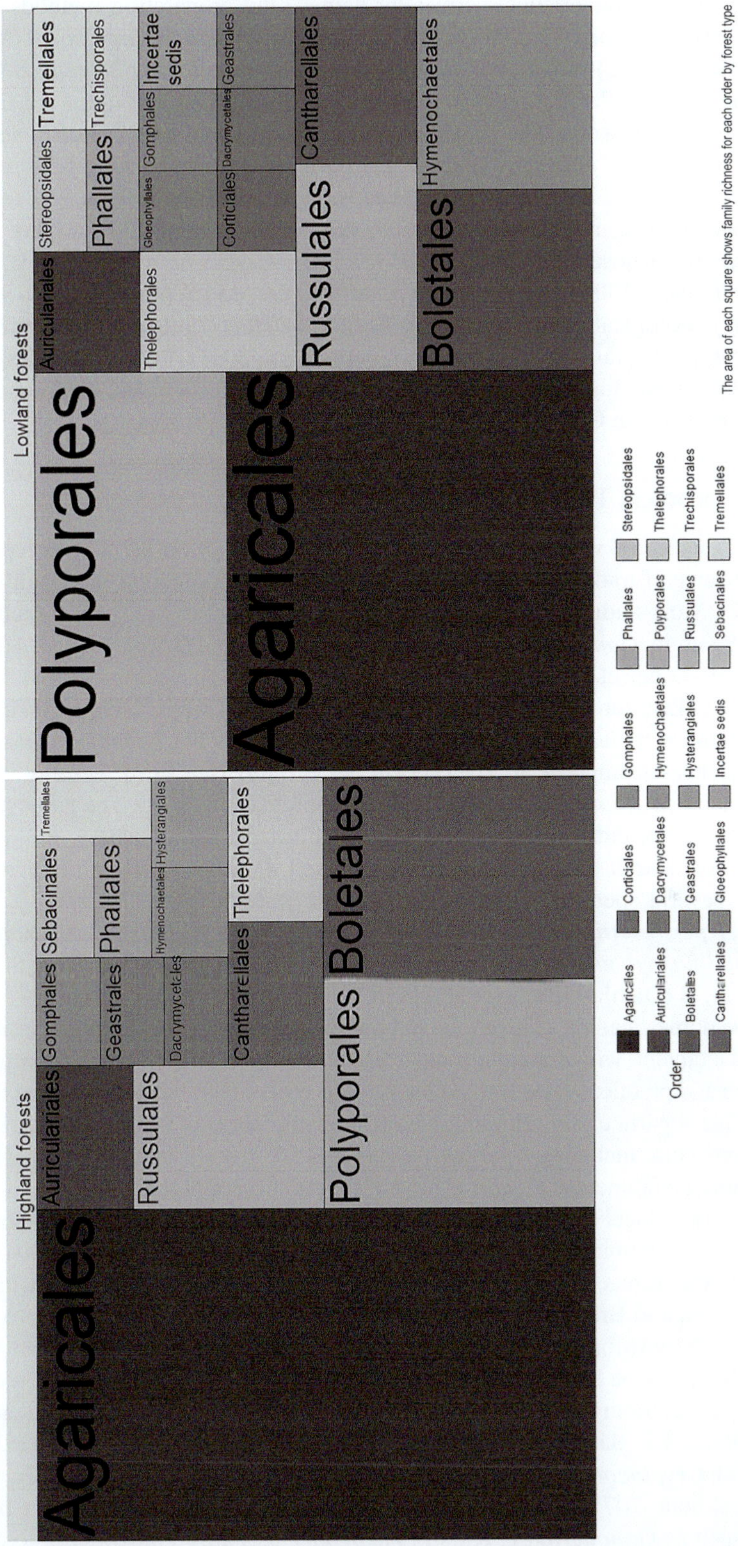

FIGURE 6.3 Treemap of the composition and abundance of the main fungal orders across the Colombian ecosystem types.

role in transmitting traditional knowledge about these edible fungi, including the seasons of their appearance, identifying characteristics, vernacular names, and preparation methods (Peña-Cañón and Henao-Mejía 2014). Reported edible species include *Auricularia fuscosuccinea* (Mont.) Henn., *Cantharellus lateritius* (Berk.) Singer, *Crepidotus palmarum* Singer, *Favolus tenuiculus* P. Beauv., *Hericium erinaceus* (Bull.) Pers., *Lentinus crinitus* (L.) Fr., *Macrocybe titans* (H.E. Bigelow & Kimbr.) Pegler, Lodge, Nakasone, *Macrolepiota colombiana* Franco-Mol., *Oudemansiella canarii* (Jungh.) Höhn., *Porphyrellus indecisus* (Peck.) E.-J. Gilbert, and *Ramaria* spp. (Peña-Cañón and Henao-Mejía 2014; Pérez and Piragauta 2006; Henao and Ruíz 2006).

Beyond their dietary use, rural communities in the Andean region also employ certain macrofungi for medicinal purposes. For example, *Psilocybe cubensis* (Earle) Singer is used to treat boils (Pérez and Piragauta 2006), while *Bovista plumbea* Pers. and *Scleroderma citrinum* Pers. are applied to treat dermatological conditions and as healing agents for burns and wounds (Henao and Ruíz 2006; Pérez and Piragauta 2006; González-Cuellar et al. 2021). These examples highlight the need for further ethnomycological research to document and preserve the traditional knowledge associated with fungal use in Colombia.

6.5.2 Biotechnological Potential

In Colombia, 179 species of wild fungi used for human consumption have been reported, underscoring the significance of traditional knowledge associated with indigenous and rural communities (Vargas et al. 2022). Most studies have focused on commercially cultivated species such as *Pleurotus ostreatus* (Jacq.) P. Kumm, *Ganoderma lucidum* (Curtis) P. Karst, and *Grifola frondosa* (Dicks.) Gray, among others (Torres-López et al. 2011). To date, only three studies have documented the domestication and large-scale cultivation of wild macrofungi. The species involved include *Auricularia fuscosuccinea* (Mont.) Henn, *Crepidotus palmarum* Singer (Niño et al. 2017), *Favolus rugulosus* Palacio & R.M. Silveira (Sánchez-Ocampo et al. 2022), *Lentinus crinitus* (L.) Fr., *Pleurotus tubarius* (Pat.) Pegler, and *Fabisporus sanguineus* (L.) Zmitr. (Dávila-Giraldo et al. 2022). The latter study provides detailed information on the nutritional composition of cultivated basidiomata.

Within the same research line, a recent study assessed the nutritional value of wild-collected basidiomata from several species, including *Auricularia fuscosuccinea*, *Bovista* sp., *Ganoderma australe* (Fr.) Pat., *Irpex rosettiformis* C.C. Chen & Sheng H. Wu, *Lentinus* sp., *Pleurotus floridanus* Singer, Pegler, *Phellinus gilvus* (Schwein.) Pat., *Trametes coccinea* (Fr.) Hai J. Li & S.H. He, and *T. elegans* (Spreng.) Fr. B (Dávila-Giraldo et al. 2023).

Research on the domestication, cultivation, and nutritional composition of wild Basidiomycota in Colombia remains incipient, with major gaps regarding edible species. The social appropriation of traditional knowledge and practices related to edible fungi is essential to fully understand their potential as food sources. This is particularly relevant given that a wide range of bioproducts can be generated from wild Basidiomycota, including food ingredients and additives, basidiomata as sources of proteins, fibers, vitamins, and minerals, as well as protein formulations and functional products (Mortimer et al. 2021). Such bioproducts could contribute to diversifying local and national economies through the sustainable use of Colombia's Basidiomycota diversity (Cuesta and Castro-Ríos 2017).

Colombia, however, ranks among the countries with the fewest fungal DNA barcodes, lagging behind Argentina and Brazil (Barcode of Life Data System 2023). According to the National Center for Biotechnology Information (NCBI), a total of 62,820 nucleotide sequences belong to the Ascomycota and 14,117 to the Basidiomycota (data revised in February 2024, https://www.ncbi.nlm.nih.gov). Within the Basidiomycota, several species stand out by the number of sequences reported: *Blastosporella zonata* T.J. Baroni & Franco-Mol (4,808), *Malassezia pachydermatis* (Weidman) C.W. Dodge (4,478), *Cryptococcus neoformans* (San Felice) Vuill (911), and *Cryptococcus tetragattii* F. Hagen & Boekhout (707). In contrast, fewer than one hundred sequences have been reported for other genera such as *Ganoderma* P. Karst, *Leucocoprinus* Pat., *Neofavolus* Sotome & T. Hatt. *Russula* Pers. and *Trametes* Fr.

Various strains of native fungi from Colombian forests have been cultivated for the production and extraction of exopolysaccharides, as well as ligninolytic and cellulolytic enzymes. These include *Schizophyllum radiatum* Fr. *Lentinus crinitus* (L.) Fr. (Dávila-Giraldo et al. 2023), *Fabisporus sanguineus* (L.) Zmitr., and *Cerrena* spp. (Zambrano-Forero et al. 2021). To support biotechnological applications, Colombia currently maintains 17 registered culture collections of filamentous fungi across nine departments, providing legally compliant mechanisms for the deposition and acquisition and acquisition of macrofungal strains (Ortiz-Moreno et al. 2022).

In biotechnology, colors are often used to denote major application fields: red for health (R), green for agriculture (G), white for industrial processes (W), and gray (Gy) for environmental protection. Native fungi from Colombia exhibit strong potential across all these biotechnological domains (Barcelos et al. 2018). Table 6.1 presents selected fungal genera and their biotechnological potential as reported by different authors worldwide.

TABLE 6.1
Traditional Knowledge and Biotechnological Potential of Wild Basidiomycota

Relevant Genera	Traditional Use	Biotechnological Application*				References
		R	G	W	Gy	
Agaricus	Food	**R**		W		Henao and Ruíz (2006); Pérez and Piragauta (2006); Arya and Rusevska (2022); Mago et al. (2023); Peña-Cañón et al. (2023):Díaz-Montes et al. (2021)
Amanita		**R**	**G**		**Gy**	Al-Rajhi et al. (2023)
Amauroderma		**R**		W		Loi et al. (2021)
Auricularia	Food	**R**				Pérez and Piragauta (2006); Vargas et al. (2022); Shevchuk et al. (2023)
Boletus		**R**				Tian et al. (2025)
Calvatia		**R**		W		Kshirsagar et al. (2025)
Cantharellus	Food	**R**	**G**			Pérez and Piragauta (2006); Ghobad-Nejhad et al. (2023)
Coprinellus				W	**Gy**	Tešanović et al. (2017); Shalini et al. (2009)
Coprinus		**R**		W	**Gy**	Tešanović et al. (2017); Wang et al. (2018); Nowakowski et al. (2020)
Coriolopsis				W	**Gy**	Munck et al. (2018)
Favolus	Food	**R**		W	**Gy**	Arya and Rusevska (2022); Abd and Aty (2023); Peña-Cañón et al. (2023)
Lactarius	Food	**R**	**G**			González-Cuellar et al. (2021); Kostić et al. (2023)
Laetiporus	Food	**R**		W	**Gy**	Abd and Aty (2023); Peña-Cañón et al. (2023)
Lentinula	Food	**R**		W		Pérez and Piragauta (2006), Barrios-Gutiérrez et al. (2023); Mago et al. (2023); Peña-Cañón et al. (2023)
Lentinus	Food	**R**		W	**Gy**	Maestre-Reyna et al. (2015); Torres-Morales et al. (2022); Peña-Cañón et al. (2023)
Lenzites				W	**Gy**	Lakshmi et al. (2018); Arya and Rusevska (2022)
Lepiota		**R**	**G**			Arya and Rusevska (2022)
Lepista	Food	**R**				Arya and Rusevska (2022); Peña-Cañón et al. (2023)
Leucoagaricus			**G**		**Gy**	Maya-Yescas et al. (2021)
Leucopaxillus		**R**		W		Malya et al. (2020); Wang et al. (2022)

(Continued)

TABLE 6.1 (*Continued*)
Traditional Knowledge and Biotechnological Potential of Wild Basidiomycota

Relevant Genera	Traditional Use	Biotechnological Application*				References
		R	G	W	Gy	
Lopharia					**Gy**	Wang et al. (2010)
Macrocybe	Food	**R**				Torres-Morales et al. (2022); Peiris et al. (2024)
Macrolepiota	Food	**R**		W		Henao and Ruíz (2006); Pérez and Piragauta (2006); González-Cuellar et al. (2021); Adamska and Tokarczyk (2022); Eyupoglu et al. (2025)
Marasmiellus		**R**		W	**Gy**	Chenthamarakshan et al. (2017); Kardena et al. (2020); Yao et al. (2025)
Marasmius		**R**		W	**Gy**	Gröbe et al. (2011); Jeon and Lim (2017); Dimopoulou et al. (2022); Barrios-Gutiérrez et al. (2023).
Neonothopanus				W	**Gy**	Mogilnaya et al. (2021); Ventura et al. (2021)
Oudemansiella	Food	**R**		W	**Gy**	Sanjuan (1999); Pérez and Piragauta (2006); Iark et al. (2019); Balaraju et al. (2010); Dulay (2023)
Oxyporus			**G**			Lee et al. (2009)
Panaeolus		**R**				Nkadimeng et al. (2020)
Phellinus		**R**		W		Chen et al. (2016); Mago et al. (2023)
Phylloporus		**R**		W		Liu et al. (2024)
Pleurotus	Food	**R**	**G**	W	**Gy**	Sufiate et al. (2017); Wang et al. (2018); González-Cuellar et al. (2021); Abd and Aty (2023); Barrios-Gutiérrez et al. (2023); Mago et al. (2023); Peña-Cañón et al. (2023)
Polyporus		**R**		W	**Gy**	Loi et al. (2016)
Psilocybe		**R**				Lowe et al. (2021)
Pycnoporus		**R**		W	**Gy**	Barrios-Gutiérrez et al. (2023); Loi et al. (2021)
Schizophyllum	Food	**R**		W	**Gy**	Sanjuan (1999); González-Cuellar et al. (2021); Mago et al (2023); Loi et al. (2021); Peña-Cañón et al. (2023); Zhao et al. (2024)
Stereum				W		Barrios-Gutiérrez et al. (2023)
Suillus	Food	**R**	**G**			Pera and Parladé (2005); Venditti et al. (2017)
Trametes		**R**		W		Mago et al. (2023); Barrios-Gutiérrez et al. (2023)
Tricholoma		**R**			**Gy**	Li et al. (2022); Cao et al. (2010)

* The colors of biotechnology represent its main application fields: red for health (R), green for agriculture (G), white for industrial processes (W), and gray for environmental protection (Gy).

6.6 MAIN THREATS AND CHALLENGES

Colombian forests face serious threats, including the expansion of the agricultural frontier, land-use conversion, illegal mining, urbanization, and climate change. These factors have significantly reduced both the extent and ecological quality of forested areas. In response, the country has promoted conservation, ecological restoration, and sustainable development strategies through programs such as the Andean Forests Program, implemented in collaboration with international organizations (Báez et al. 2016).

Due to their particular characteristics, highland ecosystems such as Andean forests are highly sensitive to environmental changes. Their location in areas of steep terrain subject to high rainfall favors intense erosion processes (Ohl and Bussmann 2004; Sabogal et al. 2015; Armenteras et al. 2017). Duque et al. (2021) suggest that Mature Andean forests as globally important carbon sinks and future carbon refuges. Conversely, tropical lowlands, such as dry forest, face threats of both natural and anthropogenic origin. Natural threats include wildfires, extreme drought, and climate change, while anthropogenic pressures stem from unsustainable livestock farming, human settlement, and the extraction and exploitation of natural resources (Tamayo-Quintana and Torres-Romero 2022).

Because the composition and distribution of fungal species correlate with environmental and biotic factors such as climate, water, and soil chemistry, altitude, plant community composition, and habitat quality, these ecosystem-level threats jeopardize fungal diversity, making conservation actions urgent (Geml et al. 2022; Mueller et al. 2022). One essential conservation tool is the assessment and reporting of the conservation status of fungal species. However, such evaluations require a robust baseline of taxonomic and ecological documentation (Dahlberg and Mueller 2011). In this context, species checklists are of particular importance, as they enable long-term monitoring of the conservation status of both ecosystems and fungal populations, and serve as essential resource for curating specimens in biological collections (Eppink and Van Den Bergh 2007; González and Jiménez 2013; Zotti et al. 2013).

Effectively documenting this diversity requires the use of integrative taxonomy (Cao et al. 2021), for which molecular studies are pivotal. Characterizing the actual genetic diversity of Colombian fungi remains a major challenge, as DNA sequences are necessary to improve our understanding of fungal taxonomy, phylogenetic affinities, and ecological interactions. The scarcity of fungal genetic data in public repositories is attributed not only to limited financial resources but also to two major bottlenecks: a global shortage of trained mycologists (Lücking 2020) and the lack of widespread academic training in molecular biology within the country.

Furthermore, advancing mycological research in a megadiverse country like Colombia faces institutional and legal hurdles. Navigating the Access and Benefit-Sharing (ABS) frameworks established under the Nagoya Protocol presents significant challenges for researchers seeking to legally access genetic resources and conduct molecular or biotechnological studies (Silvestri 2016).

Despite these limitations, the opportunities for harnessing the country's fungal diversity are immense. Establishing a solid baseline of knowledge on fungal diversity is a fundamental first step. This documentation is not only critical for conservation planning but also essential for investigating the traditional and biotechnological potential of these resources, ensuring their sustainable use for future generations.

6.7 ACKNOWLEDGMENTS

Thanks to Sebastián Serna-Muñoz for his advice on generating the treemaps graphs.

REFERENCES

Abd, E. & A.-A Aty. 2023. Valorization of agro-industrial wastes using fungi for industrial enzyme production. *International Journal of Frontline Research in Science and Technology* 2:1–23. https://doi.org/10.56355/ijfrst.2023.2.2.0055

Adamska, I. & G. Tokarczyk. 2022. Possibilities of using *Macrolepiota procera* in the production of pro-health food and in medicine. *International Journal of Food Science* 1:1–27. https://doi.org/10.1155/2022/5773275

Al-Rajhi, A.M.H., M.M. Alawlaqi, T.M. Abdel Ghany & H. Moawad. 2023. *Amanit*a sp. from the subtropical region of Saudi Arabia as a source of chitinase enzyme and its antifungal activity. *BioResources* 18:2928–2939. https://doi.org/10.15376/biores.18.2.2928-2939

ANLA. 2020. Reporte Análisis Regional: Actualización del Reporte de Alertas de Análisis Regional de la Zona Hidrográfica Caribe-Guajira. *Autoridad Nacional de Licencias Ambientales (ANLA)*. https://www.anla.gov.co/documentos/biblioteca/07-01-2021-anla-aarzh-caribe-guajira.pdf

Armenteras, D., J.M. Espelta, N. Rodríguez & J. Retana. 2017. Deforestation dynamics and drivers in different forest types in Latin America: Three decades of studies (1980–2010). *Global Environmental Change* 46:139–147. https://doi.org/10.1016/j.gloenvcha.2017.09.002

Arya, A. & K. Rusevska (eds). 2022. *Biology, Cultivation and Applications of Mushrooms*. Springer, Singapore.

Báez, S., A. Malizia, J. Carilla & O. Osinaga. 2016. Propuestas Andinas #14: Respuestas de los bosques andinos a los cambios ambientales globales. *Proyecto EcoAndes, CONDESAN, Bosques Andinos. Recuperado*. https://condesan.org/recursos/propuestas-andinas-14-respuestas-de-los-bosques-andinos-a-los-cambios-ambientales-globales/

Bahram, M. & T. Netherway. 2021. Fungi as mediators linking organisms and ecosystems. *FEMS Microbiology Reviews* 46:fuab058. https://doi.org/10.1093/femsre/fuab058

Balaraju, K.K., S. Park, S. Jahagirdar & V. Kaviyarasan. 2010. Production of cellulase and laccase enzymes by *Oudemansiella radicata* using agro wastes under solid-state and submerged conditions. *Research in Biotechnology* 1:21–28.

Ball, M.C. 1988. Ecophysiology of mangroves. *Trees* 2:129–142. https://doi.org/10.1007/BF00196018

Barcelos, M., F. Lupki, G. Campolina, D. Lee & G. Molina. 2018. The colors of biotechnology: General overview and developments of white, green and blue areas. *FEMS Microbiology Letters* 365:21. https://doi.org/10.1093/femsle/fny239

Barcode of Life Data System. 2023. *BOLD: Barcode of Life Data System*. https://www.boldsystems.org/

Barrios-Gutiérrez, S., M. Vélez-Mercado & J. Rodrigues. 2023. Oxidative machinery of basidiomycetes as potential enhancers in lignocellulosic biorefineries: A lytic polysaccharide monooxygenases approach. *Bioresource Technology* 386:129481. https://doi.org/10.1016/j.biortech.129481.

Begon, M., C.R. Townsend & J.L. Harper. 2006. *Ecology: From Individuals to Ecosystems*. Blackwell Publishing, Oxford, England.

Benjumea-Aristizábal, D.C., A. Builes-Vélez, N. Gómez-Montoya & C.A. López-Quintero. 2023. New records of macrofungi from Colombia and the department of Antioquia. *Funga Latina* 1: Article A9. https://doi.org/10.5281/zenodo.10267771

Bernal, B., G. Sidman & T. Pearson. 2017. *Assessment of Mangrove Ecosystems in Colombia and Their Potential for Emissions Reductions and Restoration*. Winrock International, Nairobi, Kenya.

Boiko, S. 2021. The local populations of the fungus *Schizophyllum commune* Fr. as drivers of its biodiversity. *Folia Forestalia Polonica Series A-Forestry* 634:308–316. https://doi.org/10.2478/ffp-2021-0031

Bolaños-Rojas, A.C., V. Bononi, A. Gugliotta & J. Muñoz. 2016. New records of *Ganoderma multiplicatum* (Mont.) Pat. (Polyporales, Basidiomycota) from Colombia and its geographic distribution in South America. *Check List* 12:948–1948. https://doi.org/http://dx.doi.org/10.15560/12.4.1948

Bolaños-Rojas, A.C., J. Londoño-Caicedo, A. Cortés & V. Motato-Vásquez. 2024. Phylogenetic diversity, host specificity, and distribution of the wood-decaying fungus *Phellinotus teixeirae* in western Colombia's seasonally dry tropical forest. *Forests* 156:1008. https://doi.org/10.3390/f15061008

Bolaños-Rojas, A.C. & E. Soto-Medina. 2011. Macrohongos comestibles y medicinales comunes en la vegetación de la Universidad del Valle, Colombia. *Revista de Ciencias* 15:31–38. https://doi.org/10.25100/rc.v15i0.515

Buriticá-Céspedes, P., M. Salazar-Yepes & V. Pardo-Cardona. 2014. Pucciniales (Fungi), Royas de Colombia. *Revista Facultad Nacional de Agronomía Medellín* 67(suppl. 1):S1–S93.

Candelier, H. 1893. *Rio-Hacha et les Indiens Goajires*. Ulan Press, New York. https://archive.org/details/riohachaetlesin00candgoog (consultado el 2 de marzo de 2015)

Cannon, P., B. Aguirre-Hudson & M. Catherine. 2018. Definition and diversity. In K.J. Willis (ed.), *State of the World's Fungi*, pp. 4–11. Royal Botanic Gardens, Kew.

Cao, B., D. Haelewaters & N. Schoutteten. 2021. Delimiting species in Basidiomycota: A review. *Fungal Diversity* 109:181–237. https://doi.org/10.1007/s13225-021-00479-5

Cao, Y.-R., Z. Liu, G.-L. Cheng, X.-B. Jing & H. Xu. 2010. Exploring single and multi-metal biosorption by immobilized spent *Tricholoma lobayense* using multi-step response surface methodology. *Chemical Engineering Journal* 164:183–195. https://doi.org/10.1016/j.cej.2010.08.055

Chardón, C. & R. Toro. 1930. Mycological exploration of Colombia. *Journal of the Department of Agriculture of Porto Rico* 144:195–369. https://doi.org/10.46429/jaupr.v14i4.14223

Charria-Girón, E., A.M. Vasco-Palacios, B. Moncada & Y. Marin-Felix. 2023. Colombian fungal diversity: Untapped potential for diverse applications. *Microbiology Research* 14:2000–2021. https//doi.org/10.3390/microbiolres1404013

Chaves, M.E., R. Gómez-S., W. Ramírez & C. Solano (eds.) 2021. *Colombia's National Biodiversity and Ecosystem Assessment: Summary for Policymakers. Alexander von Humboldt Biological Resources Research Institute; United Nations Environment Programme World Conservation Monitoring Centre (UNEP-WCMC)*. Federal Ministry for the Environment, Nature Conservation, Nuclear Safety and Consumer Protection of Germany.

Chen, H., T. Tian, H. Miao & Y.Y. Zhao. 2016. Traditional uses, fermentation, phytochemistry and pharmacology of *Phellinus linteus*: A review. *Fitoterapia* 113:6–26. https://doi.org/10.1016/j.fitote.2016.06.009

Chenthamarakshan, A., N. Parambayil & N. Miziriya. 2017. Optimization of laccase production from *Marasmiellus palmivorus* LA1 by Taguchi method of design of experiments. *BMC Biotechnology* 17:12. https://doi.org/10.1186/s12896-017-0333-x.

Corrales, A., H. Terry & M. Smith. 2018. Ectomycorrhizal associations in the tropics–biogeography, diversity patterns and ecosystem roles. *New Phytologist* 2204:1076–1091. https://doi.org/10.1111/nph.15151

Cuesta, M.-C. & K. Castro-Ríos. 2017. Mushrooms as a strategy to reduce food insecurity in Colombia. *Nutrition & Food Science* 47:817–828. https://doi.org/10.1108/NFS-03-2017-0039

Dahlberg, A. & G. Mueller. 2011. Applying IUCN red-listing criteria for assessing and reporting on the conservation status of fungal species. *Fungal Ecology* 42:147–162. https://doi.org/10.1016/j.funeco.2010.11.001

Dávila-Giraldo, L., C. Pérez Jaramillo, J. Méndez & W. Murillo-Arango. 2023. Nutritional value and antioxidant, antimicrobial and cytotoxic activity of wild macrofungi. *Microorganisms* 11:1158. https://doi.org/10.3390/microorganisms11051158

Dávila-Giraldo, L., C. Zambrano-Forero, O. Torres-Arango, J. Betancur & W. Murillo-Arango. 2022. Integral use of rice husks for bioconversion with white-rot fungi. *Biomass Conversion and Biorefinery* 12:2981–2991. https://doi.org/10.1007/s13399-020-00940-6

De Almeida, R., R. Lücking, A.M. Vasco-Palácios, E. Gaya & M. Diaz-Granados. 2022. Catalogue of fungi of Colombia. *Revista de la Academia Colombiana de Ciencias Exactas, Físicas y Naturales* 46:1050–1053. https://doi.org/10.18257/raccefyn.1842

Díaz-Montes, E., J. Yáñez-Fernández & R. Castro-Muñoz. 2021. Dextran/chitosan blend film fabrication for bio-packaging of mushrooms (*Agaricus bisporus*). *Journal of Food Processing and Preservation* 45:e15489. https://doi.org/10.1111/jfpp.15489

Dimopoulou, M., A. Kolonas, S. Mourtakos, O. Androutsos & O. Gortzi. 2022. Nutritional composition and biological properties of sixteen edible mushroom species. *Applied Sciences* 12:8074. https://doi.org/10.3390/app12168074

Dryflor, K.B., A. Delgado-Salinas, K.G. Dexter, R. Linares-Palomino & A. Oliveira-Filho. 2016. Plant diversity patterns in Neotropical dry forests and their conservation implications. *Science* 353:1383–1387. https://doi.org/10.1126/science.aaf5080

Dulay, R.M. 2023. *Oudemansiella* (Physalacriaceae) mushrooms: A status review on the distribution, cultivation, composition and bioactivity profile. *Studies in Fungi* 8:13. https://doi.org/10.48130/SIF-2023-0013

Dumont, K.P., P. Buriticá & E. Forero. 1978. Los hongos de Colombia–I. *Caldasia* 12:159–164. http://www.jstor.org/stable/43405904

Dumont, K.P. & M.I. Umaña. 1978. Los hongos de Colombia V: *Laterna triscapa* y *Calostoma cinnabarina* en Colombia. *Caldasia* 12:349–352.

Duque, A., M.A. Peña, F. Cuesta, S. González-Caro, P. Kennedy, O.L. Phillips, M. Calderón-Loor, C. Blundo, J. Carilla, L. Cayola, W. Farfán-Ríos, A. Fuentes, R. Grau, J. Homeier, M.I. Loza-Rivera, Y. Malhi, A. Malizia, L. Malizia, J.A. Martínez-Villa, J.A. Myers, O. Osinaga-Acosta, M. Peralvo, E. Pinto, S. Saatchi, M. Silman, J.S. Tello, A. Terán-Valdez & K.J. Feeley. 2021. Mature Andean forests as globally important carbon sinks and future carbon refuges. *Nature Communications* 12:2138. https://doi.org/10.1038/s41467-021-23955-7

Eppink, F. & J. Van Den Bergh. 2007. Ecological theories and indicators in economic models of biodiversity loss and conservation: A critical review. *Ecological Economics* 61:284–293. https://doi.org/10.1016/j.ecolecon.2006.01.013

Etter, A., C. McAlpine & H. Possingham. 2008. Historical patterns and drivers of landscape change in Colombia since 1500: A regionalized spatial approach. *Annals of the Association of American Geographers* 981:2–23.

Eyupoglu, V., M.B. Akin, S. Kaya, O. Çaylak, A. Berisha & S. Çetinkaya. 2025. Effective removal of methylene blue dye from aqueous solution using *Macrolepiota procera* Mushroom: Experimental and theoretical studies. *Journal of Molecular Liquids* 418:126714. https://doi.org/10.1016/j.molliq.2024.126714

Franco-Molano, A.E., T.J. Baroni & L.J.J. van de Peppel. 2024. *Arthromyces pulverulentus* sp. nov. (Basidiomycota, Agaricales, Lyophyllaceae), a new insect associated conidia producing species from the cloud forests of Colombia. *Phytotaxa* 662:271–278. https://doi.org/10.11646/phytotaxa.662.3.6

Franco-Molano, A.E., A. Corrales & A.M. Vasco-Palacios. 2010. Macrohongos de Colombia II. Listado de especies de los órdenes Agaricales, Boletales, Cantharellales y Russulales (Agaricomycetes, Basidiomycota). *Actualidades Biológicas* 32:89–113.

Franco-Molano, A.E. & E. Uribe-Calle. 2000. Hongos agaricales y boletales de Colombia. *Bioya Colombiana* 1:25–43. http://revistas.humboldt.org.co/index.php/biota/article/view/55

Franco-Molano, A.E., A.M. Vasco-Palacios, C.A. López-Quintero & T. Boekhout. 2005. *Macromicetes de la región del Medio Caquetá. Guía de campo*, p. 211. Medellín: Multimpresos.

Fuhrmann, O. & E. Mayor. 1914. Voyage d'exploration scientifique en Colombie par Dr O Fuhrmann et Dr Eug. Mayor. *Mémoires de la Société des Sciences Naturelles de Neuchâtel* 5:1087.

Fungal Conservation Network. 2024. *Contribution of Fungi to the Global Biodiversity Framework.* https://doi.org/10.5281/zenodo.14680634

García-Lemos, A. & A.C. Bolaños-Rojas. 2010. Macrohongos presentes en el Bosque Seco Tropical de la Región del Valle del Cauca, Colombia. *Revista de Ciencias* 14:45–54. https://doi.org/10.25100/rc.v14i0.653

Gaya, E., V. Motato-Vásquez & R. Lücking. 2022. Diversity of fungi of Colombia. In R.F. De Almeida, R. Lücking, A. Vasco-Palacios, E. Gaya & M. Diazgranados (eds.), *Catalogue of Fungi of Colombia,* pp. 45–57. Royal Botanic Gardens, Kew, UK.

Gaya, E., A.M. Vasco-Palacios, N. Vargas-Estupiñan, R. Lücking, J. Carretero, T. Sanjuan, B. Moncada, B. Allkin, A.C. Bolaños-Rojas, C. Castellanos-Castro, L.F. Coca, A. Corrales, T. Cossu, L. Davis, J. dSouza, A. Dufat, A.E. Franco-Molano, F. García, N. Gómez-Montoya, F.E. González-Cuellar, D. Hammond, A. Herrera, M.M. Jaramillo-Ciro, C. Lasso-Benavides, M.P. Mira, J. Morley, V. Motato-Vásquez, Y. Niño-Fernández, M.L. Ortiz-Moreno, E.R. Peña-Cañón, M. Ramírez-Castrillón, T. Rojas, J. Ruff, D. Simijaca, H.J.M. Sipman, E. Soto-Medina, G. Torres, P.A. Torres-Andrade, T. Ulian, K. White & M. Diazgranados. 2021. *ColFungi: Colombian Resources for Fungi Made Accessible.* Royal Botanic Gardens, Kew.

Geml, J., A. Arnold & T. Semenova-Nelsen. 2022. Community dynamics of soil-borne fungal communities along elevation gradients in Neotropical and Palaeotropical forests. *Molecular Ecology* 31:2044–2060. https://doi.org/10.1111/mec.16368

Ghobad-Nejhad, M., D. Bálint, C. Bao-Kai & S. Jing. 2023. Basidiomycete fungi: From biosystematics and biodiversity to biotechnology. *Frontiers in Microbiology* 14:1128319. https://doi.org/10.3389/fmicb.2023.1128319

Gibertoni, T., P. Medeiros & A. Soares. 2016. The distribution of polypore fungi in endemism centres in Brazilian Amazonia. *Fungal Ecology* 20:1–6. https://doi.org/10.1016/j.funeco.2015.09.012

Gómez-Montoya, N., C. Ríos-Sarmiento, B. Zora-Vergara, C. Benjumea-Aristizábal, D.J. Santa-Santa, M. Zuluaga-Moreno & A.E. Franco-Molano. 2022. Macrofungal diversity (Basidiomycota) of Colombia: List of species. *Actualidades Biológicas* 44:e7. https://doi.org/10.17533/udea.acbi.v44n116a07

González, D. & J. Jiménez. 2013. Colecciones microbianas: Importancia, establecimiento y Regulación. *Hechos Microbiológicos* 4:23–33. https://doi.org/10.17533/udea.hm.20093

González-Cuellar, F.E., C.M. Lasso-Benavides, B.Y. Adrada-Gómez, O.L. Sanabria-Diago & A.M. Vasco-Palacios. 2021. Estudio etnomicológico con tres comunidades rurales Ubicadas en la Zona Andina del Departamento del Cauca, Colombia. *Boletín de Antropología* 36:147–164. https://doi.org/10.17533/udea.boan.v36n62a08

Gröbe, G., R. Ullrich & M. Pecyna. 2011. High-yield production of aromatic peroxygenase by the agaric fungus *Marasmius rotula. AMB Express* 1:31. https://doi.org/10.1186/2191-0855-1-31

Guo, W., H. Wu & Z. Zhang. 2017. Comparative analysis of transcriptomes in Rhizophoraceae provides insights into the origin and adaptive evolution of mangrove plants in intertidal environments. *Frontiers in Plant Science* 8:795. https://doi.org/10.3389/fpls.2017.00795

Guzmán, G. & L. Varela. 1978. Los hongos de Colombia III. Observaciones sobre los hongos, líquenes y mixomicetes de Colombia. *Caldasia* 12:309–338.

Henao, L. & A. Ruíz. 2006. Investigación y gestión local de robledales alrededor del uso tradicional de macromicetos en la cordillera Oriental colombiana. *Memorias del I Simposio de Robles y Ecosistemas Asociados. Bogotá. Fundación Natura y Pontificia Universidad Javeriana.* Fundación Natura Colombia, Bogotá, Colombia.

Higuita, H.D., F.C. Naranjo, A.M.C. Arias, Y.M.M. Figueroa & J.S.M. Serna. 2019. 50 años documentando algas, hongos y plantas de Colombia. *Herbario Universidad de Antioquia (HUA) 1969–2019.* Publicaciones ViD, Medellín.

Hyde, K.D., J. Xu, S. Rapior, R. Jeewon, S. Lumyong & A.G.T. Niego. 2019. The amazing potential of fungi: 50 ways we can exploit fungi industrially. *Fungal Diversity* 97:1–136. https://doi.org/10.1007/s13225-019-00430-9

Iark, D., A.J. dos Reis Buzzo, J.A.A. Garcia, V.G. Côrrea, C.V. Helm & R.C.G. Corrêa. 2019. Enzymatic degradation and detoxification of azo dye congo red by a new laccase from *Oudemansiella canarii. Bioresource Technology* 289:121655. https://doi.org/10.1016/j.biortech.2019.121655

Jeon, S.-J. & S.-J. Lim. 2017. Purification and characterization of the laccase involved in dye decolorization by the white-rot fungus *Marasmius scorodonius. Journal of Microbiology and Biotechnology* 27:1120–1127. https://doi.org/10.4014/jmb.1701.01004

Kardena, E., A.N.K. Pratama, S.H. Suhardi, K. Hassan & A.J. Effendi. 2020. Decolorisation of batik waste water by *Marasmiellus palmivorus* using modified fixed bed reactor. *Malaysian Applied Biology* 49:55–61. https://doi.org/10.55230/mabjournal.v49i2.1523

Kirk, P.M., P.F. Cannon, D.W. Minter & J.A. Stalpers. 2008. *Dictionary of the Fungi*, 10th ed. CABI, Wallingford, United Kingdom.

Kostić, M., M. Ivanov, Â. Fernandes, R.C. Calhelha, J. Glamočlija, L. Barros, M. Soković & A. Ćirić. 2023. A comparative study of *Lactarius* mushrooms: Chemical characterization, antibacterial, antibiofilm, antioxidant and cytotoxic activity. *Journal of Fungi (Basel)* 9:70. https://doi.org/10.3390/jof9010070

Krauss, K.W., C.E. Lovelock, K.L. McKee, L. Lopez-Hoffman, S.M.L. Ewe & W.P. Sousa. 2008. Environmental drivers in mangrove establishment and early development: A review. *Aquatic Botany* 89:105–127. https://doi.org/10.1016/j.aquabot.2007.12.014

Kshirsagar, Y., M. Borde & C. Manoharachary. 2025. Biodiversity, ecological interaction, and bioprospecting potential of gasteroid Fungi. In C. Manoharachary, H.B. Singh, S.K. Singh & Y.P. Sharma (eds.), *Developments in Applied Microbiology and Biotechnology: Biodiversity, Bioengineering, and Biotechnology of Fungi*, pp. 27–40. Academic Press. https://doi.org/10.1016/B978-0-443-13856-0.00002-6

Lakshmi, K., C. Aiswarya, P. Nayana, P.K. Dash & P. Nambisan. 2018. Enhancement of extracellular laccase production from *Lenzites elegans* KSG32: Taguchi orthogonal array experimental design methodology. *International Journal of Engineering Development and Research* 6:369–374.

Lee, S.O., H.Y. Kim, G.J. Choi, H.B. Lee, K.S. Jang, Y.H. Choi & J.C. Kim. 2009. Mycofumigation with *Oxyporus latemarginatus* EF069 for control of postharvest apple decay and *Rhizoctonia* root rot on moth orchid. *Journal of Applied Microbiology* 1064:1213–1219. https://doi.org/10.1111/j.1365-2672.2008.04087.x

Li, M., Y. Zhang, Q. Lu, Y. Gao, T. Ye, C. Wang & D. Xing. 2022. Structure, bioactivities and applications of the polysaccharides from *Tricholoma matsutake*: A review. *Food Science and Technology* 42:e44922. https://doi.org/10.1590/fst.44922

Liu, M., L. Liu, G. Zhang, G. Wang, R. Hou, T. Zhang & X. Tian. 2024. Study on optimization of liquid fermentation medium and antitumor activity of the mycelium on *Phylloporia lonicerae*. *Journal of Microbiology and Biotechnology* 34:1898. https://doi.org/10.4014/jmb.2405.05004

Loi, M., F. Fanelli, P. Zucca & V.C. Liuzzi. 2016. Aflatoxin B1 and M1 degradation by Lac2 from *Pleurotus pulmonarius* and redox mediators. *Toxins* 8:245. https://doi.org/10.3390/toxins8090245

Loi, M., O. Glazunova, T. Fedorova, A.F. Logrieco & G. Mulè. 2021. Fungal laccases: The forefront of enzymes for sustainability. *Journal of Fungi* 7:1048. https://doi.org/10.3390/jof7121048

Lombana, P., J.A. Monterroza-Álvarez, L.F. Chamorro-Quiroz, A.E. Franco-Molano & I.R. Payares-Díaz. 2016. Nuevos registros de macromicetos para Colombia. *Actualidades Biológicas* 38:181–189. https://doi.org/10.17533/udea.acbi.v37n105a05

Lowe, H., N. Toyang & B. Steele. 2021 The therapeutic potential of psilocybin. *Molecules* 26:2948. https://doi.org/10.3390/molecules26102948

Lu, Y.Z., J. Liu, J. Chen, X.Y. Zeng & J.C. Kang. 2023. Insights on fungal diversity of ascomycetes and basidiomycetes: Taxonomy and interactions with their host. *Frontiers in Microbiology* 14:1245204. https://doi.org/10.3389/fmicb.2023.1245204

Lücking, R. 2020. Three challenges to contemporaneous taxonomy from a licheno-mycological perspective. *Megataxa* 1:78–103. https://doi.org/10.11646/megataxa.1.1.16

Lücking, R., M.C. Aime & B. Robbertse. 2021. Fungal taxonomy and sequence-based nomenclature. *Nature Microbiology* 6:540–548.

Luna-Fontalvo, J.A., C. Abaunza, A. Barrios, K. Ramírez-Roncallo, R.J. Guerrero & M.A. Negritto. 2023. New records of agaricoid macrofungi (Agaricales, Basidiomycota) in an urban fragment of tropical dry forest from Colombian Caribbean region. *Check List* 19:371–379. https://doi.org/10.15560/19.3.371

Luna-Fontalvo, J.A., A. Barrios, C. Abaunza, K. Ramírez-Roncallo & M.A. Negritto. 2021. New records of *Agaricaceae* (Agaricales, Basidiomycota) from Colombia. *Check List* 17:1383–1393. https://doi.org/10.15560/17.5.1383

Maestre-Reyna, M., W.-C. Liu, W.-Y. Jeng, C.-C. Lee, C.-A. Hsu & T.-N. Wen. 2015. Structural and functional roles of glycosylation in fungal laccase from *Lentinus* sp. *PLoS ONE* 10:e0120601. https://doi.org/10.1371/journal.pone.0120601

Mago, P., R. Sharma, I. Hafeez, I. Nawaz, M. Joshi & R. Mehrotra. 2023. Mushroom-based cosmeceuticals: An upcoming biotechnology sector. *Biosciences Biotechnology Research Asia* 20:381–394. http://dx.doi.org/10.13005/bbra/3097

Malya, I.Y., J. Wu, E. Harada, M. Toda & C.N. D'Alessandro-Gabazza. 2020. Plant growth regulators and Axl and immune checkpoint inhibitors from the edible mushroom *Leucopaxillus giganteus*. *Bioscience, Biotechnology, and Biochemistry* 847:1332–1338. https://doi.org/10.1080/09168451.2020.1743170

Maya-Yescas, M.E., S. Revah & S. Le Borgne. 2021. Growth of *Leucoagaricus gongylophorus* Möller (Singer) and production of key enzymes in submerged and solid-state cultures with lignocellulosic dubstrates. *Biotechnology Letters* 43:845–854. https://doi.org/10.1007/s10529-020-03057-y

Meyer, V., E.Y. Basenko, J.P. Benz, G.H. Braus, M.X. Caddik & M. Csukai. 2020. Growing a circular economy with fungal biotechnology: A white paper. *Fungal Biology and Biotechnology* 7:5. https://doi.org/10.1186/s40694-020-00095-z

Mogilnaya, O., N. Ronzhin, E. Posokhina & V. Bondar. 2021. Extracellular oxidase from the *Neonothopanus nambi* fungus as a promising enzyme for analytical applications. *The Protein Journal* 40:731–740. https://doi.org/10.1007/s10930-021-10010-z

Mortimer, P.E., E. Boa, K.D. Hyde & H. Li. 2021. Macrofungi as food. In *Encyclopedia of Mycology*, pp. 405–417. Elsevier, Ámsterdam.

Motato-Vásquez, V., L.K. Vinasco-Díaz, M. Londoño-Caicedo & A.C. Bolaños-Rojas. 2025. Hidden treasures of Colombia's Pacific mangrove: New fungal species and records of macrofungi (Basidiomycota). *Journal of Fungi* 11:459. https://doi.org/10.3390/jof11060459

Mueller, G.M., K.M. Cunha, T.W. May, J.L. Allen, J.R. Westrip, C. Canteiro & A. Dahlberg. 2022. What do the first 597 global fungal red list assessments tell us about the threat status of fungi. *Diversity* 14:736. https://doi.org/10.3390/d14090736

Mueller, G.M., R.E. Halling, J. Carranza, M. Mata & J.P. Schmit. 2006. Saprotrophic and ectomycorrhizal macrofungi of Costa Rican oak forests. In M. Kappelle (ed.), *Ecology and Conservation of Neotropical Montane Oak Forests*, pp. 233–250. Springer Berlin Heidelberg, Germany.

Munck, C., E. Thierry, S. Gräßle, S. Hui-Chen & A. Yien-Ting. 2018. Biofilm formation of filamentous fungi *Coriolopsis* sp. on simple muslin cloth to enhance removal of triphenylmethane dyes. *Journal of Environmental Management* 214:261–266. https://doi.org/10.1016/j.jenvman.2018.03.025

Niño, Y.M., E.R. Peña & L.G. Enao. 2017. Spawn production and isolation for *Auricularia fuscosuccinea* (Mont.) Henn. and *Crepidotus palmarum* traditionally used in Pauna (Boyaca, Colombia). *Revista Colombiana de Ciencias Hortícolas* 11:151–158.

Niskanen, T., R. Lücking, A. Dahlberg, E. Gaya, L.M. Suz & V. Mikryukov. 2023. Pushing the frontiers of biodiversity research: Unveiling the global diversity, distribution and conservation of fungi. *Annual Review of Environment and Resources* 48:149–176. https://doi.org/10.1146/annurev-environ-112621-090937

Nkadimeng, S.M., C. Steinmann & J.N. Eloff. 2020. Effects and safety of *Psilocybe cubensis* and *Panaeolus cyanescens* magic mushroom extracts on endothelin-1-induced hypertrophy and cell injury in cardiomyocytes. *Scientific Reports* 10:22314. https://doi.org/10.1038/s41598-020-79328-5

Nowakowski, P., S.K. Naliwajko, R. Markiewicz-Żukowska, M.H. Borawska & K. Socha. 2020. The two faces of *Coprinus comatus*—functional properties and potential hazards. *Phytotherapy Research* 34:2932–2944. https://doi.org/10.1002/ptr.6741

Ocando-Rincón, H.A. 2014. *El psicoanalista y el ascenso de los ángeles caídos*, p. 560. Bloomington, Palibrio. ISBN 978-980-12-6870-3.

Ohl, C. & R. Bussmann. 2004. Recolonisation of natural landslides in tropical mountain Forests of southern Ecuador. *Feddes Repertorium: Zeitschrift für Botanische Taxonomie und Geobotanik* 115:248–264. https://doi.org/10.1002/fedr.200311041

Ortiz-Moreno, M.L., B. Moncada, A.M. Vasco-Palacios, R.F. de Almeida & E. Gaya. 2022. Fungi in Colombian and International biological collections. In R.F. De Almeida, R. Lücking, A. Vasco-Palacios, E. Gaya & M. Diazgranados (eds.), *Catalogue of Fungi of Colombia*, pp. 189–207. Royal Botanic Gardens, Kew, UK.

Palacio, M., Y. Gutiérrez, A.E. Franco-Molano & R. Callejas-Posada. 2015. Nuevos registros de macrohongos (Basidiomycota) para Colombia procedentes de un bosque seco tropical. *Actualidades Biológicas* 37:319–339. https://doi.org/10.17533/udea.acbi.331466

Peiris, T.M., M. Perera, H.H. Munasinghe, K.M. Thambugala, B.P. Dharmasena, P. Suttiprapan & R. Cheewangkoon. 2024. The treasured giants: A current overview on agricultural, nutritional, bioactive, and economic potential of *Macrocybe* species (Agaricales, Basidiomycota). *Frontiers in Cellular and Infection Microbiology* 14:1493532. https://doi.org/10.3389/fcimb.2024.1493532

Peña-Cañón, E.R. & L.G. Henao-Mejía. 2014. Conocimiento y uso tradicional de hongos silvestres de las comunidades campesinas asociadas a bosques de roble (*Quercus humboldtii*) en la zona de influencia de la Laguna de Fúquene, Andes Nororientales. *Etnobiología* 12:28–40. https://revistaetnobiologia.mx/index.php/etno/article/view/169

Peña-Cañón, E.R., Y. Niño-Fernández & L.G. Henao-Mejía. 2023. Importancia cultural de hongos silvestres comestibles en Cuatro Municipios de Boyacá (Colombia). *Ciencia en Desarrollo* 14:31–46. https://doi.org/10.19053/01217488.v14.n2.2023.15082

Pera, J. & J. Parladé. 2005. Inoculación controlada con hongos ectomicorrícicos en la producción de planta destinada a repoblaciones forestales: estado actual en España. *Investigación Agraria, Sistemas y Recursos Forestales* 14:419–433.

Pérez, E. & M. Piragauta. 2006. *Estudio etnomicológico entre campesinos de los municipios de Arcabuco y Moniquirá, departamento de Boyacá.* Trabajo de grado, Universidad Pedagógica y Tecnológica de Colombia.

Pinzón-Osorio, C.A., J. Pinzón-Osorio & N. Ladino-Ospina. 2022. Nuevos registros del género *Geastrum* (Agaricomycetes, Basidiomycota) para Colombia. *Hoehnea* 49:1–8. https://doi.org/10.1590/2236-8906-hoehnea-2022-4056

Pulido-LM, M. 1983. *Estudios en Agaricales Colombianos: Los Hongos de Colombia IX. Biblioteca José Jerónimo Triana.* Instituto de Ciencias Naturales, Bogotá.

Quiroga, L.P.A. 2011. La expedición Helvética. Viaje de exploración científica por Colombia en 1910 de los profesores Otto Fuhrman y Eugéne Mayor. *Historia Crítica* 44:219–220.

Robledo, G.L., M. Palacio, C. Urcelay, A.M. Vasco-Palacios, E. Crespo, O. Popoff & D.H. Costa-Rezende. 2020. Mystery unveiled: *Diacanthodes* Singer—a lineage within the core polyporoid clade. *Systematics and Biodiversity* 186:538–556. https://doi.org/10.1080/14772000.2020.1776784

Ruiz-Roa, W., L.G. Henao-M., E.R. Peña-Cañón, L. Amézquita & J. Cipamocha. 2008. *Reconocimiento y valoración de hongos silvestres comestibles de bosques de roble (Quercus humboldtii) en Paipa, Boyacá.* Informe de proyecto, Universidad Pedagógica y Tecnológica de Colombia.

Runnel, K., L. Tedersoo, F.S. Krah, M. Piepenbring, J.F. Scheepens, H. Hollert, S. Johann, N. Meyer & C. Bässler. 2025. Toward harnessing biodiversity–ecosystem function relationships in fungi. *Trends in Ecology & Evolution* 402:180–191. https://doi.org/10.1016/j.tree.2024.10.004

Sabogal, C., C. Besacier & D. McGuire. 2015. Restauración de bosques y paisajes: conceptos, enfoques y desafíos que plantea su ejecución. *Unasylva* 66:3–10.

Saldarriaga, Y., F. Pineda, G. García, L.F. Velásquez & G. Guzmán. 1988. Nuevos registros de Agaricales en Colombia. *Scientia Fungorum* 3:333–342.

Sánchez-Ocampo, S., M. Palacio, C. Ríos-Sarmiento & N. Gómez-Montoya. 2022. *Favolus rugulosus* en Colombia: producción de micelio y basidiomas en diferentes condiciones nutricionales. *Lilloa* 59 (suplemento):427–444. https://doi.org/10.30550/j.lil/2022.59.S/2022.10.01

Sanjuan, T. 1999. *La Diversidad del género Cordyceps en hormigas del bosque húmedo tropical de Colombia.* Bachelor's Thesis, Universidad Nacional de Colombia. Santafé de Bogotá, Colombia.

Shalini, S., C. Tyagi, D. Dutt & J. Upadhyaya. 2009. Production of high levels of cellulase-poor xylanases by wild strains of white-rot fungus *Coprinellus disseminatus* in solid-state fermentation. *New Biotechnology* 26:165–170. https://doi.org/10.1016/j.nbt.2009.09.004

Shevchuk, Y., K. Kuypers & G.E. Janssens. 2023. Fungi as a source of bioactive molecules for the development of longevity medicines. *Ageing Research Reviews* 87:101929. https://doi.org/10.1016/j.arr.2023.101929

SiB Colombia. 2025. *Biodiversidad en Cifras, Sistema de Información sobre Biodiversidad de Colombia.* https://gitlab.com/sib-colombia/administracion-contenidos/-/tree/master/

Silvestri, L.-C. 2016. Acceso a recursos genéticos y distribución de beneficios en Colombia: desafíos del régimen normativo. *Investigación y Desarrollo* 24:1–25. https://doi.org/10.14482/indes.24.1.8682

Simijaca, D., G. Muller, A.M. Vasco-Palacios, R. Almeida, R. Lücking, E. Gaya & M. Díaz-Granados. 2022. Fungal conservation in Colombia. In *Catalogue of Fungi of Colombia*, pp. 175–187. Royal Botanical Gardens, Kew.

Sufiate, B.L., F.E. de Freitas Soares, S.S. Moreira, A. de Souza Gouveia, T.S.A. Monteiro, L.G. de Freitas & J.H. de Queiroz. 2017. Nematicidal action of *Pleurotus eryngii* metabolites. *Biocatalysis and Agricultural Biotechnology* 12:216–219. https://doi.org/10.1016/j.bcab.2017.10.009

Tamayo-Quintana, A. & F. Torres-Romero. 2022. Amenazas y riesgos de origen natural y antrópico que pueden afectar un proceso de restauración ecológica en bosque seco tropical: estudio de caso Central Hidroeléctrica El Quimbo (Huila, Colombia). *Gestión y Ambiente* 25:10.

Tedersoo, L., V. Mikryukov, A. Zizka, M. Bahram, N. Hagh-Doust & T. Aslan. 2022. Global patterns in endemicity and vulnerability of soil fungi. *Global Change Biology* 28:6696–6710. https://doi.org/10.1111/gcb.16398.

Tešanović, K., B. Pejin & F. Šibul. 2017. A comparative overview of antioxidative properties and phenolic profiles of different fungal origins: Fruiting bodies and submerged cultures of *Coprinus comatus* and *Coprinellus truncorum*. *Journal of Food Science and Technology* 54:430–438. https://doi.org/10.1007/s13197-016-2479-2.

Tian, J., Z. Zhang, Y. Shang, T. Yang & R. Zhou. 2025. Isolation, structures, bioactivities, and applications of the polysaccharides from *Boletus* spp.: A review. *International Journal of Biological Macromolecules* 285:137622. https://doi.org/10.1016/j.ijbiomac.2024.137622

Tomlinson, P.B. 1986. *The Botany of Mangroves.* Cambridge University Press, Cambridge.

Torres-López, A.M., J.C. Quintero-Díaz & L. Atehortua-Garcés. 2011. Efecto de nutrientes sobre la producción de biomasa del hongo medicinal *Ganoderma lucidum*. *Revista Colombiana de Biotecnología* 13:103–109.

Torres-Morales, G., M. Flórez-Pulido, N. Vargas, R. Peña-Cañón & M. Fernández-Lucer. 2022. *Especies de Plantas y hongos útiles y de interés para las comunidades de tres lugares biodiversos de Colombia. Royal Botanical Garden (Kew) and Instituto de Investigaciones en Recursos Biológicos Alexander von Humboldt*, p. 111.

United Nations General Assembly. 2015. *Transforming Our World: The 2030 Agenda for Sustainable Development.* United Nations, New York. https://sdgs.un.org/2030agenda

Vargas, N., N. Gómez-Montoya, R. Peña-Cañón & G. Torres-Morales. 2022. Useful Fungi of Colombia. In R.F. De Almeida, R. Lücking, A. Vasco-Palacios, E. Gaya & M. Diazgranados (eds.), *Catalogue of Fungi of Colombia*, pp. 151–163. Royal Botanic Gardens, Kew, UK.

Vasco-Palacios, A.M. & B. Moncada. 2022. Two centuries of mycological history in Colombia. In R.F. De Almeida, R. Lücking, A. Vasco-Palacios, E. Gaya & M. Diazgranados (eds.), *Catalogue of Fungi of Colombia*, pp. 32–43. Royal Botanic Gardens, Kew, UK.

Vasco-Palacios, A.M., E.R. Peña-Cañón, O.L. Benavides & L.R. 2025. Hongos comestibles y desarrollo sostenible en Colombia. *Lilloa* 62 (Suplemento 1):223–251. https://doi.org/10.30550/j.lil/1920

Vasco-Palacios, A.M., S.C. Suaza, M. Castaño-Betancur & A.E. Franco-Molano. 2008. Conocimiento etnoecológico de los hongos entre los indígenas Uitoto, Muinane y Andoke de la Amazonía Colombiana. *Acta Amazonica* 38:17–30. https://doi.org/10.1590/s0044-59672008000100004

Venditti, A., C. Frezza, F. Sciubba, M. Serafini & A. Bianco. 2017. Primary and secondary metabolites of and European edible mushroom and its nutraceutical value: *Suillus bellinii* (Inzenga) Kuntze. *Natural Product Research* 3116:1910–1919. https://doi.org/10.1080/14786419.2016.1267731

Ventura, F.F., D.M. Soares, K. Bayle, A.G. Oliveira, E.J. Bechara, R.S. Freire & C.V. Stevani. 2021. Toxicity of metal cations and phenolic compounds to the bioluminescent fungus *Neonothopanus gardneri. Environmental Advances* 4:100044. https://doi.org/10.1016/j.envadv.2021.100044.

Vergara-González, O. 1986. *Etnografía de los guajiros (Wayuu).* Instituto Colombiano de Antropología, Bogotá.

Villalobos, S., M. Mengual & L.G. Henao-Mejía. 2017. Uso de los hongos *Podaxis pistillaris*, *Inonotus rickii* y *Phellorinia herculeana* (Basidiomycetes) por la Etnia Wayuu en la Alta Guajira Colombiana. *Etnobiología* 15:64–73. https://revistaetnobiologia.mx/index.php/etno/article/view/142

Wang, Q., M. Su, W. Zhu, X. Li, Y. Jia, P. Guo & X. Tian. 2010. Growth inhibition of *Microcystis aeruginosa* by white-rot fungus *Lopharia spadicea. Water Science and Technology* 62:317–323. https://doi.org/10.2166/wst.2010.214.

Wang, Y., B. Zhang, N. Chen, C. Wang, S. Feng & H. Xu. 2018. Combined bioremediation of soil co-contaminated with cadmium and endosulfan by *Pleurotus eryngii* and *Coprinus comatus. Journal of Soils and Sediments* 18:2136–2147. https://doi.org/10.1007/s11368-017-1762-9

Wang, Z.W., S. Guo, L.P. Hou & Y.T. Li. 2022. Optimization of liquid fermentation process of *Leucopaxillus giganteus* by response surface. *Journal of Shanxi Agricultural Sciences* 50:424–432. https://doi.org/10.3969/j.issn.1002-2481.2022.03.19

Webster, J. & R. Weber. 2007. *Introduction to Fungi*, 3rd ed. Cambridge University Press, The Edinburgh Building, Cambridge, UK.

Yao, L., S. Li, J. Zhang, Z. Li & J. Lv. 2025. Ten previously undescribed norsesquiterpenoids from the fungus *Marasmiellus candidus* and their anti-inflammatory activities. *Mycology* 1–12. https://doi.org/10.1080/21501203.2025.2551650

Zambrano-Forero, C.J., L.R. Dávila-Giraldo, L.O.B. Jaimes, J.J.M. Arteaga, G.L. Robledo & W. Murillo-Arango. 2021. The lignocellulolytic effect from newly wild white rot fungi isolated from Colombia. *International Journal of Environment and Waste Management* 274:440–454. https://doi.org/10.1504/IJEWM.2021.115379

Zambrano-Forero, C.J., L.R. Dávila-Giraldo, V. Motato-Vásquez, P.X. Villanueva, I.S. Rondón-Barragán & W. Murillo-Arango. 2023. Diversity and distribution of macrofungi (Ascomycota and Basidiomycota) in Tolima, a Department of the Colombian Andes: An annotated checklist. *Biodiversity Data Journal* 11:e104307. https://doi.org/10.3897/BDJ.11.e104307

Zhao, M., D. Li, L. Jie, J. Fang & L. Changhong. 2024 Pressure-tolerant survival mechanism of *Schizophyllum commune* 20R-7-F01 isolated from deep sediments 2 kilometers below the seafloor. *Frontiers in Marine Science* 11:1471465. https://doi.org/10.3389/fmars.2024.1471465

Zotti, M., A.M. Persiani, E. Ambrosio, A. Vizzini, G. Venturella, D. Donnini & G.I. Zervakis. 2013. Macrofungi as ecosystem resources: Conservation versus exploitation. *Plant Biosystems—An International Journal Dealing with All Aspects of Plant Biology* 1471:219–225. https://doi.org/10.1080/11263504.2012.753133.

7 Myxomycetes and Tropical Forests

Steven L. Stephenson

7.1 INTRODUCTION

The myxomycetes (also called "myxogastrids" or "plasmodial slime molds") are a group of eumycetozoans commonly found in most terrestrial ecosystems. Because many species produce fruiting bodies large enough to be observed with the naked eye, myxomycetes can be collected and curated in much the same manner as macrofungi. In fact, myxomycetes were long thought to be fungi but actually belong to a totally different phylum (the Amoebozoa). Their fruiting bodies resemble those of certain macrofungi (especially some small basidiomycetes), and the two groups of organisms occur in many of the same habitats. However, the fruiting bodies of macrofungi and myxomycetes are structurally very different, and their life cycles have little in common except that sexual reproduction involves the production of resistant spores.

7.2 LIFE CYCLE

The primary vegetative (or amoeboflagellate) stage in the myxomycete life cycle is a fundamental element of the soil biota, where these organisms play an essential but very much underappreciated role in nutrient cycles. Amoeboflagellates consume bacteria and thus make available nutrients locked up in bacterial biomass that would otherwise be unavailable to plants. Because they are microscopic, amoeboflagellates cannot be studied directly in the field. The amoeboflagellate is the first of two trophic stages in the life cycle. The second trophic stage, the plasmodium, is derived from the amoeboflagellate stage. Some plasmodia are large enough to be seen with the naked eye, but this tends to be the case for those produced by members of the traditional orders Physarales. Most of the time a plasmodium occurs within a particular substrate (e.g., a decaying log or ground litter) and emerges to an exposed position only just before fruiting. Plasmodia of the smallest species of myxomycetes give rise to just a single (very small) fruiting body, but the more usual situation is for a plasmodium to give rise to a group of fruiting bodies. Some very large fruitings may consist of thousands of individual fruiting bodies. Many of these are small but visible to the naked eye. Fruiting bodies are often colorful and sometimes quite ornate (Stephenson and Stempen 1994).

7.3 HISTORY OF THE STUDY OF MYXOMYCETES

For most of the history of their study, records of myxomycetes consisted of specimens that had developed under natural conditions in the field. This changed in the 1930s, when Gilbert and Martin (1933) first described the moist chamber culture technique. The technique involves culturing myxomycetes in the laboratory from samples of litter, bark, or other types of dead plant material collected in the field. Plasmodia and fruiting bodies of myxomycetes appear in the cultures, and once they are mature, fruiting bodies can be harvested and curated in the same manner as specimens collected in the field. Since the introduction of the moist chamber culture technique, it has become an important component of surveys for myxomycetes, and it was used in all of the major survey efforts mentioned later in his chapter.

DOI: 10.1201/9781003647140-7

7.4 TROPICAL FORESTS

The tropical zone (or tropics) of the earth, as defined by latitude, extends from the Tropic of Cancer (23°25′ N) in the Northern Hemisphere to the Tropic of Capricorn (23°25′ S) in the Southern Hemisphere. Tropical forests cover nearly half of the land area in the tropical zone, which represents a total of about 1.73 billion hectares (Poker and MacDicken 2016). Tropical forests vary considerably in structure and composition throughout the world, but they are best expressed as tropical rainforests (Figure 7.1). These forests are characterized by abundant precipitation, high average temperatures, nutrient-poor soil, and high levels of biodiversity (or species richness). The range of microhabitats available for myxomycetes in tropical rainforests exceeds that of any other type of ecosystem (Stephenson et al. 2004a; Liu et al. 2015).

7.5 MICROHABITATS FOR MYXOMYCETES

Myxomycetes are known to be associated with a wide range of microhabitats in tropical forests. Myxomycetes are terrestrial organisms, but a few species have been reported from aquatic microhabitats (Lindley et al. 2007; Winsett and Stephenson 2013), and one species of *Didymium* was described as new to science from a small stream in Brazil (Gottsberger and Nannenga-Bremekamp 1971).

7.5.1 Coarse Woody Debris

In boreal, temperate, and tropical forests, coarse woody debris (also referred to as decaying wood) represents the primary microhabitat for many species of myxomycetes (Figure 7.2). These lignicolous (wood loving) myxomycetes include some of the more common species in such genera as

FIGURE 7.1 Tropical forest in Costa Rica.

Source: Photo by Carlos Rojas

FIGURE 7.2 Fruiting bodies of *Stemonitopsis typhina* on coarse woody debris in a forest in a tropical forest in Vietnam.

Source: Photo by Yura Novozhilov

Arcyria, *Hemitrichia*, *Lycogala*, *Stemonitis*, and *Trichia*. Coarse woody debris is either decorticated or still has bark present. Some species of myxomycetes tend to be consistently associated with bark, which is more nutrient-rich than wood. The assemblage of species associated with coarse woody debris (with or without bark) is different from the assemblage associated with the bark surface of still-living trees and woody shrubs. These corticolous (bark loving) species are mostly very small and almost always recorded with the use of moist chamber cultures. The latter assemblage includes species in such genera as *Echinostelium*, *Licea*, *Macbrideola*, and *Paradiacheopsis*.

7.5.2 Ground Litter

The leaves and other plant parts (e.g., seeds, fruits, small twigs, and old inflorescences) that fall to the ground accumulate over time to form what is known as ground litter or the litter layer (Figure 7.3). Most of this layer is made up of decaying leaves. Certain species of myxomycetes are typically found fruiting on litter. Some of these produce fruiting bodies (or more likely groups of fruiting bodies) that can be observed and collected in the field. Species of *Diachea*, *Diderma*, and *Didymium* are often encountered on litter. Although small twigs are usually regarded as components of the litter layer on the forest floor, they tend to support a different assemblage of myxomycetes and are best considered a separate microhabitat.

7.5.3 Aerial Litter

The term "aerial litter" is applied to dead parts (mostly leaves) of plants that are still attached and have not fallen to the ground (Figure 7.4). In tropical rainforests, rain occurs almost daily, and this means that everything remains relatively moist. A film of water is often present on plant surfaces, including both living and dead leaves. On the forest floor, dead leaves require a period when there is no additional rain to dry out, and this does not happen for extended periods of time. In contrast,

FIGURE 7.3 Litter layer of a tropical forest in northern Thailand.

Source: Photo by Steve Stephenson

FIGURE 7.4 Areal litter in a forest in Costa Rica.

Source: Photo by Steve Stephenson

dead leaves that are still attached but well above the forest floor are capable of drying out rather quickly, which makes them a more favorable microhabitat for the establishment and development of myxomycetes. This has been supported by the results obtained in a number of studies. For example, Schnittler and Stephenson (2000) reported that for a primary tropical forest in Costa Rica, 137 cultures prepared with samples of ground litter yielded an average of 0.7 species of myxomycetes per moist chamber culture, whereas 100 cultures prepared with samples of aerial litter produced an average of 1.4 species. Interestingly, the aerial rhizomorphs of certain fungi can trap some of the leaves that fall well above the ground, thus creating a special type of aerial litter (Stephenson 2020).

7.5.4 Bark Surface of Living Trees

It is now known that a diverse assemblage of myxomycetes is associated with the bark surface of living trees (Figure 7.5). However, most of the species involved are inconspicuous or sporadic in their occurrence and difficult, if not impossible, to detect in the field. When Gilbert and Martin (1933) introduced the moist chamber culture technique as it applies to myxomycetes, it allowed these corticolous (bark loving) myxomycetes to be studied for the first time. Since its introduction, the technique has been used with considerable success by many researchers (e.g., Blackwell and Gilbertson 1980; Härkönen 1981; Stephenson 1989; Schnittler and Stephenson 2000). More than 100 species of corticolous myxomycetes have been reported from the bark microhabitat as specimens obtained with the use of moist chamber cultures (Mitchell 1980). Some of these also are known to occur in other microhabitats, but others appear to be restricted to the bark of living trees. Prominent examples include various species of *Echinostelium*, *Licea* and *Macbrideola*.

Both the number of species and the relative abundance of corticolous myxomycetes appear to be higher (sometimes much higher) for the bark of temperate trees than for the bark of tropical

FIGURE 7.5 Bark surface of a living tree in a tropical forest in northern Thailand.

Source: Photo by Steve Stephenson

trees (Stephenson et al. 2008). Just why the corticolous myxomycetes are less common on the bark of tropical trees is not known, although the fact that many species of tropical trees have relatively smooth bark is likely to be a factor. Presumably, a smooth bark surface would represent a less effective "spore trap" for air-borne spores of myxomycetes if most myxomycetes are introduced to a particular microhabitat as the result of being dispersed by air currents, which is likely to be the case.

7.5.5 Inflorescences

Other microhabitats found in tropical forests but not (or only rarely) in the forests of the boreal and temperate zones include the inflorescences of large herbaceous plants (Schnittler and Stephenson 2002), lianas (Wrigley de Basanta et al. 2008; Ko Ko et al. 2010b), and the cover of epiphylls that sometimes develops on the upper surface of evergreen leaves (Schnittler 2001). Schnittler and Stephenson (2002) characterized the assemblage of myxomycetes associated with large Neotropical herbaceous plants at a number of study sites in Costa Rica, Ecuador, and Puerto Rico. They recorded 653 species representing 31 different taxa of myxomycetes. When they compared these results with those obtained from 696 moist chamber cultures prepared with various other types of litter, 13 taxa of myxomycetes occurred more often on inflorescences. Moreover, six species, such as *Physarum compressum*, *P. didermoides*, and *Didymium iridis*), displayed a strong preference for this microhabitat, and three of these were new records for the Neotropics. Correspondence analysis of the data set compiled for inflorescences indicated that the assemblage of myxomycetes was relatively consistent across all of the various study sites. This suggests that inflorescences support a distinct ecological assemblage of species of myxomycetes. The actual substrates upon which myxomycete fruiting bodies occurred were the rapidly decaying floral parts enclosed by the massive, still-living bracts. Birds visit the individual flowers in the inflorescences and likely serve as spore vectors, carrying spores from one inflorescence to another.

7.5.6 Lianas

Woody vines (lianas) are a conspicuous feature of most tropical forests, where their leaves can constitute a large proportion of the total leaf surface area of the entire forest. Lianas, which reach the canopy by using other plants as support, include representatives from many diverse groups of plants, ranging from ferns to angiosperms. Some flowering plant families such as the Bignoniaceae and Vitaceae are particularly rich in species of lianas. Larger examples of lianas have an outer bark-like layer that can be collected and processed in the same way as the bark of trees. Wrigley de Basanta et al. (2008) compiled data on the myxomycetes recorded on samples of lianas obtained from tropical forests in Australia, Cuba, Ecuador, Mexico, Peru, and Puerto Rico. Moist chamber cultures prepared with these samples yielded several hundred collections of myxomycetes representing 65 species in 21 genera, and at least 87% of a total of 202 moist chamber cultures produced some evidence (either plasmodia or fruiting bodies) of these organisms. *Arcyria cinerea*, *Diderma hemisphaericum*, *Didymium squamulosum*, *Physarum pusillum*, and *Stemonitis fusca* var. *nigrescens* were among the more consistently abundant and widespread myxomycetes associated with liana microhabitat. The cultures also produced a number of noteworthy collections, including *Perichaena dictyonema*, *Physarum hongkongense*, *Stemonitis foliicola*, *Stemonaria gracilis*, *Ceratiomyxa fruticulosa* (not usually known from any type of aerial microhabitat), and *Willkommlangea reticulata*.

7.5.7 Living Leaves

Another distinctive feature of tropical evergreen forests is the occurrence of a specialized group of liverworts and lichens on the upper surface of living, often leathery leaves of understory plants. Recent studies of these assemblages of foliicolous liverworts and lichens indicate a surprisingly high species diversity of these organisms in this microhabitat. For example, Lücking (1999) reported more than 250 species of lichenized fungi from a single locality in an Amazonian rainforest, with more than 80 species occurring on a single, medium-sized palm leaf. During a study of myxomycete ecology carried out in Costa Rica, Schnittler (2001) attempted to screen all available microhabitats for the occurrence of myxomycetes. As part of this effort, ten moist chamber cultures were prepared with small pieces of living leaves covered by liverworts from the cloud forest at Monteverde. All ten cultures yielded myxomycetes, although often with only one to five tiny fruiting bodies present. Eleven species of myxomycetes were recovered, with *Arcyria afroalpina* as a new record for the Neotropics. Three species (*Arcyria cinerea*, *Didymium iridis* and *D. squamulosum*) were recorded with a high frequency (between 59 and 66%) in the moist chamber cultures. These data suggest that these three species may occur regularly on epiphyll-covered leaf surfaces as populations of amoeboflagellates. The unusually small numbers of fruiting bodies found in the cultures, along with their atypically small size, suggest a microhabitat poor in the nutrients necessary to support a plentiful supply of bacteria for myxomycetes to feed upon. Lowland rainforests with a high annual rainfall seem to provide the best conditions for the growth of epiphyllous myxomycetes, but the frequent occurrence of rain would seem likely to "flush" spores and perhaps even amoeboflagellates from the leaves. However, rain also would mean that a film of water is likely to be maintained on the leaf surface. The only other data on the myxomycetes associated with epiphylls is unpublished and was obtained by the author for a few samples collected in northern Queensland, Australia. Two of the species recorded were *Arcyria cinerea* and *D. squamulosum*. This is one aspect of the distribution and ecology of myxomycetes that is interesting enough to warrant additional study.

7.5.8 Soil

Soil is now known to be one of the primary habitats for myxomycetes. Myxomycetes occur in soil as amoeboflagellates, and they exist in considerable numbers. Feest and Madelin (1985) were the first to obtain a body of meaningful data on the presence and abundance of myxomycetes in soils. They used a culture-based method in which abundance was measured by the number of plasmodium-forming units (PFUs), as determined by using the most-probable number (MPN) method. A PFU was defined as an amoeboflagellate or cyst capable of giving rise to a plasmodium on a culture plate. More recently, data have been obtained using molecular methods. Limited data exist for the soils of tropical forests, and they did not determine the species composition of the assemblages of myxomycetes present. Gao et al. (2019, 2022) used DNA sequence analysis to examine community structure of dark-spored myxomycetes in soils at several study sites in subtropical/tropical forests of southern China, but specific species were not identified. However, other studies have indicated that the majority of the amoeboflagellates in soil belong to the traditional order Physarales, with species of *Didymium* especially prominent. There is little evidence to suggest that myxomycetes occur in soil as plasmodia or that fruiting bodies are produced in soil. Feest and Madelin (1985) did not initially observe plasmodia in the culture plates they prepared with samples of soil. Plasmodia appeared only later when they presumably developed from some of the amoeboflagellates (PFU's) present. Myxomycete fruiting bodies only occasionally are found on bare soil but commonly occur on ground litter. It seems likely that many fruitings collected from ground litter were ultimately derived from soil-inhabiting amoeboflagellates. The distribution and ecology of myxomycetes in tropical soils are clearly understudied, and more research needs to be carried out.

7.5.9 Canopy Soil

Trees in tropical forests tend to be characterized by the presence of an assemblage of vascular and nonvascular epiphytes on their branches and to some extent their trunks. The presence of this cover of epiphytes contributes to the development of a layer of accumulated organic matter derived from decaying epiphytes, partially decomposed tree bark, insect frass, and intercepted litter. This microhabitat, which has been referred to as "canopy soil" (Stephenson and Landolt 2011, 2015), is known to represent an ecologically important subsystem of the forests but remains understudied for most groups of organisms (Orlovich et al. 2013). Stephenson and Landolt (1998) first noted the presence of dictyostelids (cellular slime molds) in canopy soil. Their studies carried out at a number of localities throughout the world revealed that they were not uncommon in this microhabitat. They also noticed that the plasmodia of myxomycetes occasionally turned up in the primary isolation plates prepared for dictyostelids.

The occurrence of myxomycetes in the canopy soil microhabitat would not seem particularly surprising, since their amoeboflagellate cells presumably occupy a niche similar to that of the amoeboid cells of dictyostelids. Moreover, myxomycetes are often exceedingly common in the aerial litter microhabitat of tropical forests (e.g., Black et al. 2004) and their wind-dispersed spores seemingly would have no problem reaching above-ground microhabitats. Animal vectors also might well play a role, since a number of different kinds of invertebrates are known to be capable of transporting the spores of myxomycetes. However, canopy soil as a microhabitat for myxomycetes has been studied so little that we don't know how important it is for these organisms.

7.5.10 Dung

Myxomycetes are known to occur in the dung of herbivorous mammals (Cox 1981; Eliasson 2013; Calaça et al. 2020; Stephenson et al. 2022), and some species are largely or completely restricted to this microhabitat. However, these coprophilous (dung-loving) myxomycetes appear to be almost completely absent from tropical rainforests, although sometimes present in seasonal tropical forests where there is a definite dry season. Coprophilous myxomycetes very rarely appear on the dung of small mammals, regardless of the type of habitat. Instead, the dung is very quickly colonized by filamentous fungi. Worldwide, there are relatively few large mammals that naturally occur in tropical rainforests, which limits the availability of large masses of dung. Even if the latter were available, the dung rapidly decomposes due to the constant warm and moist conditions. It does not persist long enough to provide a suitable microhabitat for the establishment and development of myxomycetes. In contrast, dung can be an important microhabitat for myxomycetes in deserts.

7.6 RECORDS OF MYXOMYCETES FROM TROPICAL FORESTS

The earliest published records of myxomycetes were largely confined to western Europe, but eventually myxomycetes began to be reported from North America and the rest of the world. The first collections of myxomycetes from the tropics were made in the early nineteenth century by Europeans whose primary focus was on other groups of organisms, particularly fungi. Since myxomycetes were considered to be fungi at the time, it is not surprising that the records of myxomycetes appeared along with fungi in species lists and publications. During this period, specimens collected in other parts of the world were often sent to Europeans for identification. Obviously, collecting a specimen is one thing, but knowing what it is is quite a different thing.

Northern Queensland in Australia falls within the tropical zone, and the first reports of myxomycetes from the tropical forests of this region were those of Berkeley (1881). Since then, tropical forests of mainland Australia have received relatively little attention. The few published reports include Black et al. (2004) and McHugh et al. (2009), although species listed as occurring in tropical forests were listed by Stephenson (2021). Tropical forests are found on three of the Australian offshore islands, and published reports of myxomycetes from these islands are reported later in this chapter.

The first published report of myxomycetes from the tropical zone of Africa appears to be that of Bresadola and Saccardo (1899) for the Congo. However, the first significant report from tropical Africa was that of Farquharson and Lister (1916) for Nigeria. Duthie (1917) made an effort to list all species of myxomycetes known from Africa prior to the publication of his paper. These included records from both tropical and non-tropical regions of the continent. However, numerous records were from Cameroon, Central Africa, East Africa, Nigeria, and West Africa, all of which are within the tropical zone. More recent reports of myxomycetes from Africa include those of Härkönen (1981), Härkönen and Saarimäki (1991), Ukkola (1998, 2000), and Ndiritu et al. (2009a, 2009b).

As reported by Lado and Wrigley de Basanta (2008), the first reports of myxomycetes from the Neotropics were by Bertero (1828) and Montagne (1837). Since then, a number of reports of myxomycetes from tropical forests in various areas within the region have been published. These include Costa Rica (Schnittler and Stephenson 2000; Rojas et al. 2010a, 2010c, 2015a, 2018; Lado and Rojas 2018); Cuba (Camino et al. 2008); Ecuador (Stephenson and Mitchell 1994; Schnittler et al. 2002; Stephenson et al. 2004b; Lado et al. 2003, 2017, 2018), Peru (Stephenson and Mitchell 1994; Rojas and Stephenson 2012, 2013), and Puerto Rico (Novozhilov et al. 2001). Additional reports published earlier than 2008 were listed by Lado and Wrigley de Basanta (2008).

Early reports of the myxomycetes associated with tropical forests of Asia are few. The paper by Raciborski (1898), who listed a number of species from Java, possibly represents the earliest example. The twenty-first century has witnessed an increased interest in the myxomycetes, with numerous studies being carried out. The islands of the Philippines (Alfaro et al. 2015; Dagamac et al. 2011, 2012, 2014, 2015; Macabago et al. 2017, 2020; Macabago and Stephenson 2020, 2021) and Southeast Asia (Kylin et al. 2013; Tran et al. 2006, 2008, 2014; Ko Ko et al. 2010a, 2011, 2012, 2013, 2015) have received particular interest. Reports prior to this time period include Sanderson (1922), Reynolds (1981), and Ing and Spooner (1998).

7.7 MAJOR SURVEY EFFORTS IN TROPICAL FORESTS

In 1997 the National Science Foundation of the United States funded a three-year project titled "Studies of Neotropical Myxomycetes" and directed by the author. Funding for the project was extended for an additional three years in 2001. The overall objective of the project was to carry out biodiversity surveys for myxomycetes in four localities in the Neotropics. These were the Luquillo Experimental Forest in Puerto Rico, the Maquipucuna Cloud Forest Reserve in Ecuador, the Guanacaste Conservation Area in Costa Rica, and a portion of the state of Tlaxcala in eastern Central Mexico. Later, the El Eden Ecological Reserve on the Yucatan Peninsula was added as an additional locality. The surveys carried out yielded the largest body of data yet assembled on the myxomycetes associated with tropical forests (Stephenson et al. 2003, 2004a). Moreover, several microhabitats such as tropical lianas, inflorescences of large tropical herbaceous plants, the cover of epiphylls that often occur on the leaves of tropical plants, and what has since been referred to as "canopy soil" not previously investigated for myxomycetes were subjected to detailed study (Schnittler and Stephenson 2000, 2002; Schnittler 2001).

In 2003, two of the most ambitious survey projects for myxomycetes ever carried out received funding. The first of these was a project titled "Global Biodiversity of Eumycetozoans" funded by the Planetary Biodiversity Inventories Program (PBI) of the National Science Foundation and directed by the author. The purpose of the PBI program was to provide groups of scientists with the funding necessary to do the fieldwork to develop a worldwide, species-level inventory of a major group of organisms. In the first year of the program, the eumycetozoans were one of the four groups of organisms selected for funding. As the name suggests, the PBI project was directed toward all three groups of eumycetozoans—protosteloid amoebae, dictyostelids, and myxomycetes. However, the myxomycetes, as the most species-rich group, were the primary focus.

The funding for the PBI project extended over six years, ending in 2009. However, some of the data obtained were not published until years later. During the project, there were numerous expeditions to countries around the world, including areas of tropical forests in Australia (Black et al. 2004), Cambodia (Ko Ko et al. 2015), Cuba (Camino et al. 2008), Guatemala (Rojas et al. 2010b, 2020b), Kenya (Ndiritu et al. 2009a, 2009b), Laos (Ko Ko et al. 2012), Madagascar (Wrigley de Basanta et al. 2013), Myanmar (Ko Ko et al. 2013), Peru (Rojas et al. 2011b; Rojas and Stephenson 2013), the Philippines (Macabago et al. 2012, 2016; dela Cruz et al. 2014; Dagamac et al. 2011, 2012, 2014, 2017), South Africa (Winsett et al. 2008), and Thailand (Ko Ko et al. 2010a). The one thing all of these had in common was that they represented or contained areas where myxomycetes were an understudied group. Several thousand collections or records of myxomycetes were compiled, and papers reporting these data were published as noted above.

The second major project that began in 2003 was the Myxotropic project directed by Carlos Lado and funded by the Spanish government. The objective of the Myxotropic project was similar to that of the Global Diodiversity of Eumycetozoans project except that the region of study was limited to the Neotropics. Particular emphasis was placed on deserts, which had been largely neglected previously (Estrada-Torres et al. 2009). However, surveys also were carried out in other types of tropical ecosystems. Specimens were collected from localities in Argentina, Bolivia, Chile, Mexico, and Peru. Elevations of the actual collecting sites ranged from near sea level to more than 4,500 m. The data from this project greatly increased our knowledge of the assemblages of myxomycetes associated with extreme environments in the Neotropics. Overall, the Myxotropic project yielded the largest set of records and specimens of myxomycetes available for any single region of the world (Lado et al. 2011, 2013, 2014, 2016, 2019). Of particular interest was the special ecological group of succulenticolous (cactus loving) myxomycetes associated with succulent plants in xeric habitats (Wrigley de Basanta et al. 2009, 2011; Garcia-Martin et al. 2018).

A project titled "Taxonomic and ecological diversity of the mycobiota of tropical forests in Vietnam" began in 2010 and is still ongoing at the time this chapter was written. This project is based at the Komarov Botanical Institute of the Russian Academy of Sciences and funded by the Ministry of Education and Science of the Russian Federation. Yuri Novozhilov is the supervisor for the myxomycete component, which has the objective of characterizing the biodiversity of myxomycetes in the lowland and montane tropical forests in nature preserves and national parks throughout Vietnam. The set of data being compiled exceeds that available for any country in the tropical region of the world (Novozhilov et al. 2017, 2018, 2020; Fedorova et al. 2020, 2021).

Much of the other data available on the myxomycetes of tropical forests was obtained from more limited studies carried out by numerous individuals. Studies carried out in the Neotropics were listed by Farr (1976) and Lado and Wrigley de Basanta (2008). More recent studies include Rojas et al. (2010c, 2011a, 2014), Rojas and Stephenson (2012), and Walker et al. (2015). Ndiritu et al. (2009a) provided information on published records for all of Africa, including those from the tropical zone. In his monograph on the myxomycetes of Australia, Stephenson (2021) referred to those species known from the tropical region of Australia and listed the source of each record. Unfortunately, no comprehensive checklist or review of published records of myxomycetes yet exists for the tropical forests of Asia, although many of the more significant studies were listed by Dagamac et al. (2017).

Tropical forests (albeit sometimes limited in extent) also occur on some isolated oceanic islands in the tropics, and surveys for myxomycetes have been carried out on some of these islands. Examples include Ascension Island (Stephenson 2009), Christmas Island (Stephenson and Stephenson 2019), Cocos Island (Rojas and Stephenson 2008), Keeling Island (Stephenson, unpub. data), La Reuion Isand (Adamonyte et al. 2011), and the Seychelles (Kryvomaz et al. 2017, 2020a, 2020b).

7.8 MYXOMYCETE BIOGEOGRAPHY

The data compiled from the surveys described above and other earlier, more limited surveys have yielded the information necessary to compare the assemblages of myxomycetes associated with tropical forests and temperate forests as well as to examine the differences that exist for tropical forests in different parts of the world. In the first effort to compare the assemblages of myxomycetes associated with the tropical forests and temperate forests, Stephenson et al. (1993) used the data collected from two regions in the mid-Appalachians of eastern North America and two regions of India to analyze differences in the species present. The two regions in eastern North America and one region in India were in the temperate zone, whereas the second region in India was in the tropics. The coefficient of community indices calculated for all pairwise combinations of the four regions indicated that the one in the tropics (southern India) had the lowest overall similarity to the others. Moreover, there were differences among the four regions in the relative proportions of the members of the traditional orders of myxomycetes. For example, members of the traditional order Physarales made up 63% of all records in southern India, much higher than the values for the other orders. Differences among the regions also existed for the substrates (woody versus non-woody) upon which fruiting occurred.

In a later more comprehensive study, Dagamac et al. (2017) compared the assemblages of myxomycetes associated with tropical forests in the Neotropics (study areas in Costa Rica and Ecuador) and the Palaeotropics (the Philippines, Thailand, and Vietnam). The data came from surveys carried out in eight different areas, with four in each region of the world. Altogether, the surveys from these eight areas included a total of 7,585 records of myxomycete records (2,844 from Neotropical forests and 4,741 from Asian Palaeotropical forests). Both specimens collected in the field and those obtained in moist chamber cultures were considered. A total of 239 taxa (species and varieties) were recorded. Neotropical forests were less diverse (150 taxa) than Asian Palaeotropical forests (196 taxa). In both sets of data, numerous taxa were represented by only a single record (singleton). For the Neotropics there were 40 singletons (representing 27% of all taxa), while for the Palaeotropics the corresponding figures were 43 singletons and 22%. According to the Chao1 estimator, the eight surveys ranged from 70% to 88% complete.

Rojas and Stephenson (2020a) compared assemblages of myxomycetes from boreal, temperate, and tropical forests. The data for boreal forests were from Denali National Park in the United States, the data for temperate forests were from the Great Smoky Mountains National Park in the United States, and the data for tropical forests were from the La Selva Biological Station in Costa Rica. The data used consisted primarily of records (a total of 3,558) for specimens collected in the field. Forty-six percent of the records in the entire set of data were from temperate forests, 29% from boreal forests, and 23% from tropical forests. A total of 208 species were recorded for the three different types of forests, with 69% temperate, 49% boreal, and 40% tropical. The coefficient of community index calculated for pairwise combinations of three sets of data was lowest (0.38) for the boreal-tropical pair and highest (0.56) for the temperate-tropical pair. The taxonomic diversity indices were 2.68 and 2.83 for the boreal and tropical datasets, but 3.76 for the temperate dataset.

7.9 PATTERNS OF BIODIVERSITY

Tropical forests are characterized by their high species diversity for most groups of organisms, but this may or may not be the case for myxomycetes. Some studies have suggested that myxomycetes achieve their highest species diversity in temperate forests. In one study, dela Cruz et al. (2014) collected samples of forest floor litter, aerial litter, and woody twigs from three study sites in mid-latitude temperate deciduous forests in northwest Arkansas in the United States and three study sites in lowland evergreen tropical forests in the Philippines. Thirty samples of each substrate type were collected in each of the six study sites. These samples were used to prepare moist chamber cultures. The tropical forest samples yielded a total of 1,597 collections, while the temperate forest samples produced 1,089 collections. Although the number of collections for tropical forests was appreciably higher than the number recorded for temperate forests, the numbers of different species recorded were similar (82 for temperate and 80 for tropical). The number of genera represented by the species recorded in the temperate data set was slightly higher (23) than the number (20) for the tropical data set.

Unquestionably, the substrates potentially available to myxomycetes in a particular type of habitat represent a major factor in the distribution of these organisms. For example, some species are almost always collected from coarse woody debris, while others are invariably associated with ground litter. However, something more than the simple availability of a suitable substrate is sometimes apparent. This suggests that there are other yet to be determined more subtle factors involved. Rojas et al. (2020a) examined this aspect of myxomycete ecology by carrying out a project involving translocated and non-translocated substrates. Samples of dead leaves on the forest floor were collected from a locality in central Costa Rica (tropical climate) and another locality in northwest Arkansas in the United States (temperate climate) and then were placed in fine-mesh nylon bags, autoclaved, and then left out in the field for four months. Half of the bags prepared with sterile material from each country were placed out in the country where they were collected, and the other half were placed in the other country. When recollected, all samples were processed for myxomycetes with the use of the moist chamber culture technique. The cultures yielded a total of 156 records of myxomycetes representing 16 species, but only 13 records and six species were isolated from the samples left out in Costa Rica. One hundred percent of cultures placed out in Arkansas produced myxomycetes, whereas only 68% of those placed out in Costa Rica were positive for these organisms. Differences of this magnitude were unexpected, especially the fact that the origin of the leaves (i.e., the country in which they were collected) did not seem to be a factor. Clearly, something else was involved. The authors suggested that the xeric conditions in the temperate forest in Arkansas, where there would have been no film of water covering the surface of the leaves on the ground, may have played a role in favoring the establishment of myxomycete propagules. Presumably, the presence of a film of water on the leaf surfaces in the moist forest in Costa Rica negatively impacted the establishment of myxomycete propagules. This would not have been the case in the moist tropical forest in Costa Rica. However, it seems likely that other not yet determined factors also were involved.

7.10 ECOLOGICAL FACTORS

It has been well established that ecological factors affect the distribution patterns of particular species of myxomycetes. Some of these are easily discernable and can be quantified to at some extent, but this is not the case for others.

7.10.1 Temperature and Moisture

Temperature and moisture are the two main factors that determine the distribution of living organisms—including myxomycetes—on earth and account for the fundamental differences that exist in boreal, temperate, and tropical forests. Differences in these two factors exist across a given landscape, and these differences are reflected in differences in the composition and structure of the particular type of tropical forest present. Temperature and moisture affect myxomycetes in two fundamental ways. First, they directly affect the establishment, growth, and development of the myxomycetes associated with various types of substrates. (As used herein, the term "substrate" refers to microhabitat in a practical and not conceptional sense.) Second, they largely determine what types of plants and, by extension, the substrates available for myxomycetes exist at a particular locality.

7.10.2 Substrate pH

One notable but less apparent example of a factor that affects the distribution of myxomycetes is the pH of the substrate upon or within which the trophic stages (amoeboflagellates and plasmodia) occur. Some species, especially members of the traditional order Physarales, do not tolerate low pH conditions. However, other myxomycetes, such as members of the traditional Stemonitales, are invariably associated with a low pH. The data obtained in numerous studies in which the moist chamber culture technique was used clearly indicate that most species of myxomycetes display at least some degree of substrate specificity. This is reflected in the differences in the assemblages of species typically associated with the different microhabitats described earlier in this chapter.

7.10.3 Phenology

The term "phenology" refers to the study of recurring natural events that are correlated with climatic conditions. In temperate and boreal regions of the world, life cycle events in all groups of organisms are highly correlated with season, especially the length of the effective growing season when conditions are most favorable for growth, development, and reproduction. In the mid-latitude temperate deciduous forests of eastern North America, fresh fruiting bodies of myxomycetes are rarely evident before May (Stephenson 1988), with *Ceratiomyxa fruticulosa* (traditionally considered a myxomycete almost a member of a sister group of eumycetozoans) and *Lycogala epidendrum* among the first species to appear.

Rojas et al. (2021) studied the phenology of myxomycetes over a 30-month period in a premontane tropical forest in Turrialba, Costa Rica. At the species level, *Hemitrichia calyculata* was the only species observed during every month, closely followed by *Arcyria cinerea*, *A. denudata,* and *Physarum compressum*, all three of which were recorded on most visits. Sporadic fruiting was observed for species such as *Tubifera microsperma*, *P. tenerum*, *P. bogoriense*, *P. melleum*, and *Metatrichia vesparia*. Their data suggested that phenological patterns can be observed at the species level, but that in the Neotropics, under favorable conditions, myxomycete fruiting bodies are practically always present. However, species assemblages vary temporally. This study was limited to one small area. More comprehensive studies of phenology in myxomycetes have yet to be carried out.

7.10.4 Elevation

The relationship of diversity and abundance of myxomycetes in relation to increasing elevation in the tropics has been investigated in several studies. Stephenson et al. (1999) sampled (using moist chamber cultures) the myxomycetes associated with the litter layer in five types of tropical forests that occurred along an elevation gradient that extended from <600 m to more than 750 m. They found that overall diversity of myxomycetes was highest in the forest at the lowest elevation and lowest in the forest at the highest elevation. These authors hypothesized, based on their data, that diversity of myxomycetes in tropical forests decreased with increasing elevation.

Rojas and Stephenson (2008) examined patterns of diversity of myxomycetes (primarily data from moist chamber cultures but also including some field collections) on Cocos Island. Samples were obtained from six study sites located along an elevation gradient that extended from five to 575 m. A general pattern of decreasing diversity with increasing elevation was suggested by their data. These authors discussed the factors that could be responsible for this pattern, including differences in resource availability and the possible effect of wind on spore dispersal in myxomycetes at higher elevations.

Rojas et al. (2016) designed an experiment to test the hypothesis that myxomycete diversity decreases with increasing elevation. They obtained samples of ground litter, aerial litter, and woody twigs from 40 study sites along each of four different elevational transects that extended from 100 to 1,000 m. Sets of samples were collected during both a rainy season and a dry season and processed in moist chamber cultures. When subjected to statistical analysis, their data revealed little evidence of a pattern of diversity related to elevation. They concluded that elevation may not be as important a factor as previously reported.

7.11 TROPICAL SPECIES OF MYXOMYCETES

Many species of myxomycetes appear to be cosmopolitan, occurring wherever suitable conditions for their growth and fruiting exist. Arcyria cinerea, *Arcyria denudata*, *Collaria arcyrionema*, *Craterium leucocephalum*, *Cribraria cancellata*, *Diderma hemisphaericum*, *Didymium clavus*, *D. iridis*, *D. nigripes*, *D. squamulosum*, *Fuligo septica*, *Hemitrichia calyculata*, *H. serpula*, *Lamproderma scintillans*, *Lycogala epidendrum*, *Metatrichia vesparia*, *Perichaena chrysosperma*, *Physarum album*, *P. cinereum*, *P. pusillum*, *P. viride*, *Stemonitis axifera*, *S. fusca*, *Stemonitopsis typhina*, and *Trichia favoginea* fall into this category. These species have been reported from the majority of the field surveys ever carried out. However, other species display a distribution that is centered largely in tropical forests although they also can be found in temperate forests. For example, *Hemitrichia calyculata* and *H. serpula* are often exceedingly abundant in tropical forests but occasionally can be collected in temperate forests. The same situation applies to *Comatricha tenerrima*, *Cribraria microcarpa*, *C. minutissima*, *C. tenella*, *Diachea bulbillosa*, *Diderma chondrioderma*, *Didymium intermedium*, *Lycogala exiguum*, *Physarella oblonga*, *Physarum aeneum*, *Physarum compressum*, *P. crateriforme*, *P. echinosporum*, *P. fulgens*, *P. melleum*, *P. nucleatum*, *P. oblatum*, *P. polycephalum*, *P. roseum*, *P. stellatum*, *P. superbum*, *Stemonaria longa*, *Stemonitis herbatica*, *S. pallida*, and *Tubifera microsperma*.

Species that appear to be restricted to tropical forests include *Alwisia bombarda*, *Ceratiomyxa morchella*, *C. sphaerosperma*, *Craterium paraguayense*, *Erionema aureum*, *Physarum javanicum*, *P. nicaraguense*, and *Trichamphora pezizoidea*. Whether species described from material collected in tropical forests (e.g., *Arcyria annulata*) but known from only a few records can be considered having a distribution limited to the tropics remains problematic.

7.12 GENERAL OBSERVATIONS

One factor that limits the occurrence of myxomycetes in tropical forests is the fact that as soon as fruiting bodies are produced, they are subject to being colonized by filamentous fungi. The same moist conditions that favor myxomycetes also favor the development of fungi. Under moist conditions, fruiting bodies are likely to be colonized before the spores can be dispersed. Once established, the hyphae of the fungi surround the fruiting body and invade the spore mass, in some cases actually penetrating individual cells. The spores of the myxomycete are never dispersed, so there is no chance of successful reproduction for the species involved. Interestingly, some of the fungi that colonize the fruiting bodies of myxomycetes appear to be restricted to this substrate (Rogerson and Stephenson 1993).

In those tropical forests located in regions of the world where there are distinct "rainy" and "dry" seasons, there is a noticeable difference in the relative abundance of fruiting bodies of myxomycetes in the field. However, fruiting bodies appearing on samples of plant material collected in the field during the two different seasons and then processed in moist chamber cultures may not show the same degree of difference. This does not hold true if samples collected during the rainy season are still moist when placed in moist chamber cultures. These samples are likely to be overrun by filamentous fungi. Allowing the sample to dry out for a couple of days before placing them in moist chambers largely prevents the excessive development of fungi.

Myxomycete plasmodia are more commonly observed in the field during the rainy season, and some plasmodia can be quite large. It is not unusual for a plasmodium to be associated with the fruiting bodies of wood-decay fungi. Some species of myxomycetes found in the tropics are capable of forming huge plasmodia: *Physarella oblonga* is a good example. Where masses of moist decaying leaves occur on the forest floor, such as at the base of a slope, small fruitings develop on leaf surfaces. These sometimes consist of only one or a few sporocarps. Documenting the occurrence of these species can be done only with very careful collecting.

7.13 RECENT DEVELOPMENTS

It seems likely that some of the morphologically identified species (morphospecies) of myxomycetes reported from both tropical and non-tropical regions of the world are represented by one or more "biospecies" (defined as reproductively isolated units) restricted to one region or the other. The results obtained from intensive studies (e.g., Clark and Haskins 2013; Feng and Schnittler 2015, 2017) of several cosmopolitan species have shown that they consist of several biospecies with a more geographically restricted distribution. This is the case for truly cosmopolitan morphospecies such as *Arcyria cinerea* that have been recorded in virtually every survey for myxomycetes ever carried out.

Until recently, classification of myxomycetes was based almost entirely on easily observable morphological features. However, the past decade has witnessed an unparalleled advance in molecular-based classification and the development of a phylogeny that truly reflects relationships among taxa. This has resulted in changes in the nomenclature used for myxomycetes. For example, Garcia et al. (2022) proposed a new classification for the traditional family Trichiales, which included 13 new combinations in the genera *Arcyria*, *Hemitrichia*, *Ophiotheca*, *Oligonema*, *Gulielmina*, and *Perichaena*. The genus *Ophiotheca* was resurrected from older literature, and *Gulielmina* was described as a genus new to science. Prikhodko et al. (2023) reconstructed the phylogeny of the family Didymiaceae in which a number of traditionally recognized species were transferred to new genera. These recent changes are available only in the primary literature and have not yet made their way to monographic treatments of the myxomycetes. As such, the nomenclature used in this chapter reflects the more traditional classification still widely in use.

One other aspect of the classification of myxomycetes should be noted. As already mentioned, it has become apparent that some of the easily recognized morphospecies are actually complexes of a number of biologically distinct taxonomic entities. For example, Leontyev et al. (2023) described 15 new species from the genus *Lycogala*. The majority of these were new species separated out from the single species *L. epidendrum*, the first species of myxomycete formally described and easily one of the most common and widespread species. The new species are morphologically very similar to *L. epidendrum* but are clearly different when minor morphological features and genetic sequences are considered. In an earlier paper, Leontyev et al. (2015) revised the complex of species that collectively had been recognized as the single species *Tubifera ferruginosa*.

7.14 ACKNOWLEDGMENTS

Studies of myxomycetes in tropical forests were supported by grants 0701464, 0431371, 0102895, 0329102, 0139547, and 0316284 from the National Science Foundation of the United States.

REFERENCES

Adamonyte, G., S.L. Stephenson, A. Michaud, E.-H. Seraoui, M. Meyer, Y.K. Novozhilov & T.I. Krivomaz. 2011. Myxomycete species diversity on the island of La Réunion (Indian Ocean). *Nova Hedwigia* 92:523–549.

Alfaro, J.R.D., D.L. Alcayde, J.B. Agbulos, N.H.A. Dagamac & T.E.E. dela Cruz. 2015. The occurrence of myxomycetes from a lowland montane forest and agricultural plantations of Negros Occidental, Western Visayas, Philippines. *Fine Focus* 1:7–20.

Berkeley, M.J. 1881. Australian fungi, II. *Journal of the Linnean Society of London, Botany* 18:383–389.

Bertero, C. 1828. Variedades Ciencias Naturales. *El Mercurio Chileno* 4:194–195.

Black, D.R., S.L. Stephenson & C.A. Pearce. 2004. Myxomycetes associated with the aerial litter microhabitat in tropical forests of northern Queensland, Australia. *Systematics and Geography of Plants* 74:129–132.

Blackwell, B. & R.L. Gilbertson. 1980. Sonoran Desert myxomycetes. *Mycotaxon* 11:139–149.

Bresadola, J. & P.A. Saccardo. 1899. Fungi Congoenses. *Bulletin de la Société Royale de Botanique de Belgique* 38:163.

Calaça F.J.S., J.C. Araúja, G. Cacialli, N.C. Silva, C. Rojas & S. Xavier-Santos. 2020. Fimicolous myxomycetes: Overview of their global distribution and scientific production. *Biologia.* http://doi.org/10.2478/s11756-0-20-00578-9

Camino, M., S.L. Stephenson, T. Krivomaz, D. Wrigley de Basanta, C. Lado & A. Estrada-Torres. 2008. Biodiversity surveys for myxomycetes in the mountains of central Cuba. *Revista Mexicana de Micologia* 27:39–51.

Clark, J. & E.F. Haskins. 2013. The nuclear reproductive cycle in the myxomycetes: A review. *Mycosphere* 4:233–248.

Cox, J.J. 1981. Notes on coprophilous myxomycetes from the western United States. *Mycologia* 73:741–747.

Dagamac, N.H.A., T.E.E. dela Cruz, M.V.B. Pangilinan & S.L. Stephenson. 2011. List of species collected and interactive database of myxomycetes (plasmodial slime molds) for Mt. Arayat National Park, Pampanga, Philippines. *Mycosphere* 2:449–455.

Dagamac, N.H.A., T.E.E. dela Cruz & S.L. Stephenson. 2014. The occurrence of myxomycetes at different elevations in Mt. Arayat National Park, Pampanga, Philippines. *Nova Hedwigia* 98:187–196.

Dagamac, N.H.A., Y.K. Novozhilov, S.L. Stephenson, C. Lado, C. Rojas, T.E.E. de la Crux, M. Unterseer & M. Schnittler. 2017. Biogeographical assessment of myxomycete assembles from Neotropical and Asian Paleotropical forests. *Journal of Biogeography* 44:1523–1536.

Dagamac, N.H.A., M.A.D. Rea-Maminta, N.S. Batungbacal, S.H. Jung, C.R.T. Bulang, A.G.R. Cayago & T.E.E. dela Cruz. 2015. Diversity of plasmodial slime molds (myxomycetes) on coastal, mountain, and community forests of Puerto Galera, Oriental Mindoro, Philippines. *Journal of Asia Pacific Biodiversity* 8:322–329.

Dagamac, N.H.A., S.L. Stephenson & T.E.E. dela Cruz. 2012. Occurrence, distribution and diversity of myxomycetes (plasmodial slime moulds) along two transects in Mt. Arayat National Park, Pampanga, Philippines. *Mycology* 3:119–126.

dela Cruz, T.E.E., M.A.D. Rea, H.T.M. Tran, T.W. Ko Ko & S.L. Stephenson. 2014. A comparative species listing of myxomycetes from tropical (Philippines) and temperate (United States) forests. *Mycosphere* 5:299–311.

Duthie, V.A. 1917. African myxomycetes. *Transactions of the Royal Society of South Africa* 6:297–310.

Eliasson, U. 2013. Coprophilous myxomycetes: Recent advances and future research directions. *Fungal Diversity* 59:85–90.

Estrada-Torres, A., D. Wrigley de Basanta, E. Conde & C. Lado. 2009. Myxomycetes associated with dryland ecosystems of the Tehuacán-Culcatlán Valley Biosphere Reserve, Mexico. *Fungal Diversity* 36:17–56.

Farquharson, C.O. & G. Lister. 1916. Notes on South Nigerian myxomycetes. *Journal of Botany* 54:121–123.

Farr, M.L. 1976. *Flora Neotropica Monograph No. 16 (Myxomycetes)*. New York Botanical Garden, New York.

Fedorova, N.A., Y.K. Novozhilov & V.I. Gmoshinskiy. 2020. Diversity of slime moulds (Myxomycetes = Myxogastrea) in mountain tropical forests of the Phia Oac Reserve (northern Vietnam) revealed by moist chamber cultures. *Mikologiya i Fitopatologiya* 54:288–298. https://doi.org/10.31857/S0026364820040030

Fedorova, N.A., Y.K. Novozhilov, O.N. Shchepin & D.H. Nguyen. 2021. Substrate assemblages of Myxomycetes (Myxogastria) on the bark of living trees and aerial litter in Vietnamese forests. *Journal of Tropical Science and Technology (Tọp chí Khoa học và Công nghệ nhiệt đới, sổ). Joint Russian-Vietnamese Tropical Research and Technological Centre* 24:41–52.

Feest, A. & M.F. Madelin. 1985. A method for the enumeration of myxomycetes in soils and its application to a wide range of soils. *FEMS Microbiology Ecology* 1:103–109. https://doi.org/10.1111/j.1574–6968.1985.tb01137.x

Feng, Y. & M. Schnittaler. 2015. Sex or no sex? Group I introns and independent marker genes reveal the existence of three sexual but reproductively isolated biospecies in *Trichia varia* (Myxomycetes). *Organisms Diversity & Evolution.* https://doi.org/10.1007/s13127-015-0230x

Feng, Y. & M. Schnittler. 2017. Molecular or morphological species? Myxomycete diversity in a deciduous forest of northeastern Germany. *Nova Hedwigia* 104:359–380.

Gao, Y., H. Song, F. Zhou, S. Chen, G. He, J. Yan, Q. Sun, H. Long, Z. Zhai, D. Hu & H. Hu. 2022. Community of soil-inhabiting myxomycetes shares similar assembly mechanisms with fungi, and is affected by bacterial community in subtropical forests of China. *Soil Biology and Biochemistry.* https//doi.org/10.1016/j.soilbio.2022.108854

Gao, Y., X. Zhang, G. He, O.N. Shchepin, S. Yan & S. Chen. 2019. Influence of forest type on dark-spored myxomycete community in subtropical forest soil, China. *Soil Biology and Biochemistry.* https://doe.org/10.106/j.soilbio.2019.107606

Garcia, I., J.C. Zamora, M. Ryberg & C. Lado. 2022. Phylogeny and evolution of a morphological structures in a highly diverse linage of fruiting-forming amoebae, order Trichiales (Myxomycetes, Amoebozoa). *Molecular Phylogenetics and Evolution* 177:107609. https://doi.org/10.1016/

Garcia-Martin, J.M., J. Mosquera & C. Lado. 2018. Morphological and molecular characterization of a new succulenticolous *Physarum* (Myxomycetes, Amoebozoa) with unique polygonal spores linked in chains. *European Journal of Protistology* 63:13–25.

Gilbert, H.C. & G.W. Martin. 1933. Myxomycetes found on the bark of living trees. *University of Iowa Studies in Natural History* 15:3–8.

Gottsberger, G. & N.E. Nannenga-Bremekamp. 1971. A new species of *Didymium* from Brazil. *Proceedings of the Koninklijke Nederlandse Akademie van Wetenschappen. Series C. Biological and Medical Sciences* 74:264–269.

Härkönen, M. 1981. Gambian myxomycetes developed in moist chamber cultures. *Karstenia* 21:21–25.

Härkönen, M. & T. Saarimäki. 1991. Tanzanian myxomycetes: First survey. *Karstenia* 31:41–54.

Ing, B. & B.M. Spooner. 1998. Myxomycetes from Bruni Darussalam. *Kew Bulletin* 53:829–834.

Ko Ko, T.W., T.T.M. Hanh, S.L. Stephenson, D.W. Mitchell, C. Rojas, K.D. Hyde, S. Lumyong & A.H. Bahkali. 2010a. Myxomycetes of Thailand. *Sydowia* 62:243–260.

Ko Ko, T.W., Z.Z.W. Ko Ko, W.C. Rosing & S.L. Stephenson. 2013. Myxomycetes from Myanmar. *Sydowia* 65:267–276.

Ko Ko, T.W., W.C. Rosing & S.L. Stephenson. 2015. First records of myxomycetes from Cambodia. *Austrian Journal of Mycology* 24:31–37.

Ko Ko, T.W., S.L. Stephenson, K.D. Hyde & S. Lumyong. 2010b. Patterns of occurrence of myxomycetes on lianas. *Fungal Ecology* 3:302–310.

Ko Ko, T.W., S.L. Stephenson, K.D. Hyde & S. Lumyong. 2011. Influence of seasonality on the occurrence of myxomycetes. *Chiang Mai Journal of Science* 38:71–84.

Ko Ko, T.W., H.T.M. Tran, M.E. Clayton & S.L. Stephenson. 2012. First records of myxomycetes from Laos. *Nova Hedwigia* 96:73–81.

Kryvomaz, T., A. Michaud & S.L. Stephenson. 2017. First survey of myxomycetes on Mahe Island in the Seychelles. *Nova Hedwigia* 104:65–84.

Kryvomaz, T., A. Michaud & S.L. Stephenson. 2020a. An annotated checklist of myxomycetes from the Seychelles Islands, Indian Ocean. *Karstenia* 58:215–240.

Kryvomaz, T., A. Michaud & S.L. Stephenson. 2020b. Myxomycete biodiversity on five islands of the Seychelles. *Zootaxa* 4851:201–244.

Kylin, H., D.W. Mitchell, E.H. Seraoui & B. Buyck. 2013. Myxomycetes from Papua New Guinea and New Caledonia. *Fungal Diversity* 59:33–44.

Lado, C., A. Estrada-Torres & C. Rojas. 2018. New records of genera and species of myxomycetes (Amoebozoa) from the Neotropics. *Check List* 14:509–518.

Lado, C., A. Estrada-Torres, S.L. Stephenson, D. Wrigley de Basanta & M. Schnittler. 2003. Biodiversity assessment of myxomycetes from two tropical forest reserves in Mexico. *Fungal Diversity* 12:67–110.

Lado, C., A. Estrada-Torres, D. Wrigley de Basanta, M. Schnittler & S.L. Stephenson. 2017. A rapid biodiversity assessment of myxomycetes from a primary tropical moist forest of the Amazon Basin in Ecuador. *Nova Hedwigia* 104:283–321.

Lado, C. & C. Rojas. 2018. Diversity patterns, ecological associations and future research on Costa Rican myxomycetes. *Mycology*. http://doi.org/10.1080/21501203.2018.1481153

Lado, C. & D. Wrigley de Basanta. 2008. A review of Neotropical myxomycetes (1828–2008). *Anales Jardín Botánico de Madrid* 65:211–254.

Lado, C., D. Wrigley de Basabta & A. Estrada-Torres. 2011. Biodiversity of myxomycetes from the Monte Desert of Argentina. *Anales Jardín Botánico de Madrid* 68:61–95.

Lado, C., D. Wrigley de Basanta, A. Estrada-Torres & E. Garcia-Carvjl. 2014. Myxomycete diversity in steppe and bordering areas in Argentina. *Anales Jardín Botánico de Madrid* 71:e006. http://doi.org/10.2989/aibm.2394

Lado, C., D. Wrigley de Basanta, A. Estrada-Torres & S.L. Stephenson. 2013. The biodiversity of myxomycetes in central Chile. *Fungal Diversity* 59:3–32.

Lado, C., D. Wrigley de Basanta, A. Estrada-Torres & S.L. Stephenson. 2016. The myxomycete diversity of the coastal deserts of Peru with emphasis on the lomas formations. *Anales del Jardín Botánico de Madrid* 73(1):e032. ISSN: 0211-1322. http://dx.doi.org/10.3989/ajbm.2436

Lado, C., D. Wrigley de Basanta, A. Estrada-Torres, S.L. Stephenson & I. Treviño. 2019. Diversity of Myxomycetes in arid zones of Peru part II: The cactus belt and transition zones. *Anales del Jardín Botánico de Madrid* 76(2):e083.

Leontyev, D., Y. Ishchenko & M. Schnittler. 2023. Fifteen new species from the myxomycete genus *Lycogala*. *Mycologia* 115:524–560.

Leontyev, D., M. Schnittler & S.L. Stephenson. 2015. A critical revision of the *Tubifera ferruginosa* complex. *Mycologia* 107:959–985.

Lindley, L.A., S.L. Stephenson & F.W. Spiegel. 2007. Protostelids and myxomycetes isolated from aquatic habitats. *Mycologia* 99:504–509.

Liu, Q.-S., S.Z. Yan & S.-L. Chen. 2015. Species diversity of myxomycetes associated with different terrestrial ecosystems, substrata (microhabitats) and environmental factors. *Mycological Progress* 14:27. http://doi.org/10.1007/s11557-015-1048-9

Lücking, R. 1999. Factors influencing the diversity and distribution of foliicolous, lichenized fungi in tropical rain forests. *Sixth International Mycological Congress, Tel Aviv, Israel* 1:7.

Macabago, S.A.B., N.H.A. Dagamac, T.E.E. dela Cruz & S.L. Stephenson. 2017. Implications of the role of dispersal on the occurrence of litter-inhabiting myxomycetes in different vegetation types after a disturbance: A case study in Bohol Islands, Philippines. *Nova Hedwigia* 104:221–236.

Macabago, S.A.B., N.H.A. Dagamac, T.E.E. dela Cruz & S.L. Stephenson. 2020. Myxomycetes of the Caramoan Islands and an update on the myxomycetes of the Bicol Peninsula in the Philippines. *Karstenia* 58:157–167.

Macabago, S.A.B., T.E. dela Cruz & S.L. Stephenson. 2012. First records of myxomycetes from Lubang Island, Occidental Mindoro, Philippines. *Sydowia* 64:109–118.

Macabago, S.A.B. & S.L. Stephenson. 2020. Myxomycetes of Coron Island and additions to the myxomycetes of the Palawan group of islands in the Philippines. *Karstenia* 58:157–167.

Macabago, S.A.B. & S.L. Stephenson. 2021. Does the Wallace Line apply to myxomycetes? *Pacific Science* 75:531–541.

Macabago, S.A.B., S.L. Stephenson & T.E.E. dela Cruz. 2016. Diversity and distribution of myxomycetes in coastal and mountain forests of Lubang Island, Occidental Mindoro, Philippines. *Mycosphere* 7:18–29.

McHugh, R., D.W. Mitchell, M.H. Brims & S.L. Stephenson. 2009. New additions to the Myxomycota of Australia. *Australian Mycologist* 28:56–64.

Mitchell, D.W. 1980. *A Key to Corticolous Myxomycetes*. The British Mycological Society, Cambridge, England.

Montagne, C. 1837. Centurie de plantes cellulaires exotiques nouvelles. *Ser. II. Annales des Sciences Naturelles. Paris* 8:345–370.

Ndiritu, G.G., F.W. Spiegel & S.L. Stephenson. 2009b. Rapid biodiversity assessment of myxomycetes in two regions of Kenya. *Sydowia* 61:287–319.

Ndiritu, G.G., K.E. Winsett, F.W. Spiegel & S.L. Stephenson. 2009a. African myxomycetes II. *Mycotaxon* 107:353–356.

Novozhilov, Y.K., D.A. Erastova, O.N. Shchepin, M. Schnittler, A.V. Aleksandrova, E.S. Popov & A.N. Kuznetsov. 2017. Myxomycetes associated with monsoon lowland tropical forests in southern Vietnam. *Nova Hedwigia* 104:143–182. https://doi.org/10.1127/nova_hedwigia/2016/0395

Novozhilov, Y.K., M. Schnittler, A.W. Rollins & S.L. Stephenson. 2001. Myxomycetes in different forest types of Puerto Rico. *Mycotaxon* 77:285–299.

Novozhilov, Y.K., O.N. Shchepin, A.V. Alexandrova, E.S. Popov & N.H.A. Dagamac. 2018. Altitudinal patterns of myxomycete diversity across tropical forests of Southern Vietnam. *Protistology* 12:73–80. https://doi.org/10.21685/1680–0826-2018-12-2-2

Novozhilov, Y.K., O.N. Shchepin, M. Schnittler, N.H.A. Dagamac, A.V. Alexandrova, E.S. Popov & A.N. Kuznetsov. 2020. Myxomycetes associated with mountain tropical forests of Bidoup Nui Ba and Chu Yang Sin national parks (Dalat Plateau, southern Vietnam). *Nova Hedwigia* 110:185–224. https://doi.org/10.1127/nova_hedwigia/2019/0560

Orlovich, D.A., S.J. Draffin, R.A. Daly & S.L. Stephenson. 2013. Piracy in the high trees: Ectomycorrhizal fungi from an aerial "canopy soil" microhabitat. *Mycologia* 105:52–60.

Poker, J. & K. MacDicken. 2016. Tropical forest resources: Facts and tables. In L. Pancel & M. Köhl (eds.), *Tropical Forestry Handbook*, pp. 3–45. Springer, Berlin, Heidelberg. https://doi.org/10.1007/978-3-642-54601-3_7

Prikhodko, I.S., O.N. Shchepin, N.A. Bortnikova, Y.K. Novozhilov, V.I. Gmoshinskiy, G. Moreno, A. Lopez-Villalba, S.L. Stephenson & M. Schnittler. 2023. A three-gene phylogeny supports taxonomic rearrangements in the family Didymiaceae (Myxomycetes). *Mycological Progress* 22:11. http://doi.org/10.1007/s11557-022-01858-1

Raciborski, M. 1898. Uber die javanischen schleimpilze. *Hedwigia* 37:50–55.

Reynolds, D.R. 1981. Southeast Asian myxomycetes II. Philippines. *Philippine Journal of Biology* 10:127–150.

Rogerson, C.T. & S.L. Stephenson. 1993. Myxomyceticolous fungi. *Mycologia* 85:456–469.

Rojas, C., A. Rojas & C. Lado. 2018. Myxomycete diversity in Costa Rica. *Mycosphere* 9:227–255.

Rojas, C., P.A. Rojas & S.L. Stephenson. 2021. Phenology of myxomycetes in Turrialba, Costa Rica. *Karstenia* 59:1–12.

Rojas, C., A.W. Rollins & S.L. Stephenson. 2014. Distribution of myxomycetes among the microhabitats available for these organisms in tropical forests. In J.K. Misra, J.P. Tewari, S.K. Deshmukh & C. Vágvölgyi (eds.), *Fungi From Different Substrates*, pp. 126–143. CRC Press, Boca Raton, Florida.

Rojas, C., M. Schnittler & S.L. Stephenson. 2010a. A review of the Costa Rican myxomycetes (Amebozoa). *Brenesia* 73–74:39–57.

Rojas, C. & S.L. Stephenson. 2008. Myxomycete ecology along an elevation gradient on Cocos Island, Costa Rica. *Fungal Diversity* 29:117–127.

Rojas, C. & S.L. Stephenson. 2012. Rapid assessment of the distribution of myxomycetes in a southwestern Amazon forest. *Fungal Ecology* 5:726–733.

Rojas, C. & S.L. Stephenson. 2013. Effect of forest disturbance on myxomycete assemblages in the southwestern Peruvian Amazon. *Fungal Diversity* 59:45–53.

Rojas, C. & S.L. Stephenson. 2020a. Comparative analysis on datasets of myxomycetes associated with boreal, temperate, and tropical regions of North America. *Karstenia* 58:190–200.
Rojas, C., S.L. Stephenson, A. Estrada-Torres, R. Valverde & O. Morales. 2010b. New records of myxomycetes from high-elevation areas of Mexico and Guatemala. *Mycosphere* 1:73–82.
Rojas, C., S.L. Stephenson & M. Pavlich. 2011b. New additions to the myxobiota of Peru. *Mycosphere* 2:583–592.
Rojas, C., S.L. Stephenson & P.A. Rojas. 2020a. Myxomycete colonization on translocated and non-translocated dead leaves from temperate and tropical forests. *Studies in Fungi* 5:462–470. https://doi.org/10.5943/sif/5/1/27
Rojas, C., S.L. Stephenson, R. Valverde & A. Estrada-Torres. 2011a. Biogeographical aspects of high-elevation myxomycete assemblages in the northern Neotropics. *Fungal Ecology* 3:1–15.
Rojas, C., R. Valverde & E. Calvo. 2016. Does elevation influence the distributional patterns of tropical myxomycetes? A case study in Costa Rica. *Mycology* 7:45–52.
Rojas, C., R. Valverde, S. Somerille, A.W. Rollins & S.L. Stephenson. 2020b. Myxomycetes within forest ecotones in temperate and tropical forests. *Uniciencia* 35:299–311.
Rojas, C., R. Valverde & S.L. Stephenson. 2015a. New additions to the myxobiota of Costa Rica. *Mycosphere* 6:709–715.
Rojas, C., R. Valverde, S.L. Stephenson & M.J. Vargas. 2010c. Biogeographical and ecological patterns of Costa Rican myxomycetes. *Fungal Ecology* 3:139–147.
Rojas, C., J.M. Zúñiga & S.L. Stephenson. 2015b. Ecological niche modeling for Costa Rican myxomycetes. *Current Research in Environmental & Applied Mycology* 5:153–159.
Sanderson, A.R. 1922. Notes on Malayan mycetozoa. *Transactions of the British Mycological Society* 53:829–834.
Schnittler, M. 2001. Foliicolous liverworts as a microhabitat for Neotropical Myxomycetes. *Nova Hedwigia* 72:259–270.
Schnittler, M., C. Lado & S.L. Stephenson. 2002. Rapid biodiversity assessment of a tropical myxomycete assemblage—Maquipucuna Cloud Forest Reserve, Ecuador. *Fungal Diversity* 9:135–167.
Schnittler, M. & S.L. Stephenson. 2000. Myxomycete biodiversity in four different forest types in Costa Rica. *Mycologia* 92:626–637.
Schnittler, M. & S.L. Stephenson. 2002. Inflorescences of Neotropical herbs as a newly discovered microhabitat for myxomycetes. *Mycologia* 94:6–20.
Stephenson, S.L. 1988. Distribution and ecology of myxomycetes in temperate forests. I. Patterns of occurrence in the upland forests of southwestern Virginia. *Canadian Journal of Botany* 66:2187–2207.
Stephenson, S.L. 1989. Distribution and ecology of myxomycetes in temperate forests. II. Patterns of occurrence on bark surface of living trees, leaf litter, and dung. *Mycologia* 81:608–621.
Stephenson, S.L. 2009. First records of myxomycetes from Ascension Island. *Sydowia* 61:105–115.
Stephenson, S.L. 2020. An unusual ecological association between higher fungi and myxomycetes. *Neotropical Naturalist* 4:1–5.
Stephenson, S.L. 2021. *Secretive Slime Moulds: Myxomycetes of Australia.* CSIRO Publishing, Australia.
Stephenson, S.L., T.F. Elliott, K. Elliott & K. Varnes. 2022. Myxomycetes associated with Australian vertebrate dung. *Australian Zoologist* https://doi.org/10.7882/AZ.2022.041
Stephenson, S.L., A. Estrada-Torres, M. Schnittler, C. Lado, D. Wrigley de Basanta & N. Ogata. 2003. Distribution and ecology of myxomycetes in the forests of Yucatan. In A. Gómez-Pompa, A.M. Allen, S. Fedick & J. Jimenez (eds.), *Lowland Maya Area: Three Millennia at the Human-Wildland Interface*, pp. 241–259. Haworth Press, New York.
Stephenson, S.L., I. Kalyanasundaram & T.N. Lakhanpal. 1993. A comparative biogeographical study of myxomycetes in the mid-Appalachians of eastern North America and two regions of India. *Journal of Biogeography* 20:645–657.
Stephenson, S.L. & J.C. Landolt. 1998. Dictyostelid cellular slime molds in canopy soils of tropical forests. *Biotropica* 30:657–661.
Stephenson, S.L. & J.C. Landolt. 2011. Dictyostelids from aerial "canopy soil" microhabitats. *Fungal Ecology* 4:191–195.
Stephenson, S.L. & J.C. Landolt. 2015. Occurrence of myxomycetes in the aerial "canopy soil" microhabitat. *Mycosphere* 6:83–86.
Stephenson, S.L., J.C. Landolt & D.L. Moore. 1999. Protostelids, dictyostelids, and myxomycetes in the litter microhabitat of the Experimental Forest, Puerto Rico. *Mycological Research* 103:209–214.
Stephenson, S.L. & D. Mitchell. 1994. Notes on tropical myxomycetes. I. Collections from Ecuador and Peru. *Micologia Neotropical Aplicada* 7:17–21.

Stephenson, S.L., M. Schnittler & C. Lado. 2004b. Ecological characterization of a tropical myxomycete assemblage—Maquipucuna Cloud Forest Reserve, Ecuador. *Mycologia* 96:488–497.

Stephenson, S.L., M. Schnittler, C. Lado, A. Estrada-Torres, D. Wrigley de Basanta, J.C. Landolt, Y.K. Novozhilov, J. Clark, D.L. Moore & F.W. Spiegel. 2004a. Studies of Neotropical mycetozoans. *Systematics and Geography of Plants* 74:87–108.

Stephenson, S.L., M. Schnittler & Y.K. Novozhilov. 2008. Myxomycete diversity and distribution from the fossil record to the present. *Biodiversity and Conservation* 17:285–301.

Stephenson, S.L. & H. Stempen. 1994. *Myxomycetes: A Handbook of Slime Molds*. Timber Press, Portland, Oregon.

Stephenson, S.L. & B.C. Stephenson. 2019. Distribution and ecology of myxomycetes on Christmas Island, Indian Ocean. *Phytotaxa* 416:138–148.

Tran, D.Q., H.T.N. Nguyen, H.T.M. Tran & S.L. Stephenson. 2014. Myxomycetes from three lowland tropical forests in Vietnam. *Mycosphere* 5:662–672.

Tran, H.T.M., S.L. Stephenson, K.D. Hyden & O. Mongkolporn. 2006. Distribution and occurrence of myxomycetes in tropical forests of northern Thailand. *Fungal Diversity* 22:227–242.

Tran, H.T.M., S.L. Stephenson, K.D. Hyden & O. Mongkolporn. 2008. Distribution and occurrence of myxomycetes on agricultural ground litter and forest floor litter in Thailand. *Mycologia* 100:181–190.

Ukkola, T. 1998. Tanzania myxomycetes to the end of 1998. *Publications in Botany of the University of Helsinki* 27:1–45.

Ukkola, T. 2000. A checklist of Tanzanian myxomycetes. *Karstenia* 40:189–194.

Walker, L.M., C. Rojas & S.L. Stephenson. 2015. The myxomycetes of the La Selva Biological Station. *Austrian Journal of Mycology* 24:99–111.

Winsett, K.E., J.C. Cavender & S.L. Stephenson. 2008. Myxomycetes of South Africa. *Abstracts of Oral and Poster Presentations, Sixth International Congress on the Systematics and Ecology of Myxomycetes* 1:66.

Winsett, K.E. & S.L. Stephenson. 2013. Myxomycetes isolated from submerged plant material collected in the Big Thicket National Preserve, Texas. *Mycosphere* 4:227–231.

Wrigley de Basanta, D., C. Lado, A. Estrada-Torres & S.L. Stephenson. 2009. Description and life cycle of a new *Didymium* (Myxomycetes) from arid areas of Argentina and Chile. *Mycologia* 101:707–716

Wrigley de Basanta, D., C. Lado, A. Estrada-Torres & S.L. Stephenson. 2013. Biodiversity studies of myxomycetes in Madagascar. *Fungal Diversity* 59:55–83.

Wrigley de Basanta, D., S.L. Stephenson, C. Lado, A. Estrada-Torres & A.M. Nieves-Rivera. 2008. Lianas as a microhabitat for myxomycetes in tropical forests. *Fungal Diversity* 28:109–125.

8 Entomopathogenic Fungi and Tropical Forests

Leela Maya Rizal

8.1 HISTORY OF THE STUDY OF ENTOMOPATHOGENIC FUNGI

Entomopathogenic fungi (EPF) have been recognized for their insect-killing capabilities since the early 1800s, with the earliest documented observation of this attributed to Agostino Bassi in 1835 (Steinhaus 1957; Lacey et al. 2015). Bassi demonstrated that the white muscardine disease in silkworms was caused by a fungal pathogen, later named *Beauveria bassiana* in his honor (Steinhaus 1957; Lacey et al. 2015). This foundational work laid the groundwork for the germ theory of disease and sparked interest in the use of fungi as biological control agents. In the nineteenth and early twentieth centuries, entomopathogenic fungi were studied primarily for their natural epizootics in insect populations. Steinhaus (1949) advanced the field by cataloging fungal pathogens in insects and promoting their use in integrated pest management (IPM). His work emphasized the ecological roles of these fungi and their potential for environmentally sustainable pest control.

8.2 CLASSIFICATION OF ENTOMOPATHOGENIC FUNGI

The fungi most commonly associated with insect pathogenicity belong to the orders Hypocreales (class Sordariomycetes, phylum Ascomycota) and Entomophthorales (class Entomophthoromycetes, phylum Zygomycota) (Kepler et al. 2015; Chandler 2017). The Hypocreales include such genera as *Beauveria*, *Metarhizium*, *Cordyceps*, and *Isaria*, which are known for their broad host ranges and adaptability to diverse environments, including tropical forest soils and canopies. These fungi reproduce via asexual conidia, parasexual hyphal fusion, and sexual recombination, allowing them to persist and evolve under varying ecological conditions (Castrillo et al. 2004; Chu et al. 2017). In contrast, members of the Entomophthorales exhibit biotrophic relationships with their insect hosts and are less amenable to mass production due to their limited saprophytic capabilities. Their specificity and ecological stability make them ideal candidates for conservation biological control, particularly in tropical forest systems where maintaining native fungal-insect dynamics is crucial (Mora et al. 2018).

The development of fungal biopesticides accelerated in the late twentieth century, with commercial formulations of *Beauveria bassiana*, *Metarhizium anisopliae*, and *Isaria fumosorosea* registered for use in various countries, including Australia, South Africa, and the United States. These products are valued for their environmental safety, minimal risk to humans, and compatibility with IPM strategies. Recent advances in molecular biology, genomics, and ecological modeling have deepened our understanding of fungal virulence, host specificity, and environmental persistence (Rizal et al. 2022; Rizal et al. 2024; Rizal et al. 2025). Studies have revealed that entomopathogenic fungi can also function as endophytes, colonizing plant tissues and contributing to plant health and defense, an especially relevant trait in tropical agroforestry systems (Moreira et al. 2025). Despite these advances, challenges remain. Many fungal strains lose virulence under field conditions or fail to infect target pests consistently. This has prompted a shift toward exploring native fungal diversity and ecological adaptations, particularly in under-studied regions such as tropical Australia,

DOI: 10.1201/9781003647140-8

where unique fungal-insect-plant interactions may yield more effective biocontrol agents (Barta et al. 2025).

8.3 LIFE CYCLE

In tropical forest ecosystems, entomopathogenic fungi of the order Hypocreales exhibit remarkable ecological flexibility, growing saprophytically on organic matter, endophytically within plants, or parasitically within insects, depending on resource availability and environmental conditions (Abd-Elsalam et al. 2025). Their life cycle is complex, involving asexual, parasexual, and sexual reproductive pathways (Rizal et al. 2025). Asexual reproduction, the most common mode, involves mitotic division of haploid hyphae to produce conidia across all substrates (Minarni et al. 2025). Parasexual reproduction occurs primarily on organic matter, where compatible haploid hyphae undergo anastomosis, plasmogamy, and karyogamy to form a diploid heterokaryon that produces haploid conidia mitotically (Rizal et al. 2025). Sexual reproduction, less frequent and observed only in insect hosts, involves mating-type idiomorphs (MAT1–1 and MAT1–2) that initiate pheromone signaling and the formation of sexual stromata bearing perithecia (Cai et al. 2022; Rizal et al. 2025).

Conidia are passively dispersed by wind, rain, insects, or human activity (Golan and Pringle 2017), and infection begins upon contact with the insect cuticle (Ortiz-Urquiza and Keyhani 2013). Their hydrophobic nature facilitates adhesion to the epicuticle, followed by germ tube formation and development of appressoria that secrete cuticle-degrading enzymes such as chitinases, proteases, and lipases (Zibaee and Ramzi 2018). Upon penetration into the hemocoel, hyphae transition into blastospores, wall-less, yeast-like cells that evade immune detection and rapidly multiply (Li et al. 2020). These blastospores secrete secondary metabolites that induce host toxicosis (Pedrini 2022). After host death, the fungus re-emerges from the cadaver, producing conidiogenous cells and conidia to complete its life cycle (González-Hernández et al. 2020). In tropical forest systems, these fungi not only regulate insect populations but also persist in soil and plant tissues, contributing to ecosystem resilience. Their ability to switch between ecological roles—saprophyte, endophyte, and pathogen—underpins their adaptability to the dynamic and resource-rich environments of tropical forests (Lodge and Cantrell 1995).

8.4 OCCURRENCE AND ENVIRONMENTS INHABITED BY ENTOMOPATHOGENIC FUNGI IN TROPICAL FORESTS

Entomopathogenic fungi (EPF), particularly those in the order Hypocreales, are widespread across terrestrial ecosystems and have been isolated from soils in tropical, temperate, and even polar regions (Meyling and Eilenberg 2007). In tropical forest systems, their occurrence is closely tied to the region's high humidity, stable temperatures, and rich organic substrates, which support both fungal survival and insect host availability (Evans 1982). While these fungi can tolerate temperatures as low as 2°C, their optimal growth typically occurs above 20°C, making tropical environments particularly conducive to their proliferation (Rizal et al. 2022; Rizal et al. 2024). The diversity and distribution of EPF in tropical forests are influenced more by vegetation type and organic matter content than by soil physicochemical properties or geographic location (Quesada-Moraga et al. 2007). Studies from temperate regions have shown inconsistent correlations between fungal diversity and environmental variables, but in tropical systems, dense vegetation and high organic input appear to enhance EPF richness. For instance, in tropical Australia, climate, vegetation type, and crop species significantly affect EPF distribution (Aung et al. 2008). Species of *Metarhizium* were found to be more abundant in undisturbed urban garden soils that were often watered than in dry grassland soils in Queensland, suggesting that moisture and the land type can shape fungal community structure (Rizal 2023; Rizal et al. 2024). In addition, annual rainfall also plays a role in EPF occurrence. In Tasmania, rainfall patterns were linked to the distribution of species of *Metarhizium*

(Rath et al. 1992), indicating that moisture availability may influence fungal persistence and dispersal. In tropical forests, EPF are commonly found in leaf litter, decaying wood, rhizospheres, and insect cadavers, where they function as saprotrophs, endophytes, or pathogens depending on environmental conditions and host interactions (Sharma et al. 2023). In Australia, half of the fungal isolates collected for mosquito biocontrol were obtained from agricultural soils using insect baiting techniques, such as *Galleria mellonella* (Lepidoptera) and *Tenebrio molitor* (Coleoptera) (Darbro et al. 2011). Forty percent were recovered directly from field-collected insect cadavers, primarily representing species of *Metarhizium* and *Beauveria* (Korosi et al. 2019). However, similar baiting efforts in non-agricultural tropical habitats remain limited. Exploring undisturbed forest soils using diverse insect baits may reveal novel EPF species adapted to native insect hosts and microhabitats, offering new opportunities for conservation biocontrol and ecological research. Overall, tropical forests provide a rich and varied environment for EPF, supporting their complex life cycles and ecological roles. Their occurrence in these systems reflects a dynamic interplay between climate, vegetation, host availability, and land use, underscoring the need for targeted exploration and characterization of native fungal communities.

8.5 TROPICAL SPECIES OF ENTOMOPATHOGENIC FUNGI

Tropical forest ecosystems host a rich diversity of entomopathogenic fungi (EPF), many of which play critical roles in regulating insect populations, promoting plant health, and contributing to nutrient cycling. The ecological roles of tropical EPF extend beyond insect mortality (Gielen et al. 2024). They form endophytic relationships with forest trees, enhance plant defense responses, and contribute to soil health through decomposition and nutrient solubilization (Wong and Wong 2025). Their presence is often an indicator of forest ecosystem resilience, and their conservation is increasingly recognized as vital in the face of deforestation and climate change. Recent surveys in tropical regions such as Malaysia, Thailand, Vietnam, Indonesia, and parts of tropical Australia have revealed novel EPF species and high fungal diversity in undisturbed urban garden soil and forest soils compared to agricultural lands (Bunsak et al. 2023; Lin et al. 2017). The EntomoFun 1.0 database has catalogued over 1,700 records of EPF, insect associations globally, with many entries from tropical zones, highlighting the ecological importance and biogeographic richness of these fungi, particularly abundant in the orders Hypocreales and *Entomophthorales*, with genera such as *Beauveria*, *Metarhizium*, *Cordyceps*, *Ophiocordyceps*, *Isaria*, *Purpureocillium*, and *Paecilomyces* frequently reported from tropical soils and insect cadavers (Rajula et al. 2020). Among the most studied tropical EPF are the species listed below.

Beauveria bassiana: A generalist pathogen with a wide host range, commonly found in tropical agricultural and forest soils. It also colonizes tropical crops such as cocoa, banana, and coffee as an endophyte (Bara and Laing 2020).

Metarhizium anisopliae: Widely distributed in tropical regions, this species is used in biocontrol formulations and has been isolated from diverse insect hosts and soil types. It forms endophytic associations and can transfer insect-derived nitrogen to plants (Meyling and Eilenberg 2007).

Metarhizium acridum: A host-specific pathogen used in tropical biocontrol programs, particularly against locusts and grasshoppers in Africa and Australia (Brunner-Mendoza et al. 2019).

Cordyceps militaris: Found in tropical and subtropical forests, this species is known for its striking colorful fruiting bodies and insecticidal properties (Krishna et al. 2025).

Ophiocordyceps unilateralis: Popular for its "zombie-ant" behavior manipulation, this fungus is native to tropical forests in Southeast Asia, South America and Australia where it infects ants and alters their behavior to optimize spore dispersal (van Roosmalen and de Bekker 2024).

Isaria fumosorosea: Common in tropical agroecosystems, this fungus infects soft-bodied insects and is used in commercial biocontrol products (Mascarin 2015).

Purpureocillium lilacinum: Found in tropical soils, this fungus has both entomopathogenic and nematophagous properties, making it useful in IPM (Nguyen et al. 2025).

Paecilomyces spp: Frequently isolated from tropical forest soils and insect cadavers, though their taxonomy is still being refined (Stauderman et al. 2012).

8.6 BIOGEOGRAPHY AND SPECIES DISCOVERY

Tropical regions such as Southeast Asia, tropical Australia, and sub-Saharan Africa remain underexplored in terms of entomopathogenic fungal diversity, despite their ecological richness and high insect host density (Rizal et al. 2024). These regions offer unique microhabitats, ranging from dense forest canopies to decomposing leaf litter and insect cadavers, that support a wide array of fungal taxa, many of which are yet to be described. Soil baiting techniques using insect larvae (e.g., *Galleria mellonella* and *Tenebrio molitor*) have proven effective in isolating novel fungal strains, particularly in undisturbed forest soils (Rizal et al. 2022; Rizal et al. 2024). However, most surveys have focused on agricultural landscapes, leaving native forest ecosystems largely uncharacterized (Rizal et al. 2024). The use of curated databases such as EntomoFun 1.0 has begun to map global EPF-insect associations, revealing significant biogeographic gaps in tropical zones (De Wint et al. 2024). Expanding fungal biodiversity surveys in these regions is essential not only for species discovery but also for identifying locally adapted strains with potential for conservation biocontrol and ecological restoration.

8.7 COMMERCIAL AND ECOLOGICAL RELEVANCE OF ENTOMOPATHOGENIC HYPOCREALES IN TROPICAL FORESTS

In tropical forest ecosystems, entomopathogenic fungi from the order Hypocreales play a dual role as natural regulators of insect populations and as promising agents for sustainable pest management. The most extensively studied genera *Metarhizium* (Clavicipitaceae), *Beauveria*, *Cordyceps*, and *Isaria* (Cordycipitaceae), and *Ophiocordyceps* (Ophiocordycipitaceae) are known for their virulence and ecological versatility (Humber 2008). Several species, including *Beauveria bassiana*, *Cordyceps militaris*, and *Metarhizium anisopliae*, have been developed into commercial biopesticides and are used globally, including in tropical regions such as Australia and parts of Africa (Lacey 2017; Rajula et al. 2020). These fungi are particularly well-suited to tropical environments due to their ability to thrive in humid, biodiverse soils and their capacity to switch between saprophytic and pathogenic lifestyles. Species of *Beauveria* and *Metarhizium*, for example, persist in tropical soils as saprotrophs and become facultative pathogens upon contact with susceptible insect hosts. Their entomopathogenic phase begins when conidia penetrate the insect cuticle, often leading to natural epizootics under favorable conditions, high humidity, optimal temperature, and dense insect populations (Lacey 2017).

Beyond insect control, some entomopathogenic fungi also function as plant endophytes, particularly in tropical agroforestry systems. *Beauveria bassiana*, for instance, colonizes plant tissues and can enhance plant health by suppressing pathogens or stimulating defense responses (Vidal and Jaber 2015). This multifunctionality makes them valuable components of integrated pest and crop management strategies in tropical landscapes. Despite their promise, only a few species consistently demonstrate high virulence and field efficacy. Many fail to maintain pathogenicity across generations or under variable tropical field conditions (Ortiz-Urquiza and Keyhani 2015). This highlights the need for ecological studies of native fungal diversity in tropical forests, where locally adapted strains may offer greater reliability and effectiveness. While global research has mapped fungal-insect associations in temperate zones, tropical regions like Australia remain underexplored (Korosi et al. 2019). Expanding our understanding of these native entomopathogens could unlock new avenues for biocontrol tailored to tropical forest systems.

8.8 CRYPTIC DIVERSITY AND IDENTIFICATION OF ENTOMOPATHOGENIC FUNGI IN TROPICAL FORESTS

Entomopathogenic fungi, particularly those in the genera *Beauveria* and *Metarhizium*, exhibit extensive cryptic diversity, which poses significant challenges for species identification and ecological interpretation especially in biodiverse tropical forest systems (Rizal et al. 2025). Morphological similarities among species are common, and reproductive structures such as perithecia and ascospores often lack distinct diagnostic features (Mongkolsamrit et al. 2020; Rizal et al. 2024). This issue is compounded by the occurrence of both sexual and asexual states in many species, which are phenotypically divergent and difficult to associate with one another (Rizal et al. 2025). In tropical forests, where fungal diversity is exceptionally high and ecological niches are complex, the limitations of morphology-based identification are particularly pronounced. The sexual state of these fungi is characterized by stromata, perithecia, asci, and ascospores, while the asexual state is defined by conidiogenous cells and conidia (Mongkolsamrit et al. 2020; Rizal et al. 2025). Some species, such as *Metarhizium brittlebankisoides* and *M. campsosterni*, are known only from their sexual forms, whereas others like *Beauveria bassiana* are predominantly described in their asexual state, with sexual forms rarely observed in nature (Rizal et al. 2025; Mei et al. 2020). Although the recent study by Mei et al. (2020) and Rizal et al. (2025) have had success in generating the sexual stromata in the laboratory conditions in China and Australia, respectively.

Molecular tools have significantly advanced fungal taxonomy, with internal transcribed spacer (ITS2) DNA barcoding being widely used. However, ITS2 lacks sufficient resolution for many entomopathogenic taxa, particularly in *Beauveria* and *Metarhizium*, where closely related species cannot be reliably separated (Rehner and Buckley 2005; Sung et al. 2007; Rizal et al. 2024). This has led to the adoption of multilocus phylogenetic analysis, which incorporates multiple gene regions to improve species delimitation and resolve relationships between sexual and asexual states (Rehner et al. 2011). In tropical forest contexts, multilocus phylogenetics has been instrumental in clarifying species boundaries and linking morphologically distinct life stages. For example, *Cordyceps bassiana* has been shown to represent the sexual form of *Beauveria bassiana*, with similar associations confirmed for *B. brongniartii*, *C. brongniartii*, *B. sobolifera*, *C. sobolifera*, *B. sungii* and *C. scarabaeicola* (Shrestha et al. 2014). These findings are particularly relevant in tropical forests, where fungal-insect interactions are diverse and often host-specific, and where accurate identification is essential for ecological and biocontrol applications.

Despite its advantages, multilocus phylogenetics is not without limitations. Poor node support and ambiguous clade separation have been reported in species such as *B. majiangensis* and *B. asiatica*, as well as *B. araneola* and *B. medogensis*, raising concerns about the reliability of species assignments based solely on genetic data (Chen et al. 2018; Khonsanit et al. 2020). In tropical forest systems, where sampling is often limited and species richness is high, integrating molecular data with detailed morphological and ecological observations is critical to avoid taxonomic inflation and misclassification (Rizal et al. 2024). The cryptic diversity of entomopathogenic fungi in tropical forests underscores the need for comprehensive, integrative approaches to fungal identification. Accurate species delimitation is essential not only for understanding fungal ecology and evolution but also for selecting effective biocontrol agents that are adapted to local environmental conditions and insect hosts.

8.9 RECENT IDENTIFICATION SYSTEMS

Recent advances in genomic and phylogenetic tools have fundamentally transformed the identification of entomopathogenic fungi, especially in complex environments like tropical forests where cryptic species diversity is high (Rizal et al. 2025). Traditional identification methods, such as morphological analysis and ITS2 DNA barcoding, have proven inadequate for resolving closely related species within genera like *Beauveria* and *Metarhizium*, due to high morphological variability and insufficient genetic divergence at the ITS2 locus (Rizal et al. 2024; Imoulan et al. 2017). This limitation is particularly evident in species complexes such as *Metarhizium anisopliae*, where even multilocus phylogenies sometimes fail to resolve species boundaries (Rehner and Kepler 2017). To overcome these challenges, researchers now employ a combination of multilocus phylogenetic analyses, whole-genome sequencing, and population genomics (Rizal et al. 2025; Rizal et al. 2024; Mei et al. 2020; Kobmoo et al. 2021). Multilocus approaches, using markers such as TEF1-α, RPB1, and β-tubulin, have revealed previously unrecognized lineages within *Beauveria* and *Metarhizium*, but whole-genome sequencing provides even greater resolution (Rizal et al. 2024). For example, principal component analyses (PCA) of genome-wide SNP data can identify distinct genetic clusters among isolates that appear identical in multilocus phylogenies (Rizal et al. 2025). Rizal et al. (2024) found out that the isolates assigned to the same lineage by multilocus genotyping were often separated into multiple, statistically robust genetic clusters by whole-genome data and they were morphologically different (Figure 8.1), indicating the presence of hidden diversity below the resolution of traditional markers. This integrative approach, combining whole-genome data, multilocus phylogenies, and morphological analysis, enables the detection of cryptic species that are reproductively isolated, even when they coexist in the same habitat (Rizal et al. 2024; Rizal et al. 2025; Mei et al. 2020). In *B. bassiana* from Australian soils, for instance, four distinct genetic groups were identified, each representing a separate gene pool with no evidence of recombination or gene flow between them (Rizal et al. 2025). High FST values, concordance between nuclear and mitochondrial phylogenies, and vegetative incompatibility tests all supported the recognition of these groups as cryptic species. Moreover, the presence of mito-nuclear discordance in some global lineages highlights the complexity of fungal evolution and the need for integrated nuclear and mitochondrial analyses (Rizal et al. 2025).

FIGURE 8.1 Morphology of entomopathogenic fungi collected from Australian subtropical soil. (A) *Beauveria bassiana*, (B) *B. australi,* (C) *M. campsosterni*, (D) *Metarhizium robertsii*, (E) *M. guizhounse*, (F) *M. pinghaense.*

Importantly, combining genomic data with reproductive biology experiments such as crossing tests to assess sexual or parasexual recombination, provides further evidence for species boundaries (Rizal et al. 2025). This is crucial for understanding the ecology, host specificity, and biocontrol potential of these fungi. For example, some genetic clusters may represent new species with unique ecological roles or enhanced efficacy as biological control agents against specific insect pests.

8.10 VIRULENCE AND INTERACTION OF ENTOMOPATHOGENIC FUNGI WITH THEIR INSECT HOSTS IN TROPICAL FORESTS

In tropical forest ecosystems, entomopathogenic fungi play a critical role in regulating insect populations through complex pathogenic interactions. Pathogenicity refers to the inherent ability of a fungus to breach host defenses, while virulence describes the speed and intensity of this invasion (Lacey 1997). In tropical environments, where insect diversity and density are high, fungi such as *Beauveria* and *Metarhizium* exhibit sophisticated mechanisms to infect and overcome host immune responses (Lacey 2017; Rizal et al. 2022; Rizal 2023).

Virulence is mediated by a suite of enzymes and secondary metabolites secreted during infection. These include cuticle-degrading enzymes and immunomodulatory compounds that trigger host responses such as heat shock proteins (HSPs), cytochrome P450 monooxygenases, and antimicrobial peptides (Mannino et al. 2021). Insects in tropical forests often respond with early ecdysis (moulting) to shed infected cuticle layers and activate melanizations via the prophenoloxidase system for wound healing and defense (Zdybicka-Barabas et al. 2025).

Once the fungal appressorium penetrates the cuticle, the insect's cellular immune system deploys hemocytes to encapsulate and digest the invading hyphae (Wang et al. 2023). This is followed by nodulation and phagocytosis, which aim to prevent fungal entry into the hemocoel. In response, the fungi secrete additional enzymes and toxic metabolites to suppress these defenses (Chu et al. 2016), highlighting a dynamic arms race between pathogen and host.

In tropical systems, the virulence of fungal isolates is influenced by multiple factors, including host species, life stage, and environmental conditions such as temperature and humidity (Rizal et al. 2022; Vandenberg et al. 1998). Bioassay variability is common, even under controlled laboratory conditions (Uma Devi et al. 2008), emphasizing the need for field-based validation in tropical habitats.

Importantly, virulence can be enhanced by passaging fungal isolates through their target insect hosts, a method shown to acclimate fungal enzymes to host immune barriers and increase pathogenicity (Humber 2016). For example, *Metarhizium brunneum* exhibited significantly increased virulence after a single passage through its target host (Eckard et al. 2014). This technique is particularly relevant in tropical forests, where host-specific adaptation may improve biocontrol efficacy against native pest species.

Given the ecological complexity of tropical forests, understanding the molecular and immunological interactions between entomopathogenic fungi and their insect hosts is essential for developing effective, environmentally compatible biocontrol strategies (Eckard et al. 2017). Future research should prioritize native fungal strains and host-pathogen dynamics under field conditions to optimize virulence and persistence in tropical landscapes.

8.11 HOST SPECIFICITY AND ENVIRONMENTAL REQUIREMENTS IN TROPICAL FOREST SYSTEMS

Host-pathogen specificity is a cornerstone of effective biological control, particularly in ecologically sensitive environments such as tropical forests (Imoulan et al. 2017). While fungi in the genera *Beauveria* and *Metarhizium* are known to infect over 700 insect species globally (Shahid et al. 2012), only a limited number of isolates demonstrate high virulence against specific hosts. This specificity is exemplified by *Metarhizium acridum*, the active ingredient in the Australian biopesticide Green Guard®, which targets acridid insects such as locusts and grasshoppers (Milner

2000; de Faria and Wraight 2007). Similarly, *Beauveria* bassiana formulations like Ostrinil® and Boverin® are tailored to specific pests such as *Ostrinia nubilalis* and *Leptinotarsa decemlineata* (Safavi et al. 2010).

In tropical forest ecosystems, where insect diversity is exceptionally high and environmental conditions are variable, understanding host specificity must be coupled with an assessment of fungal environmental tolerances. Laboratory screenings often show promising results under controlled conditions (Rizal et al. 2022; Rizal et al. 2024), but field applications frequently fail due to uncontrolled abiotic factors such as temperature, humidity, and solar radiation (Athanassiou et al. 2017; Ibarra-Cortés et al. 2018). These factors are particularly influential in tropical systems, where microclimatic variation can significantly affect fungal germination, growth, and pathogenicity (Ortiz-Urquiza and Keyhani 2015; Wang et al. 2024).

Despite this, many biocontrol programs prioritize virulence over environmental adaptability (Niassy et al. 2012; Onsongo et al. 2019, 2022), overlooking the fact that fungal isolates from different tropical habitats may have distinct temperature and moisture requirements (Rizal et al. 2022). Screening protocols typically involve testing conidial germination, radial growth, and sporulation on media such as SDA, PDA, or OMA across a temperature gradient (Vandenberg et al. 1998). For tropical and subtropical isolates, optimal performance is generally observed at 30°C, with moderate growth at 20–25°C and minimal or no growth below 15°C or above 35°C (Acheampong et al. 2020). Interestingly, some isolates exposed to suboptimal temperatures (10–15°C) can recover and grow rapidly when returned to optimal conditions (25–30°C), suggesting a degree of thermal resilience (Rizal et al. 2022). However, selecting isolates solely based on performance at optimal temperatures may exclude strains better adapted to fluctuating tropical microclimates. In tropical forest systems, where insect hosts and fungal pathogens co-evolve under complex environmental pressures, biocontrol success depends on matching fungal isolates not only to target hosts but also to the specific ecological conditions of deployment. Future research should prioritize the characterization of native fungal strains, their host ranges, and their environmental tolerances to enhance the reliability and sustainability of biological control in tropical landscapes.

REFERENCES

Abd-Elsalam, K.A., R.K. Hassan, T.E. Abdelkhalek, H. Almoammar & A.A. El-Sayed. 2025. Hidden biocontrol agents: The world of insect-pathogenic fungi. In K.A. Abd-Elsalam & S.F. Abu Qamar (eds.), *Fungal endophytes Volume I*, pp. 1–20. Springer Publishing. https://doi.org/10.1007/978-981-97-7312-1_1

Acheampong, M.A., C.A. Coombes, S.D. Moore & M.P. Hill. 2020. Temperature tolerance and humidity requirements of select entomopathogenic fungal isolates for future use in citrus IPM programmes. *Journal of Invertebrate Pathology* 174:107436.

Athanassiou, C.G., N.G. Kavallieratos, C.I. Rumbos & D.C. Kontodimas. 2017. Influence of temperature and relative humidity on the insecticidal efficacy of *Metarhizium anisopliae* against larvae of *Ephestia kuehniella* (Lepidoptera: Pyralidae) on wheat. *Journal of Insect Science* 17:22.

Aung, O.M., K. Soytong & K.D. Hyde. 2008. Diversity of entomopathogenic fungi in rainforests of Chiang Mai Province, Thailand. *Fungal Diversity* 30:15–22.

Bara, G.T. & M.D. Laing. 2020. Entomopathogens: Potential to control thrips in avocado, with special reference to *Beauveria bassiana*. *Horticultural Reviews* 47:325–368.

Barta, M., D. Takov & D. Pilarska. 2025. Entomopathogenic fungi associated with cereal leaf beetles and their comparative pathogenicity in the laboratory: Possible sustainable control approaches. *Cereal Research Communications* 1–18.

Brunner-Mendoza, C., M.D.R. Reyes-Montes, S. Moonjely, M.J. Bidochka & C. Toriello. 2019. A review on the genus *Metarhizium* as an entomopathogenic microbial biocontrol agent with emphasis on its use and utility in Mexico. *Biocontrol Science and Technology* 29:83–102.

Bunsak, A., W. Homhaul, W. Pongprasert, K. Ruangrit & S. Intanon. 2023. Occurrence of entomopathogenic fungi from natural ecosystems in Phetchabun, Thailand and their virulence against brown planthopper. *Current Applied Science and Technology* 10:55003.

Cai, Y., Y. Nie, Y. Gao & B. Huang. 2022. Natural populations of the entomopathogenic fungus *Beauveria bassiana* in Chinese forest ecosystems are diverse and reveal equal frequencies of mating types within phylogenetic species. *Fungal Ecology* 56:101–139.

Castrillo, L.A., M.H. Griggs & J.D. Vandenberg. 2004. Vegetative compatibility groups in indigenous and mass-released strains of the entomopathogenic fungus *Beauveria bassiana*: Likelihood of recombination in the field. *Journal of Invertebrate Pathology* 86:26–37.

Chandler, D. 2017. Basic and applied research on entomopathogenic fungi. In *Microbial Control of Insect and Mite Pests*, pp. 69–89. Academic Press, Cambridge, MA.

Chen, W.H., M. Liu, Z.X. Huang, G.M. Yang, Y.F. Han, J.D. Liang & Z.Q. Liang. 2018. *Beauveria majiangensis*, a new entomopathogenic fungus from Guizhou, China. *Phytotaxa* 333:243–250.

Chu, Z.J., H.H. Sun, S.H. Ying & M.G. Feng. 2017. Vital role for cyclophilin B (CypB) in asexual development, dimorphic transition and virulence of *Beauveria bassiana. Fungal Genetics and Biology* 105:8–15.

Chu, Z.J., Y.J. Wang, S.H. Ying, X.W. Wang & M.G. Feng. 2016. Genome-wide host-pathogen interaction unveiled by transcriptomic response of diamondback moth to fungal infection. *PLoS One* 11:e0152908.

Darbro, J.M., R.I. Graham, B.H. Kay, P.A. Ryan & M.B. Thomas. 2011. Evaluation of entomopathogenic fungi as potential biological control agents of the dengue mosquito, *Aedes aegypti* (Diptera: Culicidae). *Biocontrol Science and Technology* 21:1027–1047.

de Faria, M.R. & S.P. Wraight. 2007. Mycoinsecticides and mycoacaricides: A comprehensive list with worldwide coverage and international classification of formulation types. *Biological Control* 43:237–256.

De Wint, F.C., S. Nicholson, Q.Q. Koid, S. Zahra, G. Chestney-Claassen, J.S.S. Seelan, J. Xie, S. Xing, T.M. Fayle & D. Haelewaters. 2024. Introducing a global database of entomopathogenic fungi and their host associations. *Scientific Data* 11:1418.

Eckard, S., M.A. Ansari, S. Bacher, T.M. Butt, J. Enkerli & G. Grabenweger. 2014. Virulence of in vivo and in vitro produced conidia of *Metarhizium brunneum* strains for control of wireworms. *Crop Protection* 64:137–142.

Eckard, S., S. Bacher, J. Enkerli & G. Grabenweger. 2017. A simple in vitro method to study interactions between soil insects, entomopathogenic fungi, and plant extracts. *Entomologia Experimentalis et Applicata* 163:315–327.

Evans, H.C. 1982. Entomogenous fungi in tropical forest ecosystems: An appraisal. *Ecological Entomology* 7:47–60.

Gielen, R., K. Ude, A. Kaasik, K. Põldmaa, T. Teder & T. Tammaru. 2024. Entomopathogenic fungi as mortality agents in insect populations: A review. *Ecology and Evolution* 14:e70666.

Golan, J.J. & A. Pringle. 2017. Long-distance dispersal of fungi. *Microbiology Spectrum* 5:10–128.

González-Hernández, G.A., I. Padilla-Guerrero, A. Martínez-Vázquez & J.C. Torres-Guzmán. 2020. Virulence factors of the entomopathogenic genus *Metarhizium. Current Protein and Peptide Science* 21:324–330.

Humber, R.A. 2008. Evolution of entomopathogenicity in fungi. *Journal of Invertebrate Pathology* 98:262–266.

Humber, R.A. 2016. Seeking stability for research and applied uses of entomopathogenic fungi as biological control agents. *Journal of Asia-Pacific Entomology* 19:1019–1025.

Ibarra-Cortés, K.H., H. González-Hernández, A.W. Guzmán-Franco, L.D. Ortega-Arenas, J.A. Villanueva-Jiménez & A. Robles-Bermúdez. 2018. Interactions between entomopathogenic fungi and *Tamarixia radiata* (Hymenoptera: Eulophidae) in *Diaphorina citri* (Hemiptera: Liviidae) populations under laboratory conditions. *Journal of Pest Science* 91:373–384.

Imoulan, A., M. Hussain, P.M. Kirk, A. El Meziane & Y.J. Yao. 2017. Entomopathogenic fungus *Beauveria*: Host specificity, ecology and significance of morpho-molecular characterization in accurate taxonomic classification. *Journal of Asia-Pacific Entomology* 20:1204–1212.

Kepler, R.M., T.A. Ugine, J.E. Maul, M.A. Cavigelli & S.A. Rehner. 2015. Community composition and population genetics of insect pathogenic fungi in the genus *Metarhizium* from soils of a long-term agricultural research system. *Environmental Microbiology* 17:2791–2804.

Khonsanit, A., J.J. Luangsa-ard, D. Thanakitpipattana, W. Noisripoom, T. Chaitika & N. Kobmoo. 2020. Cryptic diversity of the genus *Beauveria* with a new species from Thailand. *Mycological Progress* 19:291–315.

Kobmoo, N., N. Arnamnart, W. Pootakham, C. Sonthirod, A. Khonsanit, W. Kuephadungphan, R. Suntivich, O.V. Mosunova, T. Giraud & J.J. Luangsa-Ard. 2021. The integrative taxonomy of *Beauveria asiatica* and *B. bassiana* species complexes with whole-genome sequencing, morphometric and chemical analyses. *Persoonia–Molecular Phylogeny and Evolution of Fungi* 47:136–150.

Korosi, G.A., B.A. Wilson, K.S. Powell, G.J. Ash, A. Reineke & S. Savocchia. 2019. Occurrence and diversity of entomopathogenic fungi (*Beauveria* spp. and *Metarhizium* spp.) in Australian vineyard soils. *Journal of Invertebrate Pathology* 164:69–77.

Krishna, K.V., B. Balasubramanian, S. Park, S. Bhattacharya, J. Kadanthottu Sebastian, W.C. Liu, M. Pappuswamy, A. Meyyazhagan, H. Kamyab, S. Chelliapan & A. Malaviya. 2025. Conservation of endangered *Cordyceps sinensis* through artificial cultivation strategies of *C. militaris*, an alternate. *Molecular Biotechnology* 67:1382–1397.

Lacey, L.A. (ed.) 1997. *Manual of Techniques in Insect Pathology*. Academic Press.

Lacey, L.A. 2017. Entomopathogens used as microbial control agents. In *Microbial Control of Insect and Mite Pests*, pp. 3–12. Academic Press, Cambridge, MA.

Lacey, L.A., D. Grzywacz, D.I. Shapiro-Ilan, R. Frutos, M. Brownbridge & M.S. Goettel. 2015. Insect pathogens as biological control agents: Back to the future. *Journal of Invertebrate Pathology* 132:1–41.

Li, M., Q. Meng, H. Zhang, R. Ni, G. Zhou, Y. Zhao, P. Wu, R. Shu, Q. Qin & J. Zhang. 2020. Vegetative development and host immune interaction of *Ophiocordyceps sinensis* within the hemocoel of the ghost moth larva, *Thitarodes xiaojinensis*. *Journal of Invertebrate Pathology* 170:107–331.

Lin, G.L.E., J.M. Salim, M.F.A. Halim & W.A. Azmi. 2017. Entomopathogenic fungi isolated from the soil of Terengganu, Malaysia as potential bio-pesticides against the red palm weevil, *Rhynchophorus ferrugineus*. *Journal of Sustainability Science and Management* 12:71–79.

Lodge, D.J. & S. Cantrell. 1995. Fungal communities in wet tropical forests: Variation in time and space. *Canadian Journal of Botany* 73:1391–1398.

Mannino, M.C., B. Davyt-Colo & N. Pedrini. 2021. Toxic secondary metabolites and virulence factors expression by entomopathogenic fungi during insect infection and potential impact as a tool for pest management. In *Microbes for Sustainable Insect Pest Management: Hydrolytic Enzyme & Secondary Metabolite–Volume 2*, pp. 121–134. Springer International Publishing, Cham.

Mascarin, G.M. 2015. *Production by solid-state and liquid fermentation and formulation of virulent strains of the fungal entomopathogens Beauveria bassiana and Isaria fumosorosea against whiteflies*. Ph.D. Dissertation. Universidade de São Paulo, Brazil.

Mei, L., M. Chen, Y. Shang, G. Tang, Y. Tao, L. Zeng, B. Huang, Z. Li, S. Zhan & C. Wang. 2020. Population genomics and evolution of a fungal pathogen after releasing exotic strains to control insect pests for 20 years. *The ISME Journal* 14:1422–1434.

Meyling, N.V. & J. Eilenberg. 2007. Ecology of the entomopathogenic fungi *Beauveria bassiana* and *Metarhizium anisopliae* in temperate agroecosystems: Potential for conservation biological control. *Biological Control* 43:145–155.

Milner, R.J. 2000. Current status of Metarhizium as a mycoinsecticide in Australia. *Biocontrol News and Information* 21:47N–50N.

Minarni, E.W., N.K. Wulansari & M.L. Rostaman. 2025. Optimisation of *Lecanicillium saksenae*: Propagation on various agro-waste media for future bioinsecticide development. *Journal of Natural Pesticide Research* 15:100–173.

Mongkolsamrit, S., A. Khonsanit, D. Thanakitpipattana, K. Tasanathai, W. Noisripoom, S. Lamlertthon, W. Himaman, J. Houbraken, R.A. Samson & J. Luangsa-Ard. 2020. Revisiting *Metarhizium* and the description of new species from Thailand. *Studies in Mycology* 95:171–251.

Mora, M.A.E., A.M.C. Castilho & M.E. Fraga. 2018. Classification and infection mechanism of entomopathogenic fungi. *Arquivos do Instituto Biológico* 84.

Moreira, C., T. Mendes-Pereira, F. Pereira, I. Cardoso, E. Mizubuti & S. Elliot. 2025. Ecosystem service providers in the rhizosphere: *Metarhizium* (Hypocreales, Clavicipitaceae) species associated with coffee and non-crop plants in an agroforestry system. *Preprint*. https://doi.org/10.22541/au.176474731.11643460/vl

Nguyen Thi, H., X.H. Nguyen Thi, N. Huynh, A.N. Phan, T.H. Dinh & A.D. Do. 2025. Antagonistic activity of *Purpureocillium lilacinum* PL2 against root-knot nematode (*Meloidogyne incognita*) infection on pepper plants. *Journal of Phytopathology* 173:e70132.

Niassy, S., N.K. Maniania, S. Subramanian, M.L. Gitonga, R. Maranga, A.B. Obonyo & S. Ekesi. 2012. Compatibility of *Metarhizium anisopliae* isolate ICIPE 69 with agrochemicals used in French bean production. *International Journal of Pest Management* 58:131–137.

Onsongo, S.K., B.M. Gichimu, K.S. Akutse, T. Dubois & S.A. Mohamed. 2019. Performance of three isolates of *Metarhizium anisopliae* and their virulence against *Zeugodacus cucurbitae* under different temperature regimes, with global extrapolation of their efficiency. *Insects* 10:270.

Onsongo, S.K., S.A. Mohamed, K.S. Akutse, B.M. Gichimu & T. Dubois. 2022. The entomopathogenic fungi *Metarhizium anisopliae* and *Beauveria bassiana* for management of the melon fly *Zeugodacus cucurbitae*: Pathogenicity, horizontal transmission, and compatibility with cuelure. *Insects* 13:859.

Ortiz-Urquiza, A. & N.O. Keyhani. 2013. Action on the surface: Entomopathogenic fungi versus the insect cuticle. *Insects* 4:357–374.

Ortiz-Urquiza, A. & N.O. Keyhani. 2015. Stress response signalling and virulence: Insights from entomopathogenic fungi. *Current Genetics* 61:239–249.

Pedrini, N. 2022. The entomopathogenic fungus *Beauveria bassiana* shows its toxic side within insects: Expression of genes encoding secondary metabolites during pathogenesis. *Journal of Fungi* 8:488.

Quesada-Moraga, E., J.A. Navas-Cortés, E.A. Maranhao, A. Ortiz-Urquiza & C. Santiago-Álvarez. 2007. Factors affecting the occurrence and distribution of entomopathogenic fungi in natural and cultivated soils. *Mycological Research* 111:947–966.

Rajula, J., A. Rahman & P. Krutmuang. 2020. Entomopathogenic fungi in Southeast Asia and Africa and their possible adoption in biological control. *Biological Control* 151:104399.

Rath, A.C., T.B. Koen & H.Y. Yip. 1992. The influence of abiotic factors on the distribution and abundance of *Metarhizium anisopliae* in Tasmanian pasture soils. *Mycological Research* 96:378–384.

Rehner, S.A. & E. Buckley. 2005. A *Beauveria* phylogeny inferred from nuclear ITS and EF1-α sequences: Evidence for cryptic diversification and links to *Cordyceps* teleomorphs. *Mycologia* 97:84–98.

Rehner, S.A. & R.M. Kepler. 2017. Species limits, phylogeography and reproductive mode in the *Metarhizium anisopliae* complex. *Journal of Invertebrate Pathology* 148:60–66.

Rehner, S.A., A.M. Minnis, G.H. Sung, J.J. Luangsa-ard, L. Devotto & R.A. Humber. 2011. Phylogeny and systematics of the anamorphic, entomopathogenic genus *Beauveria. Mycologia* 103:1055–1073.

Rizal, L.M., M.J. Furlong & G.H. Walter. 2022. Responses of diamondback moth to diverse entomopathogenic fungi collected from non-agricultural habitats: Effects of dose, temperature and starvation. *Fungal Biology* 126:648–657.

Rizal, L.M. 2023. *Hidden Diversity, Virulence and Thermal Requirements of Subtropical Entomopathogenic Fungi.* PhD Thesis, School of Biological Sciences, The University of Queensland. https://doi.org/10.14264/f262961

Rizal, L.M., J.P. Hereward, D.R. Brookes, M.J. Furlong & G.H. Walter. 2024. Hidden diversity within *Beauveria* and *Metarhizium*: Comparing morphology, barcoding, multilocus phylogenies and whole-genome sequences. *Fungal Ecology* 67:101–304.

Rizal, L.M., J.P. Hereward, A.R. McTaggart, M.J. Furlong & G.H. Walter. 2025. Genomic analyses and mating tests among sympatric isolates of *Beauveria bassiana* provide evidence for a diverse cryptic species complex. *Mycologia* 9:1–19.

Safavi, S., A. Kharrazi, G.R. Rasoulian & A. Bandani. 2010. Virulence of some isolates of entomopathogenic fungus, *Beauveria bassiana* on *Ostrinia nubilalis* (Lepidoptera: Pyralidae) larvae. *Journal of Agricultural Science and Technology* 12:13–21.

Shahid, A.A., Q.A. Rao, A. Bakhsh & T. Husnain. 2012. Entomopathogenic fungi as biological controllers: New insights into their virulence and pathogenicity. *Archives of Biological Sciences* 64:21–42.

Sharma, A., S. Sharma & P.K. Yadav. 2023. Entomopathogenic fungi and their relevance in sustainable agriculture: A review. *Cogent Food & Agriculture* 9:2180857.

Shrestha, B., M.W. Hyun, J. Oh, J.G. Han, T.H. Lee, J.Y. Cho, H. Kang, S.H. Kim & G.H. Sung. 2014. Molecular evidence of a teleomorph-anamorph connection between *Cordyceps scarabaeicola* and *Beauveria sungii* and its implication for the systematics of *Cordyceps sensu stricto*. *Mycoscience* 55:231–239.

Stauderman, K., P. Avery, L. Aristizábal & S. Arthurs. 2012. Evaluation of *Isaria fumosorosea* (Hypocreales: Cordycipitaceae) for control of the Asian citrus psyllid, *Diaphorina citri* (Hemiptera: Psyllidae). *Biocontrol Science and Technology* 22:747–761.

Steinhaus, E.A. 1957. Microbial diseases of insects. *Annual Review of Microbiology* 11:165–182.

Steinhaus, H. 1949. Sur la division pragmatique. *Econometrica: Journal of the Econometric Society* 17:315–319.

Sung, G.H., N.L. Hywel-Jones, J.M. Sung, J.J. Luangsa-Ard, B. Shrestha & J.W. Spatafora. 2007. Phylogenetic classification of *Cordyceps* and the clavicipitaceous fungi. *Studies in Mycology* 57:5–59.

Uma Devi, K., J. Padmavathi, C. Uma Maheswara Rao, A.A.P. Khan & M.C. Mohan. 2008. A study of host specificity in the entomopathogenic fungus *Beauveria bassiana* (Hypocreales, Clavicipitaceae). *Biocontrol Science and Technology* 18:975–989.

van Roosmalen, E. & C. de Bekker. 2024. Mechanisms underlying *Ophiocordyceps* infection and behavioral manipulation of ants: Unique or ubiquitous? *Annual Review of Microbiology* 78.

Vandenberg, J.D., M. Ramos & J.A. Altre. 1998. Dose–response and age- and temperature-related susceptibility of the diamondback moth (Lepidoptera: Plutellidae) to two isolates of *Beauveria bassiana* (Hyphomycetes: Moniliaceae). *Environmental Entomology* 27:1017–1021.

Vidal, S. & L.R. Jaber. 2015. Entomopathogenic fungi as endophytes: Plant–endophyte–herbivore interactions and prospects for use in biological control. *Current Science* 46–54.

Wang, H., Z. Lu, N.O. Keyhani, J. Deng, X. Zhao, S. Huang, Z. Luo, K. Jin & Y. Zhang. 2023. Insect fungal pathogens secrete a cell wall-associated glucanase that acts to help avoid recognition by the host immune system. *PLoS Pathogens* 19:e1011578.

Wang, L., N.O. Keyhani, Y. Xia & J. Xie. 2024. The potential and limitations of entomopathogenic fungi as biocontrol agents for insect pest management. *Entomologia Generalis* 44.

Wong, C.K.F. & W.L. Wong. 2025. Discovering the ecological roles of entomopathogenic fungi in tropical rainforests. *Microbiology Australia* 46:125–129.

Zdybicka-Barabas, A., S. Stączek, M. Kunat-Budzyńska & M. Cytryńska. 2025. Innate immunity in insects: The lights and shadows of phenoloxidase system activation. *International Journal of Molecular Sciences* 26:1320.

Zibaee, A. & S. Ramzi. 2018. Cuticle-degrading proteases of entomopathogenic fungi: From biochemistry to biological performance. *Archives of Phytopathology and Plant Protection* 51:779–794.

9 Dictyostelids Associated With Tropical Forests

Pu Liu, Zhaojuan Zhang, Jing Zhao, Yu Li, and Yue Zou

9.1 INTRODUCTION

Tropical forests (Figure 9.1) represent one of the most ecologically important ecosystems on Earth, encompassing extensive geographical regions and playing pivotal roles in the maintenance of global biodiversity, carbon cycle regulation, and climatic modulation (Crowther et al. 2015; Lewis et al. 2015). The spatial distribution and ecological boundaries of these forests are primarily determined by a complex interplay of climatic variables, floristic characteristics, and edaphic factors (Gillespie et al. 2006; Zhu et al. 2015; Ali et al. 2019). Among the edaphic factors, the soil microbial community—including unique organisms such as dictyostelid cellular slime molds (dictyostelids)—plays a crucial yet often overlooked role in maintaining forest health and nutrient cycling.

The prevailing climatic conditions in a typical tropical forest ecosystems are consistently elevated temperatures and substantial precipitation, which collectively foster the development of dense cover of vegetation and exceptional species richness for most of the groups of organisms present. Nevertheless, contemporary climate change has emerged as a critical anthropogenic threat to all ecosystems, with a demonstrated potential to disrupt existing species distribution patterns and compromise biodiversity integrity (Zhou et al. 2013). For example, studies have shown that global warming is causing shifts in species composition, particularly in Andean forests, where forest communities at intermediate and higher elevations are shifting toward species from lower and warmer elevations (Fadrique et al. 2018).

This phenomenon, known as thermophilization, underscores the sensitivity of tropical forests to climatic changes. The vegetation of tropical forests is incredibly diverse, with primary forests being irreplaceable for sustaining tropical biodiversity. These forests are under threat from human-driven land-use changes such as agriculture and logging, which lead to biodiversity loss and ecosystem degradation (Gibson et al. 2011). The structural integrity of these forests is crucial for their role as carbon sinks, yet degradation from activities such as selective logging and fire significantly reduces their carbon storage capacity (Heinrich et al. 2023). Protecting these forests is vital for maintaining their ecological functions and mitigating climate change impacts. Soil properties in tropical forests also play a critical role in their ecological dynamics. Soil fertility, influenced by local geochemical properties, is a major driver of net primary productivity and carbon allocation in these ecosystems (Chave et al. 2010; Bukombe et al. 2022). The interaction between soil characteristics and vegetation can significantly affect the distribution and health of tropical forests, with soil microorganisms such as dictyostelids serving as important mediators of these processes.

9.2 DICTYOSTELIDS IN TROPICAL FORESTS: BIODIVERSITY, FUNCTION, AND BIOMONITORING

Tropical forests represent this planet's most biodiverse terrestrial ecosystem, harboring over half of Earth's species while covering less than 10% of its land surface (Trew and Maclean 2021; Pillay et al. 2022). These vital ecosystems function as global climate regulators, storing approximately

DOI: 10.1201/9781003647140-9

FIGURE 9.1 Representative tropical forest habitats in China. (A, B) Mature tropical evergreen broadleaf forests in Bawangling National Nature Reserve, Hainan Province, (C–F) Tropical evergreen broadleaf forest ecosystems in Xishuangbanna National Nature Reserve, Yunnan Province.

Source: Photos by Pu Liu

250 billion tons of carbon in their biomass and soils. This is equivalent to 90 years of anthropogenic carbon emissions at current rates (Malhi and Grace 2000). Beyond their climate services, tropical forests support critical ecological processes including nutrient cycling, hydrological regulation, and evolutionary diversification, while providing essential resources for nearly 1.6 billion people globally (McKenna and Farrell 2006; Toberman et al. 2011; Ashford et al. 2013).

The extraordinary microbial diversity of tropical forest soils, particularly dictyostelid cellular slime molds, plays underappreciated yet crucial role in maintaining these ecosystems. Dictyostelids contribute significantly to bacterial population control, organic matter decomposition, and nutrient mobilization, their distribution patterns further serve as sensitive indicators of forest health (Stephenson and Landolt 2011; Rukseree et al. 2018).

The purpose of this chapter is to discuss the importance of tropical forests, as well as the life history and ecology of dictyostelids. The primary objectives are to (1) systematically review and synthesize existing literature on the geographic distribution and species richness of dictyostelids across tropical forests, identifying major research locations and regional species patterns; (2) describe the life cycle and ecological characteristics of dictyostelids, emphasizing their adaptations to tropical environments; (3) summarize and evaluate the research methodologies commonly used in studies of tropical dictyostelids, including sampling strategies, cultivation techniques, and molecular tools; and (4) identify gaps in current knowledge and propose future research directions for understanding the diversity and ecological roles of dictyostelids in tropical ecosystems.

By integrating classic and contemporary approaches, the information presented in this chapter seeks to enhance the understanding of dictyostelid biology within the context of tropical forest conservation. The synthesis will provide a foundation for using dictyostelids as model organisms in microbial ecology and as potential indicators of ecosystem health in tropical regions.

9.3 LIFE CYCLE, TAXONOMY AND ECOLOGY OF DICTYOSTELIDS

Dictyostelids are a group of protists commonly referred to as social amoebae due to their ability to form multicellular structures through collective behavior. These organisms exhibit a unique asexual life cycle characterized by alternating phases of unicellular growth and multicellular development (Romeralo et al. 2010; Schaap 2016) (Figure 9.2). During the vegetative stage, dictyostelids exist as free-living, amoeboid cells called myxamoebae. These cells proliferate via binary fission while feeding on bacteria, which they detect through chemotaxis toward folic acid secreted by their prey (Fets et al. 2010; Nasser et al. 2013). Upon nutrient depletion, individual myxamoebae initiate a coordinated aggregation process, forming a multicellular pseudoplasmodium (also termed a "slug" due to its motile, elongated morphology) (Raper 1984; Romeralo et al. 2013). The pseudoplasmodium exhibits migratory behavior, with movement patterns varying among species—some strains form a stalk during migration, while others do not. Following migration, the pseudoplasmodium undergoes morphogenesis to develop into a fruiting body (sorocarp) (Figure 9.3). The sorocarp culminates in a spore-containing structure called a "sorus," where asexual spores are encased in a droplet. Under favorable environmental conditions, these spores germinate, releasing new myxamoebae and thereby completing the life cycle (Hehmeyer 2019).

The current taxonomy used for dictyostelids recognizes 12 genera—three traditional genera (*Acytostelium*, *Dictyostelium*, and *Polysphondylium*) and nine new genera (*Cavenderia*, *Coremiostelium*, *Hagiwaraea*, *Heterostelium*, *Raperostelium*, *Rostrostelium*, *Speleostelium*, *Synstelium*, and *Tieghemostelium*). The latter nine were first recognized in 2018 (Sheikh et al. 2018).

Ecologically, dictyostelids are widespread in terrestrial environments, particularly in forest soils, where they contribute to microbial balance. While their primary habitat is the soil-humus layer, they also colonize diverse microenvironments, including caves (Landolt et al. 2006; Stephenson et al. 2007) and animal dung (Suthers 1985). As predators of bacteria, they likely play a regulatory role in soil microbial communities, although their exact ecological impact remains under investigation (Romeralo et al. 2011). Their ubiquity and adaptive life cycle underscore their importance as model organisms for studying microbial cooperation and ecosystem dynamics, particularly in the context of tropical forest ecosystems where they contribute to nutrient cycling and soil health maintenance.

As noted earlier, many groups of terrestrial organisms achieve their greatest biodiversity in tropical forests, and current data on dictyostelids suggest they follow this trend. Despite significant efforts to thoroughly document the species found in various tropical forests, some areas of the world still require further research. This is especially the case for the Amazon Basin in South America, as well as regions in Africa and Southeast Asia. There is a lack of comprehensive data on dictyostelids

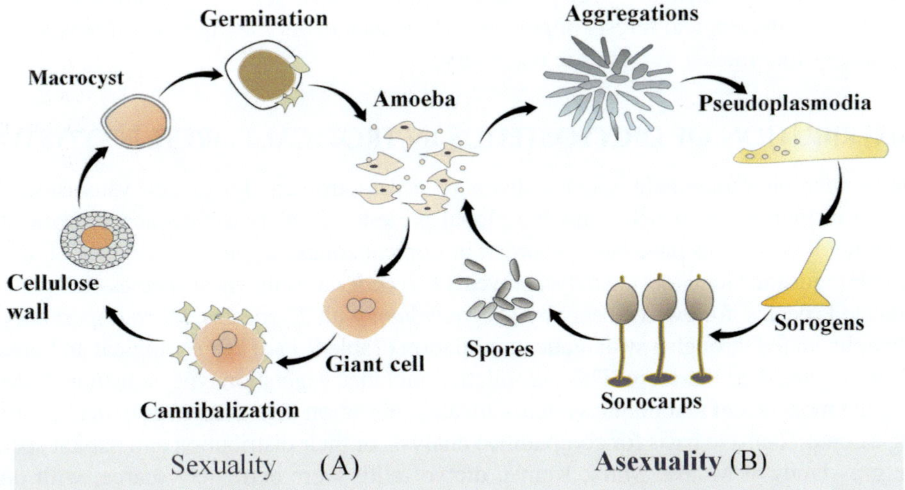

FIGURE 9.2 Life cycle of dictyostelid social amoebae. (A) sexual macrocyst formation, (B) asexual fruiting body development.

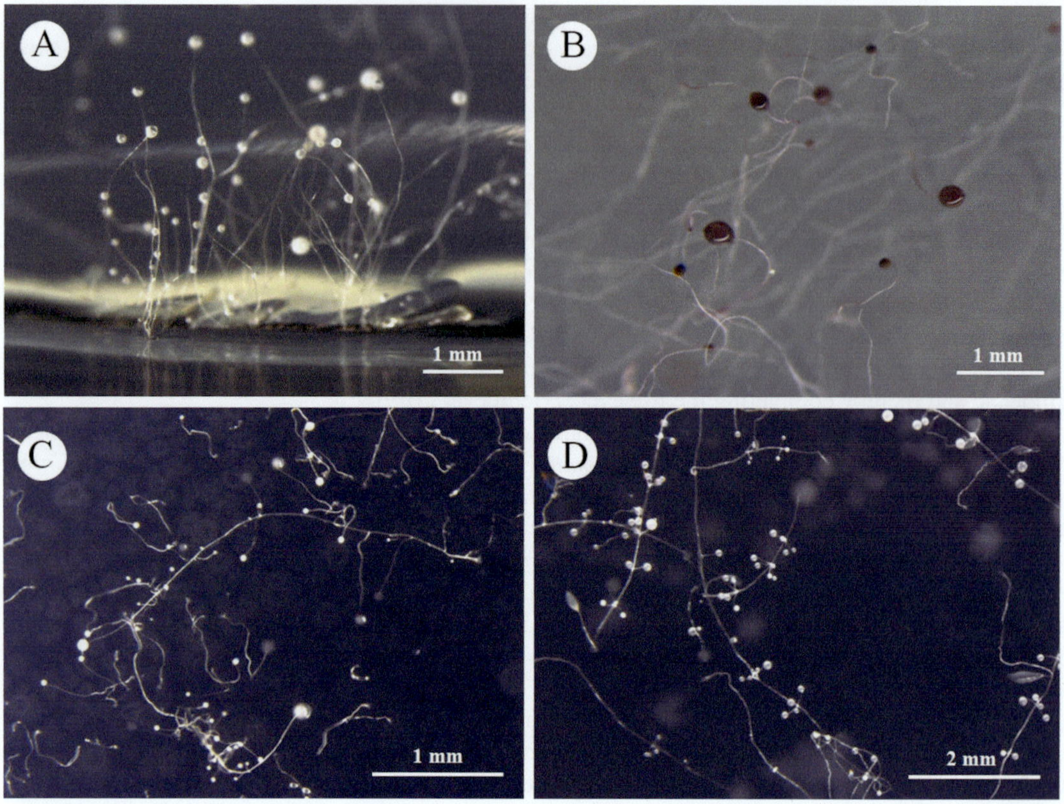

FIGURE 9.3 Morphological characteristics of fruiting bodies (sorocarps) in four species of dictyostelids collected from tropical forests in the Jianfengling National Nature Reserve (A–C) and the Bawangling National Nature Reserve (D), Hainan Island, China. (A) *Dictyostelium giganteum*, (B) *Dictyostelium purpureum*, (C) *Cavenderia aureostipes*, (D) *Heterostelium pallidum*. Scale bars: A–C = 1 mm; D = 2 mm.

in these places, making it difficult to fully document the species assemblage. There is a considerable amount of data for the tropical forests of northern Queensland, yet the high rate of newly identified species indicates that further surveys are necessary.

9.4 DISTRIBUTION OF DICTYOSTELIDS IN TROPICAL FOREST ECOSYSTEMS

Dictyostelids exhibit remarkable species diversity across tropical forest ecosystems worldwide as illustrated in Figure 9.4, which maps the global presence of dictyostelids across tropical forest regions, a total of 94 species have been recorded in tropical zones, accounting for 49% of all species from Index Fungorum. Genus-level analysis reveals *Cavenderia* as the most species-rich genus, with 25 documented species, followed closely by *Dictyostelium* with 22 species. More importantly, a pattern further elucidated through a systematic compilation (Table 9.1) of key ecological and taxonomic attributes from published literature. This compilation includes vegetation type, substrate, Latin name (following the most recent taxonomic system), locality, elevation (m), and geographical coordinates, providing a comprehensive basis for comparative analysis of their distribution and habitat specificity.

In the grasslands of Maasai Mara, Kenya, dictyostelids were extremely scarce, with only one clone isolated from ten samples. In montane forests on Mt. Kenya and Aberdare, occurrence was higher (8/13 samples positive), but diversity remained low, with all isolates belonging to the genus *Dictyostelium*. These results indicate markedly lower abundance in Kenyan grasslands compared to those in central and western North America (Landolt et al. 2011).

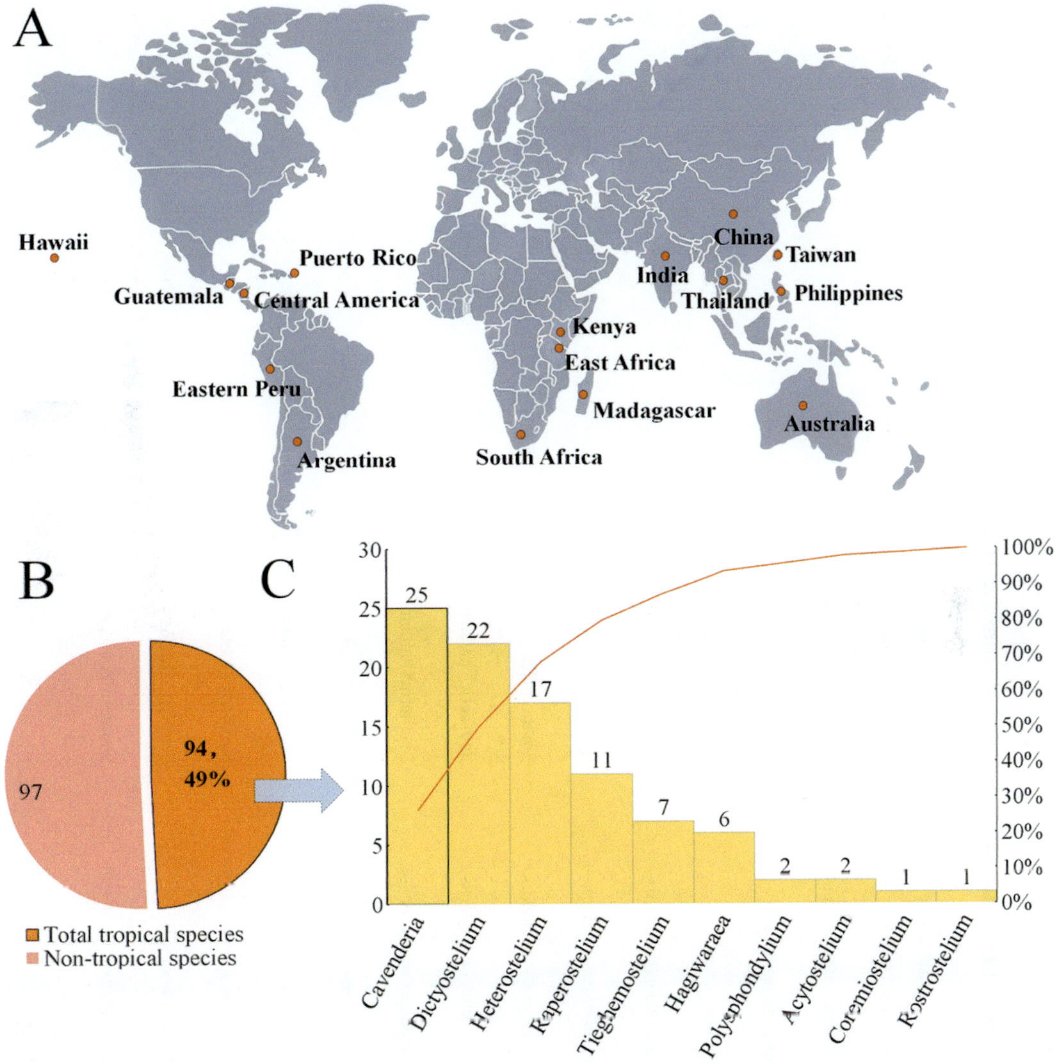

FIGURE 9.4 Global distribution and taxonomic composition of dictyostelid communities in tropical forests. (A) Worldwide collection localities of dictyostelid isolates across tropical forest ecosystems. Sampled countries are highlighted in orange circles. (B) Pie chart illustrating the relative proportion of tropical forest species within the Index Fungorum Redhead and Norvell (2012) (accessible at http://www.indexfungorum.org/names/Names.asp) dictyostelid diversity. The tropical specialist fraction is emphasized in orange. (C) Bar plot showing the distribution of tropical species across major genera. Taxa identified as dominant (based on species richness) are marked with bold black outlines.

In Vadell's doctoral dissertation, he provided detailed morphological characterizations of various species and reported a series of taxonomic novelties (Vadell 2003). Subsequently, in 2007, he described 13 new dictyostelid taxa isolated from soils in the Atlantic Forest of the Iguazú region, Misiones, Argentina. These included seven new species, comprising one *Heterostelium*, one *Tieghemostelium*, two *Cavenderia*, and three *Dictyostelium* species. Additionally, 14 dictyostelid species previously recorded in the Northern Hemisphere (Tikal, Guatemala) were identified, several of which constituted first records for southern South America. A comparative analysis of this dictyostelid assemblage with those from Northern Hemisphere forests, particularly Tikal, suggests distinct evolutionary histories and/or differential natural selection between these biogeographic regions (Vadell and Cavender 2007).

TABLE 9.1
Species of Dictyostelids Recorded from Tropical Forest Ecosystems as Compiled from Current Literature.

Vegetation type	Substrate	Latin name	Locality	Elevation (m)	Coordinates	References
Tropical cloud forest	Soil	*Cavenderia helicoidea*	Chiang Mai Province, Doi Inthanon National Park	—	18°35′32″N, 98°29′12″E	Cavender et al. (2022)
Tropical cloud forest	Soil	*Cavenderia parvibrachiata*	Doi Inthanon National Park	—	18°35′32″N, 98°29′12″E	
Tropical cloud forest	Soil	*Cavenderia protumula*	Chiang Mai Province, Doi Inthanon National Park	—	18°35′32″N, 98°29′12″E	
Tropical cloud forest	Soil	*Cavenderia ungulata*	Chiang Mai Province, Doi Inthanon National Park	—	18°35′32″N, 98°29′12″E	
—	Rubber tree leaf litter	*Dictyostelium dichotomum*	Southern Thailand	—	8°17′S, 99°22′ E	Seephueak and Petcharat (2014)
—	Rubber tree leaf litter	**Cavenderia macrocarpum*	Southern Thailand	—	8°17′N, 99°22′ E	
—	Rubber tree leaf litter	**Tieghemostelium menorah*	Southern Thailand	—	8°09′ S, 99°40′ E	
—	Rubber tree leaf litter	**Cavenderia microsporum*	Southern Thailand	—	8°09′ S, 99°40′ E	
—	Rubber tree leaf litter	*Dictyostelium minutum*	Southern Thailand	—	8°15′ S, 99°35′ E	
—	Rubber tree leaf litter	*Dictyostelium mucoroides*	Southern Thailand	—	7°57′ S, 99°59′ E	
—	Rubber tree leaf litter	*Dictyostelium rosarium*	Southern Thailand	—	8°10′ S, 100°08′ E	
—	Rubber tree leaf litter	**Heterostelium multicystogenum*	Southern Thailand	—	6°53′ S, 100°23′ E	
—	Rubber tree leaf litter	**Heterostelium pallidum*	Southern Thailand	—	7°00′ S, 100°27′ E	
—	Rubber tree leaf litter	*Polysphondylium violaceum*	Southern Thailand	—	7°00′ S, 100°27′ E	

Montane tropical forest	Soil	*Cavenderia aureostabilis*	Iguaziu National Park	1000	19°07′200’ N, 98°44′044′E	Vadell et al. (2018)
Tropical cloud forest	Soil	*Cavenderia bhumiboliana*	Thailand, Doi Inthanon National Park,	2500	18°35′32″N, 98°29′12″E	
Tropical cloud forest	Soil	*Cavenderia protodigitata*	Thailand, Mushroom Research Center	2500	18°35′32″N, 98°29′12″E	
Evergreen tropical forest	Soil	*Cavenderia pseudoaureostipes*	Thailand, Doi Inthanon National Park	800	20°08′48.3″N, 99°51′09.2″E	
Montane tropical forest	Soil	*Cavenderia subdiscoidea*	Thailand: Doi Inthanon National Park	800	19°07′200″N, 98°44′044″E	
Tropical forest	Soil/Water	**Heterostelium ampliverticillatum*	L’Isalo National Park	720	22°22’ S, 45°11’ E	Cavender et al. (2016)
Tropical forest	——	**Heterostelium cumulocystum*	L’Isalo National Park	720	22°29′S, 45°23′E	
Tropical forest	——	**Heterostelium pseudoplasmodiofascium*	Andringitra National Park	922	22°08’ S, 46°54’ E	
Subhumid tropical forest	Soil/Humus	*Cavenderia basinodulosa*, *Hagiwaraea lavandula*, *Raperostelium stabile*	Ranomafana National Park, MGR-09–05	914	21°15′31″S, 47°25′15″E	Cavender et al. (2021)
Subhumid tropical forest	Soil/Humus	*Cavenderia deminutiva*	Ranomafana National Park, MGR-09–06	1001	21°15′44″S, 47°25′16″E	
Subhumid tropical forest	Soil/Humus	*Dictyostelium* sp. A, *Heterostelium versatile*	Ranomafana National Park, MGR-09–52	1222	21°17′12″S, 47°19′09″E	
Subtropical woodland (bushveld)	Soil	*Cavenderia fulva*	Kruger National Park	530	25°05′3″S, 31°14′46″E	Cavender et al. (2018)
Remnant subtropical forest (bushvelt)	Soil	*Hagiwaraea tenebrica*	Zululand	300	28°49′5″S, 31°44′12″E	
Subtropical woodland (bushveld)	Soil	*Heterostelium irregularibrachiatum*	Kruger National Park	530	25°05′34″S, 31°14′46″E	
Subtropical woodland	Soil	*Raperostelium cymosum*	Kruger National Park	530	25°05′34″S, 31°14′46″E	
Subtropical forest	Surface soil /Humus	*Raperostelium crispum*	Birds of Eden Park	220	33°57′53″S, 23°28′60″E	

(*Continued*)

TABLE 9.1 (*Continued*)
Species of Dictyostelids Recorded from Tropical Forest Ecosystems as Compiled from Current Literature.

Vegetation type	Substrate	Latin name	Locality	Elevation (m)	Coordinates	References
Subtropical forest	Soil	*Cavenderia minima*	Birds of Eden Park near Tsitsikamma National Park in the Western Cape	220	33°57′53″S, 23°28′60″E	
Tropical semievergreen forest	Soil	*Dictyostelium rnucoroides*	East Africa	below 1,200	——	Cavender (1969)
Tropical semievergreen forest	Soil	*Dictyostelium purpureum*	East Africa	below 1,200	——	
Tropical semievergreen forest	Soil	**Hagiwaraea lavandulum*	East Africa	below 1,200	——	
Tropical semievergreen forest	Soil	* *Heterostelium pallidum*	East Africa	Below 1,200	——	
Tropical semievergreen forest	Soil	*Acytostelium leptosomum*	East Africa	below 1,200	——	
Tropical semievergreen forest	Soil	**Hagiwaraea lavandulum*	East Africa	below 1,200	——	
——	Soil	*Heterostelium colligatum*	Xishuangbanna Tropical Botanical Garden, Yunnan Province, China	570	101°25′E, 21°41′N	Liu et al. (2019b)
Tropical evergreen rainforest	Soil	*Dictyostelium gloeosporum*	Limu Mountain in Hainan Province, China	——	108°37′–111°05′E, 3°30′–20°18′N	Liu and Li (2012)
Tropical evergreen rainforest	Soil	*Dictyostelium sphaerocephalum*	Wuzhi Mountain in Hainan Province, China	——	108°37′–111°05′E, 3°30′–20°18′N	
Broadleaf rainforest	Soil	*Dictyostelium purpureum* var. *pseudosessile*	China. Hainan Province, Qixianling Hot Springs National Forestry Park	350	109°41′E, 18°42′N	Liu et al. (2019c)
vergreen broad leaved forests and bamboo thickets	Humus and fermentation layer	*Dictyostelium macrocephalum*	Taiwan, China	——	——	Hagiwara et al. (1985)
Broad ieaved forest and deciduous tree	Soil	**Heterostelium pallidum*	Hengchun Garden, Pingtung Taiwan, China	——	——	Yeh and Chien (1983)

—	Soil	*Dictyostelium macrocephalum* *Dictyostelium purpureum*	Checheng, Taiwan, China	—	—	Hagiwara et al. (1992)
—	Soil	*Acytostelium leptosomum complex*, **Cavenderia aureostipes*, **Coremiostelium polycephalum*, *Dictyostelium macrocephalum*, *Dictyostelium purpureum*, **Hagiwaraea lavandula*, **Heterostelium pallidum complex*, *Polysphondylium violaceum*, **Raperostelium monochasioides*	Maopitou, Taiwan, China	—	—	
—	Soil	**Coremiostelium polycephalum*, **Cavenderia aureostipes*, *Dictyostelium macrocephalum*, *Dictyostelium brefeldianum complex*, *Dictyostelium purpureum*, **H. pallidum complex*, **Hagiwaraea lavandula*, **Raperostelium monochasioides*	Oluanpi, Taiwan, China	—	—	
Tropical rain forests	Soil	**Cavenderia boomerangispora*	Australia: Queensland: School for Field Studies on the Atherton Tablelands near Yungaburra	—	17°12′19″S, 145°40′47″E	Landolt et al. (2008)
Tropical rain forests	Soil	**Heterostelium flexuosum*	Australia:Victoria:Jehosaphat Gulley in Kinglake National Park 50 km NE of Melbourne	—	37°32′13″S, 145°21′06″E	
Tropical rain forests	Soil	** Heterostelium granulosum*	Australia: Western Australia: Shannon National Park 300 km S of Perth	—	34°46′59″S, 116°11′30″E	
Tropical rain forests	Canopy soil	**Cavenderia myxobasis*	Australia: Queensland: Australian Canopy Crane S of Cape Tribulation	—	16°06′13″S, 145°26′48″E	

(*Continued*)

TABLE 9.1 ***(Continued)***
Species of Dictyostelids Recorded from Tropical Forest Ecosystems as Compiled from Current Literature.

Vegetation type	Substrate	Latin name	Locality	Elevation (m)	Coordinates	References
Tropical rain forests	Soil	*_Hagiwaraea radiculata_	Australia: Queensland: Mt Lewis 35 km S of Julatten	—	16°33′12″S, 145°16′25″E	
Tropical rain forests	Canopy soil	*_Heterostelium rotatum_	Australia: Queensland: Australian Canopy Crane S of Cape Tribulation	—	16°06′13″S, 145°26′48″E	
Tropical rain forests	Canopy soil	*_Heterostelium australicum_	Australia: Queensland: Australian Canopy Crane S of Cape Tribulation	—	16°06′13″S, 145°26′48″E	
Tropical rain forests	Soil	*_Heterostelium stolonicoideum_	Australia: Queensland: School for Field Studies near Yungaburra	—	17°12′19″S, 145°40′47″E	
Tropical deciduous forests	Soil	_Dictyostelium mucoroides_	Costa Rica	1000–2000	—	Cavender and Raper (1968)
Tropical deciduous forests	Soil	_Dictyostelium purpureum_	Costa Rica	1000–2000	—	
Tropical deciduous forests	Soil	*_Cavenderia mucoroides_	Costa Rica	1000–2000	—	
Tropical deciduous forests	Soil	_Polysphondylium violaceum_	Costa Rica	1000–2000	—	
Tropical deciduous forests	Soil	*_Heterostelium pallidum_	Costa Rica	1000–2000	—	
Tropical deciduous forests	Soil	*_Coremiostelium polycephalum_	Costa Rica	1000–2000	—	
Tropical deciduous forests	Soil	*_Hagiwaraea vinaceofuscum_	Costa Rica	1000–2000	—	
Semievergreen tropical rainforests	Soil	*_Tieghemostelium simplex_	Tikal, Petén Province, Guatemala	15–200	19°36′54″N, 99°13′17″W	Cavender et al. (2013)
Semievergreen tropical rainforests	Soil	*_Tieghemostelium unicornutum_	Tikal, Petén Province, Guatemala	15–200	19°36′54″N, 99°13′17″W	

Semievergreen tropical rainforests	Soil	**Raperostelium capillare*	Bladen Nature Reserve, Toledo District, Belize	100	16°34′17″N, 88°42′30″W	
Semievergreen tropical rainforests	Soil	**Raperostelium filiforme*	Tikal, Petén, Guatemala	15–200	19°36′54″N, 99°13′17″W	
Semievergreen tropical rainforests	Soil	**Raperostelium maeandriforme*	Tikal, Petén Province, Guatemal	15–200	19°36′54″N, 99°13′17″W	
Semievergreen tropical rainforests	Soil	**Raperostelium reciprocatum*	Bladen Nature Reserve, Toledo District, Belize	100	16°34′17″N, 88°42′30″W	
Semievergreen tropical rainforests	Soil	**Tieghemostelium angelicum*	Bladen Nature Reserve, Toledo District, Belize	100	16°34′17″N, 88°42′30″W	
Semievergreen tropical rainforests	Soil	**Tieghemostelium montium*	Bladen Nature Reserve, Toledo District, Belize	100	16°34′17″N, 88°42′30″W	
Semievergreen tropical rainforests	Soil	**Tieghemostelium dumosum*	Tikal, Petén Province, Guatemal	15–200	19°36′54″N, 99°13′17″W	
Semievergreen tropical rainforests	Soil	**Raperostelium reciprocatum* var. *transitum*	Tikal, Petén Province, Guatemal	15–200	19°36′54″N, 99°13′17″W	
Tropical evergreen; deciduous; thorn	Soil	**Heterostelium pallidum*, *Polysphondylium violaceum*, **Cavenderia aureostipes*, **Raperostelium minutum*, *Dictyostelium sphaerocephalum*, *Dictyostelium mucoroides*, *Dictyostelium giganteum*, *Dictyostelium purpureum*, **Coremiostelium polycephalum*, *Acytostelium subglobosum*, **Hagiwaraea vinaceofusca*, **Raperostelium tenue*	Corbett, Uttar Pradesh; Banipur, Karnataka; Bondla, Goa; Delhi, I; Guindy, Tamil Nadu; Bangalore, Karnataka; Delhi, II	400, 1000, 600, 200, 100, 600, 200	——	Cavender and Lakhanpal (1986)
——	Soil/Litter	*Dictyostelium mucoroides*	Hawaiian Island	1500–2000	——	Swanson et al. (2004)
——	Soil/Litter	**Cavenderia aureostipes*	Hawaiian Island	1500–2000	——	
——	Soil/Litter	*Dictyostelium giganteum*	Hawaiian Island	1500–2000	——	
——	Soil/Litter	**Coremiostelium polycephalum*	Hawaiian Island	1500–2000	——	
——	Soil/Litter	**Heterostelium pallidum*	Hawaiian Island	1500–2000	——	

(*Continued*)

TABLE 9.1 (*Continued*)
Species of Dictyostelids Recorded from Tropical Forest Ecosystems as Compiled from Current Literature.

Vegetation type	Substrate	Latin name	Locality	Elevation (m)	Coordinates	References
—	Soil/Litter	*Dictyostelium capitatum*	Hawaiian Island	1500–2000	—	
—	Soil/Litter	*Dictyostelium implicatum*	Hawaiian Island	1500–2000	—	
—	Soil/Litter	**Heterostelium tenuissimum*	Hawaiian Island	1500–2000	—	
—	Soil/Litter	*Acytostelium subglobosum*	Hawaiian Island	1500–2000	—	
—	Soil/Litter	**Raperostelium minutum*	Hawaiian Island	1500–2000	—	
—	Soil/Litter	* *Raperostelium tenue*	Hawaiian Island	1500–2000	—	
—	Soil/Litter	*Dictyostelium* sp. 1	Hawaiian Island	1500–2000	—	
—	Soil/Litter	*Dictyostelium* sp. 2	Hawaiian Island	1500–2000	—	
Montane mesic forest, Coastal wet shrubland, Montane wet forest, Lowland mesic forest/ Shrubland	Soil/Litter	**Cavenderia aureostipes*, *Dictyostelium gigantium*, *Dictyostelium macrocephalum*, *Dictyostelium brefeldianum*, **Coremiostelium polycephalum*, *Dictyostelium purpureum*, **Heterostelium pallidum*	Hawai'i Island	—	—	Landolt and Wong (1998)
Lowland mesic forest/ Shrubland, Agricultural mulch pile, Lowland mesic forest	Soil/Litter	*Dictyostelium brefeldianum*, *Dictyostelium mucoroides* var. *stoloniferum*, **Coremiostelium polycephalum*, Undescribed *Dictyostelium* (NAH 8), **Heterostelium pallidum*, *Polysphondylium violaceum*	O'ahu Island	—	—	
Lowland mesic shrubland, Coastal mesic forest, Montane mesic forest, Montane wet mixed community	Soil/Litter	*Dictyostelium brefeldianum*, *Dictyostelium purpureum*, Undescribed *Dictyostelium* (KU 14), *Dictyostelium mucoroides*, **Heterostelium pallidum*	Kaua'i Island	—	—	
Secondary tabonuco forest, Palo colorado forest	Litter	*Acytostelium leptosomum*	Luquillo Mountains in northeastern Puerto Rico	600–750	18°20′N, 65°49′W	Stephenson et al. (1999)

Tabonuco forest, Secondary tabonuco forest	Litter	*Dictyostelium firmibasis*	Luquillo Mountains in northeastern Puerto Rico	below 600	18°20′N, 65°49′W
Tabonuco forest	Litter	**Raperostelium minutum*	Luquillo Mountains in northeastern Puerto Rico	below 600	18°20′N, 65°49′W
Secondary tabonuco forest, Palm forest	Litter	**Raperostelium monochasioides*	Luquillo Mountains in northeastern Puerto Rico	350–450	18°20′N, 65°49′W
Tabonuco forest, Secondary tabonuco forest, Palm forest	Litter	*Dictyostelium mucoroides*	Luquillo Mountains in northeastern Puerto Rico	350–450	18°20′N, 65°49′W
Secondary tabonuco forest	Litter	*Dictyostelium purpureum*	Luquillo Mountains in northeastern Puerto Rico	350–450	18°20′N, 65°49′W
Secondary tabonuco forest	Litter	**Hagiwaraea rhizopodium*	Luquillo Mountains in northeastern Puerto Rico	350–450	18°20′N, 65°49′W
Palo colorado forest	Litter	**Raperostelium tenue*	Luquillo Mountains in northeastern Puerto Rico	600–750	18°20′N, 65°49′W
Palm forest	Litter	*Dictyostelium* sp. A	Luquillo Mountains in northeastern Puerto Rico	350–450	18°20′N, 65°49′W
Secondary tabonuco forest	Litter	*Dictyostelium* sp. B	Luquillo Mountains in northeastern Puerto Rico	350–450	18°20′N, 65°49′W
Secondary tabonuco forest	Litter	*Dictyostelium* sp. C	Luquillo Mountains in northeastern Puerto Rico	350–450	18°20′N, 65°49′W
Secondary tabonuco forest	Litter	*Dictyostelium* sp. D	Luquillo Mountains in northeastern Puerto Rico	350–450	18°20′N, 65°49′W

(*Continued*)

TABLE 9.1 (_Continued_)
Species of Dictyostelids Recorded from Tropical Forest Ecosystems as Compiled from Current Literature.

Vegetation type	Substrate	Latin name	Locality	Elevation (m)	Coordinates	References
Tabonuco forest, Secondary tabonuco forest, Palm forest	Litter	*_Heterostelium pallidum_	Luquillo Mountains in northeastern Puerto Rico	below 600, 350–450	18°20′N, 65°49′W	
Tropical rain forests	Soil	*_Rostrostelium ellipticum_, _Acytostelium subglobosum_, *_Polysphondylium laterosorum_, *_Hagiwaraea lavandula_, _Dictyostelium macrocephalum_, *_Cavenderia mexicana_, *_Raperostelium monochasioides_, _Dictyostelium mucoroides_, _Dictyostelium purpureum_, *_Hagiwaraea rhizopodia_, _Dictyostelium_ sp.A, _Dictyostelium_ sp. B, *_Heterostelium pallidum_	Eastern Peru	150	4°30′S, 73°30′W	Landolt and Stephenson (1991)
Tropical lowland evergeen rain forests, Semi-deciduous rain forests, Montane rain forest	Soil	_Dictyostelium mucoroides_, *_Heterostelium pallidum_, _Dictyostelium purpureum_, _Polysphondylium violaceum_, _Dictyostelium_ sp., *_Coremiostelium polycephalum_, *_Hagiwaraea rhizopodia_, *_Hagiwaraea lavandula_, _Acytostelium subglobosum_, *_Cavenderia bifurcata_, *_Hagiwaraea vinaceofusca_, *_Tieghemostelium lacteum_, *_Cavenderia multistipes_, *_Hagiwaraea coeruleostipes_, _Dictyostelium intermedium_, *_Dictyostelium mucoroides_	Singapore, Malaysia, Indonesia, Thailand, Philippines (Southeast Asian)	—	—	Cavender (1976)

Atlantic subtropical rainforest	Lightly acidic soil (pH 5.7)	**Cavenderia macrocarpa*	Macuco trail, Yacaratia and Yry-apútrails, Sites II, III and IV.	100	25°28′S, 56°1′W	Vadell and Cavender (2007)
	Soil-litter (pH 5.8–6.6)	*Dictyostelium brevicaule*	Iguazú, both Regional and National Parks	100–250	25°28′S, 56°1′W	
	Soil-litter (pH 5.8)	*Dictyostelium dichotomum*	Macuco Trial, Iguazú National Parks	100	25°28′S, 56°1′W	
	Soil-litter (pH 6)	**Tieghemostelium menorah*	Macuco Trail, Iguazú National Park Misiones, Argentina	180	25°28′S, 56°1′W	
	Soil-litter (pH 6.2, pH 6.3)	**Cavenderia nanopodia*	Site I, Upper-lower trail, Iguazú National Park, Misiones, Argentina; Site II, Yy-apú area.	150	25°28′S, 56°1′W	
	Soil-litter (pH 6)	*Dictyostelium vermiforme*	Yry-apú area Iguazú Regional Park, Misiones, Argentina	200	25°28′S, 56°1′W	
	Soil-litter (pH 6)	**Heterostelium arachnoideum*	Site I, Upper-Lower trail, Site II, Macuco trail and Yacaratia Trail of the Iguazú National Park; Yryapú area, site V, Iguazú Regional Park, Misiones, Argentina	200	25°28′S, 56°1′W	
Semievergreen tropical rain forest	Surface humus layers	*Dictyostelium citrinum*, **Cavenderia medusoides*, **Cavenderia granulophora*	Tikal, Petén Province, Guatemala	150–200	—	Vadell et al. (1995)
—	Soil	**Coremiostelium polycephalum*, *Dictyostelium aureum*, *Dictyostelium giganteum*, *Dictyostelium leptosomum* *Dictyostelium mucoroides*, **Heterostelium tenuissimum*, **Raperostelium minutum*	Kenya	—	—	Landolt et al. (2011)

(*Continued*)

TABLE 9.1 (*Continued*)
Species of Dictyostelids Recorded from Tropical Forest Ecosystems as Compiled from Current Literature.

Vegetation type	Substrate	Latin name	Locality	Elevation (m)	Coordinates	References
Thick tropical virgin lowland rainforests that host large dipterocarp trees	Soil	*Dictyostelium discoideum*, *Dictyostelium mucoroides*, *Dictyostelium purpureum*, **Heterostelium pallidum*, *Polysphondylium violaceum*, **Cavenderia aureostipes*, **Polysphondylium laterosorum*	Lubang Group of Islands	—	13°49′15.6″N, 120°09′54.8″E	Yulo and dela Cruz (2011)
Forest trails	Forest and mossy soils	**Polysphondylium laterosorum*, **Heterostelium pallidum*	Subic Bay Natural Forest Reserve, Zambales, Philippines	—	—	dela Cruz et al. (2011)

Note: The table includes details on vegetation type, substrate, Latin name (an asterisk* indicates a taxon name under the new classification system of Sheikh et al. (2018), locality, elevation (m), and geographical coordinates.

In the semievergreen tropical rain forest soils of Tikal, Guatemala (Vadell et al. 1995), a diverse assemblage of dictyostelids was documented, comprising 34 distinct taxa. Ten of these taxa represent newly discovered species, with several others recorded for the first time in the American tropics. This survey constitutes the most extensive single-forest collection of dictyostelids reported to date. Among the newly described species are three yellow-pigmented, medium to large forms: *Dictyostelium citrinum*, *Cavenderia medusoides*, and *C. granulophora*.

In a comprehensive survey across Southeast Asia (Cavender 1976), encompassing forest sites in Singapore, Malaysia, Thailand, Indonesia, and the Philippines, an analysis of soil samples revealed a dictyostelid community composition of 16 species. Five species—*Dictyostelium mucoroides*, *D. purpureum*, *Heterostelium pallidum*, *Polysphondylium violaceum*, and an unidentified *Dictyostelium* sp.—were ubiquitous and dominant. Four exhibited intermediate prevalence, while the remaining seven were rare and geographically restricted. Notably, the dominant species and overall diversity closely mirrored patterns observed in Neotropical forests, indicating a consistent pantropical distribution for core dictyostelid taxa, with endemicity primarily confined to rare species.

During a 1989 survey in the tropical rainforests of eastern Peru, soil samples from both primary and secondary forests were analyzed for dictyostelids. This study identified 11 dictyostelid species, including nine new records for the Amazon Basin. Nine isolates were further classified into two putative novel species, tentatively named *Dictyostelium* sp. A and *Dictyostelium* sp. B. The overall dictyostelid density was found to be low, averaging 63 clones per gram of wet soil (Landolt and Stephenson 1991).

In the Luquillo Experimental Forest in Puerto Rico, dictyostelids were studied in five forest types (tabonuco forest, secondary tabonuco forest, palo colorado forest, palm forest and elfin forest) that occurred at different elevations in a relatively small area. The secondary tabonuco forest exhibited the highest species richness and abundance of dictyostelids among five forest types studied. A general pattern of decreasing dictyostelid diversity with increasing elevation was observed, with the lowest levels found in the high-elevation elfin forest (Stephenson et al. 1999).

The dictyostelid communities across the Hawaiian Islands, including Hawai'i, O'ahu, and Kaua'i, were found to be species-poor and exhibited low clone densities, especially during dry conditions. This overall depauperate state—relative to both subtropical and neotropical mainland sites—was primarily attributed to limited dispersal and colonization success due to the lack of an efficient windborne dispersal mechanism, leading to a high frequency of unproductive sites (Landolt and Wong 1998). However, a chronosequence study on the island of Hawai'i revealed that early colonization can be rapid, with both density and diversity peaking in the youngest geological substrates (≥67 years). It was hypothesized that this initial successional peak is transient and becomes suppressed over time by the arrival and establishment of predators, particularly nematodes, which subsequently reduce dictyostelid populations (Swanson et al. 2004).

Tropical forests in Peninsular India, along with five vegetational-climatic zones of the West-Central Himalaya, were found to be characterized by 12 species of dictyostelid slime molds, with populations showing greater similarity in species composition and diversity to those of East Africa than to Southeast Asia (Cavender and Lakhanpal 1986).

North American tropical forests harbor a distinct assemblage of cellular slime molds, including species with crampon bases and the ability to form macrocysts not typically found in temperate deciduous forests, with species diversity and population composition closely linked to vegetation type and soil environment (Cavender and Raper 1968). Central America's seasonal rainforests in Belize and Guatemala yielded ten isolates of small dictyostelids, nine of which were identified as distinct new species based on detailed morphological and behavioral characteristics. Phylogenetic analysis confirmed these new species belong to dictyostelid group 3, refining the taxonomic understanding of this group (Cavender et al. 2013). In the northernmost portion of Queensland, Australia, new species of dictyostelid cellular slime molds were discovered in both ground soil and the canopy soil of tropical forests (Landolt et al. 2008).

Hainan Province from tropical China yielded the discovery of *Dictyostelium pseudosessile*, a dictyostelid variant characterized by violet sori, sorophores that widen at their midpoint, and simple recurved bases, as confirmed by its complete 50-hour life cycle documentation and phylogenetic analysis (Zou et al. 2025; Liu et al. 2019b), while two other species (*D. gloeosporum* and *D. sphaerocephalum*) were isolated from soil samples collected from Limu Mountain and Wuzhi Mountain in Hainan Province, China (Liu and Li 2012). Xishuangbanna, Yunnan Province in China, served as the discovery site for *Heterostelium colligatum*, a dictyostelid whose life cycle stages harbor distinct, dynamic, and species-specific symbiotic bacterial communities dominated by Proteobacteria, Actinobacteria, and Firmicutes (Liu et al. 2019a). Several species of dictyostelids have been reported from Taiwan Province in China, including *D. macrocephalum*, *D. purpureum*, and members of the *H. pallidum* complex, among others (Yeh and Chien 1983; Hagiwara et al. 1985; Hagiwara et al. 1992). These findings were primarily documented through field collections in regions such as Kenting, Maopitou, and Oluanpi in southern Taiwan.

Southern Thailand rubber tree plantations yield ten species of dictyostelids (seven *Dictyostelium* and three *Polysphondylium*), with significantly lower densities in herbicide-treated litter (15.0–32.5 clones/g) than in untreated areas (122.5–255.0 clones/g) (Seephueak and Petcharat 2014). Northern Thailand exhibits remarkable dictyostelid diversity, with 40 soil samples yielding 15 taxa, including nine newly described species—five species of *Cavenderia* (*C. aureostabilis*, *C. bhumiboliana*, *C. protodigitata*, *C. pseudoaureostipes*, and *C. subdiscoidea*) characterized through integrated morphological and phylogenetic analyses within the family Cavenderiaceae (Vadell et al. 2018), followed by four additional novel species identified via morphological distinctiveness, molecular phylogenetics, and DNA sequence data, collectively expanding the known biodiversity of this group in Southeast Asia (Cavender et al. 2022). Samples from Amnat Charoen Province in Thailand yielded five dictyostelid genera, dominated by *Dictyostelium*, as revealed by 18S rDNA phylogenetics (Rukseree et al. 2018).

East Africa's tropical forest soils exhibit lower species richness of dictyostelids compared to Central America, particularly among rare members of the group, potentially due to simpler forest structure or Central America's evolutionary centrality. With increasing altitude, East African forests show a shift toward *Dictyostelium minutum* dominance and reduced diversity, mirroring altitudinal patterns in Central America (Cavender 1969). However, despite these regional variations, broader surveys across the African continent reveal significant diversity. Continental Africa and Madagascar have documented approximately 25 dictyostelid species each, with numerous new forms identified. Nevertheless, research remains highly uneven, concentrated mainly in East and South Africa, while West, Central, and North Africa are critically understudied—the latter with only two species recorded (Cavender et al. 2010). Importantly, the existing research confirms that different regions host distinct dictyostelid communities. For instance, South Africa harbors a unique assemblage in its tropical and subtropical forests, as evidenced by the discovery of six novel species among the 18 recorded species there. All of these species are exclusively found in "subtropical forest" or "tropical woodland" habitats, underscoring that this region represents a unique component of the continent's mycological diversity (Cavender et al. 2018).

A total of 13 distinct dictyostelid species were identified in Madagascar, among which three species—*Heterostelium ampliverticillatum*, *H. cumulocystum*, and *H. pseudoplasmodiofascium*—were specifically isolated from tropical forest habitats (Cavender et al. 2016), while a 2009 survey across 18 localities in Madagascar revealed exceptional dictyostelid diversity, documenting 30 species (including 17 novel taxa, predominantly within the genus *Heterostelium*), surpassing continental Africa's known richness, with higher biodiversity in mesic habitats and restricted distributions among new species (Cavender et al. 2021).

Lubang Island, Occidental Mindoro, Philippines, yielded eight dictyostelid species from coastal and lowland mountain forest soils, including five previously recorded taxa (*Dictyostelium discoideum*, *D. mucoroides*, *D. purpureum*, *Heterostelium pallidum*, *Polysphondylium violaceum*), two new records for the Philippines (*Cavenderia aureostipes*, *P. laterosorum*), and one putative novel species (*Dictyostelium* sp. L08–09) (Yulo and dela Cruz 2011). Meanwhile, a parallel survey in the Subic Bay Natural Forest Reserve, Zambales, Philippines, documented population densities ranging from 821 to 3,150 clones per gram of soil and identified two species, *P. laterosorum* and *H. pallidum*, from forest and mossy soils (dela Cruz et al. 2011). Together, these two studies significantly advanced the understanding of Philippine dictyostelid diversity, the former represented the first national documentation in over three decades, while the latter provided the first record of these organisms within the Subic Bay protected area.

9.5 RESEARCH SITES AND METHODS

Research on dictyostelids in tropical forests has been conducted across diverse geographical regions, including Hawai'i, Africa, China, Thailand, India, and Central America, each offering unique ecological conditions. In Hawai'i, soil and litter samples were collected from multiple vegetation communities on the islands of Hawai'i, O'ahu, and Kaua'i, representing variations in geological age and human impact (Landolt and Wong 1998). In China, studies focused on the Xishuangbanna Tropical Botanical Garden (Yunnan Province) and Qixianling Hot Springs National Forestry Park (Hainan Province), both characterized by tropical monsoon climates with annual temperatures of 21.4°C and 23.0°C, respectively (Liu et al. 2019b). Sampling in Thailand targeted tropical cloud forests and seasonal rainforests in northern regions, with soil/humus layers collected for analysis (Vadell et al. 2018; Cavender et al. 2022). The Shimla region in India, featuring temperate forests at elevations up to 2700 m, and Belize's Bladen Nature Reserve dominated by broadleaf evergreen rainforest, and Guatemala's Tikal National Park, characterized by seasonal rainforest, were also key study sites (Cavender and Lakhanpal 1986; Cavender et al. 2013). In East Africa, samples were collected from diverse forest types, including thorn-scrub, coastal semideciduous, and tropical semievergreen forests, reflecting the continent's ecological heterogeneity (Cavender 1969). These sites were selected to capture the broad ecological diversity of tropical forests, ensuring comprehensive coverage of dictyostelid habitats.

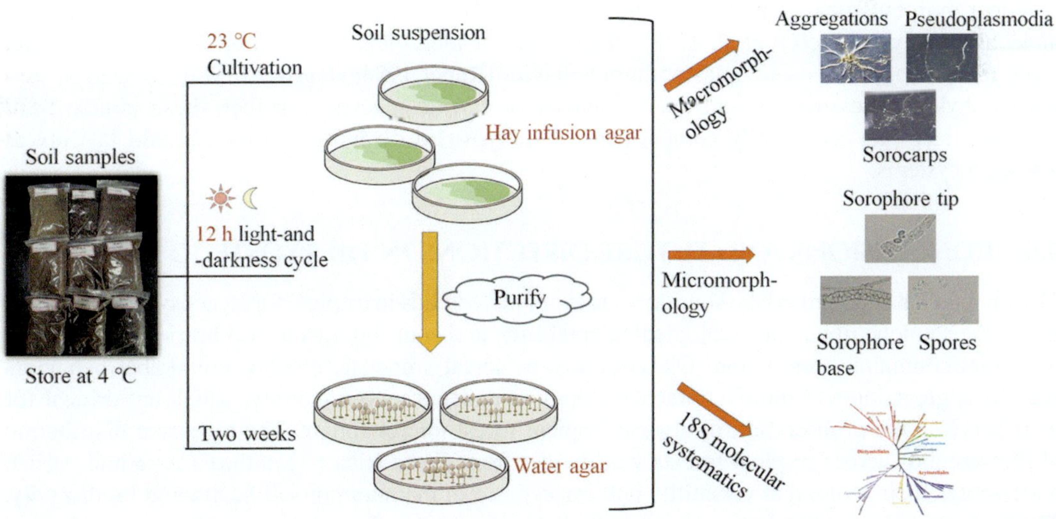

FIGURE 9.5 The experimental workflow for the isolation and characterization of dictyostelids from soil samples, comprising sample processing, morphological analysis (including both macro- and micro-morphological observations), and molecular identification based on 18S rRNA gene sequencing. The phylogenetic tree shown in the figure is derived from Zhang et al. (2024b).

Standardized methods were employed across studies to ensure consistency in sample collection and processing (Figure 9.5). Soil and litter samples, typically weighing 10–100 g, were collected from the surface humus layer and placed in sterile plastic bags. The samples were numbered and recorded in the laboratory soil sample database upon arrival, then stored at 4°C for subsequent processing (Zhang et al. 2023; Zhang et al. 2024a; Liu et al. 2025; Zhang et al. 2025a). When this was possible, to enable comprehensive analysis, a portion of each soil sample was reserved for determining soil microbial community composition and physicochemical properties. These subsamples were carefully separated and stored under appropriate conditions to preserve their integrity for downstream analyses (Liu et al. 2019a; Zhang et al. 2025b).

The isolation of dictyostelids followed the protocols of Cavender and Raper (1965a), involving soil-water dilutions (e.g., 1:10 or 1:25) and plating onto hay infusion agar with *Escherichia coli* as a food source (Raper 1984). Plates were incubated at 20–25°C (typically maintained at 23°C) under diffuse light and monitored daily for aggregations and fruiting bodies formation over a 2–3 weeks period. Isolates were purified on non-nutrient agar and preserved for taxonomic and molecular studies, with spores stored at −80°C or in silica gel (Cocucci and Sussman 1970; Liu et al. 2019c).

Morphological observations of dictyostelids are divided into two major categories: macroscopic and microscopic morphology. Macroscopic morphological observations are conducted under a stereoscopic dissecting microscope to document the location of each early aggregating clone and the developing sorocarp(s), with all positions marked on the lid of the Petri dish. Key macroscopic features noted including the growth habit, branching pattern, and color of the sorocarp (fruiting body); the length, color, and size of the sorus; the size and morphology of cell aggregates; and the formation of pseudoplasmodia. These structures can be photographed, measured, and recorded for further analysis. For microscopic morphology, observations are performed on slides using a light microscope equipped with 10× ocular and 10×, 40×, and 100× (oil) objectives. This allows for detailed examination of spore shape and size, the presence or absence of granules in spores, the thickness and number of cell lines in the sorophores, as well as the characteristics and width of sorophore tips and bases. Images can be captured with a color microscope camera and archived in a specimen identification file.

These morphological approaches, combined with DNA extraction and 18S rDNA sequencing for molecular analysis (Sheikh et al. 2018), enable species identification based on sorocarp structure, spore morphology, and pseudoplasmodium behavior (Raper 1984; Hagiwara 1989), as well as supported phylogenetic analyses and the delineation of cryptic species. Together, these standardized protocols facilitate cross-study comparisons of dictyostelid diversity and distribution in tropical forest ecosystems.

9.6 CONCLUSIONS AND FUTURE DIRECTIONS IN DICTYOSTELID RESEARCH

This chapter has highlighted the significant role of dictyostelids in tropical forest ecosystems, emphasizing their unique life cycle, ecological adaptability, and contributions to soil nutrient cycling and microbial community regulation. Dictyostelids, as social amoebae, exhibit remarkable behaviors such as aggregation and multicellular development under resource scarcity, which are critical for their survival and proliferation in diverse tropical forest microhabitats. The extensive distribution of dictyostelids across tropical forests worldwide, from Costa Rica to Southeast Asia and Africa, underscores their ecological versatility and importance in maintaining soil health and biodiversity. Their presence in various forest types, including rainforests, cloud forests, and seasonal forests, further demonstrates their adaptability to different environmental conditions.

Despite the progress made in understanding dictyostelid taxonomy and ecology, several limitations persist. Current research is hindered by insufficient data from key regions such as the Amazon Basin, Africa, and Southeast Asia, where comprehensive surveys are lacking. Moreover, the dynamic nature of tropical forests, coupled with the challenges of sampling and identifying dictyostelid species, limits our ability to fully document their diversity and ecological roles. The reliance on morphological characteristics for species identification also poses challenges, as cryptic species and subtle variations may be overlooked.

Future research should prioritize expanding taxonomic and ecological studies to under-explored regions, employing molecular techniques to enhance species identification and phylogenetic analyses. Investigating the ecological functions of dictyostelids, particularly their interactions with bacteria and other microorganisms, will provide deeper insights into their role in nutrient cycling and ecosystem stability. Furthermore, the impact of global climate change on dictyostelid distribution and behavior warrants urgent attention, as shifting temperatures and precipitation patterns may alter their habitat suitability and ecological contributions. Developing conservation strategies that incorporate dictyostelids as bioindicators of soil health and forest integrity could also aid in preserving tropical forest ecosystems. Collaborative efforts among researchers, policymakers, and local communities will be essential to address these challenges and advance our understanding of these fascinating organisms.

9.7 ACKNOWLEDGMENTS

Studies of dictyostelids in tropical forests were supported by funding supplied by the National Natural Science Foundation of China (No. 32370020) and the National Natural Science Foundation of Jilin Province (No. 20260101006JJ).

REFERENCES

Ali, A., S.L. Lin, J.K. He, F.M. Kong, J.H. Yu & H.S. Jiang. 2019. Elucidating space, climate, edaphic, and biodiversity effects on aboveground biomass in tropical forests. *Land Degradation & Development* 30:918–927.

Ashford, O.S., W.A. Foster, B.L. Turner, E.J. Sayer, L. Sutcliffe & E.V.J. Tanner. 2013. Litter manipulation and the soil arthropod community in a lowland tropical rainforest. *Soil Biology & Biochemistry* 62:5–12.

Bukombe, B., M. Bauters, P. Boeckx, L.N. Cizungu, M. Cooper, P. Fiener, L.K. Kidinda, I. Makelele, D.I. Muhindo, B. Rewald, K. Verheyen & S. Doetterl. 2022. Soil geochemistry—and not topography—as a major driver of carbon allocation, stocks, and dynamics in forests and soils of African tropical montane ecosystems. *New Phytologist* 236:1676–1690.

Cavender, J.C. 1969. The occurrence and distribution of Acrasiae in forest soils. II. East Africa. *American Journal of Botany* 56:993–998.

Cavender, J.C. 1976. Cellular slime molds of Southeast Asia. II. occurrence and distribution. *American Journal of Botany* 63:71–73.

Cavender, J.C. & T.N. Lakhanpal. 1986. Distribution of dictyostelid cellular slime molds in forest soils of India. *Mycologia* 78:56–65.

Cavender, J.C., J.C. Landolt, G.G. Ndiritu & S.L. Stephenson. 2010. Dictyostelid cellular slime moulds from Africa. *Mycosphere* 1:147–152.

Cavender, J.C., J.C. Landolt, M. Romeralo, A. Perrigo, E.M. Vadell & S.L. Stephenson. 2016. New species of *Polysphondylium* from Madagascar. *Mycologia* 108:80–109.

Cavender, J.C., J.C. Landolt, E.M. Vadell, A.L. Perrigo, S.L. Stephenson, D.W. De Basanta, C. Lado & P. Liu. 2021. Distribution and ecology of dictyostelids in Madagascar. *Phytotaxa* 505:176–186.

Cavender, J.C. & K.B. Raper. 1965a. The Acrasieae in nature. I. Isolation. *American Journal of Botany* 52:294–296.

Cavender, J.C. & K.B. Raper. 1968. The occurrence and distribution of Acrasiae in forests of subtropical and tropical America. *American Journal of Botany* 55:504–513.

Cavender, J.C., E.M. Vadell, J.C. Landolt, S.L. Stephenson, S.L. Baldauf, J. Roux, N. Cavender & K.E. Winsett. 2018. New dictyostelid cellular slime molds from South Africa. *Phytotaxa* 383:233–251.

Cavender, J.C., E.M. Vadell, J.C. Landolt, K.E. Winsett, S.L. Stephenson, A.W. Rollins & M. Romeralo. 2013. New small dictyostelids from seasonal rainforests of Central America. *Mycologia* 105:610–635.

Cavender, J.C., E.M. Vadell, A.L. Perrigo, J.C. Landolt, S.L. Stephenson & P. Liu. 2022. Four new species of dictyostelids from soil systems in northern Thailand. *Journal of Fungi* 8:593.

Chave, J., D. Navarrete, S. Almeida, E. Alvarez, L.E.O.C. Aragao, D. Bonal, P. Chatelet, J.E. Silva–Espejo, J.Y. Goret, P. von Hildebrand, E. Jimenez, S. Patino, M.C. Penuela, O.L. Phillips, P. Stevenson & Y. Malhi. 2010. Regional and seasonal patterns of litterfall in tropical South America. *Biogeosciences* 7:43–55.

Cocucci, S.M. & M. Sussman. 1970. RNA in cytoplasmic and nuclear fractions of cellular slime mold amebas. *Journal of Cell Biology* 45:399–407.

Crowther, T.W., H.B. Glick, K.R. Covey, C. Bettigole, D.S. Maynard, S.M. Thomas, J.R. Smith, G. Hintler, M.C. Duguid, G. Amatulli, M.N. Tuanmu, W. Jetz, C. Salas, C. Stam, D. Piotto, R. Tavani, S. Green, G. Bruce, S.J. Williams, S.K. Wiser, M.O. Huber, G.M. Hengeveld, G.J. Nabuurs, E. Tikhonova, P. Borchardt, C.F. Li, L.W. Powrie, M. Fischer, A. Hemp, J. Homeier, P. Cho, A.C. Vibrans, P.M. Umunay, S.L. Piao, C.W. Rowe, M.S. Ashton, P.R. Crane & M.A. Bradford. 2015. Mapping tree density at a global scale. *Nature* 525:201–205.

dela Cruz, T.E.E., K.A.A. Santiago, C.S.P. Ramirez, J.M.O. Torres, N.H.A. Dagamac, J.M. Yap, Ching & P.R.J. Yulo. 2011. Occurrence of cellular slime molds (dictyostelids) in subic bay natural forest reserve, Zambales, Philippines. *Philippine Journal of Systematic Biology* 5:99–104.

Fadrique, B., S. Baez, A. Duque, A. Malizia, C. Blundo, J. Carilla, O. Osinaga–Acosta, L. Malizia, M. Silman, W. Farfan–Rios, Y. Malhi, K.R. Young, C.F. Cuesta, J. Homeier, M. Peralvo, E. Pinto, O. Jadan, N. Aguirre, Z. Aguirre & K.J. Feeley. 2018. Widespread but heterogeneous responses of Andean forests to climate change. *Nature* 564:207–212.

Fets, L., R. Kay & F. Velazquez. 2010. *Dictyostelium*. *Current Biology* 20:R1008–R1010.

Gibson, L., T.M. Lee, L.P. Koh, B.W. Brook, T.A. Gardner, J. Barlow, C.A. Peres, C.J.A. Bradshaw, W.F. Laurance, T.E. Lovejoy & N.S. Sodhi. 2011. Primary forests are irreplaceable for sustaining tropical biodiversity. *Nature* 478:378–381.

Gillespie, T.W., B.R. Zutta, M.K. Early & S. Saatchi. 2006. Predicting and quantifying the structure of tropical dry forests in South Florida and the neotropics using spaceborne imagery. *Global Ecology and Biogeography* 15:225–236.

Hagiwara, H. 1989. *The Taxonomic Study of Japanese Dictyostelid Cellular Slime Molds*. National Science Museum, Tokyo.

Hagiwara, H., C. Chien & Z. Yeh. 1992. Dictyostelid cellular slime molds of Taiwan. *Bulletin of the National Science Museum. Tokyo, Series B* 18:39–52.

Hagiwara, H., Z. Yeh & C. Chien. 1985. *Dictyostelium macrocephalum*, a new dictyostelid cellular slime mold from Taiwan. *Bulletin of the National Science Museum, Tokyo, Series B* 11:103–108.

Hehmeyer, J. 2019. Two potential evolutionary origins of the fruiting bodies of the dictyostelid slime moulds. *Biological Reviews* 94:1591–1604.

Heinrich, V.H.A., C. Vancutsem, R. Dalagnol, T.M. Rosan, D. Fawcett, C.H.L. Silva Junior, H.L.G. Cassol, F. Achard, T. Jucker, C.A. Silva, J. House, S. Sitch, T.C. Hales & L.E.O.C. Aragao. 2023. The carbon sink of secondary and degraded humid tropical forests. *Nature* 615:436–442.

Landolt, J.C., J.C. Cavender, A.W. Rollins, G.G. Ndiritu & S.L. Stephenson. 2011. Dictyostelid cellular slime molds of Kenya. *Inoculum* 62:28.

Landolt, J.C., J.C. Cavender, S.L. Stephenson & E.M. Vadell. 2008. New species of dictyostelid cellular slime moulds from Australia. *Australian Systematic Botany* 21:50–66.

Landolt, J.C. & S.L. Stephenson. 1991. Cellular slime molds from tropical rain forests of eastern Peru. *Cryptogamic Botany* 2/3:258–260.

Landolt, J.C., S.L. Stephenson & M.E. Slay. 2006. Dictyostelid cellular slime molds from caves. *Journal of Cave and Karst Studies* 68:22–26.

Landolt, J.C. & G.J. Wong. 1998. Dictyostelid cellular slime molds from Hawai'i 1. *Pacific Science* 52:98–103.

Lewis, S.L., D.P. Edwards & D. Galbraith. 2015. Increasing human dominance of tropical forests. *Science* 349:827–832.

Liu, P., J.G. Hou, Y. Zou, S.L. Stephenson, X. Huo, X. Hu & Y. Li. 2019a. Developmental features and associated symbiont bacterial diversity in essential life cycle stages of *Heterostelium colligatum*. *European Journal of Protistology* 68:99–107.

Liu, P. & Y. Li. 2012. New records of dictyostelids from China. *Nova Hedwigia* 94:429–436.

Liu, P., Z.J. Zhang, J.H. Ge, Y. Zou, Z. Li, S. Li, Y. Li & S.L. Stephenson. 2025. Dictyostelids: The second major group of slime molds. *Mycosphere* 16:423–516.

Liu, P., Y. Zou, J. Hou, S.L. Stephenson & Y. Li. 2019b. *Dictyostelium purpureum* var. *pseudosessile*, a new variant of dictyostelid from tropical China. *BMC Evolutionary Biology* 19:78.

Liu, P., Y. Zou, W. Li, Y. Li, X. Li, S. Che & S.L. Stephenson. 2019c. Dictyostelid cellular slime molds from Christmas Island, Indian Ocean. *Msphere* 4:27.

Malhi, Y. & J. Grace. 2000. Tropical forests and atmospheric carbon dioxide. *Trends in Ecology & Evolution* 15:332–337.

McKenna, D.D. & B.D. Farrell. 2006. Tropical forests are both evolutionary cradles and museums of leaf beetle diversity. *Proceedings of the National Academy of Sciences of the United States of America* 103:10947–10951.

Nasser, W., B. Santhanam, E.R. Miranda, A. Parikh, K. Juneja, G. Rot, C. Dinh, R. Chen, B. Zupan, G. Shaulsky & A. Kuspa. 2013. Bacterial discrimination by dictyostelid amoebae reveals the complexity of ancient interspecies interactions. *Current Biology* 23:862–872.

Pillay, R., M. Venter, J. Aragon-Osejo, P. Gonzalez-del-Pliego, A.J. Hansen, J.E. Watson & O. Venter. 2022. Tropical forests are home to over half of the world's vertebrate species. *Frontiers in Ecology and the Environment* 20:10–15.

Raper, K.B. 1984. *The Dictyostelids*. Princeton University Press, Princeton.

Redhead, S.A. & L.L. Norvell. 2012. MycoBank, Index Fungorum, and Fungal Names recommended as official nomenclatural repositories for 2013. *Ima Fungus* 3:44–45.

Romeralo, M., A. Skiba, A. Gonzalez–Voyer, C. Schilde, H. Lawal, S. Kedziora, J.C. Cavender, G. Gloeckner, H. Urushihara & P. Schaap. 2013. Analysis of phenotypic evolution in Dictyostelia highlights developmental plasticity as a likely consequence of colonial multicellularity. *Proceedings of the Royal Society B–Biological Sciences* 280:20130976.

Romeralo, M., F.W. Spiegel & S.L. Baldauf. 2010. A fully resolved phylogeny of the social amoebas (Dictyostelia) based on combined SSU and ITS rDNA sequences. *Protist* 161:539–548.

Romeralo, M., J. Moya–Larano & C. Lado. 2011. Social amoebae: Environmental factors influencing their distribution and diversity across south-western Europe. *Microbial Ecology* 61:154–165.

Rukseree, K., K. Wonganun, P. Palittapongarnpim & P. Ajawatanawong. 2018. Phylogenetic diversity of 18S rDNA sequences of dictyostelids from Amnat Charoen Province, Thailand. *Mycosphere* 9:202–214.

Schaap, P. 2016. Evolution of developmental signalling in dictyostelid social amoebas. *Current Opinion in Genetics & Development* 39:29–34.

Seephueak, P. & V. Petcharat. 2014. The biodiversity of dictyostelid cellular slime molds in rubber tree leaf litter in Southern Thailand. *Mycosphere* 5:805–813.

Sheikh, S., M. Thulin, J.C. Cavender, R. Escalante, S.I. Kawakami, C. Lado, J.C. Landolt, V. Nanjundiah, D.C. Queller, J.E. Strassmann, F.W. Spiegel, S.L. Stephenson, E.M. Vadell & S.L. Baldauf. 2018. A new classification of the dictyostelids. *Protist* 169:1–28.

Stephenson, S.L. & J.C. Landolt. 2011. Dictyostelids from aerial "canopy soil" microhabitats. *Fungal Ecology* 4:191–195.

Stephenson, S.L., J.C. Landolt & D.L. Moore. 1999. Protostelids, dictyostelids, and myxomycetes in the litter microhabitat of the experimental forest, Puerto Rico. *Mycological Research* 103:209–214.

Stephenson, S.L., M.E. Slay, C.A. Slay & A.E. Tuggle. 2007. Cave crickets (Orthoptera: Rhaphidophoridae) as vectors of dictyostelids (Protista: Dictyosteliida). *Entomological News* 118:292–295.

Suthers, H.B. 1985. Ground–feeding migratory songbirds as cellular slime mold distribution vectors. *Oecologia* 65:526–530.

Swanson, A.R., J.D. Shadwick, D.E. Hemmes & F.W. Spiegel. 2004. Ecological succession of dictyostelid slime molds on the island of Hawai'i. *Systematics and Geography of Plants* 74:67–79.

Toberman, H., C. Chen & Z. Xu. 2011. Rhizosphere effects on soil nutrient dynamics and microbial activity in an Australian tropical lowland rainforest. *Soil Research* 49:652–660.

Trew, B.T. & I.M.D. Maclean. 2021. Vulnerability of global biodiversity hotspots to climate change. *Global Ecology and Biogeography* 30:768–783.

Vadell, E.M. 2003. *Contribución a la sistemática y ecología de los dictiostélidos del Parque Nacional Iguazú, Misiones, Argentina*. Universidad de Buenos Aires, Facultad de Ciencias Exactas y Naturales.

Vadell, E.M. & J.C. Cavender. 2007. Dictyostelids living in the soils of the Atlantic forest, Iguazú region, Misiones, Argentina: Description of new species. *Mycologia* 99:112–124.

Vadell, E.M., J.C. Cavender, J.C. Landolt, A.L. Perrigo, P. Liu & S.L. Stephenson. 2018. Five new species of dictyostelid social amoebae (Amoebozoa) from Thailand. *BMC Evolutionary Biology* 18:198.

Vadell, E.M., M.T. Holmes & J.C. Cavender. 1995. *Dictyostelium citrinum*, *D. medusoides* and *D. granulophorum*: Three new members of the Dictyosteliaceae fromforest soils of Tikal, Guatemala. *Mycologia* 87:551–559.

Yeh, Z.Y. & C.Y. Chien. 1983. Cellular slime molds in Taiwan I: Four newly recorded species. *Biol Bull Nat Taiwan Normal University* 18:69–86.

Yulo, P.R.J. & T.E.E. dela Cruz. 2011. Cellular slime molds isolated from Lubang Island, Occidental Mindoro, Philippines. *Mycosphere* 2:575–582.

Zhang, Z.J., L. He, Y.Q. Sun, Z. Li, Y.K. Yang, C. Zhai, S.L. Stephenson, X. Xie, Y. Li & P. Liu. 2024a. New species and records expand the checklist of cellular slime molds (dictyostelids) in Jilin Province, China. *Journal of Fungi* 10:834.

Zhang, Z.J., J.L. Jin, Y.Q. Sun, Y.B. Wang, J. Zhao, S.N. Guo, Y. Zou, W.H. Chen, S.L. Stephenson, Y. Li, J.Z. Qiu & P. Liu. 2025a. The overlooked amoebae of an agroecosystem of black soil land in China: Five new species of dictyostelids. *Applied and Environmental Microbiology* 91:e00025–25.

Zhang, Z.J., M. Li, S.H. Zhang, Y. Qin, J. Zhao, Y. Li, S.L. Stephenson, J. Qiu & P. Liu. 2024b. Diversity of cellular slime molds (Dictyostelids) in the Fanjing Mountain nature reserve and geographical distribution comparisons with other representative aature reserves in different climate zones of China. *Microorganisms* 12:1061.

Zhang, Z.J., Y.K. Yang, Y.J. Du, Y. Zou, S.L. Stephenson, Y. Li & P. Liu. 2025b. The community diversity and metabolic function of symbiont bacteria associated with soil amoebae (dictyostelids) in free–living habitats. *Applied Soil Ecology* 206:105820.

Zhang, Z.J., Y.K. Yang, J. Zhao, Y. Li, S.L. Stephenson, J.Z. Qiu & P. Liu. 2023. Environmental factors influencing the diversity and distribution of dictyostelid cellular slime molds in forest and farmland soils of western China. *Microbiology Spectrum* 11:e0173223.

Zhou, X.G., Y.L. Fu, L.Y. Zhou, B. Li & Y.Q. Luo. 2013. An imperative need for global change research in tropical forests. *Tree Physiology* 33:903–912.

Zhu, H., H. Wang, B.G. Li, S.S. Zhou & J.H. Zhang. 2015. Studies on the forest vegetation of Xishuangbanna. *Plant Science Journal* 33:641–726.

Zou, Y., S.N. Guo, Z.M. Zhu, Z.J. Zhang, Z. Li, J.X. Li, J.H. Ge, S.L. Stephenson, P. Liu & Y. Li. 2025. Integrative taxonomy and cryptic diversity in dictyostelids: A 42% expansion of taxa known from China with conservation implications. *Mycosphere* 16:3177–3256.

10 Mushrooms in the Topical Forests of Southeast Asia

Jaturong Kumla and Nakarin Suwannarach

10.1 INTRODUCTION

Macrofungi, commonly known as mushrooms, are the fruiting bodies of members of the phyla Ascomycota (commonly referred to as ascomycetes) and Basidiomycota (commonly known as basidiomycetes) that are large enough to be visible to the naked eye. Mushrooms exhibit diverse sizes, colors, and shapes and may develop either above the ground (epigeous) or below it (hypogeous). Many of the more common forms are recognized as gilled fungi (agarics), boletes, polypores, jelly fungi, puffballs, coral fungi, earth stars, stinkhorns, and cup-like forms on the basis of their shapes. The majority of mushrooms belong to the phylum Basidiomycota, while some belong to Ascomycota (Mueller et al. 2007; Petersen 2012; Kinge et al. 2020).

Mushrooms have several different functions in nature. Some are saprotrophs that decompose organic matter and recycle nutrients, others are parasites that obtain nutrition from living hosts, and still others are mycorrhizal symbionts that form mutually beneficial associations with plant roots. In addition, mushrooms interact with numerous other organisms, including insects, which may serve as dispersal agents, food sources, or substrates for breeding, thereby contributing to ecosystem functioning and biodiversity maintenance across various environments, including tropical rainforests (Cunningham and Yang 2012; Niego et al. 2023; Stallman et al. 2024). Beyond their ecological significance, mushrooms have been valued by human societies for millennia as sources of food and traditional medicine, with documented uses spanning diverse cultures worldwide. In recent decades, expanding scientific research has revealed their substantial potential for applications in modern medicine. Recent investigations also have focused on biotechnological uses, including enzyme production and biomaterial development, agricultural applications such as biofertilizers and biocontrol agents, bioremediation of contaminated environments, and pharmaceutical development of novel therapeutic compounds with antimicrobial, anticancer, and immunomodulatory properties. This growing body of data positions mushrooms as vital contributors to both natural ecosystem processes and human innovation, bridging traditional knowledge with cutting-edge scientific advancement.

10.2 EXTENT OF THE REGION

Southeast Asia is a biogeographically diverse region encompassing 11 countries—Brunei Darussalam, Cambodia, Indonesia, Laos, Malaysia, Myanmar, the Philippines, Singapore, Thailand, Timor-Leste, and Vietnam. The region includes both mainland and insular territories, extends from the eastern margin of the Indian subcontinent to the western Pacific Ocean, and falls predominantly within the tropical zone between approximately 28°N and 11°S latitude (Myers et al. 2000; Wikramanayake et al. 2002; Woodruff 2010; Stibig et al. 2014). Southeast Asia encompasses approximately 4.5 km². As noted above, the region encompasses 11 nations that span both mainland and maritime territories. The mainland portion includes Myanmar (676,578 km²), Thailand (513,120 km²), Vietnam (331,212 km²), Laos (236,800 km²), Cambodia (181,035 km²), and the small city-state of Singapore (728 km²). The maritime or insular region consists of Indonesia (1,904,569 km²), the largest archipelagic nation in the world with over 17,000 islands; the Philippines (300,000 km²), an archipelago

DOI: 10.1201/9781003647140-10

of more than 7,600 islands; Malaysia (330,803 km²), uniquely divided between the Malay Peninsula and northern Borneo; Brunei Darussalam (5,765 km²), located on the island of Borneo; and Timor-Leste (14,874 km²), occupying the eastern half of Timor Island.

10.3 GENERAL FEATURES

Characterized by consistently warm temperatures, high humidity, and abundant rainfall throughout most of the year, Southeast Asia harbors some of the world's most biologically diverse ecosystems, including vast tropical rainforests, mangrove forests, peatlands, and coral reefs. The region lies at the intersection of two major biogeographic realms, the Indomalayan and Australasian, and this has resulted in distinctive patterns of species distribution and endemism. Its complex geological history, diverse topography (ranging from lowland plains to mountain ranges exceeding 4,000 meters), and varied climatic conditions have collectively shaped an exceptional level of biodiversity, establishing Southeast Asia as one of the world's foremost hotspots for conservation and scientific investigation.

10.4 FORESTS OF SOUTHEAST ASIA

Southeast Asia harbors an extraordinary diversity of mushrooms that play critical roles in ecosystem functioning. Despite their ecological significance and potential for food, medicine, and biotechnology, mushrooms of this region remain incompletely documented and are seriously understudied when compared to plants and animals.

Tropical forests in Southeast Asia represent some of the most biodiverse and structurally complex ecosystems on Earth, covering approximately 2.5 km² across the region, though this area has declined significantly due to deforestation in recent decades. These forests are characterized by high temperatures, averaging 25–27 °C year-round, abundant rainfall typically exceeding 2,000 mm annually, and consistently high humidity, creating optimal conditions for luxuriant plant growth and exceptional species richness. The region supports several distinct forest types, each with unique ecological characteristics. Lowland dipterocarp forests, dominated by members of the family Dipterocarpaceae, are the most extensive and species-rich, forming towering canopies that reach heights of 40–50 m, with emergent trees exceeding 70 m. Montane forests occur at higher elevations, typically above 1,000 m, where cooler temperatures and frequent cloud cover support distinctive assemblages of flora and fauna, including moss-covered trees and specialized epiphytic communities. Peat swamp forests develop on waterlogged, acidic soils rich in accumulated organic matter, forming specialized habitats adapted to prolonged inundation by water. Mangrove forests occur along coastal areas, creating vital transitional zones between terrestrial and marine ecosystems. In addition, seasonal forests occur in areas with pronounced dry seasons, particularly in parts of mainland Southeast Asia, and these tropical forests provide essential ecosystem services including carbon storage, water regulation, soil protection, and climate moderation. The complex three-dimensional structure of these forests, with multiple canopy layers, abundant deadwood, and diverse microhabitats, creates innumerable ecological niches that support not only remarkable plant and animal diversity but also an exceptionally rich fungal flora. The warm, humid conditions combined with abundant organic substrates make Southeast Asian tropical forests ideal environments for mushroom diversity, supporting saprotrophic species that decompose leaf litter and woody debris, mycorrhizal species that form symbiotic relationships with the dominant tree species, and parasitic fungi that regulate forest dynamics through their effects on host organisms.

10.5 MUSHROOM DIVERSITY

Global estimates of mushroom diversity vary considerably among researchers. Hawksworth (2001) proposed that approximately 140,000 mushroom species exist worldwide, whereas Mueller et al. (2007) suggested a more conservative range of 53,000 to 110,000 species. Despite these projections, only around 14,000 species have been formally described to date (Lu et al. 2020). A 2021 assessment identified approximately 2,786 mushroom taxa as edible species based on documented reports worldwide. Asia harbors the greatest diversity of edible mushrooms, with 1,493 recorded species, followed by Europe with 629 species, North America with 487 species, Africa with 351 species, South America with 204 species, Central America with 100 species, and Oceania with the fewest at 19 species (Li et al. 2021). Early estimates suggest that tropical Asia harbors between 10,000 and 25,000 species of mushrooms (Mueller et al. 2007). Mycologists have long recognized that Southeast Asia, in particular, exhibits an exceptionally high diversity of mushrooms and represents one of the world's most important centers of fungal diversity, encompassing microfungi, yeasts, and mushrooms. Despite this richness, the region remains among the least explored areas in terms of mycological research. The region's tropical climate, characterized by consistently high temperatures, humidity, and abundant rainfall throughout the year, provides optimal conditions for the growth of mushroom mycelia and the formation of fruiting bodies. These environmental factors, together with the region's complex geological history and diverse array of forest types (including lowland dipterocarp forests, montane forests, peat swamp forests, and limestone forests) have fostered exceptional mushroom diversity. Nevertheless, the true extent of this diversity remains largely unknown due to the scarcity of systematic surveys and comprehensive taxonomic studies. Many areas, particularly the remote forests of Laos, Myanmar, and parts of Indonesia, remain virtually unexplored mycologically, suggesting that countless species are likely yet to be discovered.

10.6 MUSHROOMS OF THAILAND

Thailand, situated in the heart of mainland Southeast Asia, harbors exceptional mushroom diversity owing to its tropical climate, varied topography, and diverse forest ecosystems. The country encompasses multiple ecological zones, from lowland tropical rainforests in the south to montane forests in the north, seasonal deciduous forests in the northeast, and mangrove ecosystems along extensive coastlines. This environmental heterogeneity, combined with year-round warm temperatures and substantial rainfall, particularly during the southwest monsoon season from May to October, creates optimal conditions for the growth of mushroom mycelia and the formation of fruiting bodies.

In Thailand, the study of mycology began to develop during the twentieth century. The first formal investigation of mushrooms in the country, focusing on the gasteromycetous fungi, was published by Dissing (1963). Jones et al. (1994) compiled a list of wild mushrooms collected by Thai villagers. Subsequently, a total of 285 mushroom and other fungal species were reported in Thailand in 2001, of which 98 were identified as edible, 28 as non-edible, and 159 species had unknown edibility (Ruksawong and Flegel 2001). Throughout the twentieth century and beyond, numerous studies have focused on Thai mushrooms. Nevertheless, taxonomic research on these fungi faced significant challenges due to reliance on foreign expertise, linguistic barriers, and limited public awareness of fungal diversity. Since 2005, Thai mushroom research has undergone a transformation through the integration of molecular phylogenetics with traditional morphological taxonomy. Numerous studies have characterized various mushroom genera, including *Agaricus* (Zhao et al. 2011; Chen et al. 2012), *Lepiota* (Sysouphanthong et al. 2011, 2012), *Lactarius* (Le et al. 2007a, 2007b), *Lentinus* (Karunarathna et al. 2011), and *Marasmius* (Wannathes et al. 2009a, 2009b), primarily from specimens collected in northern Thailand. These investigations, employing DNA sequencing and phylogenetic analyses, have revealed that Thai fungal diversity far exceeds previous estimates. Many collections have come from previously unexplored habitats and geographic areas, particularly in protected forest reserves and national parks.

The "Checklist of Mushrooms (Basidiomycetes) in Thailand," published in 2011, documented 1,978 species of basidiomycetes (Chandrasrikul et al. 2011) throughout the country and has become a frequently referenced resource in numerous scientific publications. In this checklist, the order Agaricales was the most abundant (42.3% of all species), followed by Polyporales (18.2%), Boletales (13.9%), and Russulales (12.7%). Despite its widespread use, the checklist continues to include many species originally described from North America and Europe. Moreover, some entries may represent incorrectly identified taxa, as the documentation lacks preserved herbarium specimens, precise morphological descriptions, and molecular genetic data necessary for reliable species verification. Remarkably, Hyde et al. (2018) reported that among collected specimens, over 93% species of *Agaricus*, 83% of *Amanita*, and 87% of *Lactarius* subgenus *Russularia* represented species previously unknown to science. This extraordinarily high proportion of undescribed species demonstrates that Thailand's mycological diversity remains vastly underexplored and that traditional morphology-based identifications had obscured the true extent of tropical fungal speciation. Numerous new mushroom species and new records for the country have been collected and described, especially from northern Thailand, which serves as one of the country's most significant hotspots of fungal diversity. Furthermore, an updated checklist of mushrooms in Thailand has recently been revised, incorporating new taxonomic discoveries, molecular phylogenetic data, and Thai specimens accumulated over the past decade (Suwannarach et al. 2022, 2025).

Wild mushrooms in Thailand typically produce fruiting bodies during the wet season (middle May to October) of each year, with most forming ectomycorrhizal associations with common tree species belonging to the families Dipterocarpaceae, Fagaceae, and Magnoliaceae. In northern and northeastern Thailand, preliminary surveys conducted at local and roadside markets have documented a remarkable diversity of wild mushrooms, with ectomycorrhizal (ECM) species constituting the predominant group offered for sale (Figure 10.1). Based on observations of fruiting bodies in forest habitats, the most abundant mushroom families in northern Thailand are the Russulaceae, Boletaceae, and Amanitaceae (Sanmee 2004; Butkrachang et al. 2007; Watling 2013). Although survey results indicate a high frequency of the Russulaceae, accurate species identification has yet to be confirmed. Commonly encountered genera include *Amanita*, *Astraeus*, *Cantharellus*, *Lactarius*, *Termitomyces*, and *Russula*, as well as various species of the bolete group of mushrooms (Sanmee et al. 2003; Sanmee 2004). Yuwa-Amornpitak et al. (2006) studied the diversity of ECM mushrooms associated with the root tips of Dipterocarpaceae in Thailand and found that the Thelephoraceae was the most abundant family, followed by the Russulaceae, Sclerodermataceae, Cortinariaceae, and Boletaceae. The diversity of *Lactarius* in northern Thailand has been reported (Le 2007a, 2007b; van de Putte et al. 2010). The edible species of *Amanita* include *A. hemibapha*, *A. pseudoprinceps*, A. *rubromarginata*, and *A. subhemibapha*, all of which belong to *Amanita* section *Caesareae* (Kumla et al. 2023). In addition, three species of *Astraeus* (*As. asiaticus*, *As. odoratus*, and *As. sirindhorniae*) are consumed throughout the country (Suwannarach et al. 2022). Meanwhile, *Boletus griseipurpureus* is a popular edible ECM mushroom in southern Thailand (Aung-aud-chariya et al. 2012). Several edible boletes in Thailand have been studied, most of which were originally classified under *Boletus sensu lato* but have since been reclassified into genera such as *Aureoboletus*, *Retiboletus*, *Sutorius*, and *Tylopilus* (Seehanan and Petcharat 2008; Thongklam 2008; Pukahuta et al. 2009).

Among the most highly prized boletes in Thai markets are species of *Phlebopus*, particularly *P. portentosus* and *P. spongiosus*, which produce large, robust fruiting bodies that can reach an impressive size (Kumla et al. 2022a; Suwannarach et al. 2022). Interestingly, three species of *Tuber* (*T. lannaense*, *T. magnatum*, and *T. thailandicum*) have been collected from northern Thailand and identified based on both morphological characteristics and molecular data, marking the first discovery of truffles in Southeast Asia (Suwannarach et al. 2015, 2016, 2017). Furthermore, studies on mushroom diversity have been carried out in several national parks and community forests. For example, Wongchalee and Pukahute (2012) surveyed a dry dipterocarp forest in Phu Phan National Park, Sakon Nakhon Province, northeastern Thailand, between 2008 and 2009 and found that species of *Russula* were the most commonly collected, followed by *Boletus*, *Lactarius*, and *Amanita*, respectively. Duengkae (2018) surveyed the Phukhieo-Nam Nao Forest

FIGURE 10.1 Edible wild mushrooms in local and roadside markets of northern Thailand.

Source: Photos by Nakarin Suwannarach

Complex in northeastern Thailand between 2013 and 2016 and identified 359 mushroom species belonging to 54 families and 143 genera. A large number of species were found in the genera *Amanita*, *Russula*, and *Marasmius*, respectively. Yuwa-Amornpitak and Yeunyaw (2020) reported that species of Russulaceae (*Russula*) were the most frequently collected mushrooms, followed by Boletaceae (*Boletus*), Amanitaceae (*Amanita*), and Tricholomataceae (*Termitomyces*) in a dry dipterocarp forest in Na Si Nuan Sub-district, western Thailand. Recently, a study on mushroom diversity conducted in Namtok Phlio National Park, Chanthaburi Province (eastern Thailand) in 2023 by Surawut et al. (2025) revealed that the collected specimens were classified into 2 phyla, 3 classes, 9 orders, 18 families, and 23 genera. Most species belonged were basidiomycetes (89.3%). The family Polyporaceae (21.4%) was the most abundant, followed by Marasmiaceae (14.3%), Omphalotaceae (3.57%), and Sarcoscyphaceae (3.57%).

10.7 MUSHROOMS OF SINGAPORE

Singapore, despite being one of the world's smallest nations, has a mycological research history that dates back to the colonial period, when early collections and observations were documented by British naturalists. Fungal biodiversity in the country has been relatively understudied since the time of E. J. H. Corner, the first mycologist in Singapore and former Assistant Director of the Singapore Botanic Gardens, who collected numerous specimens and made extensive observations during his tenure from 1929 to 1945. However, most of his work was published long after he left Singapore, and the majority of the specimens he collected were taken with him for further study. Nonetheless, at least 100 of his voucher specimens are still deposited in the Herbarium of the Singapore Botanic Gardens (SING), and a comprehensive list of his fungal work can be found in his obituary (Watling 2001).

More recent contributions include a literature-based checklist of fungi in Singapore compiled by Turner et al. (1994), and the recording of species of bioluminescent fungi by Rodda (2012) and Lee (2022). Contemporary research has revealed that Singapore's mushrooms include representatives from major groups such as polypores, agarics, gasteromycetes, and various ascomycetes, with saprotrophic species dominating due to the abundance of decaying wood and leaf litter in tropical environments. Notable genera recorded include *Cookeina*, *Favolaschia*, *Ganoderma*, *Marasmius*, *Microporus*, and *Mycena* among many others and listed in the annotated checklist of macrofungi recorded in Singapore (Tham and Watling 2017a, 2017b, 2017c, 2017d). In 2023, "A Guide to Macrofungi in Singapore" was published, and this documented a total of 165 species from 103 genera and 51 families (Lee and Choong 2023). Some ECM mushrooms, including species belonging to the genera *Amanita*, *Lactarius*, and *Russula*, as well as various members of the bolete group such as *Boletellus*, *Heimioporus*, and *Tylopilus* have been recorded. This comprehensive guide provides detailed descriptions, photographs, and ecological information for each species, representing a significant advancement in the systematic documentation of Singapore's macrofungal diversity.

10.8 MUSHROOMS OF CAMBODIA

Mushroom biodiversity in Cambodia has been increasingly documented in recent years, particularly through comprehensive surveys conducted in protected forests. A landmark study by Kim et al. (2017) titled "A Checklist of Mushrooms of Cambodia" represents the most extensive documentation of the country's mushroom diversity, based on collections made between 2009 and 2014. Mycologists collected 3,950 mushroom specimens from seven major protected areas, including the Central Cardamom Protected Forest (Koh Kong Province), Kirirom National Park (Kampong Speu Province), Seima Biodiversity Conservation Area (Kratie Province), Mondulkiri Protected Forest (Mondulkiri Province), Phnom Bokor National Park (Kampot Province), Western Siem Pang Protected Forest (Stung Treng Province), and Ream National Park (Sihanoukville Province). From these collections, 302 species representing 165 genera were identified. The most species-rich families were the Polyporaceae (26.2%), followed by the Marasmiaceae (9.1%), Ganodermataceae (7.7%), Hymenochaetaceae (7.5%), and Mycenaceae (7.1%), which together accounted for 57.5% of all recorded specimens. Dominant genera occurring in both western and eastern Cambodia included *Ganoderma*, *Mycena*, *Marasmius*, *Microporus xanthopus*, *Phellinus*, and *Russula*.

10.9 MUSHROOMS OF LAOS

Mycological research in Laos (Lao People's Democratic Republic) has been historically limited, with systematic surveys gaining momentum only in recent decades through collaborative efforts between Laos institutions and international partners. Nevertheless, recent studies have begun to reveal the country's rich mushroom diversity. A focused survey in the Phousabous National Protected Area from 2017 to 2018 recorded 284 species of mushrooms belonging to 142 genera, 56 families, 20 orders, and 7 classes through combined morphological and molecular studies (Hong et al. 2019). Among these, the basidiomycetes dominated with 274 species, whereas the ascomycetes contributed just 10 species. The most species-rich families were the Boletaceae (16.06%), Polyporaceae (10.12%), Russulaceae (9.94%), Amanitaceae (7.68%), and Agaricaceae (4.54%) which together made up about three-quarters of all specimens identified. In 2021, a comprehensive survey in the Dong Hua Sao National Biodiversity Conservation Area documented 103 species of mushrooms, including 92 species of basidiomycetes classified into 67 genera, 32 families, 11 orders, and 3 classes; 10 species of ascomycetes in 6 genera, 4 families, 3 orders, and 2 classes (Lee et al. 2021). Among these mushrooms, the most species-rich families are the Marasmiaceae (17.3%), Polyporaceae (13.1%), Mycenaceae (8.9%), Agricaceae (8.3%), Xylariaceae (6.5%), Auriculariaceae (4.8%), Ganodermataceae (3.6%), Dacrymycetaceae (3.0%), Meruliaceae (3.0%), and Russulaceae (3.0%). Together these comprised 71.4% of the total specimens identified. A book titled "An Introduction to the Edible, Poisonous, and Medicinal Fungi of Northern Laos" was published in 2018, highlighting a survey in Xieng Khouang Province that recorded 142 species, 90 genera and approximately 80 edible mushrooms, including species of *Amanita, Boletus, Cantharellus, Lactarius, Lentinus, Pleurotus, Russula*, and *Termitomyces* (Læssøe et al. 2018). Other research has described new taxa from the family Agaricaceae in Laos, especially lepiotaceous mushrooms in genera such as *Chlorophyllum, Clarkeinda, Macrolepiota, Pseudolepiota*, and *Xanthagaricus* (Sysouphanthong et al. 2021). Łuczaj et al. (2021) surveyed wild mushrooms sold in the open-air markets of Luang Prabang in 2018 and 2019 and found that species of *Russula* were the most frequently found, followed by *Termitomyces* and *Auricularia*.

10.10 MUSHROOMS OF MALAYSIA

Malaysia is a critical global region for fungal biodiversity, with its tropical rainforests harboring an estimated 70% of mushrooms yet to be officially described (Jones 2007). While the study of Malaysian mycology traces its origins to early published records such as a Latin book on East Malaysian mushrooms from 1879 (Beccari and Cesati 1879), comprehensive research remains a modern effort. A landmark contribution to documenting Malaysian fungal diversity was "Checklist of Fungi of Malaysia" by Lee et al. (2012), which was based on five years of work (2006–2010) assembling information from the published literature scattered across numerous sources. This checklist recorded more than 3,000 species of fungi, including 1,820 species of basidiomycetes across 358 genera and 9 families. The Polyporaceae was the most abundant family, comprising 327 species in 42 genera, followed by the Marasmiaceae with 186 species in 13 genera, and the Boletaceae with 162 species in 18 genera. Paul et al. (2019) carried out a survey of macrofungi in the Rainforest Discovery Centre in Sandakan, Sabah, identifying 10 orders, 24 families, and 43 genera. The results showed the order Polyporales had the highest number of species (47.1%), followed by the Agaricales (35.5%). The Russulales and Xylariales each accounted for 3.6%. The remaining orders, the Boletales (2.9%), Pezizales (2.9%), Cantharellales (1.5%), and Thelephorales (1.5%), were represented by much lower totals. Other studies have reported that Malaysia harbors numerous species of wild-harvested mushrooms, including species in such genera as *Amanita, Lactarius, Ganoderma, Polyporus, Russula*, and *Termitomyces* as well as various boletes. These wild mushrooms are collected from a variety of forest types, including lowland and hill dipterocarp forests, montane forests, peat swamp forests, and secondary forests (Samsudin and Abdullah 2019; Ahmadni et al. 2024; Nordin et al. 2024).

10.11 MUSHROOMS OF MYANMAR

In Myanmar, mushroom diversity has historically been underexplored, with the most comprehensive list published by Thaung (2007) in which approximately 180 species across 24 orders, 56 families, 117 genera, and 176 species were documented. The predominant genera among the ascomycetes include *Claviceps* and *Xylaria*, while among the basidiomycetes, common genera are *Agaricus*, *Agrocybe*, *Conocybe*, *Coprinus*, *Lentinus*, *Leucoagaricus*, *Marasmius*, *Psathyrella*, and *Termitomyces*. Systematic surveys have increased in recent years, particularly through collaborative efforts between Myanmar and Japanese institutions. Since 2016, researchers from the Forest Research Institute of Myanmar and the National Museum of Nature and Science, Japan, have collected over 400 specimens of mushrooms, including many new records for the country. These findings have been reported in scientific publications as well as in the book "A Guide to Flora and Fauna in Southern Myanmar" (Hosaka and Linn 2021). Moreover, wild mushrooms such as species of *Ganoderma*, *Polyporus*, and *Termitomyces* are commonly harvested from diverse forest types, including lowland and hill dipterocarp forests, montane forests, peat swamp forests, and secondary forests, for both food and medicinal purposes (Linn et al. 2024).

An intriguing aspect of Myanmar's mycological heritage is the preservation of ancient fungal specimens in Burmese amber from the mid-Cretaceous period, approximately 99 million years ago. Recent paleomycological studies have described two new putative taxa (*Marasmiamimum cretaceum* and *Mesoagaracites burmitis*) of ancient mushrooms from the suborder Marasmiineae preserved in Burmese amber, providing rare insights into the deep evolutionary history of mushrooms (Guo et al. 2024). These fossil specimens indicate that diverse fungal lineages already existed in the ecosystems of present-day Myanmar during the age of the dinosaurs, although assigning fossil mushrooms to specific families remains challenging due to limited morphological information and convergent evolutionary traits.

10.12 MUSHROOMS OF INDONESIA

Indonesia, the world's largest archipelagic nation, consists of more than 17,000 islands spanning from Sumatra in the west to Papua in the east, harbors extraordinary mushroom diversity within its extensive tropical forests. The country exhibits remarkable ecological heterogeneity, including lowland tropical rainforests, montane forests, peat swamp forests, seasonal forests, and mangrove ecosystems. These habitats, distributed across major islands such as Sumatra, Java, Kalimantan (Indonesian Borneo), Sulawesi, Maluku, Papua, and the Lesser Sunda Islands, support a rich and largely underexplored fungal diversity. As of 2017, the National Research and Innovation Agency's Research Center for Biology had recorded 2,273 mushroom species in Indonesia. A checklist compiled by Putra et al. (2022) documented 107 species of wild mushrooms in Indonesia, including 103 species of basidiomycetes and 4 species of ascomycetes, highlighting the rich ethnomycological knowledge embedded within various Indonesian communities. Commonly recorded genera include *Coprinus*, *Polyporus*, and *Schizophyllum*. Regional surveys across Indonesia have documented considerable mushroom diversity in various protected areas and forest types. For example, research in Lore Lindu National Park revealed elevational patterns in macrofungal diversity, with 172 species from 33 families recorded, including 159 species of basidiomycetes and 13 species of ascomycetes (Yusran et al. 2021). *Marasmius* and *Hygrocybe* were the most abundant genera, while species of *Auricularia*, *Ganoderma*, *Lentinus*, *Pleurotus*, *Schizophyllum*, *Termitomyces*, and *Xylaria* were commonly utilized as food sources and in traditional medicine. A study on mushroom diversity in Jayapura Regency conducted by Yerisetouw et al. (2023) reported 47 species, among which *Psathyrella candolleana*, *Coprinellus disseminatus*, *Xylaria longipes*, *Mycena haematopus*, *Laccaria laccata*, and *Parasola plicatilis* were the most commonly found. Moreover, a study of basidiomycete mushrooms in West Bali National Park, West Java, and the Cisela Nature Forest Reserve (Bojonglarang Jayanti, Cianjur, West Java) documented 22 genera, including *Auricularia*,

Boletinellus, *Campanella*, *Coprinus*, *Favolaschia*, *Ganoderma*, *Hygrocybe*, *Inocybe*, *Lentinus*, *Marasmius*, *Marasmiellus*, *Microporus*, *Phellinus*, *Polyporus*, *Postia*, *Pycnoporus*, *Rimbachia*, *Schizophyllum*, *Tephrocybe*, *Trametes*, *Tyromyces*, and *Xylaria* (Wijaya et al. 2021). Research on mushroom diversity conducted by Sumarouw et al. (2024) in the Sanggabuana Mountain area in Karawang Regency, West Java, recorded 23 species from 17 genera, with *Ganoderma applanatum*, *Trametes* sp., *Inonotus* sp., *Microporus xanthopus*, and *Xylaria* sp. being the most commonly observed. A recent study by Nurzahra et al. (2025) collected and identified 33 species of mushrooms in the Banyak Mountain Forest Area, with the Polyporaceae being the most frequently found family, followed by the Xylariaceae, Auriculariaceae, and Ganodermataceae. However, many species of Indonesian mushrooms likely remain undiscovered or undescribed, with the potential for high levels of endemism given the archipelagic nature of the country and the historical biogeographic isolation of different regions.

10.13 MUSHROOMS OF THE PHILIPPINES

The Philippines, an archipelagic nation with diverse tropical ecosystems ranging from lowland dipterocarp forests to montane and mossy forests, harbors a rich but still incompletely documented diversity of macrofungi. Early mycological records were initiated during the American colonial period, but systematic research has been sporadic. However, recent decades have seen growing efforts to catalog and describe native mushrooms. Regional surveys across the Philippines have documented considerable mushroom diversity in various locations and for different habitat types, with several researchers focusing specifically on the different mushrooms found in many mountainous areas of the country. For example, Daep and Cajuday (2003) investigated the mushroom diversity of Mt. Malinao in Albay, recording nine species belonging to the families Auriculariaceae, Coprinaceae, Pluteaceae, and Tricholomataceae. De Leon et al. (2023) conducted a survey of mushrooms in six Aeta communities in Central Luzon, identifying 76 species distributed across 23 families and 41 genera. The Polyporaceae was the most abundant family, followed by the Marasmiaceae, Tricholomataceae, and Pleurotaceae, reflecting the rich macrofungal diversity utilized and recognized by local indigenous groups. A total of 21 species of mushrooms were collected from Mt. Makiling Forest Reserve, Laguna, comprising 20 basidiomycetes and one ascomycete, with the Polyporaceae identified as the most dominant family (Nacua et al. 2018). In 2019, field sampling of mushrooms in North Central Mindanao Protected Areas documented 185 species across 122 genera and 76 families, with the Polyporaceae identified as the most dominant family (Tadiosa and Lubos 2019). Research conducted by Brazas Jr. et al. (2020) examined basidiomycetous mushrooms in the northeastern side of the Quezon Protected Landscape, Southern Luzon, and identified 53 species distributed across 19 families and 31 genera. The family Polyporaceae exhibited the highest species richness, with most species belonging to the genera *Hexagonia* and *Polyporus*, followed by members of the Mycenaceae, Agaricaceae, and Tricholomataceae families. Kim et al. (2021) conducted a mushroom survey at Mt. Cleopatra Needle Forest Reserve on Palawan Island, based on collections from 2018 to 2019, documented 176 species across 114 genera, 55 families, and 22 orders. The most frequently recorded families were the Polyporaceae (18.9%), Marasmiaceae (11.5%), Xylariaceae (9.7%), Agaricaceae (8.3%), Auriculariaceae (4.8%), Ganodermataceae (4.2%), Hypoxylaceae (3.2%), and Sarcoscyphaceae (3.0%). Subsequently, Dulay et al. (2023) compiled a comprehensive checklist summarizing reports from 2001 to 2021 across the Philippines, documenting 447 species of macrofungi classified into 193 genera and 72 families. The most species-rich family was the Polyporaceae (72 species), followed by Agaricaceae (33 species), Hymenochaetaceae (18 species), Ganodermataceae (16 species), Psathyrellaceae (16 species), Marasmiaceae (15 species), Russulaceae (15 species), Mycenaceae (14 species), Meruliaceae (13 species), and Tricholomataceae (12 species). Among these, *Ganoderma lucidum*, *Auricularia polytricha*, *Schizophyllum commune*, and *Termitomyces* species were the most frequently reported and widely distributed throughout the country. Recent research on mushroom diversity by Bustillos et al. (2024) documented 224 specimens of macrofungal from the Mt. Arayat Protected Landscape, representing two phyla, four classes, 12 orders, 36 families, 53 genera, and 108 species, with *Ganoderma*, *Microporus*, *Schizophyllum*, and *Trametes* being the most commonly encountered genera. Despite extensive documentation efforts, significant knowledge gaps persist in Philippine mushroom diversity, as many remote forested regions, especially on the smaller surrounding islands, remain largely unexplored for macrofungi.

10.14 MUSHROOMS OF VIETNAM

In Vietnam, a major milestone in documenting mushroom diversity was achieved with the publication of the "Preliminary Checklist of Macrofungi of Vietnam" by T. T. Kiet in 1998, which synthesized data from studies conducted since 1890. The checklist documented a total of 829 species of mushrooms known from Vietnam, of which 519 species were reported as new records for the country (Kiet 1998). Among the most commonly consumed genera were *Amanita*, *Boletus*, *Cantharellus*, *Lactarius*, *Russula*, and *Termitomyces*, along with various wood-decay genera, including *Ganoderma*, *Lentinus*, *Polyporus*, and *Schizophyllum*. This checklist represents the most comprehensive documentation of Vietnam's mushroom diversity to date, although the actual species richness is undoubtedly much higher due to the limited survey coverage across the country's vast and ecologically diverse landscapes.

The most comprehensive account of mushrooms in Cuc Phuong National Park is the illustrated guide by Tran and Truong (2005), which documented 214 species, and Le and Doan (2018) reported 30 species belonging to the family Psathyrellaceae. A list of 40 species of *Xylaria* for the country was reported by Duong et al. (2011). In 2015, field sampling of mushrooms in the Lung Ngoc Hoang Natural Reserve wetland documented 63 identified species belonging to 42 genera, 27 families, 12 orders, 6 classes, and 2 phyla. Among these, the basidiomycetes dominated (98.97%), with the Agaricales being the most dominant order, followed by the Polyporales and Auriculariales (Truyen and Patacsil 2017). During the period of 2018–2019, mushroom diversity in Da Lat, southern Vietnam, was investigated by Lee et al. (2020). The study classified the collected specimens into two phyla, 13 orders, 38 families, 93 genera, and 124 species, with the most species belonging to the orders Polyporales, Agaricales, Boletales, Russulales, and Hymenochaetales. However, 175 taxa could not be identified to the species level. These were assigned to the genera *Russula*, *Cystidiodontia*, *Mycena*, and *Clavulinopsis*. Between 2019 and 2023, a comprehensive survey of the family Ganodermataceae in Yok Don National Park Tay in the Nguyen region of the central highlands of Vietnam revealed the presence of 10 species belonging to the genera *Ganoderma* and *Amauroderma* (Kien et al. 2025). In addition, Phu et al. (2024) investigated mushroom diversity in the Son Tra Nature Reserve, Da Nang, and classified the specimens into 42 genera, 23 families, 10 orders, and 3 classes of basidiomycetes. The most species belonged to the families Ganodermataceae (e.g., *Ganoderma* and *Amauroderma*) and Polyporaceae (e.g., *Microporus* and *Trametes*). Among them, species such as *Microporus xanthopus* and *Stereum hirsutum* were encountered more frequently.

10.15 MUSHROOMS OF BRUNEI DARUSSALAM AND TIMOR-LESTE

Brunei Darussalam and Timor-Leste represent the most significant knowledge gaps in Southeast Asian mycology. Despite being situated within the biodiversity-rich tropical region of the Indo-Malayan realm, their mushroom diversity remains virtually unexplored and undocumented. In Brunei, research has often focused on specific groups such as aquatic microfungi on submerged wood. This has contributed to regional checklists but a comprehensive inventory of mushrooms across its diverse habitats is still lacking. Timor-Leste, as a relatively young nation, faces even greater challenges, with fungal research is still in its infancy, with limited studies on terrestrial and marine fungi. Consequently, both nations represent critical frontiers for mycology.

10.16 AN OVERVIEW OF MYCOLOGY IN SOUTHEAST ASIA

Over the past decade, Southeast Asia has witnessed a surge in the discovery and documentation of new species of mushrooms, reflecting both increased mycological interest and the application of modern taxonomic tools. Molecular phylogenetics, combined with traditional morphological approaches, has revealed numerous species previously overlooked or misidentified using European and North American taxonomic concepts. Countries such as Thailand, Malaysia, Indonesia, Vietnam, and the Philippines have reported multiple new taxa across diverse mushroom groups, including ECM genera (*Amanita*, *Russula*, and members of the bolete group), wood-decay genera (*Ganoderma* and *Lentinus*), and termite-associated genera (*Termitomyces*). For example, recent taxonomic work has described multiple new species of *Amanita* (e.g., *A. albicarnosa*, *A. albifragilis, A. castanea*, *A. claristriata*, *A. fulvisquamea*, *A. kalasinensis*, and *A. ravicrocina*) from northern and northeastern Thailand, contributing substantially to knowledge of the genus in tropical Asia (Thongbai et al. 2016; Liu et al. 2022a, 2022b, 2023). New species of *Russula* (e.g., *R. bellissima*, *R. luteonana*, *R. orientalovirescens*, *R. pseudomodesta*, and *R. sribuabanensis*) have been described from dry dipterocarp forests in Thailand (Wisitrassameewong et al. 2022, 2025; Paloi et al. 2023a), and *R. griseocarnosa* is a new species described from Vietnam (Anh et al. 2023). In the bolete group, several new species have been reported from Thailand, including *Cacaoporus pallidicarneus* and *C. tenebrosus*, *Gyrodon suthepensis*, several species of *Sutorius* species (i.e., *S. mucosus*, *S. obscuripellis*, *S. pachypus*, *S. pseudotylopilus*, *S. rubinus*, *S. ubonensis*, and *S. vellingae*), and *Retiboletus brevibasidiatus* (Kumla et al. 2017; Chuankid et al. 2021; Vadthanarat et al. 2019b, 2021). In addition, two new species of *Erythrophylloporus* (*E. paucicarpus* and *E. suthepensis*) were described from Thailand (Vadthanarat et al. 2019a). Moreover, an investigation of *Ganoderma* in the Greater Mekong Subregion (GMS) of Southeast Asia, encompassing tropical areas of Laos, Myanmar, Thailand, and Vietnam, revealed 19 species. These were *G. adspersum*, *G. applanatum*, *G. australe*, *G. bambusicola*, *G. ellipsoideum*, *G. flexipes*, *G. gibbosum*, *G. hoehnelianum*, *G. hochiminhense*, *G. lucidum*, *G. multipileum*, *G. myanmarense*, *G. orbiforme*, *G. philippii*, *G. sichuanense*, *G. sinense*, *G. subresinosum*, *G. tropicum*, and *G. williamsianum* (Luangharn et al. 2019a, 2021; Wu et al. 2020). *Ganoderma thailandicum* was described as a new species from Thailand by Luangharn et al. (2019b). The new lentinoid taxa, *Panus sribuabanensis* and *Lentinus saisamorniae* also were discovered in Thailand (Kumla et al. 2024, 2025a). Several new species of *Volvariella* (i.e., *V. guttulosa*, *V. neovolvacea*, *V. niveosulcata*, *V. orientalis*, *V. ptilotricha*, *V. pulla*, *V. rostricystidiata*, and *V. thailandensis*) were discovered from Thailand and Vietnam (Malysheva et al. 2019; Niego et al. 2020; Kumla et al. 2022b; Malysheva and Popov 2022). New species of *Termitomyces*, namely *Te. flavus*, *Te. gilvus*, *Te. griseobulbus*, *Te. griseobrunneus*, *Te. planiperforatorius*, *Te. pseudoheimii*, and *Te. salmonicolor*, were found in Malaysia and Thailand, respectively (Seelan et al. 2020; Tang et al. 2024; Paloi et al. 2025). Recently, *Morchella rinjaniensis*, a novel tropical species of *Morchella*, was discovered in Indonesia (Retnowati et al. 2025). These discoveries highlight not only the region's extraordinary fungal diversity but also its largely unexplored ecological niches, from lowland dipterocarp and montane forests to peat swamps and limestone ecosystems. Despite these advances, many areas, particularly remote forests in Myanmar, Laos, Timor-Leste, and Brunei Darussalam, remain virtually unstudied, suggesting that a significant portion of Southeast Asia's mycobiota remains undiscovered. Continued systematic surveys, coupled with molecular identification techniques and regional collaboration, are critical for uncovering the true extent of fungal diversity and for informing conservation strategies in this globally significant biodiversity hotspot. Therefore, an understanding of the diversity of mushrooms in the tropics could help researchers to better understand global mycology.

10.17 ETHNOMYCOLOGY AND TRADITIONAL USES

Mushrooms play indispensable roles in tropical forest ecosystems as decomposers, symbionts, and pathogens, contributing significantly to nutrient cycling and forest sustainability. Beyond their ecological importance, mushrooms also hold ethnomycological and traditional significance in Southeast Asian societies. Rural and indigenous communities across Indonesia, Laos, Malaysia, Myanmar, the Philippines, Thailand, and Vietnam have developed sophisticated classification systems for identifying and categorizing wild mushrooms, distinguishing edible species from toxic ones through careful observation of morphological features, ecological associations, fruiting patterns, and seasonal availability. This knowledge, transmitted orally across generations, reflects a detailed understanding of habitat preferences and fungal ecology. Wild mushroom collection remains a vital subsistence activity and an important source of food and income for forest-dependent communities, particularly during the rainy season, when fruiting is most abundant. The economic importance of wild mushrooms extends beyond household consumption, as surplus harvests are sold in local markets or to regional traders, contributing substantially to rural livelihoods and food security throughout the region. The nutritional composition of mushrooms contributes to their cultural importance, with these mushrooms providing substantial amounts of protein, dietary fiber, essential amino acids, vitamins, minerals, and various bioactive compounds with potential health-promoting properties (Ho et al. 2020; Lu et al. 2020; Kumar et al. 2021; Lim and Shu 2022). Recent ethnopharmacological research has begun documenting these traditional uses and investigating their biochemical basis, revealing that several mushroom species contain diverse bioactive compounds including polysaccharides, terpenoids, phenolic compounds, and ergosterol derivatives with demonstrated antioxidant, immunomodulatory, and antimicrobial activities (Rathore et al. 2017; Thu et al. 2020; Kumar et al. 2021).

In Southeast Asia, ECM mushrooms are particularly prominent in dipterocarp forests, pine forests, oak woodlands, and mixed montane ecosystems. The fruiting bodies of ECM mushrooms, emerging seasonally in response to rainfall patterns and temperature fluctuations, have been recognized, collected, and consumed by indigenous and local communities throughout the region. The most commonly consumed ECM mushrooms in the region belong to the genera *Amanita*, *Astraeus*, *Cantharellus*, *Russula*, *Lactarius*, and various boletes, each contributing distinctive flavors and textures to regional cuisines (Sanmee et al. 2003; Mortimer et al. 2012; Læssøe et al. 2018; Thu et al. 2020; Kumla et al. 2023). Traditional knowledge holders recognize that different ECM mushroom species are associated with specific tree hosts and habitat types. Some species are typically found in dipterocarp forests, while others fruit abundantly in pine forests or oak-dominated montane forests. This ethnomycological knowledge not only guides safe and sustainable harvesting but also reflects generations of close ecological observation and cultural adaptation to local forest environments.

Members of the family Russulaceae, particularly species of *Russula* and *Lactarius*, rank among the most commonly collected and consumed ECM mushrooms throughout the region. In Laos, Myanmar, and northern Thailand, numerous species of *Russula* are harvested during the rainy season, with local people distinguishing between species primarily based on cap color, surface texture, and taste. Among *Russula* species, the green-cracking *Russula*, particularly *R. orientalovirescens* are widely consumed and traded throughout Southeast Asia, though they were previously misidentified as the European species *R. virescens* in local literature (Wisitrassameewong et al. 2025). Market surveys in Laos and Thailand, have revealed an extraordinarily diverse array of *Russula* species for sale, including many rare species with no reference sequences in databases and potentially new species unknown to science. Blackening *Russula* species including *R. densifolia* and *R. nigricans* are widely consumed in Laos and Thailand, though careful identification is essential as the lethal species *R. subnigricans*, which causes rhabdomyolysis and has been responsible for fatal poisoning cases, appears morphologically similar to edible blackening species (Sanmee et al. 2003; Manassila et al. 2005; Læssøe et al. 2018; Tawatsin et al. 2018). Other commonly consumed

Russula species documented from Southeast Asian markets and forests include *R. alboareolata*, *R. cyanoxantha*, *R. griseocarnosa*, *R. lepida*, and *R. xerampelina* (Sanmee et al. 2003; Manassila et al. 2005; Srikram and Supapvanich 2016; Læssøe et al. 2018; Kaewgrajang et al. 2020; Naksuwankul et al. 2022; Anh et al. 2023). The genus *Lactarius*, characterized by latex-exuding tissues when cut or broken, includes several highly valued edible species across Southeast Asia. *Lactarius volemus*, a well-known species with both deciduous and coniferous trees in northern Thailand and throughout the region (Sanmee et al. 2003; van de Putte et al. 2010; Læssøe et al. 2018). *Lactarius hatsudake* and *Lactarius pinguis* are consumed and widely distributed in Thailand and Laos (Le et al. 2007a, 2007b; Thu et al. 2020; Łuczaj et al. 2021). Traditional preparation methods vary according to species characteristics, with mild-tasting *Russula* and *Lactarius* species typically grilled, stir-fried with aromatics, or incorporated into spicy soups and salads, while species with acrid or bitter latex require special processing including parboiling and discarding cooking water, pickling in brine, or thorough drying to eliminate unpleasant flavors and potential gastric irritants.

Members of the genus *Astraeus* represent another important group of edible ECM mushrooms widely utilized across Southeast Asia, particularly in dry dipterocarp forests and pine woodlands. *Astraeus asiaticus, A. odoratus*, and related species, known as "Hed Phob" in Thai, "Hed Thao" in Lao, and by various local names across the region, produce distinctive star-shaped fruiting bodies (earthstars) with a central spore sac surrounded by rays that open and close in response to moisture conditions (Phosri et al. 2004, 2007; Srikram and Supapvanich 2016; Læssøe et al. 2018; Łuczaj et al. 2021; Suwannarach et al. 2022). The immature buttons of *Astraeus*, harvested before spore maturation, are consumed as seasonal delicacies with a unique texture and mild, nutty flavor. Traditional preparation methods vary across communities, with some groups boiling, grilling, or roasting the mushrooms, others incorporating them into curries or stir-fries, and some preserving them through drying or pickling.

The genus *Amanita*, while infamous for including some of the world's most toxic mushroom species, also comprises edible ECM species that feature in traditional Southeast Asian cuisines, though their consumption requires expert knowledge and careful preparation (Læssøe et al. 2018; Naksuwankul et al. 2022; Kumla et al. 2023). Edible *Amanita* species typically belong to *Amanita* section *Caesareae*. In northern Thailand and neighboring regions, *A. pseudoprinceps* is widely consumed, although it was previously misidentified as the European species *A. princeps* (Kumla et al. 2023). Three other edible *Amanita* species, *A. hemibapha*, *A. rubromarginata*, and *A. subhemibapha*, have also been reported from northern Thailand (Sanmee et al. 2003; Kumla et al. 2023). The cultural protocols surrounding *Amanita* collection reflect deep awareness of the risks involved, with knowledge transmission carefully controlled and novice foragers explicitly warned against collection. Some communities employ specific preparation methods including parboiling and discarding cooking water to remove potential toxins, while others consume only young specimens or specific developmental stages considered safer. The willingness to utilize potentially dangerous *Amanita* reflects both the nutritional importance of wild mushrooms in traditional food systems and the confidence in accumulated traditional knowledge, though modern public health authorities generally discourage consumption of *Amanita* species due to the severe consequences of misidentification.

Boletes, representing another major group of ECM mushrooms, contribute significantly to traditional food systems across Southeast Asian upland regions. The family Boletaceae includes numerous genera with edible species, though identification requires considerable expertise as the group also contains toxic species that can cause severe illness. In the montane forests of northern Thailand, Vietnam, and the highlands of Malaysia and Indonesia, local communities collect various bolete species associated with pine, oak, and dipterocarp trees, with *Boletus*, *Tylopilus*, *Phylloporus*, and related genera featuring in seasonal diets (Seehanan and Petcharat 2008; Thongklam 2008; Pukahuta et al. 2009; Wongchalee and Pukahute 2012; Læssøe et al. 2018). Tropical bolete, *Phlebopus portentosus* and *P. spongiosus* are popular consume in northern Thailand and southern Vietnam (Sanmee et al. 2003; Pham et al. 2012; Kumla et al. 2021, 2022a). Traditional collectors employ multiple identification criteria including cap and pore surface characteristics, staining reactions when bruised or cut, odor, taste testing of small quantities, and knowledge of toxic look-alikes to ensure safe consumption. The firm, meaty texture of boletes makes them particularly suitable for various cooking methods including grilling, stir-frying, and incorporation into soups and curries. Some communities preserve boletes through drying, creating concentrated flavor ingredients for use throughout the year, while others prefer consuming them fresh to appreciate their delicate textures and subtle flavors.

Chanterelles and related fungi, while less diverse in Southeast Asian tropical forests compared to temperate regions, occur in montane and subtropical zones where they contribute to local food systems and increasingly to commercial mushroom markets. Species of *Cantharellus*, *Craterellus*, and related genera found and collected during the early rainy season from mixed deciduous and dipterocarp forests in upland areas of Vietnam, northern Thailand, Myanmar, and the highlands of Malaysia and Indonesia. Popular edible chanterelle species in Southeast Asia include primarily *Cantharellus cibarius* and *Cantharellus minor* (Sanmee et al. 2003; Læssøe et al. 2018; Naksuwankul et al. 2022). These mushrooms are traditionally consumed fresh, stir-fried with local herbs, or added to curries and soups, where their fruity fragrance enhances the flavor of regional dishes. These mushrooms are traditionally consumed fresh, stir-fried with local herbs, or added to curries and soups, where their fruity fragrance enhances the flavor of regional dishes.

Termite-associated mushrooms of the genus *Termitomyces*, known as "Hed Kone" in Thailand, "Hed Khaen" in Laos, "Kulat Rayap" in Malaysia, "Kulat Rayap" in Malaysia, and "Jamur Rayap" in Indonesia, rank among the most highly prized edible fungi in Southeast Asia. They are celebrated for their delicate umami flavor, firm texture, and substantial size. These mushrooms emerge seasonally from termite mounds during the rainy period. Owing to their limited fruiting period and dependence on specific termite hosts, *Termitomyces* species are highly sought after and command premium prices in local and regional markets (Chang and Lee 2004; Mortimer et al. 2012; Seelan et al. 2020; Łuczaj et al. 2021). Beyond their culinary value, they also hold cultural and nutritional significance, featuring prominently in traditional dishes and local food heritage while providing essential income for rural foragers during the rainy season, typically between May and October of each year. In Southeast Asia, 18 species of the genus *Termitomyces* have been documented, collectively representing 31% of the total *Termitomyces* species known worldwide. These species include *Te. aurantiacus*, *Te. badius*, *Te. bulborhizus*, *Te. clypeatus*, *Te. entolomoides*, *Te. eurrhizus*, *Te. flavus*, *Te. floccosus*, *Te. fuliginosus*, *Te. gilvus*, *Te. globulus*, *Te. heimii*, *Te. microcarpus*, *Te. perforans*, *Te. radicatus*, *Te. robustus*, *Te. schimperi*, *Te. striatus*, and *Te. umkowaan* (Paloi et al. 2023b; Tang et al. 2024). They feature prominently in local cuisines, grilled, curried, or stir-fried with herbs, and are also valued in traditional medicine for their perceived health benefits, such as enhancing vitality and aiding digestion. Indigenous and rural communities possess detailed traditional knowledge for locating termite mounds, recognizing signs of fruiting, and harvesting sustainably without damaging the colony.

Wild edible saprobic mushrooms constitute an essential component of ethnomycological traditions and rural livelihoods throughout Southeast Asia, where they are prized for their flavor, nutritional value, and cultural significance. Across countries such as Thailand, Laos, Vietnam, Myanmar, Malaysia, Indonesia, and the Philippines, a wide variety of saprobic mushrooms, particularly species in the genera *Auricularia*, *Lentinus*, *Lentinula*, *Pleurotus*, *Schizophyllum*, and *Volvariella*, are collected from natural forests and agricultural landscapes during the rainy season (Kim et al. 2017; Dulay et al. 2023; Kien et al. 2025; Surawut et al. 2025; Yeni et al. 2025). *Auricularia auricula-judae* and related species within the genus *Auricularia* are commonly known as wood ear mushrooms and are referred to by various local names, including "Hed Hunu Dam" in Thai, "Nấm Mèo" in Vietnamese, "Kuping Tikus" or "Jamur Kuping" in Malay and Indonesian, as well as numerous other names across the region's diverse linguistic communities (Srikram and Supapvanich 2016; Lee et al. 2012, 2021; Yusran et al. 2021; Putra et al. 2022). Characterized by their ear-like morphology and rubbery texture, these mushrooms have been harvested from decaying hardwood logs and branches for centuries and are integral to traditional cuisines across the region (Læssøe et al. 2018; Bandara et al. 2019). The culinary applications of wood ear mushrooms reflect their unique textural qualities, with their crunchy yet slippery texture being highly valued in soups, stir-fries, salads, and braised dishes. Wood ear mushrooms occupy important roles in traditional medicinal systems across the region, where practitioners attribute to them properties such as enhancing blood circulation, reducing cholesterol, supporting respiratory health, and promoting general vitality, beliefs that have found partial support in modern pharmacological research.

Oyster mushrooms of the genus *Pleurotus*, comprising multiple species that fruit naturally on dead hardwood, rubber trees, bamboo, and various other woody substrates throughout Southeast Asian forests and agricultural landscapes, represent highly valued edible fungi with long histories of traditional use that predate modern cultivation (Thongklang and Luangharn 2016; Raman et al. 2020). Different *Pleurotus* species, distinguished by local collectors based on color (white, grey, pink, yellow), size, substrate preferences, seasonal occurrence, and subtle flavor differences, occupy different ecological niches and fruiting periods, collectively providing extended availability of these nutritious and flavorful mushrooms. Oyster mushrooms hold immense ethnomycological importance as food, medicine, and livelihood resources among rural and Indigenous communities. Southeast Asian countries, including Thailand, Laos, Vietnam, Myanmar, Malaysia, Indonesia, and the Philippines, harbor remarkable diversity of *Pleurotus* species, with commonly observed taxa such as *Pl. ostreatus*, *Pl. pulmonarius*, *Pl. djamor*, *Pl. cystidiosus*, and *Pl. cornucopiae* (Bao et al. 2004; Phonemany et al. 2024; Kumla et al. 2025b). In northern Thailand and Laos, oyster mushrooms are cooked in a variety of traditional dishes, including soups, stir-fries, and spicy salads. In Vietnam and Cambodia, wild oyster mushrooms are commonly fermented, sun-dried, or pickled to extend their shelf life. Beyond their culinary importance, oyster mushrooms are also valued in traditional medicine and local health practices. In many Southeast Asian communities, decoctions prepared from dried *Pleurotus* fruiting bodies are consumed to relieve fatigue, improve digestion, and promote overall vitality. These traditional uses are consistent with modern pharmacological research, which has identified oyster mushrooms as rich sources of bioactive compounds with antioxidant, immunomodulatory, and cholesterol-lowering properties.

The split-gill mushroom, *Schizophyllum commune*, recognized as one of the world's most widely distributed species, represents another important saprobic mushroom in Southeast Asian traditional food systems despite its tough and leathery texture. Across Thailand, Laos, Vietnam, Myanmar, Malaysia, and Indonesia, *Sc. commune* is collected for food, medicine, and occasionally as a source of income in rural communities (Yusran et al. 2024; Yeni et al. 2025). Ethnomycologically, *Sc. commune* is well integrated into local dietary practices (Fui et al. 2018; Yeni et al. 2025). In Thailand and Laos, it is typically stir-fried with herbs, added to soups, or mixed into spicy salads. In Indonesia and Malaysia, it is sometimes incorporated into traditional vegetable dishes or cooked with coconut milk, providing texture and subtle flavor. Ethnobotanical reports indicate that decoctions and extracts from *Sc. commune* are used to alleviate digestive issues, boost immunity, and promote general health.

Wild *Volvariella* species, particularly *Volvariella volvacea*, commonly known as the paddy straw mushroom, are among the most culturally, nutritionally, and economically significant saprobic fungi in Southeast Asia (Abella et al. 2014; Mohd-Aris and Amir 2024). Southeast Asian countries, including Thailand, Vietnam, Laos, Myanmar, Indonesia, and the Philippines, have a long history of harvesting *Volvariella* from both wild habitats and cultivated agricultural landscapes (Abella et al. 2014; Thawthong et al. 2014; Zikriyani et al. 2018; Kumla et al. 2022b). Ethnomycologically, *Volvariella* holds a central place in traditional diets. In Thailand, the mushroom is widely incorporated into soups, curries, and stir-fried dishes. In Vietnam, it is consumed fresh, grilled, or dried for later use. In Laos and Cambodia, *Volvariella* mushrooms, particularly *V. volvacea*, are commonly stir-fried with garlic, chili, and local vegetables, added to soups, or lightly grilled to highlight their natural flavor. They are also sun-dried or fermented to extend their shelf life beyond the rainy season. Local communities rely on traditional knowledge to select mature, healthy mushrooms and ensure safe, flavorful preparation. Traditional medicinal practices ascribe benefits such as better digestion, increased vitality, and overall well-being to regular consumption.

The genera *Lentinus* and *Lentinula* are among the most culturally and nutritionally significant saprobic mushrooms in Southeast Asia, valued for their culinary versatility, medicinal potential, and economic importance (Phonemany et al. 2021; Raghoonundon et al. 2021). Species such as *L. polychrous*, *L. sajor-caju*, *L. squarrosulus*, and *Lentinula edodes* (shiitake) are widely distributed across Thailand, Laos, Vietnam, Myanmar, Malaysia, and Indonesia, fruiting predominantly during the rainy season when humidity and temperature favor fungal growth (Corner 1981; Srijumpa et al. 2008; Læssøe et al. 2018; Hermawan 2021; Sysouphanthong et al. 2023). Ethnomycologically, these mushrooms are highly integrated into local food systems. In northern Thailand and Laos, *Lentinus* species and *Le. edodes* are commonly stir-fried with herbs, cooked in soups, or incorporated into curries, appreciated for their firm texture and savory flavor. *Lentinula edodes* is also collected from wild in rural Vietnam, where it is used fresh, dried, or rehydrated in soups and stews. In Indonesia and Malaysia, *L. sajor-caju* is frequently prepared with coconut milk, chilies, and local spices, while smaller species like *L. squarrosulus* are often sun-dried for preservation. *Lentinus* and *Lentinula* species are recognized for medicinal properties in traditional health systems (Raghoonundon et al. 2021; Filho et al. 2023).

The therapeutic application of mushrooms constitutes a fundamental component of traditional healing systems throughout Southeast Asia, where numerous species are employed to prevent and treat a wide range of ailments, reflecting centuries of accumulated ethnomycological knowledge. Medicinal mushrooms such as *Ganoderma lucidum* and related species, commonly known as lingzhi or reishi, have been highly revered in traditional Chinese medicine and among indigenous Southeast Asian communities for their purported immunomodulatory, anti-inflammatory, and longevity-promoting properties (Hapuarachchi et al. 2018; El Sheikha 2022; Swallah et al. 2023). These fungi are traditionally prepared as decoctions, teas, or powders and are believed to enhance overall health, improve resilience against illness, and promote vitality across all age groups. Similarly, *Lignosus rhinocerotis*, commonly referred to as the tiger milk mushroom or cendawan susu rimau, holds particular cultural and medicinal significance in Malaysia and among indigenous populations in Borneo (Lau et al. 2015; Nallathamby et al. 2018). Traditionally, this species has been used to treat respiratory ailments, improve energy levels, and support postpartum recovery, with its tuberous sclerotia considered the most potent part for therapeutic use. Beyond these well-known examples, other medicinally important genera such as *Phellinus*, *Trametes*, *Cordyceps*, and *Agaricus* play significant roles in local health practices, where they are processed into decoctions, powders, or fermented preparations to address conditions ranging from digestive and respiratory disorders to infections and even cancer (Bhambri et al. 2022; Łysakowska et al. 2023).

10.18 ECONOMIC AND BIOTECHNOLOGICAL POTENTIAL

Wild mushrooms in Southeast Asia constitute an abundant and culturally significant food resource, contributing substantially to both nutrition and livelihoods (Sanmee 2004; Watling 2013; Læssøe et al. 2018). The collection, processing, and sale of these mushrooms support household nutrition, generate rural incomes, and enhance food security, particularly in forest-dependent communities where access to animal protein may be limited. Beyond their culinary and ethnomycological value, the pharmaceutical potential of Southeast Asian wild mushrooms has attracted increasing research attention as scientists investigate traditional medicinal species for therapeutic applications validated by both indigenous knowledge and modern pharmacological screening. Medicinal mushrooms of the genus *Ganoderma*, particularly *G. lucidum* and numerous *Ganoderma* tropical species endemic to Southeast Asian forests, *Inonotus*, *Lignosus rhinocerotis*, and *Phellinus* have demonstrated remarkable arrays of bioactive compounds including triterpenes with anti-inflammatory and hepatoprotective properties, polysaccharides exhibiting immunomodulatory effects through activation of macrophages and natural killer cells, and various secondary metabolites with documented anticancer activity in cell culture and animal models (Nallathamby et al. 2018; Bhambri et al. 2022; Łysakowska et al. 2023; Swallah et al. 2023). Commercial development of these medicinal cultivation technologies and standardized extracts has created new economic opportunities while raising questions about intellectual property rights, benefit sharing with indigenous communities, and the balance between commercial exploitation and traditional knowledge protection.

The economic potential of wild edible mushroom cultivation and commercialization represents one of the most immediately achievable biotechnological applications in Southeast Asia. Numerous traditionally harvested species are well suited to domestication through the use of locally available agricultural residues (rice straw, sawdust, and corn cobs) and low-cost cultivation systems appropriate for small-scale rural enterprises. Species in genera *Auricularia*, *Ganoderma*, *Lentinus*, *Lentinula*, *Pleurotus*, and *Volvariella* already have well-established cultivation protocols and strong market demand (Figure 10.2), providing successful models for developing similar systems for other wild taxa that are still collected exclusively from natural forests (Kalaw et al. 2021; Sysouphanthong et al. 2023; Mohd-Aris and Amir 2024; Gebru et al. 2024; Phonemany et al. 2024). Expanding cultivation of these species could enhance rural livelihoods, reduce pressure on wild populations, and promote more sustainable forest resource management. The domestication of high-value species such as *Termitomyces*, which forms obligate symbiotic associations with termites, presents a major technical challenge but also a promising frontier for research due to its substantial culinary and economic value. Likewise, the cultivation of ECM species, including edible *Amanita*, *Lactarius*, *Phlebopus*, and *Russula* requires

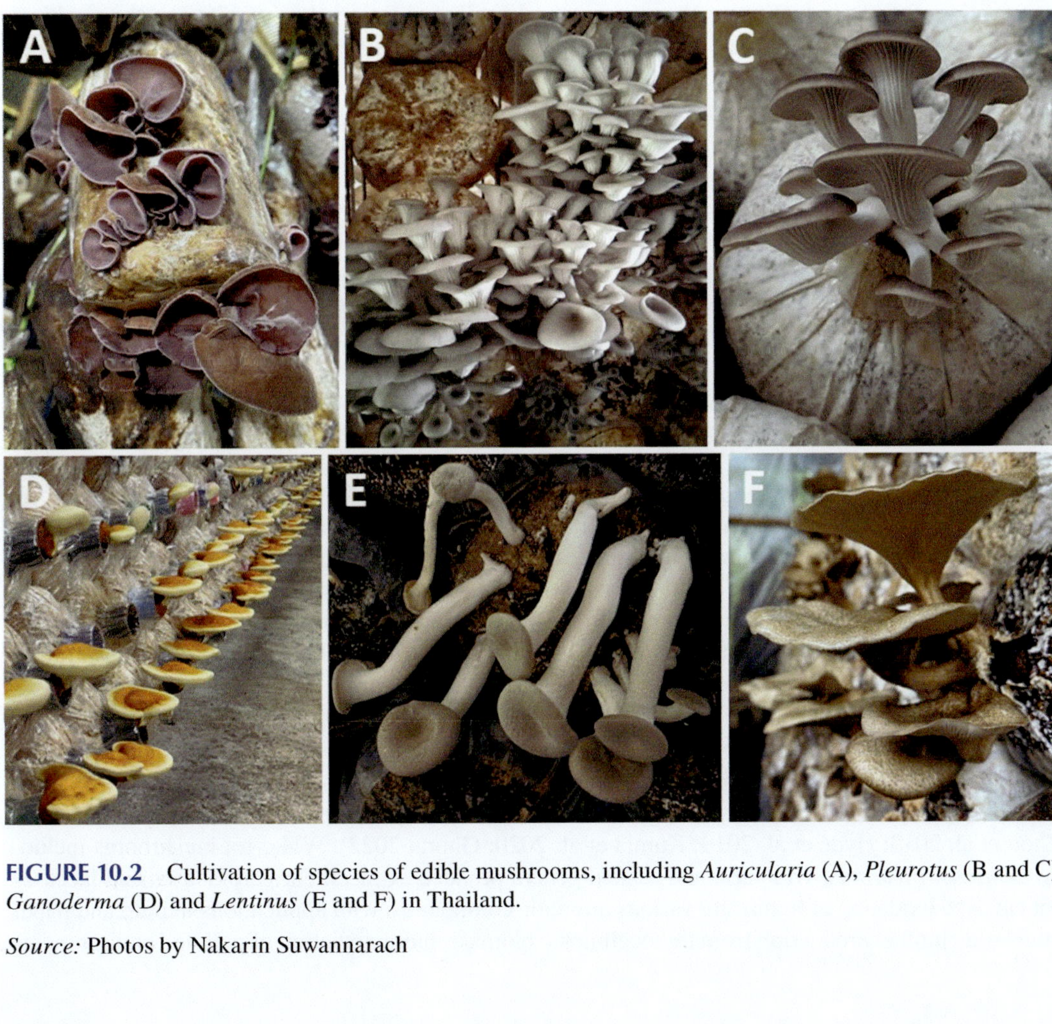

FIGURE 10.2 Cultivation of species of edible mushrooms, including *Auricularia* (A), *Pleurotus* (B and C), *Ganoderma* (D) and *Lentinus* (E and F) in Thailand.

Source: Photos by Nakarin Suwannarach

successful establishment of compatible host plant associations (Hall et al. 2003; Repáč 2011). Ongoing research programs are exploring inoculation methods, host selection, and environmental management strategies to enable plantation-based production of these valuable forest mushrooms (Kumar and Satyanarayana 2002; Repáč 2011). Continued investment in applied mycological research, technology transfer, and capacity building will be essential to transform these biological resources into viable economic opportunities that support sustainable livelihoods, biodiversity conservation, and regional bioeconomy development.

The use of ECM inoculation has become an increasingly important silvicultural practice in forest restoration and tree plantation programs in Thailand, Malaysia, and Indonesia, where seedlings of economically valuable dipterocarps and pines are inoculated with native ECM fungi to enhance growth, survival, and adaptability in degraded or reforested sites (Lee et al. 2008; Kaewgrajang et al. 2019; Imanuddin et al. 2020; Khemnark 2022). Experimental studies have demonstrated that inoculated seedlings exhibit higher biomass accumulation, better root-to-shoot ratios, and improved tolerance to drought stress compared with non-inoculated controls. Beyond their ecological function, many ECM fungi such as *Astraeus odoratus*, *Lactarius volemus*, and *Russula cyanoxantha* are also prized edible mushrooms, offering dual benefits by supporting forest regeneration and providing income for rural communities engaged in mushroom collection. The inoculation of *Shorea* and *Dipterocarpus* seedlings with *Pisolithus* and *Scleroderma* species has been shown to significantly increase root colonization, biomass accumulation, and survival rates when transplanted into degraded or nutrient-deficient soils (Turjaman et al. 2005; Lee et al. 2008). The integration of ECM fungi into reforestation strategies thus represents a sustainable approach that combines biodiversity conservation, forest productivity enhancement, and socioeconomic value, reinforcing their critical role in maintaining healthy forest ecosystems and supporting resilient seedling growth throughout Southeast Asia.

The nutraceutical and functional food potential of Southeast Asian mushrooms has catalyzed rapidly expanding market sectors, propelled by escalating consumer demand for natural health products, immune-supporting foods, and supplements grounded in traditional medicinal wisdom (Venturella et al. 2021; Rangsinth et al. 2023). Commercial products ranging from dried whole mushrooms and powdered supplements to concentrated extracts and sophisticated formulations incorporating mushroom-derived bioactive compounds, including polysaccharides, triterpenes, beta-glucans, and phenolic compounds, have successfully penetrated regional and international markets, particularly targeting health-conscious consumers in urban centers and developed nations (Łysakowska et al. 2023; Moniruzzaman et al. 2025). Species such as *Ganoderma*, *Lignosus rhinocerotis*, *Cordyceps*, and various cultivated and wild-harvested mushrooms command premium prices as functional foods and dietary supplements (Nallathamby et al. 2018; Swallah et al. 2023). Developing value chains that connect rural collectors and small-scale cultivators to urban and international markets requires investment in post-harvest handling, processing infrastructure, quality assurance systems, and market access mechanisms that ensure fair pricing while meeting consumer expectations for safety and efficacy. Certification schemes, including organic certification, fair trade verification, and geographical indication protection, could enhance the value of Southeast Asian mushroom products, while promoting sustainable harvesting and the preservation of traditional knowledge. Successful implementation, however, depends on technical support, institutional capacity, and policy frameworks, many of which are still under development across the region.

The industrial enzyme applications of Southeast Asian saprobic mushrooms represent another dimension of biotechnological potential, with wood-decay species producing sophisticated arrays of ligninolytic enzymes, cellulases, hemicellulases, and oxidoreductases capable of breaking down complex plant polymers that resist degradation by conventional chemical or biological processes (Zhou et al. 2013; Hyde et al. 2019; Kumla et al. 2020; Ghorai 2022). White-rot mushrooms including *Lentinus*, *Pleurotus*, *Trametes*, and related genera produce extracellular peroxidases and laccases that catalyze oxidation of lignin and various aromatic compounds, with applications in pulp and paper processing, biofuel production from lignocellulosic biomass, bioremediation of polluted environments contaminated with persistent organic compounds, textile processing and dye decolorization, and synthesis of fine chemicals through selective oxidation reactions (Latif et al. 2023; Torres-Farradá et al. 2024). The extreme biodiversity of Southeast Asian saprobic mushrooms suggests that novel enzyme variants with superior thermostability, pH tolerance, substrate specificity, or catalytic efficiency remain to be discovered among understudied tropical species, particularly those adapted to the high-temperature, high-humidity conditions characteristic of tropical decomposition environments.

Despite their high potential, many wild mushroom species in Southeast Asia remain underexplored, with limited documentation of their ecology, bioactive compounds, and traditional knowledge. Integrating ethnomycological wisdom with modern cultivation, biotechnological research, and metabolomic profiling can enhance both scientific understanding and commercialization. Protecting habitats, promoting sustainable harvesting, and supporting local capacity for mushroom processing can transform these mushrooms from seasonal food sources into high-value biotechnological and economic resources. Wild mushrooms in Southeast Asia exemplify a convergence of ecological, cultural, and economic value, offering opportunities for food security, health, environmental management, and rural livelihoods while sustaining the region's rich fungal biodiversity.

10.19 ACKNOWLEDGMENTS

This work was financially supported by Chiang Mai University, Thailand.

REFERENCES

Abella, E.A., G.D. Robson & R.F. Preziosi. 2014. Molecular discrimination of wild Philippine paddy straw mushroom (*Volvariella volvacea*). *Journal of Life Sciences* 8:27–34.

Ahmadni, A.S.A., N.A.S. Kamar, M.A. Hadi & R. Terhem. 2024. Assessment of macrofungi diversity in Perlis State Park, Perlis, Malaysia. *Journal of Wildlife and Biodiversity* 8:298–307.

Anh, C.N., N.M. Chi, T.T. Kiet, P.D. Long, P.T.T. Thuy, V. van Loi & B. Dell. 2023. Morphological and molecular identification of an edible *Russula* mushroom in northeast Vietnam. *Journal of Forestry Science and Technology* 15:50–59.

Aung-aud-chariya, A., P. Bangrak, B. Dell, S. Lumyong & N. Kamlangdee. 2012. Preliminary molecular identification of *Boletus griseipurpureus* Corner from Thailand and its nutritional value. *International Journal of Agricultural Technology* 8:1991–1998.

Bandara, A.R., S. Rapior, P.E. Mortimer, P. Kakumyan, K.D. Hyde & J. Xu. 2019. A review of the polysaccharide, protein and selected nutrient content of *Auricularia*, and their potential pharmacological value. *Mycosphere* 10:579–607.

Bao, D., S. Kinugasa & Y. Kitamoto. 2004. The biological species of oyster mushrooms (*Pleurotus* spp.) from Asia based on mating compatibility tests. *Journal of Wood Science* 50:162–168.

Beccari, O. & V. Cesati. 1879. *Mycetum in itinere Borneensi lectorum* (Mushroom scholars on a journey to Borneo). *Atti della Real Accademia delle Scienze Fisiche e Matematiche* (Proceedings of the Royal Academy of Physical and Mathematical Sciences), pp. 1–28. Regiae Scientiarum Academiae (Royal Academy of Sciences), Naples, Italy.

Bhambri, A., M. Srivastava, V.G. Mahale, S. Mahale & S.K. Karn. 2022. Mushrooms as potential sources of active metabolites and medicines. *Frontiers in Microbiology* 13:e837266.

Brazas, Jr.F.P., L.P. Talina, A.G.M. Revilla, R.F. Javier & E.R. Tadiosa. 2020. Diversity and taxonomy of basidiomycetous fungi at the northeastern side of Quezon Protected Landscape, Southern Luzon, Philippines. *Journal of Agricultural Science and Technology A* 10:1–11.

Bustillos, R.G., R.M.R. Dulay, S.P. Kalaw & R.G. Reyes. 2024. Diversity of macrofungi along elevation gradients in Mt. Arayat Protected Landscape, Arayat, Pampanga, The Philippines. *Studies in Fungi* 9:e017.

Butkrachang, S., E. Boonchieng, U. Sardsud, M. Sukchotiratana, A. Plikomol, G. Chairotec & P. Narongchai. 2007. Wild mushroom database of Chiang Mai community forest. *Asian Journal of Biology Education* 3:65–70.

Chandrasrikul, A., P. Suwanarit, U. Sangwanit, S. Lumyong, A. Payapanon, N. Sanoamuang, C. Pukahuta, V. Petcharat, U. Sardsud, K. Duengkae, U. Klinhom, S. Thongkantha & S. Thongklam. 2011. *Check list of Mushroom (Basidiomycetes) in Thailand*. Office of Natural Resources and Environmental Policy and Planning, Bangkok, Thailand.

Chang, Y.S. & S.S. Lee. 2004. Utilisation of macrofungi species in Malaysia. *Fungal Diversity* 15:15–22.

Chen, J., R.L. Zhao, S.C. Karunarathna, P. Callac, O. Raspé, A.H. Bahkali & K.D. Hyde. 2012. *Agaricus megalosporus*: A new species in section Minores. Cryptogamie *Mycologie* 33:145–155.

Chuankid, B., S. Vadthanarat, B. Thongbai, M. Stadler, S. Lumyong, K.D. Hyde & O. Raspé. 2021. *Retiboletus* (Boletaceae) in northern Thailand: One novel species and two first records. *Mycoscience* 62:297–306.

Corner, E.J.H. 1981. The agaric genera *Lentinus*, *Panus*, and *Pleurotus* with particular reference to Malaysian species. *Nova Hedwigia* 69:1–169.

Cunningham, A.B. & X. Yang. 2012. *Mushrooms in Forests and Woodlands: Resource Management, Values and Local Livelihoods*. Routledge, Earthscan Ltd., London, UK.

Daep, N.A. & L.A. Cajuday. 2003. Mushroom diversity at Mt. Malinao, Albay. *Philippine Society for the Study of Nature News* 2:e57.

De Leon, A., T.R.L. Evangelista, Jr., N.D. Morales, J. Undan & S. Kalaw. 2023. Assessment of utilized species of macrofungi and evaluation of mycochemical composition and bioactive properties of Phellinus robiniae collected from Benguet Province, Philippines. *CLSU International Journal of Science and Technology* 7:9–27.

Dissing, H. 1963. Studies in the flora of Thailand 25: Discomycetes and Gasteromycetes. *Dansk Botanisk Arkiv* 23:117–130.

Duengkae, K. 2018. Macrofungi communities in Phukieo—Namnao forest complex. *Thai Forest Ecological Research* 2:37–44.

Dulay, R.M.R., J.N. Batangan, S.P. Kalaw, A.M. de Leon, E.C. Cabrera, K. Kimura, F. Eguchi & R.G. Reyes. 2023. Records of wild mushrooms in the Philippines: A review. *Journal of Applied Biology and Biotechnology* 11:11–32.

Duong, L.M., Q.D. Do & T.H. Tran. 2011. Species diversity of *Xylaria* genus. In *Cuc Phuong National Park, Ninh Binh Province: Proceedings of the Fourth National Conference on Ecology and Biological Resources*. Institute of Ecology and Biological Resources, Hanoi, pp. 166–170.

El Sheikha, A.F. 2022. Nutritional profile and health benefits of *Ganoderma lucidum* "Lingzhi, Reishi, or Mannentake" as functional foods: Current scenario and future perspectives. *Foods* 11:e1030.

Filho, J.R.D.S., É.D.S. Santos, G.A. Linde, N.B. Colauto, R.A.C. Gonçalves & A.J.B. de Oliveira. 2023. *Lentinus crinitus*: Traditional use, phytochemical and pharmacological activities, and industrial and biotechnological applications. *Food Research International* 172:e113179.

Fui, F.S., F.H. Saikim, J. Kulip & J.S.S. Seelan. 2018. Distribution and ethnomycological knowledge of wild edible mushrooms in Sabah (Northern Borneo), Malaysia. *Journal of Tropical Biology and Conservation* 15:203–222.

Gebru, H., T. Belete & G. Faye. 2024. Growth and yield performance of *Pleurotus ostreatus* cultivated on agricultural residues. *Mycobiology* 52:388–397.

Ghorai, S. 2022. Edible mushroom: A potent producer of industrial enzymes. *Journal of Scientific Research* 66:148–166.

Guo, S., X. Deng, Z. Ma, N. Mao & W. Huang. 2024. Two new species of suspected mushrooms of the suborder Marasmiineae from mid-Cretaceous Burmese amber (Basidiomycota, Agaricales). *Cretaceous Research* 164:e105968.

Hall, I.R., W. Yun & A. Amicucci. 2003. Cultivation of edible ectomycorrhizal mushrooms. *TRENDS in Biotechnology* 21:433–438.

Hapuarachchi, K.K., W.A. Elkhateeb, S.C. Karunarathna, C.R. Cheng, A.R. Bandara, P. Kakumyan, K.D. Hyde, G.M. Daba & T.C. Wen. 2018. Current status of global *Ganoderma* cultivation, products, industry and market. *Mycosphere* 9:1025–1052.

Hawksworth, D.L. 2001. Mushrooms: The extent of the unexplored potential. *International Journal of Medicinal Mushrooms* 3:333–337.

Hermawan, R. 2021. Study of *Lentinus squarrosulus* from West Java on the basis of molecular and morphological data. *Journal Biota* 7:1–9.

Ho, L.H., N.A. Zulkifli & T.C. Tan. 2020. Edible mushroom: Nutritional properties, potential nutraceutical values, and its utilization in food product development. In *An Introduction to Mushroom*, pp. 1–19. IntechOpen, London, England.

Hong, M.Y., S.W. Park, D.H. Kim, V. Saysavanh & J.K. Lee. 2019. A checklist of mushrooms of *Phousabous* National Protected Area (PNPA) of Lao pdr. *Journal of Forest and Environmental Science* 35:268–271.

Hosaka, K. & W.W. Linn. 2021. Flora 3. Fungi. In N. Tanaka (ed.), *A Guide to Flora and Fauna in Southern Myanmar*, pp. 67–85. National Museum of Nature and Science, Tokyo, Japan.

Hyde, K.D., C. Norphanphoun, J. Chen, A.J. Dissanayake, M. Doilom, S. Hongsanan, R.S. Jayawardena, R. Jeewon, R.H. Perera, B. Thongbai, D.N. Wanasinghe, K. Wisitrassameewong, S. Tibpromma & M. Stadler. 2018. Thailand's amazing diversity: Up to 96% of fungi in Northern Thailand may be novel. *Fungal Diversity* 93:215–239.

Hyde, K.D., J. Xu, S. Rapior, R. Jeewon, S. Lumyong, A.G.T. Niego, P.D. Abeywickrama, J.V.S. Aluthmuhandiram, R.S. Brahamanage, S. Brooks, A. Chaiyasen, K.W.T. Chethana, P. Chomnunti, C. Chepkirui, B. Chuankid, N.I. de Silva, M. Doilom, C. Faulds, E. Gentekaki, V. Gopalan, P. Kakumyan, D. Harishchandra, H. Hemachandran, S. Hongsanan, A. Karunarathna, S.C. Karunarathnn, S. Khan, J. Kumla, R.S. Jayawardena, J.K. Liu, N. Liu, T. Luangharn, A.P.G. Macabeo, D.S., Marasinghe, D, Meeks, P.E. Mortimer, P. Mueller, S. Nadir, K.N. Nataraja, S. Nontachaiyapoom, M. O'Brien, W. Penkhrue, C. Phukhamsakda, U.S. Ramanan, A.R. Rathnayaka, R.B. Sadaba, B. Sandargo, B.C. Samarakoon, D.S. Tennakoon, R. Siva, W. Sriprom, T.S. Suryanarayanan, K. Sujarit, N. Suwannarach, T. Suwunwong, B. Thongbai, N. Thongklang, D. Wei, S.N. Wijesinghe, J. Winiski, J. Yan, E. Yasanthika & M. Stadler. 2019. The amazing potential of fungi: 50 ways we can exploit fungi industrially. *Fungal Diversity* 97:1–136.

Imanuddin, R., A. Hidayat, H.H. Rachmat, M. Turjaman, Pratiwi, F. Nurfatriani, Y. Indrajaya & A. Susilowati. 2020. Reforestation and sustainable management of *Pinus merkusii* forest plantation in Indonesia: A review. *Forests* 11:e1235.

Jones, E.B.G. 2007. Introduction to Malaysian fungal diversity. In E.B.G. Jones, K.D. Hyde & S. Vikineswary (eds.), *Malaysian Fungal Diversity*, p. 484. Mushroom Research Centre, University of Malaya and Ministry of Natural Resources and Environment, Kuala Lumpur, Malaysia.

Jones, E.B.G., A.J.S. Whalley & N.L. Hywel-Jones. 1994. A fungus foray to Chiang Mai market in Northern Thailand. *The Mycologist* 8:87–90.

Kaewgrajang, T., S. Kaewjunsri, N. Jannual & M. Nipitwattanaphon. 2020. Morphology and molecular identification of some *Lactarius* and *Russula* species. *Genomics and Genetics* 13:44–58.

Kaewgrajang, T., B. Sakolrak & U. Sangwanit. 2019. Growth response of *Dipterocarpus tuberculatus* and *Shorea roxburghii* seedlings to *Astraeus odoratus*. *Environment and Natural Resources Journal* 17:80–88.

Kalaw, S.P., R.M.R. Dulay, E.J. Damaso, J.C. Ramos, M.A.G. Del Rosario, M.D. Abon, A.M. De Leon, J.R. Undan & R.G. Reyes. 2021. Optimization of mycelial culture conditions and fructification of *Lentinus* species using rice straw and sawdust based substrates. *Studies in Fungi* 6:519–530.

Karunarathna, S.C., Z.L. Yang, R.L. Zhao, E.C. Vellinga, A.H. Bahkali, E. Chukeatirote & K.D. Hyde. 2011. Three new species of *Lentinus* from northern Thailand. *Mycological Progress* 10:389–398.

Khemnark, C. 2022. Study on inoculation of ectomycorrhizal fungi on *Pinus kesiya* in Thailand. *Thai Journal of Forestry* 6:1–17.

Kien, N.H., D.T.T. Huong, T.T.G. Thoa, N.V. Sinh & N.P.D. Nguyen. 2025. Species diversity of Ganodermataceae fungi in Yok Don National Park, Central Highlands of Vietnam. *International Journal of Environmental Sciences* 11:639–650.

Kiet, T.T. 1998. Preliminary checklist of macrofungi of Vietnam. *Feddes Repertorium* 109:257–277.

Kim, D.H., N.M. Ha, M.M.Q. Manalo, B. Manuel & J.K. Lee. 2021. Checklist of mushrooms of Mt. Cleopatra needle forest reserve in Palawan Island, Philippines. *Journal of Korean Society of Forest Science* 110:289–294.

Kim, N.K., J.H. Lee, J.W. Jo & J.K. Lee. 2017. A checklist of mushrooms of Cambodia. A checklist of mushrooms of Cambodia. *Journal of Forest and Environmental Science* 33:49–65.

Kinge, T.R., G. Goldman, A. Jacobs, G.G. Ndiritu & M. Gryzenhout. 2020. A first checklist of macrofungi for South Africa. *MycoKeys* 63:1–48.

Kumar, K., R. Mehra, R.P.F. Guiné, M.J. Lima, N. Kumar, R. Kaushik, N. Ahmed, A.N. Yadav & H. Kumar. 2021. Edible mushrooms: A comprehensive review on bioactive compounds with health benefits and processing aspects. *Foods* 10:e2996.

Kumar, S. & T. Satyanarayana. 2002. Production of inoculum of ectomycorrhizal fungi. In *Techniques in Mycorrhizal Studies*, pp. 143–166. Springer Netherlands, Dordrecht.

Kumla, J., K. Jatuwong, K. Tanruean, S. Khuna, S. Srinuanpan, S. Lumyong & N. Suwannarach. 2024. A new edible wild mushroom species, *Panus sribuabanensis* (Panaceae, Polyporales) from Northern Thailand and its nutritional composition, total phenolic content, and antioxidant activity. *Mycobiology* 52:1–12.

Kumla, J., A. Kaewnunta & N. Suwannarach. 2025a. *Lentinus saisamorniae* (Polyporaceae, Polyporales), a new edible macrofungus from northern Thailand. *Phytotaxa* 705:149–161.

Kumla, J., S. Lumyong & N. Suwannarach. 2025b. Centesimal composition, bioactive compounds, antioxidant and α-glucosidase inhibitory activities of commercial edible oyster mushrooms at different maturity stages in northern Thailand. *Foods* 14:e3511.

Kumla, J., N. Suwannarach, Y.S. Liu, K. Tanruean & S. Lumyong. 2023. Survey of edible *Amanita* in Northern Thailand and their nutritional value, total phenolic content, antioxidant and α-glucosidase inhibitory activities. *Journal of Fungi* 9:e343.

Kumla, J., N. Suwannarach & S. Lumyong. 2022a. Cultivation of edible tropical bolete, *Phlebopus spongiosus*, in Thailand and yield improvement by high-voltage pulsed stimulation. *Agronomy* 12:e115.

Kumla, J., N. Suwannarach, K. Sujarit, W. Penkhrue, P. Kakumyan, K. Jatuwong, S. Vadthanarat & S. Lumyong. 2020. Cultivation of mushrooms and their lignocellulolytic enzyme production through the utilization of agro-industrial waste. *Molecules* 25:e2811.

Kumla, J., N. Suwannarach, K. Tanruean & S. Lumyong. 2021. Comparative evaluation of chemical composition, phenolic compounds, and antioxidant and antimicrobial activities of tropical black bolete mushroom using different preservation methods. *Foods* 10:e781.

Kumla, J., N. Suwannarach, S. Vadthanarat & S. Lumyong. 2017. *Gyrodon suthepensis* (Boletales, Basidiomycota), a new ectomycorrhizal fungus from northern Thailand and its ecomycorrhizal association. *Phytotaxa* 321:181–193.

Kumla, J., N. Suwannarach, N. Wannathes & S. Lumyong. 2022b. Survey of *Volvariella* (Agaricales, Basidiomycota) including two new species, *V. neovolvacea* and *V. thailandensis*, from northern Thailand. *Diversity*14:e161.

Læssøe, T., O.S. Pedersen & P. Sysouphanthong. 2018. *An Introduction to the Edible, Poisonous, and Medicinal Fungi of Northern Laos*. Biotechnology and Ecology Institute, Ministry of Science and Technology, Vientiane, Laos.

Latif, W., C. Ciniglia, M. Iovinella, M. Shafiq & S. Papa. 2023. Role of white rot fungi in industrial wastewater treatment: A review. *Applied Sciences* 13:e8318.

Lau, B.F., N. Abdullah, N. Aminudin, H.B. Lee & P.J. Tan. 2015. Ethnomedicinal uses, pharmacological activities, and cultivation of *Lignosus* spp. (tiger's milk mushrooms) in Malaysia. *Journal of Ethnopharmacology* 169:441–458.

Le, H.T. & Q.N.T. Doan. 2018. Biodiversity investigate of ink family Coprinaceae in Cuc Phuong National Park Ninh Binh province. *Vietnam Journal of Agricultural and Rural Development* 336:137–141.

Le, H.T., J. Nuytinck, A. Verbeken, S. Lumyong & D.E. Desjardin. 2007a. *Lactarius* in Northern Thailand: 1. *Lactarius* subgenus *Piperites*. *Fungal Diversity* 24:173–224.

Le, H.T., D. Stubbe, A. Verbeken, J. Nuytinck, S. Lumyong & D.E. Desjardin. 2007b. *Lactarius* in Northern Thailand: 2. *Lactarius* subgenus *Plinthogali*. *Fungal Diversity* 27:61–94.

Lee, J.S., D. Kim, M.H. Nguyen, Y.J. Bae & P. Manilak. 2021. A checklist of mushrooms of Dong Hua Sao National Biodiversity Conservation Area (DHSNBCA) of Lao-PDR. *Journal of Forest and Environmental Science* 37:163–167.

Lee, J.S., J.Y. Park, N.K. Kim, T.B. Nguyen, M. Kim, E.J. Woo & C. Kim. 2020. A survey of macrofungal diversity in Da-Lat, southern Vietnam. *The Korean Journal of Mycology* 48:135–149.

Lee, S.M.L. 2022. The status of bioluminescent fungal species in Singapore. *Nature in Singapore* 1:e2022124.

Lee, S.M.L. & A.M.F. Choong. 2023. *A Guide to Macrofungi in Singapore*. National Parks Board. Singapore Botanic Gardens, Singapore.

Lee, S.S., S.A. Alias, E.G.B. Jones, N. Zainuddin & H.T. Chan. 2012. *Checklist of Fungi of Malaysia*. Forest Research Institute Malaysia, Malaysia.

Lee, S.S., M. Patahayah, W.S. Chong & F. Lapeyrie. 2008. Successful ectomycorrhizal inoculation of two dipterocarp species with a locally isolated fungus in Peninsular Malaysia. *Journal of Tropical Forest Science* 20:237–247.

Li, H., Y. Tian, Jr.N. Menolli, L. Ye, S.C. Karunarathna, J. Perez-Moreno, M.M. Rahman, Md.H. Rashid, P. Phengsintham, L. Rizal, T. Kasuya, Y.W. Lim, A.K. Dutta, A.N. Khalid, L.T. Huyen, M.P. Balolong, G. Baruah, S. Madawala, N. Thongklang, K.D. Hyde, P.M. Kirk, J. Xu, J. Sheng, E. Boa & P.E. Mortimer. 2021. Reviewing the world's edible mushroom species: A new evidence-based classification system. *Comprehensive Reviews in Food Science and Food Safety* 20:1982–2014.

Lim, M. & Y. Shu. 2022. *The Future Is Fungi: How Fungi Can Feed us, Heal us, Free us and Save Our World*. Thames & Hudson Australia.

Linn, W.W., D. Htwe & W.W. Than. 2024. Identification, morphology, and nutrient composition of edible fungi and insects in Myanmar. *South Florida Journal of Environmental and Animal Science* 4:e4222.

Liu, Y.S., J.K. Kumla, N. Suwannarach, P. Sysouphanthong & S. Lumyong. 2022a. Three species of *Amanita* section *Lepidella* (Amanitaceae, Agaricales) from northern Thailand. *Phytotaxa* 570:16–28.

Liu, Y.S., J.K. Liu, J. Kumla & S. Lumyong. 2022b. Two new *Amanita* species in section *Amanita* from Thailand. *Diversity* 14:e101.

Liu, Y.S., J.K. Liu, J. Kumla, N. Suwannarach & S. Lumyong. 2023. Taxonomic novelties and new records of *Amanita* subgenus *Amanitina* from Thailand. *Journal of Fungi* 9:e601.

Lu, H., H. Lou, J. Hu, Z. Liu & Q. Chen. 2020. Macrofungi: A review of cultivation strategies, bioactivity, and application of mushrooms. *Comprehensive Reviews in Food Science and Food Safety* 19:2333–2356.

Luangharn, T., S.C. Karunarathna, A.K. Dutta, S. Paloi, I. Promputtha, K.D. Hyde, J. Xu & P.E. Mortimer. 2021. *Ganoderma* (Ganodermataceae, Basidiomycota) species from the Greater Mekong Subregion. *Journal of Fungi* 7:e819.

Luangharn, T., S.C. Karunarathna, P.E. Mortimer, K.D. Hyde, N. Thongklang & J. Xu. 2019a. A new record of *Ganoderma tropicum* (Basidiomycota, Polyporales) for Thailand and first assessment of optimum conditions for mycelia production. *MycoKeys* 51:65–83.

Luangharn, T., S.C. Karunarathna, P.E. Mortimer, K.D. Hyde & J. Xu. 2019b. Additions to the knowledge of *Ganoderma* in Thailand: *Ganoderma casuarinicola*, a new record; and *Ganoderma thailandicum* sp. nov. *MycoKeys* 59:47–65.

Łuczaj, Ł., V. Lamxay, K. Tongchan, K. Xayphakatsa, K. Phimmakong, S. Radavanh, V. Kanyasone, M. Pietras & M. Karbarz. 2021. Wild food plants and fungi sold in the markets of Luang Prabang, Lao PDR. *Journal of Ethnobiology and Ethnomedicine* 17:e6.

Łysakowska, P., A. Sobota & A. Wirkijowska. 2023. Medicinal mushrooms: Their bioactive components, nutritional value and application in functional food production-a review. *Molecules* 28:e5393.

Malysheva, E., V. Malysheva, A. Alexandrova & O. Morozova. 2019. Observations on Pluteaceae in Vietnam. 1. New species and new records of *Volvariella*. *Phytotaxa* 408:233–254.

Malysheva, E. & E. Popov. 2022. Observations on Pluteaceae in Vietnam. 3. One new species of *Volvariella*. *Phytotaxa* 545:278–286.

Manassila, M., T. Sooksa-Nguan, N. Boonkerd, S. Rodtong & N. Teaumroong. 2005. Phylogenetic diversity of wild edible *Russula* from northeastern Thailand on the basis of internal transcribed spacer sequence. *Science Asia* 31:323–328.

Mohd-Aris, A. & N.F. Amir. 2024. Nutritional composition of *Volvariella volvacea* grow using different cultivation techniques and substrate utilization. *Journal of Academia* 12:132–138.

Moniruzzaman, M., D.T.C. Tuong, S. Chin, S. Anjana, A. Karthikeyan, T. Min & I.J. Kim. 2025. A review on pharmacological insights of edible and medicinal mushroom based β-glucans. *Applied Biological Chemistry* 68:e41.

Mortimer, P.E., S.C. Karunarathna, Q. Li, H. Gui, X.-Q. Yang, X.-F. Yang, J. He, L. Ye, J. Guo, H. Li, P. Sysouphanthong, D. Zhou, J. Xu & K.D. Hyde. 2012. Prized edible Asian mushrooms: Ecology, conservation and sustainability. *Fungal Diversity* 56:31–47.

Mueller, G.M., J.P. Schmit, P.R. Leacock, B. Buyck, J. Cifuentes, D.E. Desjardin, R.E. Halling, K. Hjortstam, T. Iturriaga, K-H. Larsson, D.J. Lodge, T.W. May, D. Minter, M. Rajchenberg, S.A. Redhead, L. Ryvarden, J.M. Trappe, R. Watling & Q. Wu. 2007. Global diversity and distribution of macrofungi. *Biodiversity and Conservation* 16:37–48.

Myers, N., R.A. Mittermeier, C.G. Mittermeier, G.A.B. da Fonseca & J. Kent. 2000. Biodiversity hotspots for conservation priorities. *Nature* 403:853–858.

Nacua, A.E., H.Y.M. Pacis, J.R. Manalo, C.J.M. Soriano, N.R.N. Tosoc, R. Padirogao, K.J.E. Clemente & C.C. Deocaris. 2018. Macrofungal diversity in Mt. Makiling forest reserve, Laguna, Philippines: With floristic update on roadside samples in Makiling Botanic Gardens (MBG). *Biodiversitas* 19:1579–1585.

Naksuwankul, K., A. Thongbor, C. Chantharasena, W. Khottawong, S. Parnmen, N. Nooron, S. Sikaphan, S. Leudang, P. Kongdang, M. Khamenkhetkarn, J. Chaichana, C. Ong-ardnarong & S. Polrat. 2022. Identification by morphological and local wisdom and distribution of poisonous and edible mushroom in Thailand. *Burapha Science Journal* 27:66–84.

Nallathamby, N., C.W. Phan, S.L.S. Seow, A. Baskaran, H. Lakshmanan, S.N. Abd Malek & V. Sabaratnam. 2018. A status review of the bioactive activities of tiger milk mushroom *Lignosus rhinocerotis* (Cooke) Ryvarden. *Frontiers in Pharmacology* 8:e998.

Niego, A.G.T., S. Rapior, N. Thongklang, O. Raspé, K.D. Hyde & P. Mortimer. 2023. Reviewing the contributions of macrofungi to forest ecosystem processes and services. *Fungal Biology Reviews* 44:e100294.

Niego, A.G.T., P. Sysouphanthong, N. Thongklang, K.D. Hyde, M. Phonemany, R. Phokkamsak & O. Raspé. 2020. A new species of *Volvariella* and the first record of *Volvariella pulla* (Agaricales: Incertae sedis) from Thailand. *Phytotaxa* 480:237–250.

Nordin, N.A.M., N.A.I.M. Zainudin, W.M. Syazwan, N.A. Abd Aziz, M.T. Yusof & N.S. Ramli. 2024. Unveiling the diversity and ecological roles of macrofungi in Ayer Hitam Forest Reserve, Selangor, Malaysia. *Malaysian Applied Biology* 53:59–70.

Nurzahra, V.Y., M. Muzazzinah & M. Indrowati. 2025. Diversity of macrofungi (Ascomycota and Basidiomycota) in the Banyak Mountain Forest Area, Sragen District, Indonesia. *Nusantara Bioscience* 17:39–48.

Paloi, S., J. Kumla, S.C. Karunarathna, S. Lumyong & N. Suwannarach. 2023a. Taxonomic and phylogenetic evidence reveal two new *Russula* species (Russulaceae, Russulales) from northern Thailand. *Mycological Progress* 22:e72.

Paloi, S., J. Kumla, B.P. Paloi, S. Srinuanpan, S. Hoijang, S.C. Karunarathna, K. Acharya, N. Suwannarach & S. Lumyong. 2023b. Termite mushrooms (*Termitomyces*), a potential source of nutrients and bioactive compounds exhibiting human health benefits: A review. *Journal of Fungi* 9:e112.

Paloi, S., J. Kumla, W. Phonrob, B.P. Paloi & N. Suwannarach. 2025. Unveiling the hidden diversity of *Termitomyces* (Lyophyllaceae, Agaricales) in northern Thailand: Identification of five new species and the first report of *Termitomyces acriumbonatus*. *Journal of Fungi* 11:e830.

Paul, V., P. Tang, C. Jocelyn, M. Ajik, A. Chung & N. Midin. 2019. The preliminary checklist and notes on macrofungi from the Rainforest Discovery Centre, Sandakan, Sabah. *Sepilok Bulletin* 28:1–17.

Petersen, J.H. 2012. *The Kingdom of Fungi*. Princeton University Press, Princeton, NJ.

Pham, N.D.H., H. Takahashi, T. Fukiharu, K. Shimizu, B.D. Le & A. Suzuki. 2012. *Phlebopus spongiosus* sp. nov. (Boletales, Boletinellaceae) with a sponge-like tissue. *Mycotaxon* 119:27–34.

Phonemany, M., A.G. Niego & N. Thongklang. 2024. Pleurotoid (Basidiomycetes) from Laos and Thailand: A review of their taxonomy, distribution, cultivation, nutrition and bioactivities. *Fungal Biotec Journal* 4:12–35.

Phonemany, M., B. Raghoonundon, T. Luangharn, S.M. Tang, K.D. Hyde & N. Thongklang. 2021. A mini review on the potential pharmacological properties, cultivation, and market value of edible *Lentinus* mushrooms (Polyporaceae). *Fungal Biotec Journal* 1:49–64.

Phosri, C., M.P. Martín, P. Sihanonth, A.J. Whalley & R. Watling. 2007. Molecular study of the genus *Astraeus*. *Mycological Research* 111:275–286.

Phosri, C., R. Watling, M.P. Martin & A.J.S. Whalley. 2004. The genus *Astraeus* in Thailand. *Mycotaxon* 89:453–464.

Phu, T.T., D.H. Hanh & N.T. Trung. 2024. Study on species diversity of macrofungi belonging to the phylum Basidiomycota in Son Tra Nature Reserve, Da Nang. *Vietnam Journal of Agricultural Sciences* 21:1635–1646.

Pukahuta, C., W. Palasarn & A. Pimmongkol. 2009. Taxonomic study of wild edible pore fungi found in Ubon Ratchathani. *The Microscopy Society of Thailand* 23:30–33.

Putra, I.P., N. Nurdebyandaru, M.P. Amelya & R. Hermawan. 2022. Current checklist of local names and utilization information of Indonesian wild mushrooms. *Journal of Tropical Biodiversity and Biotechnology* 7:1–14.

Raghoonundon, B., D. Gonkhom, M. Phonemany, T. Luangharn & N. Thongklang. 2021. Nutritional content, nutraceutical properties, cultivation methods and economical importance of *Lentinula*: A review. *Fungal Biotec Journal* 1:88–100.

Raman, J., K.Y. Jang, Y.L. Oh, M. Oh, J.H. Im, H. Lakshmanan & V. Sabaratnam. 2020. Cultivation and nutritional value of prominent *Pleurotus* spp.: An overview. *Mycobiology* 49:1–14.

Rangsinth, P., R. Sharika, N. Pattarachotanant, C. Duangjan, C. Wongwan, C. Sillapachaiyaporn, S. Nilkhet, N. Wongsirojkul, A. Prasansuklab, T. Tencomnao, G.P-H. Leung & S. Chuchawankul. 2023. Potential beneficial effects and pharmacological properties of ergosterol, a common bioactive compound in edible mushrooms. *Foods* 12:e2529.

Rathore, H., S. Prasad & S. Sharma. 2017. Mushroom nutraceuticals for improved nutrition and better human health: A review. *PharmaNutrition* 5:35–46.

Repáč, I. 2011. Ectomycorrhizal inoculum and inoculation techniques. In M. Rai & A. Varma (ed.), *Diversity and Biotechnology of Ectomycorrhizae*, pp. 43–63. Springer-Verlag, Berlin, Germany.

Retnowati, A., S.A. Faulina, H. Helbert, R. Riffiani, I. Saskiawan, A. Hidayat, M. Turjaman, S.S. Hakim, L.M. Rahayu, A. Aryanto, R.S. Ningrum, D. Girmansyah, T. Rianto, B. Soesmardi, I.L. Surahmat, I. Hidayat, W. Kuswanda, H.S. Alikodra, Y. Silfariani & F.A.H. Nurlhida. 2025. *Morchella rinjaniensis*: A novel species of tropical *Morchella* (Ascomycota, Pezizales, and Morchellaceae) discovered in UNESCO Rinjani-Lombok Biosphere Reserve, Indonesia. *Mycobiology* 53:367–378.

Rodda, M. 2012. Night encounters with bioluminescent mushrooms. *Gardenwise* 38:16–17.

Ruksawong, P. & T.W. Flegel. 2001. *Thai Mushrooms and Other Fungi*. BIOTEC, Thailand.

Samsudin, N.I.P. & N. Abdullah. 2019. Edible mushrooms from Malaysia; a literature review on their nutritional and medicinal properties. *International Food Research Journal* 26:11–31.

Sanmee, R. 2004. *Biodiversity, Host Range and Growth of Edible Ectomycorrhizal in Upper Northern Thailand*. PhD's Thesis, Chiang Mai University, Thailand.

Sanmee, R., B. Dell, P. Lumyong, K. Izumori & S. Lumgyong. 2003. Nutritive value of popular wild edible mushroom from Northen Thailand. *Food Chemistry* 82:527–532.

Seehanan, S. & V. Petcharat. 2008. Some species of wild boletes in Thailand. *Journal of Agricultural Technology* 4:109–118.

Seelan, S.J.S., C.S. Yee, F.S. Fui, M. Dawood & Y.S. Tan. 2020. New Species of *Termitomyces* (Lyophyllaceae, Basidiomycota) from Sabah (Northern Borneo), Malaysia. *Mycobiology* 48:95–103.

Srijumpa, N., A. PAyapanon & C. Phosri. 2008. Comparison of *Lentinula edodes* (shiitake) strains in Northern Thailand. *Thai Agricultural Research Journal* 26:255–263.

Srikram, A. & S. Supapvanich. 2016. Proximate compositions and bioactive compounds of edible wild and cultivated mushrooms from Northeast Thailand. *Agriculture and Natural Resources* 50:432–436.

Stallman, J.K., D. Haelewaters, R.A.K. Bach, M. Brann, S. Fatemi, P. Gomez-Zapata, D.R. Husbands, B. Jumbam, P.J. Kaishian, A. Moffitt & M.C. Aime. 2024. The contribution of tropical long-term studies to mycology. *IMA Fungus* 15:e35.

Stibig, H.J., F. Achard, S. Carboni, R. Raši & J. Miettinen. 2014. Change in tropical forest cover of Southeast Asia from 1990 to 2010. *Biogeosciences* 11:247–258.

Sumarouw, A.A.D., I. Yafi, F.K. Vansha & N. Noverita. 2024. Macro fungi diversity in The Sanggabuana Mountain Area, Karawang Regency, Tegalwaru District, West Java. *Journal of Tropical Biodiversity* 4:58–66.

Surawut, S., S. Nak-eiam, C. Kunsook, A. Khamchatra, N. Khamchatra, T. Jenkarn, K. Chanapon, O. Boonmee, C. Seeprasong & P. Kanjanavas. 2025. Diversity of macrofungi in the nature trail of Namtok Phlio National Park, Chanthaburi Province, Thailand. *Current Applied Science and Technology* 25:e0263544.

Suwannarach, N., J. Kumla, S. Khuna, N. Wannathes, N. Thongklang & P. Sysouphanthong. 2022. History of Thai mycology and resolution of taxonomy for Thai macrofungi confused with Europe and American names. *Chiang Mai Journal of Science* 49:654–683.

Suwannarach, N., J. Kumla, W. Lu, A. Kaewnunta, N. Thongklang, P. Sysouphanthong, Y-S. Liu, T. Luangharn, N. Wannathes, A. Walker, N. Kraisitudomsook, P. Nutaratat, S. Wongkanoun, E. Charria-Girón, S.C. Karunarathna, W. Pathom-aree & S. Lumyong. 2025. Revised and updated checklist of macrofungi in Thailand–II. *Chiang Mai Journal of Science* 52:e2025027.

Suwannarach, N., J. Kumla & S. Lumyong. 2015. A new whitish truffle, *Tuber thailandicum* from northern Thailand and its ectomycorrhizal association. *Mycological Progress* 14:e83.

Suwannarach, N., J. Kumla, J. Meerak & S. Lumyong. 2017. *Tuber magnatum* in Thailand, a first report from Asia. *Mycotaxon* 132:635–642.

Suwannarach, N., J. Kumla, S. Vadthanarat, O. Raspé & S. Lumyong. 2016. Morphology and molecular evidence support a new truffle, *Tuber lannaense*, from Thailand. *Mycological Progress* 15:827–834.

Swallah, M.S., P. Bondzie-Quaye, Y. Wu, A. Acheampong, F.L. Sossah, S.M. Elsherbiny & Q. Huang. 2023. Therapeutic potential and nutritional significance of *Ganoderma lucidum*—a comprehensive review from 2010 to 2022. *Food and Function* 14:1812–1838.

Sysouphanthong, P., K.D. Hyde, E. Chukeatirote, A.H. Bahkali & E.C. Vellinga. 2011. *Lepiota* (Agaricales) in northern Thailand–1. L. section *Stenosporae*. *Mycotaxon* 117:53–85.

Sysouphanthong, P., K.D. Hyde, E. Chukeatirote, A.H. Bahkali & E.C. Vellinga. 2012. *Lepiota* (Agaricales) in Northern Thailand-2 *Lepiota* section *Lepiota*. *Cryptogamie Mycologie* 33:25–42.

Sysouphanthong, P., W. Taliam, N. Wannasawang & N. Thongklang. 2023. The genus *Lentinus* in Thailand: Taxonomy, cultivation tests, nutritional analysis and screening for the biological activity of wild strains. *Current Research in Environmental & Applied Mycology* 13:136–161.

Sysouphanthong, P., N. Thongklang, J.K. Liu & E.C. Vellinga. 2021. Description of lepiotaceous fungal species of the genera *Chlorophyllum*, *Clarkeinda*, *Macrolepiota*, *Pseudolepiota*, and *Xanthagaricus*, from Laos and Thailand. *Diversity* 13:e666.

Tadiosa, E.R. & L.C. Lubos. 2019. Common macroscopic fungi and mosses of North Central Mindanao. *Asian Journal of Biodiversity* 10:26–137.

Tang, S.M., E. Li, H.M. Luo, C.C. Ao, T. Lv & S.H. Li. 2024. *Termitomyces flavus* sp. nov. (Lyophyllaceae, Agaricales), a new species from Northern Thailand. *Chiang Mai Journal of science* 51:e2024060.

Tawatsin, A., S. Parnmen, U. Thavara, P. Siriyasatien & P. Kongtip. 2018. Mushroom poisoning in Thailand: Incidence and intoxication to human health. *Medical Research Archives* 6:1–12.
Tham, F.Y. & R. Watling. 2017a. *Annotated Checklist of Macrofungi Recorded from Singapore: Bukit Timah*. Singapore.
Tham, F.Y. & R. Watling. 2017b. *Annotated Checklist of Macrofungi Recorded from Singapore: MacRitchie-Pierce*. Singapore.
Tham, F.Y. & R. Watling. 2017c. *Annotated Checklist of Macrofungi Recorded from Singapore: Mandai-Seletar*. Singapore.
Tham, F.Y. & R. Watling. 2017d. *Annotated Checklist of Macrofungi Recorded from Singapore: Singapore Botanic Gardens*. Singapore.
Thaung, M.M. 2007. A preliminary survey of macromycetes in Burma. *Australian Mycologist* 26:16–36.
Thawthong, A., S.C. Karunarathna, N. Thongklang, E. Chukeatirote, P. Kakumyan, S. Chamyuang, L.M. Rizal, P.E. Mortimer, J. Xu, P. Callac & K.D. Hyde. 2014. Discovering and domesticating wild tropical cultivatable mushrooms. *Chiang Mai Journal of Science* 41:731–764.
Thongbai, B., R.E. Tulloss, S.L. Miller, K.D. Hyde, J. Chen, R. Zhao & O. Raspe. 2016. A new species and four new records of *Amanita* (Amanitaceae; Basidiomycota) from Northern Thailand. *Phytotaxa* 286:211–231.
Thongklam, S. 2008. *Diersity of Boletes in Some National Park of Upper Northern Thailand*. PhD's Thesis, Chiang Mai University, Thailand.
Thongklang, N. & T. Luangharn. 2016. Testing agricultural wastes for the production of *Pleurotus ostreatus*. *Mycosphere* 7:766–772.
Thu, Z.M., K.K. Myo, H.T. Aung, M. Clericuzio, C. Armijos & G. Vidari. 2020. Bioactive phytochemical constituents of wild edible mushrooms from Southeast Asia. *Molecules* 25:e1972.
Torres-Farradá, G., S. Thijs, F. Rineau, G. Guerra & J. Vangronsveld. 2024. White rot fungi as tools for the bioremediation of xenobiotics: A review. *Journal of Fungi* 10:e167.
Tran, M.V. & B.Q. Truong. 2005. *Nấm lớn Cúc Phương* [The Macrofungi in Cuc Phuong], p. 244. Nhà Xuất Bản Nông Nghiệp, Hà Nội. [In Vietnamese].
Truyen, D.M. & F.F. Patacsil. 2017. Survey the composition and distribution of fungi species in the natural reserve Wetland Lung Ngoc Hoang, Vietnam. *Journal of Advances in Technology and Engineering Research* 3:19–26.
Turjaman, M., Y. Tamai, H. Segah, S.H. Limin, J.Y. Cha, M. Osaki & K. Tawaraya. 2005. Inoculation with the ectomycorrhizal fungi *Pisolithus arhizus* and *Scleroderma* sp. improves early growth of *Shorea pinanga* nursery seedlings. *New Forests* 30:67–73.
Turner, I.M., T.K. Tan & D.J. Metcalfe. 1994. Fungi: Zygomycetes, ascomycetes, basidiomycetes and deuteromycetes. In Y.C. Wee & P.K.L. Ng (eds.), *A First Look at Biodiversity in Singapore*. Singapore: National Council on the Environment.
Vadthanarat, S., M. Amalfi, R.E. Halling, V. Bandala, S. Lumyong & O. Raspé. 2019a. Two new *Erythrophylloporus* species (Boletaceae) from Thailand, with two new combinations of American species. *MycoKeys* 55:e29.
Vadthanarat, S., R.E. Halling, M. Amalfi, S. Lumyong & O. Raspé. 2021. An unexpectedly high number of new *Sutorius* (Boletaceae) species from northern and northeastern Thailand. *Frontiers in Microbiology* 12:e643505.
Vadthanarat, S., S. Lumyong & O. Raspé. 2019b. *Cacaoporus*, a new Boletaceae genus, with two new species from Thailand. *MycoKeys* 54:1–29.
van de Putte, K., J. Nuytinck, D. Stubbe, H. Thanh & A. Verbeken. 2010. *Lactarius volemus* sensu lato (Russulales) from northern Thailand: Morphological and phylogenetic species concepts explored. *Fungal Diversity* 45:99–130.
Venturella, G., V. Ferraro, F. Cirlincion & M.L. Gargano. 2021. Medicinal mushrooms: Bioactive compounds, use, and clinical trials. *International Journal of Molecular Sciences* 22:e634.
Wannathes, N., D.E. Desjardin, K.D. Hyde, B.A. Perry & S. Lumyong. 2009a. A monograph of *Marasmius* (Basidiomycota) from Northern Thailand based on morphological and molecular (ITS sequences). *Fungal Diversity* 37:209–306.
Wannathes, N., D.E. Desjardin & S. Lumyong. 2009b. Four new species of *Marasmius* section *Globulares* from Northern Thailand. *Fungal Diversity* 36:155–163.
Watling, R. 2001. Edred John Henry Corner (1906–1996): A pioneer in tropical mycology. *Mycological Research* 105:1533–1536.
Watling, R. 2013. Foray in Thailand. *Fungi* 6:45–46.

Wijaya, N.H., A.D. Savitri, A.T. Wahyuni, E.S. Alhadad, N. Edo, A Shabrina, M.A. Farid, S.S. Basri, A.R. Febrianti, D.A.R. Putri & I.A. Pratiwi. 2021. Wild mushrooms diversity in tropical rainforest. *Ecology Environment and Conservation* 27:622–627.

Wikramanayake, E., E. Dinerstein, C. Loucks, D. Olson, J. Morrison & J. Lamoreux. 2002. *Terrestrial Ecoregions of the Indo-Pacific: A Conservation Assessment*, p. 643. Island Press, Washington.

Wisitrassameewong, K., S. Adamčík, K. Adamčíková, S.M. Tang, N. Tangthirasunun, B. Chuankid & O. Raspé. 2025. *Russula orientalovirescens* sp. nov., a common Southeast Asian edible fungus is different from the European look-alike *R. virescens*. *PLoS One* 20:e0322545.

Wisitrassameewong, K., C. Manz, F. Hampe, B.P. Looney, T. Boonpratuang, A. Verbeken, T. Thummarukcharoen, T. Apichitnaranon, M. Pobkwamsuk, M. Caboň & S. Adamčík. 2022. Two new *Russula* species (fungi) from dry dipterocarp forest in Thailand suggest niche specialization to this habitat type. *Scientific Reports* 12:e2826.

Wongchalee, P. & C. Pukahute. 2012. Diversity of mushrooms in dry dipterocarp forest at Phuphan National Park, Sakon Nakhon Province. *Natural Science* 4:1153–1160.

Woodruff, D.S. 2010. Biogeography and conservation in Southeast Asia: How 2.7 million years of repeated environmental fluctuations affect today's patterns and the future of the remaining refugial-phase biodiversity. *Biodiversity and Conservation* 19:919–941.

Wu, S.H., C.L. Chern, C.L. Wei, Y.P. Chen, M. Akiba & T. Hattori. 2020. *Ganoderma bambusicola* sp. nov. (Polyporales, Basidiomycota) from southern Asia. *Phytotaxa* 456:75–85.

Yeni, L.F., M.A. Faturrahman, D.A. Widyastuti & L.A. Syakuran. 2025. Traditional knowledge of mushroom utilization by the Dayak Kubin Tribe: A case study in Nanga Raku Village, West Kalimantan. Biosaintifika: *Journal of Biology & Biology Education* 17:50–60.

Yerisetouw, Y., S. Sufaati, E.K. Raunsay & D.Y. Runtuboi. 2023. Diversity of mushrooms in areas sago, secondary, and swamp forest in village Khameyaka, Ebungfau District, Jayapura Regency. *Asian Journal of Natural Sciences* 2:109–126.

Yusran, Y., E. Erniwati & R. Rukmi. 2024. Ethnomycology of wild edible mushrooms by the Bunggu tribe in West Sulawesi, Indonesia. *International Journal of Design & Nature and Ecodynamics* 19:2097–2107.

Yusran, Y., E. Erniwati, D. Wahyuni, R. Ramadhanil & A. Khumaidi. 2021. Diversity of macro fungus across three altitudinal ranges in Lore Lindu National Park, Central Sulawesi, Indonesia and their utilization by local residents. *Biodiversitas* 22:199–210.

Yuwa-Amornpitak, T., T. Vichitsoonthonkul, M. Tanticharoen, S. Cheevadhanarak & S. Ratchadawong. 2006. Diversity of ectomycorrhizal fungi on Dipterocarpaceae in Thailand. *Journal of Biological Sciences* 6:1059–1064.

Yuwa-Amornpitak, T. & P. Yeunyaw. 2020. Diversity wild mushrooms in the community forest of Na Si Nuan Sub-District, Thailand. *Journal of Biochemical Technology* 11:28–36.

Zhao, R.L., S. Karunarathna, O. Raspé, L.A. Parra, J. Guinberteau, M. Moinard, A. de Kesel, G. Barroso, R. Courtecuisse, K.D. Hyde, A.K. Guelly, D.E. Desjardin & P. Callac. 2011. Major clades in tropical Agaricus. *Fungal Diversity* 51:279–296.

Zhou, X.W., W.R. Cong, K.Q. Su & Y.M. Zhang. 2013. Ligninolytic enzymes from *Ganoderma* spp: Current status and potential applications. *Critical Reviews in Microbiology* 39:416–426.

Zikriyani, H., I. Saskiawan & W. Mangunwardoyo. 2018. Utilization of agricultural waste for cultivation of paddy straw mushrooms (*Volvariella volvacea* (Bull.) Singer 1951). *International Journal of Agricultural Technology* 14:805–814.

11 Diversity, Distribution, Threats, and Conservation Status of Macrofungi Communities in Kenya
A Synthesis of Literature and Citizen Science Data

Mary Nyawira Muchane

11.1 INTRODUCTION

Fungi represent the largest component of global biological diversity and contribute significantly to the development, functioning, and sustainability of other forms of life on Earth (Hasan and Gupta 2012). The total fungal diversity is estimated at 2.0 to 3.8 million species (Hawksworth and Lücking 2017). Because of their unique characteristics such as heterotrophy and filamentous growth, fungi are classified in their own kingdom, the kingdom Fungi (Wijayawardene et al. 2024). Species of fungal can be broadly grouped into microfungi and macrofungi. Microfungi (including yeasts and molds) possess microscopic spore-producing structures and make up about 90% of the total fungal diversity. Macrofungi (mushrooms), by contrast, are fungi with conspicuous fruiting bodies (sporocarps) and extensive belowground or within substrate mycelial networks (Chang and Miles 1992). They represent about 10% of global fungal diversity, with an estimated 110,000 to 140,000 species (Hawksworth 2001; Müller et al. 2007; Hawksworth 2012).

Macrofungi are classified within the kingdom Fungi, primarily under the division Basidiomycota, with a few members belonging to the division Ascomycota. They occur in diverse forms such as mushrooms (or agarics), puffballs, coral fungi, polypores, and boletes and are distributed across various ecosystems including all terrestrial landscapes, wetlands, and even saline desert habitats (Niego et al. 2023). They represent highly diverse biological resources that play a critical role in sustaining plant communities by regulating nutrient cycling. They decompose dead plant and animal matter, thereby releasing nutrients back into the environment and making these available to plants, while also providing food resources for insects, small mammals, and soil microbes (López-Quintero et al. 2012; Ambrosio et al. 2015). Specifically, macrofungi maintain ecosystem balance by releasing carbon dioxide (CO_2) into the atmosphere during decomposition and returning essential nutrients such as nitrogen and phosphorus to the soil (Niego et al. 2023), thus contributing directly to soil fertility and plant productivity. Beyond their ecological functions, macrofungi contribute significantly to rural economies, food security, and human nutrition through their use as food, medicine, and in local trade (Chang 1996; Chadha and Sharmam 1995; Gupta 1986).

Despite their importance, the paucity of knowledge on fungal diversity and ecological roles remains a major gap in our knowledge. Addressing this gap is crucial for harnessing their ecological, economic, and cultural potential to sustain stable and functional ecosystems. To date, only

DOI: 10.1201/9781003647140-11

about 10% of fungi have been formally described, while nearly 80% of global fungal biodiversity remains poorly understood (Müller et al. 2007). Much of this undiscovered diversity is assumed to occur in tropical regions, which provide favorable environmental conditions year-round and host a higher diversity of vascular plants that create niches and microhabitats for fungi (Hawksworth 2001; Kark 2007). However, the number of species described does not reflect the actual richness, largely due to limited documentation and conservation efforts, even as these ecosystems face increasing threats from forest degradation and climate change. Advancing the understanding of macrofungal communities in tropical ecosystems is therefore critical not only for their management, protection, and conservation, but also for enhancing their sustainable utilization in supporting livelihoods, biodiversity, and ecosystem resilience.

11.2 MACROFUNGI RESEARCH IN KENYA

Kenya, one of the tropical countries in Africa, is renowned for its rich and diverse flora and fauna. Geographically, the country lies within the equatorial zone of eastern Africa, bordering the Indian Ocean between Somalia and Tanzania, with Uganda to the west, South Sudan and Ethiopia to the north, Somalia to the east and Tanzania to the south. It is located between latitudes 4°N and 4°S and longitudes 34°E and 42°E, covering an area of about 580,000 km² (225,000 sq. miles). The terrain ranges from the low-lying coastal plains along the Indian Ocean to a series of mountain ridges and plateaus, which rise above 3,000 meters (9,000 ft.) in central Kenya. The Great Rift Valley bisects the country north of Nairobi, stretching into a broad arid plain in the north. To the south, mountain plains extend before descending toward the shores of Lake Victoria in the west. The country's highest elevation is Mount Kenya at 5,199 meters.

The country comprises ten of the world's 14 biogeographical biomes, making it one of the richest biotic regions in Africa, the tropics, and globally (MEWNR 2015). This remarkable biological diversity is largely attributed to its wide altitudinal gradients, ranging from the coastal lowlands to the snow-clad peaks of Mt. Kenya at 5,199 m above sea level, as well as the presence of unique and distinctive biomes. These biomes include the coastal forests (Arabuko Sokoke and the lower Tana River), afromontane forests (Mt. Kenya, Aberdares, and Mt. Elgon), rainforest (Kakamega rainforest, the easternmost outlier of the Guineo-Congolian equatorial forests), the Somali-Maasai zone (Afro-tropical grassland and highlands biome), the Victoria Basin biome and the Sudan and Guinea Savannah biomes. Collectively, these ecosystems harbor exceptionally high levels of biodiversity, including approximately 5,245 species of animals and 7,000 to 8,000 species of plants, many of which are endemic, rare, endangered, or threatened (MEWNR 2015; Zhou et al. 2017).

Fungal diversity in Kenya, though less studied, is equally significant. The Kenya Biodiversity Atlas (2015) estimated that about 5,000 species of fungi and lichens occur in the country (MEWNR 2015). However, this number is likely much higher when extrapolated using the plant-to-macrofungi ratio of 1:33 that has been suggested for tropical regions (Srivastava 2021). Given Kenya's vast range of vegetation types, climatic zones and soil conditions, the actual diversity of fungi may be substantially underestimated.

Mycological research in Kenya dates back to the eighteenth century. In brief, the first checklist of fungi, which contained four species of agarics, appeared in 1926 (Pegler 1977). This list was later expanded through the works of Otieno (1966, 1968), Dring and Rayne (1966), and Pegler and Rayner (1969). Pegler (1977) produced a preliminary agaric flora of East Africa, a guide to the agaricoid macrofungi of East Africa that documented 398 species across Kenya, Tanzania, and Uganda. Kost (2002) later documented 162 species of macrofungi from indigenous forests in Kijabe (Aberdares), Naru Moro (Mount Kenya), Kakamega Forest and Mount Elgon, including some records from pine (*Pinus*) plantations. More recent studies have enriched this knowledge, recording 45 species from Maasai Mara, 12 species from the Coast, 224 species from the Aberdare Forest, and 445 species from Lake Olbolosaat ecosystems (Tibuhwa et al. 2011; Gateri et al. 2014; Njuguini et al. 2018; Muchane et al. 2021, 2025). Additional surveys have reported macrofungi from the Mt. Kenya, the

Tana River Basin, Elgeyo Marakwet, the Nyanza region, the Lolldaiga ecosystem, Ewaso Narok, and the Aberdare Forest. Results from these studies are available in technical reports, while macrofungal specimens are preserved at the East African Herbarium fungarium (Muchane et al. 2016; Nyaga et al. 2009; Muchane and Waithaka 2023, 2024; Muchane et al. 2019, 2021, 2022, 2025). For a comprehensive overview, see Muchane et al. (2022).

In addition, online databases have become important sources of fungal records (Haelewaters et al. 2024). For instance, iNaturalist, a global online social network, citizen science platform, and community database for recording and sharing observations of plants, animals, and fungi, has documented extensive information on macrofungi and lichen communities in Kenya. As of now, the iNaturalist Mushrooms of Kenya portal lists 7,171 macrofungi observations, representing 563 species (https://www.inaturalist.org/projects/mushrooms-of-kenya). Similarly, the GBIF database has compiled macrofungal records from various Kenyan research projects and citizen science contributions (Muchane et al. 2025).

The purpose of this review is to describe the macrofungal communities within Kenyan ecosystems and endeavors to match this information to different vegetation types across biomes. Plant species composition, soil characteristics and microclimatic conditions strongly influence the type and abundance of macrofungi present since many fungi form obligate symbiotic associations such as ectomycorrhizae while others thrive as saprobes on specific plant litter or woody debris. This highlights the need to understand fungal species alongside plant and animal diversity, as they form critical components of ecosystem functioning and resilience. Such information not only provides a framework for comparing macrofungal diversity across other tropical regions but is also vital for biodiversity conservation, sustainable land management, and for predicting the impacts of climate change on ecosystem structure and functioning.

11.3 MACROFUNGI COMMUNITIES IN KENYA

Underpinning and providing the foundation for Kenya's rich plant and animal diversity is the remarkable abundance of fungi. The actual number of species of macrofungi in Kenya is still unknown. Based on the plant-fungi ratio of 1:33 in tropical regions compared to 1:6 in temperate regions (Srivastava 2021), Muchane et al. (2022) estimated that Kenyan ecosystems may host about 264,000 species of fungi, derived from the 8,000 species of plants documented for the country (Zhou et al. 2017). Using the global assumption that macrofungi constitute approximately 10% of total fungal diversity (Rossman 1994), this translates to an estimated 26,400 species of macrofungi occurring in Kenya. Despite this potential richness, only 1,100 species of macrofungi have been documented in Kenya, with roughly 753 species identified to the level of species (Muchane et al. 2022). This suggests that nearly 96% of macrofungi in Kenya remain undocumented. However, the number of species recorded in Kenya is comparable to those reported for other tropical countries such as Colombia (1,231 species) and Nepal (2,391 species) (Vasco-Palacios and Franco-Molano 2013; Devkota and Aryal 2020), but are much higher than figures from Ethiopia (66–188 species), Cambodia (302) and Guatemala (350) (Dejene et al. 2017; Kim et al. 2017; Arzú et al. 2012; Kewessa et al. 2024). This striking discrepancy underscores a critical knowledge gap in fungal diversity within tropical regions, a gap that constrains the development of effective management, sustainable utilization, and conservation strategies.

Macrofungi are classified into two major phyla, the Basidiomycota and Ascomycota. The phylum Basidiomycota comprises a highly diverse group of fungi that includes macrofungi, such as mushrooms, polypores fungi, club fungi, puffballs, stinkhorns, earthstars, bird's nest fungi, and jelly fungi, along with some microfungi, such as yeasts, smuts, and rusts. Scientifically, members of this phylum are characterized by the sexual production of basidiospores on specialized cells called "basidia," as well as the presence of walled and septate hyphae with dolipore septa. Ecologically, members of the Basidiomycota exhibit diverse feeding habits, functioning as saprotrophs (decomposers of organic matter), symbionts (notably ectomycorrhizal partners with plants), and parasites (including pathogens of crops and trees). The phylum Basidiomycota is estimated to contain approximately 41,000 species of fungi (He et al. 2019).

Members of the phylum Basidiomycota account for the majority of species of macrofungi in Kenya, representing approximately 94% of the total documented diversity, distributed across 234 genera and 64 families (Muchane et al. 2022). Within this phylum, the class Agaricomycetes is dominant, comprising about 89% of the species. At the order level, the Agaricales is the largest, contributing nearly 70% of the documented taxa, followed by the Polyporales, with 12% of the total. In terms of families, the Agaricaceae is the most species-rich, with 77 species, followed by the Polyporaceae (67), Mycenaceae (60), Strophariaceae (40), Tricholomataceae (36), Marasmiaceae (34), and Psathyrellaceae (33). All other families have fewer than 30 recorded species (Figure 11.1). For a comprehensive species checklist and taxonomic details, refer to Muchane et al. (2022).

The phylum Ascomycota represents one of the most diverse fungal groups, with an estimated 65,000 species globally. Members include both macrofungi, such as morels and truffles, and numerous microfungi, including important plant pathogens and species that form lichens in association with algae or cyanobacteria. A defining characteristic of this phylum is the sexual production of ascospores within specialized sac-like cells called "asci," along with the presence of walled and septate hyphae with simple pores. Ecologically, members of the Ascomycota display a wide range of nutritional modes, functioning as saprotrophs; symbionts with trees, algae, and cyanobacteria; or as parasites of plants and other organisms.

In Kenya, macrofungal species of Ascomycota are relatively few compared to the Basidiomycota (Muchane et al. 2022). They are represented by 47 species, distributed across 31 genera, 24 families, 16 orders, and 10 classes (Table 11.1). Among the largest orders recorded are the Lecanorales (12 species), Xylariales (8 species), Pezizales (6 species), and Hypocreales (5 species). At the family level, the most species-rich groups include the Parmeliaceae (6 species), Xylariaceae (6 species), and Cordycipitaceae (4 species).

11.4 MACROFUNGI IN DIFFERENT BIOTROPHIC GROUPS

Macrofungi are generally categorized into three major biotrophic groups based on their mode of substrate utilization. The first group comprises saprotrophic species, which decompose organic matter such as leaf litter, soil organic material, and woody substrates. By recycling nutrients, these fungi play a critical role in maintaining soil fertility and supporting ecosystem productivity. Saprotrophic macrofungi, dominate macrofungal communities in Kenya and account for 83% (Figure 11.2) of total number of species (Tibuhwa et al. 2011; Njuguini et al. 2018; Muchane et al. 2021). The families represented by the most species include the Agaricaceae, Marasmiaceae, Mycenaceae, and Polyporaceae. Saprotrophic species typically occur in habitats that provide favorable conditions for decomposition. Forested ecosystems are particularly important habitats for saprotrophic fungi due to the abundance of organic substrates. Their ability to efficiently break down complex organic matter not only sustains nutrient cycling but also makes them relatively easy to domesticate and cultivate. As such, saprotrophic species also play a significant role in food security and traditional medicine.

TABLE 11.1

Summary of Species of Macrofungi Documented for Kenyan Ecosystems.

Phylum	Classes	Order	Families	Genus	Species
Basidiomycota	3	15	64	234	706
Ascomycota	10	16	24	31	47
Total					753

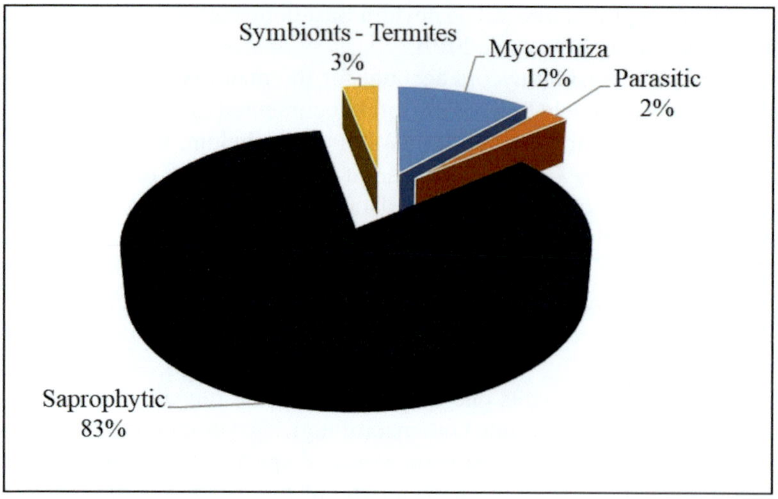

FIGURE 11.1 Quantitative distribution of macrofungi in Kenya based on mode of nutrition. Source: Adapted from Muchane et al. (2022)

The second group consists of ectomycorrhizal fungi (ECM), which form mutualistic associations with plant roots. In this symbiotic relationship, the fungi enhance the plant's uptake of essential nutrients such as nitrogen and phosphorus while receiving carbohydrates in return. This group is particularly important in forest ecosystems where they influence plant community dynamics and forest regeneration. Ectomycorrhizal (ECM) species are also prevalent in Kenya, occurring mainly in exotic plantations (*Pinus*, *Eucalyptus*, and *Acacia*) as well as in Miombo woodlands where they are associated with native trees such as *Brachystegia* and *Julbernardia* (Gateri et al. 2014). ECM species account for 12% of the total number of macrofungi, although the proportion may change depending on the study area and habitat. In exotic plantations, common ECM genera include *Suillus*, *Chroogomphus*, *Laccaria*, *Inocybe*, *Lactarius*, *Russula*, *Amanita*, *Cortinarius*, *Hebeloma*, and *Rhizopogon* (Kost 2002; Njuguini et al. 2018; Muchane and Waithaka 2023). Additional genera such as *Pisolithus* have also been reported in association with species of *Eucalyptus* and *Acacia* (Muchane et al. 2021; Muchane and Waithaka 2023). Most of these ECM fungi were introduced into Kenya during the early 1920s alongside the establishment of fast-growing trees in pine and eucalyptus plantations. Notably, pine trees are obligate hosts of ECM fungi, which are essential for their growth and survival (Smith and Read 2007). On the other hand, a considerable number of species of ECM fungi have been documented in Miombo woodlands in the coastal region, forming associations with trees in the genera *Brachystegia*, *Julbernardia*, *Isoberlinia*, *Marquesia*, and *Uapaca*. Common ECM genera in this region include *Cantharellus*, *Russula*, *Lactarius*, *Boletus*, *Lactifluus*, *Afrocastellanoa,* and *Mackintoshia* (Gateri et al. 2014). Species of ECM fungi play a particularly important role in enhancing tree nutrition by facilitating the uptake of essential minerals such as nitrogen and phosphorus, thereby improving forest health and productivity. Beyond their ecological significance, ECM fungi also contribute to local livelihoods. For example, in coastal regions of Kenya local communities actively collect edible species of ECM fungi such as *Cantharellus* and *Russula*, which are consumed as food and sold in local markets and to hotels, providing an important source of household income and food security (Gateri et al. 2014).

The third group consists of parasitic macrofungi, which colonize living or dead trees, often leading to diseases, structural weakness, or decay. Within forest ecosystems, parasitic fungi are considered a natural component as long as their populations remain below damaging thresholds. When pathogenic outbreaks occur, the fungi may directly kill trees, thereby opening forest canopies and creating light gaps that promote regeneration of shade-intolerant plant species (Molina

1994). The resulting dead wood also becomes an important nutrient source once decomposed by saprotrophic fungi, thus sustaining nutrient cycling. Although often perceived as destructive, parasitic species contribute to ecosystem functioning by regulating host populations, accelerating decomposition, and maintaining dynamic forest structure. Parasitic macrofungi represent a relatively small fraction of the fungal community, accounting for about 5% of total of all species of macrofungi. In Kenya, commonly documented genera include *Armillaria*, *Ganoderma*, and *Phellinus* (Kost 2002; Tibuhwa et al. 2011; Njuguini et al. 2018; Muchane et al. 2021). Notably, some species such as *Ganoderma applanatum* and *Phellinus gilvus* possess medicinal properties, offering bioactive compounds that could be harnessed sustainably from forest ecosystems to support the growth of pharmaceutical industries (Tapwal et al. 2013).

Other unique group consists of species of macrofungi cultivated by termites, the members of the genus *Termitomyces*. Members of this genus live symbiotically with termites of the family Termitidae and are found exclusively in certain regions of Africa and Asia (Tibuhwa et al. 2011; Vesala et al. 2017; Muchane et al. 2021). Species of *Termitomyces* (family Lyophyllaceae) in Kenya have a wide distribution with species recorded in Nyanza, Western, Central, Coastal, and Maasai Mara regions, representing about 3% of the total of all macrofungi (Tibuhwa et al. 2011; Muchane et al. 2021; Muchane et al. 2022). Although the geographical patterns of fungus-growing termites and their symbionts have not been studied systematically, reports exist of their occurrence in the savannah ecosystems of Maasai Mara and Tsavo, coastal region, the nyanza region, and the highlands of western and central Kenya (Tibuhwa et al. 2011; Gateri et al. 2014; Vesala et al. 2017; Muchane et al. 2021). Apart from assisting the proliferation of their termite hosts, species of *Termitomyces* are economically valuable natural resource, serving as an alternative to plant- and animal-derived foods and medicine. Many species of *Termitomyces* are used as food by many local Kenyan communities. According to these local communities, species of *Termitomyces* harvested from wild are cooked as stew for *ugali* (maize porridge) dishes. They are also used for medicinal purposes since they contain antioxidants, immunomodulators, antitumor substances, and substances with antimicrobials properties (Hsieh and Ju 2018).

11.5 MACROFUNGI COMMUNITIES ACROSS DIFFERENT BIOMES

Kenya has seven plant distribution regions (K1–K7) as delimited during the publication of *Flora of Tropical East Africa* (FTEA) between 1952 and 2012 (Polhill 1988). These include K1 (Northern Frontier Province) located in northern Kenya, K2 (Turkana Province) located in northwestern Kenya, K3 (Rift Valley Province) located in western Kenya, K4 (Central Province) located in central Kenya, K5 (Nyanza Province) located southwestern Kenya, K6 (Masai Province) located in southern Kenya, and K7 (Coast Province) located in the coastal region (Figure 11.2 and Table 11.2). Each of these regions represents a unique vegetation type recognizable at a regional scale and defined by geographic, climatic, and historical factors that shape its plant and fungal diversity. Since plants and fungi are ecologically interdependent, it is reasonable to assume that macrofungal diversity parallels the vegetation in these regions.

Using the iNaturalist repository of mushrooms in Kenya, macrofungi have been documented across all regions of the country (Central, Rift Valley, Nyanza, Eastern, Western, Coastal, and Northeastern Kenya) (Muchane et al. 2022). However, species richness varies considerably. Central Kenya recorded the highest diversity, accounting for 56% of the total species observed, followed by the Rift Valley (15%), Nairobi (10%), Coast (6%), Eastern (6%), Nyanza (5%), Western (3%), and Northeastern Kenya, which had the lowest representation with just 0.1% of all species. From these patterns, three major diversity hotspots can be identified. These are Central Kenya (Aberdare forests and Mt. Kenya ecosystems), Rift Valley (Mau and Cherangani forest), Nairobi highland forests and Coastal forests. This distribution likely reflects not only environmental gradients (highland to lowland transitions, rainfall variability, and vegetation types) but also differences in survey intensity and accessibility. Central Kenya, with its montane forests, is the most heavily studied region, which may partly explain the higher number of species recorded. Conversely, the arid Northeastern region remains poorly surveyed, and the very low fungal records for this region could be due to both limited sampling effort and the harsh climatic conditions that are less favorable to macrofungal growth.

Examples of macrofungi species documented across Kenya's ecological zones are presented in Table 11.2, together with their associated vegetation types and climatic conditions. Although current data remain insufficient to draw firm conclusions about macrofungal population structure, early patterns suggest regional differentiation, with certain macrofungal families restricted to specific ecological zones (Muchane et al. 2022). These preliminary trends highlight the influence of climate and vegetation gradients on fungal diversity and distribution.

For example, K1 and K2 (Arid to Semi-arid Zones) characterized by *Acacia-Commiphora* woodlands, scattered grasslands, and patches of riverine forest is characterized by species of macrofungi adapted to xeric conditions. Notably, species such as *Podaxis* and *Tulostoma* are well-represented. These fungi are adapted to dry environments, often producing long-lasting fruiting bodies with protective peridia to reduce water loss. Their presence indicates ecological specialization to withstand

TABLE 11.2

Characteristics of Different Seven Plant Distribution Regions (K1–K7) and Examples of Macrofungi Species Documented.

Region	Location	Climate	Vegetation type	Common fungi taxa
K1	Northern Frontier Province in Northern Kenya	Arid to semi-arid	Acacia–Commiphora bushland, scattered grasses and semi-desert shrubs.	*Podaxis pistillaris, Polyporaceae, Lichenized Ascomycota*
K2	Turkana Province) in Northwest Kenya	Semi-arid	Savannah-woodland bushland, grasslands, and scattered riverine forests.	*Coprinellus*, *Polyporales*, *Podaxis*, *Termitomyces*, *Ganoderma, Tulostoma, Stereum*
K3	Rift Valley Province in Western highland	Sub-humid to humid	Montane forests (Mau, Cherangani), Escarpments wooded grasslands and agricultural land.	Ectomycorrhizal fungi (*Suillus, Amanita, Russula*) in exotic pine/eucalyptus plantations, saprotrophs (Agaricaceae, *Gymnopus*, *Mycena, Marasmius*), *Termitomyces*
K4	Central Province in central Kenya	Sub-humid to humid	Aberdare Ranges, Mt. Kenya region, Moist montane and bamboo forests, afro-alpine habitats, tea/coffee farmlands.	Ectomycorrhizal fungi (*Boletus, Suillus, Lactarius*), saprotrophs (*Agaricus, Clitocybe, Gymnopus*), parasitic (*Ganoderma, Phellinus*), *Termitomyces*.
K5	Nyanza Province in Lake Victoria	Sub-humid to humid	Kakamega forest (Guineo-Congolian rainforest relic), farmlands and wetlands.	Saprotrophs (Agaricaceae, *Coprinopsis, Psathyrella, Mycena, Marasmius, Ganoderma, Trametes, Polyporus*) and *Termitomyces*
K6	Masai Province in Southern Kenya	Semi-arid	Savannah, dry forests, grasslands, acacia woodlands, and dry bush	Termitomyces, saprotrophic (Agaricaceae, *Coprinus, Panaeolus, Polypores, Psilocybe*).
K7	Coast Province in Coastal region	Sub-humid to humid	Coral rag forests, Arabuko-Sokoke forest, mangrove swamps, and humid tropical coastal belt.	Ectomycorrhizal fungi (*Amanita, Cantharellus, Lactarius, Boletus Russula*), saprotrophs (Agaricaceae, *Marasmius, Gymnopus, Mycena*), *Termitomyces* and Ascomycota

arid conditions where organic matter turnover is slow. K3, K4, and K5 (Sub-humid to humid zones) with rainforests, montane forests, and exotic plantation forests harbors a wide diversity of saprotrophic fungi, with dominant families including the Agaricaceae, Polyporaceae, and Mycenaceae. These groups thrive in moist environments with abundant organic litter and woody debris, playing critical roles in nutrient cycling and forest soil fertility. In addition, the presence of exotic pine and eucalyptus plantations supports rich ectomycorrhizal (ECM) communities, particularly from the Amanitaceae, Boletaceae, Hydnangiaceae, and Russulaceae. This reflects both the adaptability of native ECM fungi to introduced tree hosts and the establishment of new symbiotic associations, although often at the cost of reduced native fungal diversity. K6 (Semi-arid Savannah and Dry Forests) dominated by semi-arid savannah, dry forests, grasslands, acacia woodlands, and dry bushland is characterized by the relative abundance of the Lyophyllaceae (notably species of *Termitomyces* species) and families of the Agaricaceae. Members of the genus *Termitomyces* occur in close symbiosis with species of termites, reflecting the ecological importance of termite-fungus mutualism in nutrient cycling and soil fertility in semi-arid systems. Members of the Agaricaceae serve as key decomposers, further sustaining soil organic matter under nutrient-limited conditions. K7 (Coastal Zone) with Arabuko-Sokoke forests, mangrove swamps, and the humid tropical coastal belt, supports a distinctive fungal assemblage. ECM families such as the Cantharellaceae, Amanitaceae, Boletaceae, Amanitaceae, and Russulaceae dominate these ecosystems, indicating strong tree-fungus symbioses in tropical coastal forests. Mangrove ecosystems in particular, provide specialized habitats for salt-tolerant fungi, which may harbor adaptations valuable for biotechnological exploration.

These regional variations demonstrate that macrofungal diversity is shaped by ecological zonation, vegetation type, and climatic conditions. Understanding these gradients is critical for estimating levels of endemism, identifying habitat-specific species, and designing conservation strategies that integrate both macrofungi and their associated plant communities. Expanding systematic research across all Kenyan ecological zones will therefore provide the foundation for species-specific conservation planning, especially in the face of climate change and land-use transformation.

11.6 ECOLOGICAL AND ECONOMIC IMPORTANCE OF MACROFUNGI

The macrofungal community plays critical roles in ecosystem processes (nutrient cycling, crop nutrition, and carbon sequestration), contributing directly and indirectly to plant production and improvement livelihood (Gateri et al. 2014). The mode of nutrition of a fungus defines its role in the environment.

Saprotrophic fungi, which degrade recalcitrant organic substrates, play a vital role in nutrient cycling and biogeochemical processes. They secrete extracellular enzymes that break down organic matter, plant litter, and woody debris, thereby releasing essential nutrients such as phosphorus, nitrogen, potassium, calcium, magnesium, and various micronutrients into the soil. This process contributes significantly to the maintenance of soil fertility and health. In Kenya, more than 50% of all species of macrofungi species are saprotrophic, making them key drivers of nutrient cycling within natural ecosystems (Muchane et al. 2022).

Macrofungi also play a critical role in enhancing carbon sequestration in soils, thereby contributing to climate change mitigation. Ectomycorrhizal (ECM) species, in particular, have been shown to account for a significant proportion of global carbon sinks (50 to 70% of the total carbon present in leaf litter and soil), as they enhance the uptake and storage of atmospheric carbon dioxide (CO_2) through symbiotic associations with trees (Parihara et al. 2020). Similarly, saprotrophic fungi degrade recalcitrant organic matter and woody substrates, contributing to long-term carbon storage by retaining carbon in stable soil organic pools (He et al. 2017). In addition to direct carbon storage, macrofungi influence soil physical properties. Through the decomposition of organic debris and lignocellulosic substrates, they release organic compounds that facilitate the formation of soil

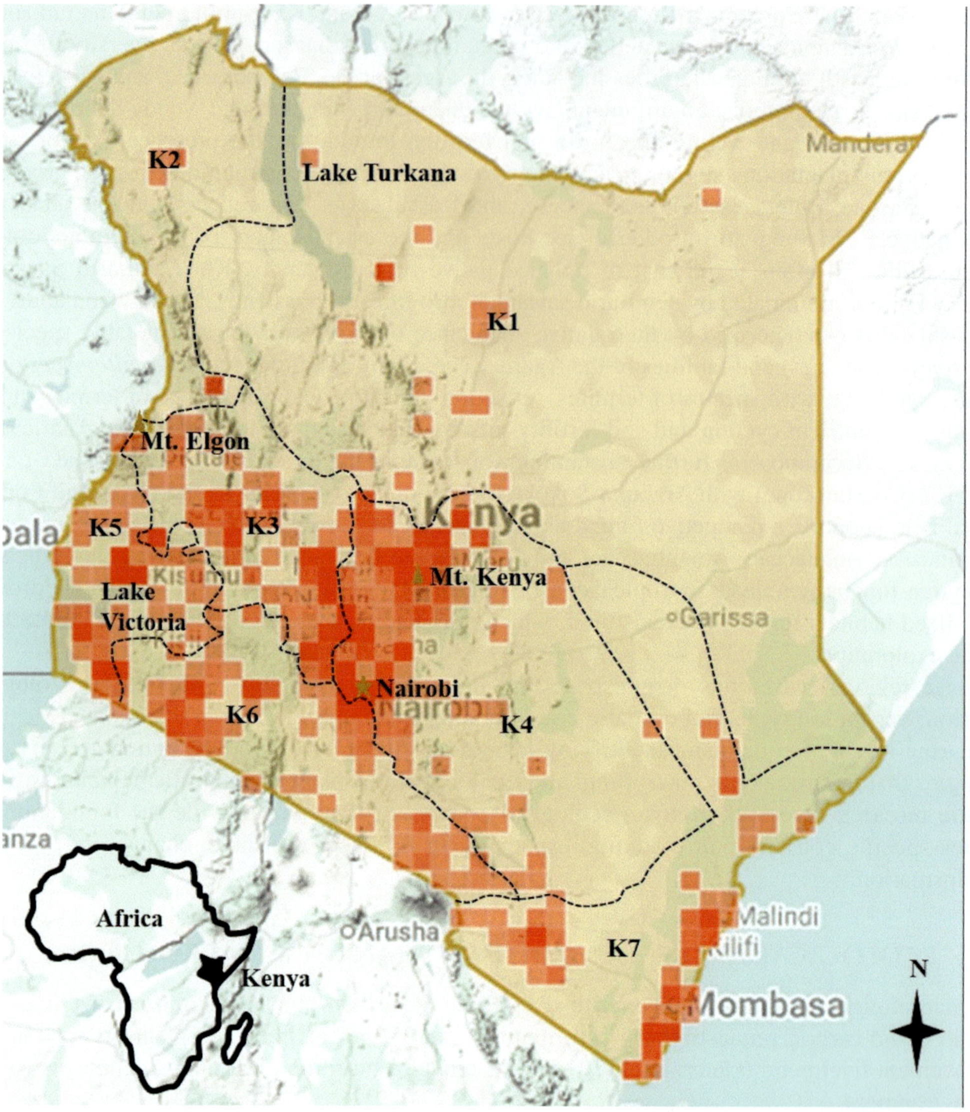

FIGURE 11.2 Kenya, its location in Africa and macrofungi diversity distribution. K1–K7 indicate key plant distribution regions as presented in *Flora of Tropical East Africa* Polhill (1988).

Source: Macrofungi distribution map generated from iNaturalist observations records in Mushrooms of Kenya project. (https://www.inaturalist.org/projects/mushrooms-of-kenya?tab=observations&subtab=map)

aggregates (Muchane 2013). Improved soil aggregation enhances soil structure and porosity, which not only promotes water infiltration and nutrient cycling but also increases soil carbon stability. Soils with well-developed, stable aggregates act as more efficient carbon reservoirs and contribute to the reduction of greenhouse gas emissions.

Although parasitic macrofungi make only small proportion of the total number of species of macrofungi, they cause huge losses to the forestry and agricultural sectors in Kenya. For example, *Armillaria* root rot, a disease caused by species of *Armillaria*, have been reported to affect *Pinus*, Eucalyptus, and *Acacia mearnsii* production in the forestry industry in Kenya (Roux et al. 2005). It has also been documented to cause huge economic losses to tea and coffee plantations (Onsando et al. 1997; Otieno 2002; Roux et al. 2005). Species of *Ganoderma* and *Phellinus* are important pathogens that infect living trees, often causing wood decay and decline in forest health. Species of *Cordyceps* species, on the other hand, parasitize insects and other arthropods, thereby playing a regulatory role in insect populations and showing great potential for use in the biocontrol of agricultural and forest pests.

Edible species of macrofungi play an important role in enhancing food security. In East Africa, 306 species of macrofungi are known to be edible, with Kenya contributing 53 species (Njuguini and Muchane 2023). In an expanded checklist of Kenya's macrofungal diversity, 210 species have been recognized as edible, although it is important to note that some of these species are very small in size and therefore do not provide a substantial meal (Muchane et al. 2022). Among these, species of *Termitomyces* (such as *T. letestui*, *T. microcarpus*, and *T. robustus*) are the most culturally and economically significant, being highly prized as seasonal delicacies and sold in local markets across Western, Nyanza, Central, and Coastal regions. Other notable edible groups include species of *Agaricus*, *Macrolepiota*, *Pleurotus*, *Cantharellus* (chanterelles), *Lactarius*, and *Russula*. Ectomycorrhizal (ECM) species, some of which were introduced into Kenya along with exotic pine plantations, also play a crucial role in improving host tree nutrition. A number of ECM fungi, such as species of *Cantharellus*, *Lactarius*, and *Russula*, are edible and may be sustainably harvested to provide food, support community livelihoods, and contribute to the restoration of degraded ecosystems. While some of these are consumed directly as food, others hold additional medicinal or nutraceutical properties, further enhancing their value to local communities. The recognition of such a large number of edible species highlights the untapped potential of wild mushrooms in contributing to food security, nutrition and rural livelihoods. However, proper taxonomic verification, sustainable harvesting practices, and food safety assessments are essential to avoid misidentification with poisonous look-alikes and to ensure long-term availability of these valuable resources.

Apart from supporting food security, macrofungi also play an important role in traditional and modern medicine. Although the exact number of species of medicinal macrofungi in Kenya remains unknown, several species are widely recognized for their therapeutic properties. For example, species of *Ganoderma* (commonly known as *reishi* or *lingzhi* mushrooms) are highly valued for their bioactive compounds, which possess antioxidant, immunomodulatory, antitumor, and antimicrobial properties. Similarly, species of *Phellinus* are known for their role in strengthening the immune system, managing inflammation, and contributing to cancer research. Recent studies in Kenya have documented at least six species of *Ganoderma* occurring in different ecosystems, including forests and savanna woodlands (Muchane et al. 2022). Other genera known for medicinal applications recorded in Kenya include *Cordyceps*, *Trametes* (*Trametes versicolor*, turkey tail mushroom), and *Inonotus*. The reliance of rural communities on fungi for ethnomedicine such as for the treatment of wounds, stomach ailments, and general immune support highlights their cultural and healthcare significance. This underscores the urgent need for systematic documentation, biochemical screening, and conservation of medicinal macrofungi species in Kenya.

11.7 MACROFUNGI THREATS IN KENYAN ECOSYSTEMS

Although fungi communities play critical roles in ecosystem processes such as nutrient cycling, soil formation, plant symbiosis, and organic matter decomposition, the ecological factors controlling their diversity and distribution patterns remain poorly understood. In their natural conditions, macrofungi communities respond rapidly to ecosystem disturbances, soil quality changes, and shifts in plant composition (Tibuhwa et al. 2011). This sensitivity makes fungi important bioindicators of ecosystem health.

Unsustainable human activities such as soil degradation, poor land management, pollution, overexploitation of plant resources, introduction of invasive plant species, reforestation with non-native tree species, fragmentation, ecosystem transformation, and land-use change alter fungi community composition and impair their ecological functioning (Kewessa et al. 2024). Such disruptions are closely linked with unstable plant communities, reduced crop yields, and overall negative impacts on environmental sustainability (Tibuhwa et al. 2011). In particular, land use change and management practices influence soil properties, including pH, carbon-to-nitrogen (C/N) ratio, soil organic carbon (SOC) content, and nutrient availability. These soil factors strongly determine the species composition and richness of macrofungi (Tibuhwa et al. 2011; Muchane et al. 2016). As a result, changes in soil fertility and structure not only affect plant productivity but also disrupt the fungi networks essential for long-term ecosystem resilience.

Current research in Kenya has shown that conversion of natural forests to monoculture plantation forests reduces macrofungi community richness by more than 50% (Muchane et al. 2016; Njuguini et al. 2018; Muchane et al. 2021). The shift to exotic plantations made up of *Pinus* and *Eucalyptus* alters soil chemistry, reduces native ectomycorrhizal fungi, and favors a limited set of generalist decomposers, thereby homogenizing the fungal community. Even more severe losses occur when forests are cleared for agriculture. Studies in Kenya show that conversion of forest ecosystems to farmland can lead to a 90 to 95% decline in macrofungal diversity (Muchane et al. 2019; Muchane et al. 2021). Such transformations eliminate ectomycorrhizal taxa, reduce saprotrophic decomposers, and diminish the soil biota crucial for nutrient cycling.

Another factor that negatively influences macrofungal communities in both forested and grassland ecosystems is overgrazing (Muchane et al. 2021). Overgrazing reduces plant growth, ground cover, and both below- and above-ground productivity through continuous removal of nutrients, particularly by grazing on palatable grasses and herbaceous plants. The reduced plant growth diminishes root biomass in the soil, which in turn lowers the availability of organic substrates required by macrofungi that decompose dead roots. Moreover, soil compaction caused by livestock trampling not only alters soil structure and aeration but also physically destroys the development of the fruiting bodies of macrofungi. Collectively, these processes lead to significant declines in macrofungal diversity, regeneration, and overall ecosystem functioning.

Since many Kenyan forests have been subjected to extensive alteration of their native plant communities whether through agricultural expansion, plantation forestry or invasive species encroachment, it is highly likely that significant portions of Kenya's macrofungal diversity have already been lost before proper documentation. This silent erosion of fungal biodiversity carries both short-term and long-term consequences for ecosystem services, restoration programs, and biodiversity conservation. Without addressing these losses, ecological restoration efforts may fail to restore essential underground fungal networks that drive nutrient availability, forest regeneration, and resilience against climate change.

11.8 CONSERVATION OF MACROFUNGI IN KENYA

Despite their ecological and socioeconomic importance, macrofungi remain one of the most neglected groups in biodiversity conservation. As already noted, their diversity is highly sensitive to anthropogenic activities such as deforestation, agricultural expansion, urbanization, and climate change, all of which degrade natural ecosystems (Buchana and May 2003). Since nearly 80% of plants directly or indirectly depend on fungi for nutrient acquisition, soil health and resilience, the conservation of plants is inherently tied to the preservation of their fungal partners. As much, a failure to address fungal diversity undermines long-term plant conservation and ecosystem restoration efforts.

In many conservation programs, fungi are often assumed to be protected as "non-target" organisms whenever plant conservation and habitat restoration activities are implemented. However, research shows that plant conservation alone may not guarantee the persistence of fungi. This is because fungi have unique ecological requirements such as specific host associations, microclimatic conditions or substrate availability that may not be restored merely by replanting vegetation. Monitoring of fungal indicator species is therefore essential for tracking

TABLE 11.3
Conservation Status of Macrofungi Species in Kenya.

Species	Edibility	Distribution in Kenya	IUCN Status	Notes
Agaricus arvensis	Edible	Highlands, grasslands	Least Concern (LC)	Widely distributed
Agaricus campestris	Edible	Highlands, grasslands	Least Concern (LC)	Widely distributed
Agaricus crocopeplus	Edible	Higlands, Charcoal	Least Concern (LC)	Limited to forest habitats
Agaricus silvaticus	Edible	Mt. Kenya, Aberdares	Least Concern (LC)	Widely distributed
Agaricus xanthodermus	Poisonous	Mt. Kenya, Aberdares	Vulnerable (VU)	Declining habitats
Amanita loosii	Edible (local use)	Central, Western	Near Threatened (NT)	Limited to forest habitats
Amanita novinupta	Edible	Coastal Region	Least of Concern (LC)	Widely distributed
Amanita phalloides	Toxic	Central Kenya	Least Concern (LC)	Pine plantation
Armillaria fumosa	Parasitic	Higlands	Least concern	Widely distributed
Astraeus hygrometricus	Ectomycorrhizal, medicinal	Coastal Region,	Vulnerable (VU)	Rare, habitat-specific
Auricularia auricula-judae	Edible & medicinal	Central, Western, Coast	Least Concern (LC)	Cultivated locally
Boletinellus merulioides	Edible	Highland, Coastal	Vulnerable (VU)	Declining habitats
Boletus edulis	Edible (ectomycorrhiza)	Coastal Region	Vulnerable (VU)	Declining habitat
Cantharellus cibarius	Edible (ectomycorrhiza)	Coastal Region	Near Threatened (NT)	Declining habitat
Coprinus comatus	Edible	Highlands (grasslands)	Least Concern (LC)	Disturbed soils
Cortinarius caperatus	Edible	Fine plantations	Least of Concern (LC)	Pine plantations
Cuphophyllus niveus	Edible	Grassland	Threatened	Declining grasslands
Cuphophyllus pratensis	Edible	Grassland	Threatened	Declining grasslands
Cuphophyllus russocoriaceus	Not edible	Grassland	Threatened	Declining grasslands
Dictyophora duplicata	Medicinal	Coast	Endangered (EN)	Habitat loss
Ganoderma applanatum	Medicinal	Central, Rift Valley	Least Concern (LC)	Locally abundant
Hygrocybe ovina	Not edible	Grassland	Vulnerable (VU)	Declining grasslands
Hygrocybe reidii	Not edible	Grassland	Vulnerable (VU)	Declining grasslands
Hygrophorus hedrychii	Not edible	Grassland	Vulnerable (VU)	Declining grasslands
Hypholoma capnoides	Not edible	Forested ecosystems	Least concern	Widely distributed
Laetiporus sulphureus	Edible	Aberdares, Mt. Elgon	Vulnerable (VU)	Forest ecosystems
Lentinus squarrosulus	Edible	Coast, Western	Least Concern (LC)	Common in humid forests
Lycoperdon perlatum	Edible, medicinal	Central, Rift Valley	Least Concern (LC)	Common in grasslands

(Continued)

TABLE 11.3 (*Continued*)
Conservation Status of Macrofungi Species in Kenya.

Species	Edibility	Distribution in Kenya	IUCN Status	Notes
Macrolepiota procera	Edible	Aberdares	Least of Concern (LC)	Grassland, Forest edges
Mycena alba	Not edible	Aberdares	Near Threatened (NT)	Forested ecosystems
Myriostoma coliforme	Not edible	Lake Olbolosat	Endangered	Declining habitat
Phallus indusiatus	Edible, medicinal	Coast, Kakamega	Vulnerable (VU)	Habitat-specific, rare
Phellinus igniarius	Medicinal	Central, Western	Data Deficient (DD)	Rarely studied
Psilocybe cubensis	Hallucinogenic	Higlands, Western Kenya	Near Threatened (NT)	Habitat loss
Psilocybe hoogshagenii	Hallucinogenic	Highlands	Rare	Declining habitats
Stropharia ambigua	Edible	Aberdares	Least concern (LC)	Widely distributed
Stropharia melanosperma	Not edible	Aberdares,	Rare	Declining grasslands
Suillus luteus	Ectomycorrhizal	Exotic pine plantations	Near Threatened (NT)	Pine plantations
Termitomyces titanicus	Edible, cultural value	Western Kenya	Near Threatened (NT)	Habitat loss

the recovery of ecosystem processes and ensuring that fungi are not silently lost during conservation interventions.

Globally, the conservation status of fungi is poor compared to plants and animals. As of 2019, only 280 species of fungi (0.02% of the known fungal community) have been formally red-listed as requiring urgent conservation action (IUCN 2019). In contrast, between 20 to 47% of plant species worldwide are considered threatened with extinction (Pitman and Jørgensen 2002). Given the often-cited plant-to-fungi ratio of 1:5, it is estimated that over 450,000 species of fungi could be currently at risk, though most remain undocumented or undescribed. This discrepancy reflects the severe knowledge gap in fungal biodiversity.

In Kenya, the situation is even more concerning. Less than 5% of species of fungi have been documented, with the vast majority of macrofungal diversity remaining unexplored. The conservation status of these species is unknown, as no systematic red-list assessments or monitoring frameworks currently exist. By extrapolating from global trends and studies from other countries, it is possible to assign provisional conservation status to at least 28 species of macrofungi documented from Kenyan ecosystems (Table 11.3). These include edible, medicinal, and ecologically significant taxa. However, this number likely represents only a fraction of the true diversity at risk.

Moreover, it is estimated that over 50% of all species of fungi may be lost due to ongoing anthropogenic pressures such as forest clearance in central highlands, degradation of coastal forests, and agricultural expansion in savannah regions. This highlights an urgent need for comprehensive fungal inventories, ecological studies, and red-listing initiatives. Such documentation is the first step in identifying endemic, threatened or endangered macrofungal species, which can then be integrated into national conservation strategies.

At present, Kenya lacks both site-specific and species-specific conservation action plans for fungi. Integrating fungi into conservation frameworks would not only safeguard unique fungal lineages but also enhance the success of broader biodiversity management programs. This is particularly critical in biodiversity hotspots such as the Aberdare ranges, Mt. Kenya, the Kakamega forest, and coastal forests, where fungi are key players in nutrient cycling, forest regeneration and food webs.

11.9 CONCLUSIONS AND FUTURE RESEARCH

This chapter provides a comprehensive synthesis of the state of macrofungi diversity, distribution, and conservation needs in Kenyan ecosystems. The findings reveal a rich and diverse macrofungi community with immense potential to sustain ecosystem health and functioning, enhance food security, and contribute to climate change mitigation. The presence of diverse saprotrophic fungi underscores their critical role in nutrient cycling and soil fertility, thereby maintaining healthy ecosystems. The diversity of ectomycorrhizal species further highlights their importance in enhancing forest productivity and carbon sequestration, while the abundance of species of *Termitomyces* reflects both the ecological role of termites and their contribution to local livelihoods through food and income.

Despite these values, this review emphasizes the need for more detailed studies across all Kenyan ecosystems, as more than 90% of species of macrofungi remain undocumented. Addressing this gap will be fundamental to promoting sustainable use, management, and conservation of fungal resources. Future research should prioritize systematic monitoring of macrofungi in already studied areas and expand surveys into unexplored habitats. To improve accuracy in species identification and uncover new taxa, a combination of morphological and molecular approaches should be integrated in macrofungi taxonomy.

Equally important is the development of conservation assessments for key species, including endemic and threatened taxa, along with the formulation of species-specific and habitat-based conservation strategies. Integrating fungi into broader biodiversity and ecosystem management frameworks will be critical to ensure their long-term survival and continued contribution to ecosystem services. Ultimately, advancing macrofungi research and conservation in Kenya will provide a foundation for sustainable ecosystem management, improved livelihoods, and resilience against environmental change.

11.10 ACKNOWLEDGMENTS

Citizen Science studies of macrofungi species in Kenya were supported by grant from the National Geographical Society.

REFERENCES

Ambrosio, E.L., E. Lancellotti, R. Brotzu & H. Salch. 2015. Assessment of macrofungal diversity in a Silver Fir plantation in Sardinia (Italy) using a standardized sampling procedure. *Italian Journal of Mycology* 44:1–17.

Arzú, R.F., O. Comandini & A.C. Rinaldi. 2012. A preliminary checklist of macrofungi of Guatemala, with notes on edibility and traditional knowledge. *Mycosphere* 3:1–21. https://doi.org/10.5943/mycosphere/3/1/1/

Buchanan, P.K. & T.W. May. 2003. Conservation of New Zealand and Australian fungi. *New Zealand Journal of Botany* 41:407–421. https://doi.org/10.1080/0028825X.2003.9512859

Chadha, K.L. & S.R. Sharmam. 1995. Mushroom research in India—history, infrastructure and achievements. In K.L. Chadha (ed.), *Advances in Horticulture*, pp. 1–8. MPH (Malhotra Publishing House, New Delhi, India.

Chang, R. 1996. Functional properties of mushrooms. *Nutrition Reviews* 54:91–93.

Chang, S.T. & P.G. Miles. 1992. Mushroom biology, new discipline. *The Mycologist* 6:64–65.

Dejene, T., J.A. Oria-de-Rueda & P. Martín-Pinto. 2017. Fungal diversity and succession following stand development in *Pinus patula* Schiede ex Schltdl. & Cham. plantations in Ethiopia. *Forest Ecology and Management* 395:9–18. https://doi.org/10.1016/j.foreco.2017.03.032

Devkota, S. & H.P. Aryal. 2020. Wild Mushrooms of Nepal. In M. Siwakoti, P.K. Jha, S. Rajbhandary & S.K. Rai (eds.), *Plant Diversity in Nepal*, pp. 41–54. Botanical Society of Nepal, Kathmandu, Nepal.

Dring, D.M. & R.W. Rayne. 1966. Some gasteromycetes from Eastern Africa. *Journal of East Africa Natural History Society* 26:114.

Gateri, M.W., B. Ndungu, A.W. Muriuki, V. Rauwl & S. Kabacia. 2014. Collection, identification and morphological characterization of indigenous mushrooms in coastal Kenya. In *Proceedings of 8th International Conference*, New Delhi, India, ICAR-Directorate of Mushroom Research. New Delhi, India, pp. 17–23.

Gupta, R.S. 1986. Mushroom cultivation. *Indian Horticulture* 31:1.

Haelewaters, D., C.A. Quandt, L. Bartrop, J. Cazabonne, M.E. Crockatt, S.P. Cunha, R. De Lange, L. Dominici, B. Douglas, E.R. Drechsler-Santos, J. Heilmann-Clausen, P.J. Irga, S. Jakob, L. Lofgren, T.E. Martin, M.N. Muchane, J.K. Stallman, A. Verbeken, A.K. Walker & S.C. Gonçalves. 2024. The power of citizen science to advance fungal conservation. *Conservation Letters* 00:e13013. https://doi.org/10.1111/conl.13013

Hasan, S. & G. Gupta. 2012. Fungal biodiversity: Evolution & distribution—a review. *International Journal of Applied Research & Studies* 1:1–8.

Hawksworth, D.L. 2001. The magnitude of fungal diversity: The 1.5 million species estimate revisited. *Mycological Research* 105:1422–1432.

Hawksworth, D.L. 2012. Global species numbers of fungi: Are tropical studies and molecular approaches contributing to a more robust estimate? *Biodiversity Conservation* 21:2425–2433. https://doi.org/10.1007/s10531-012-0335-x

Hawksworth, D.L. & R. Lücking. 2017. Fungal diversity revisited: 2.2 to 3.8 million species. *The Fungal Kingdom* 89:79–95. https://doi.org/10.1128/9781555819583.ch4

He, K., G. Chen, G. Zeng, Z. Huang, Z. Guo, T. Huang, M. Peng, J. Shi & L. Hu. 2017. Applications of white rot fungi in bioremediation with nanoparticles and biosynthesis of metallic nanoparticles. *Applied Microbiology and Biotechnology* 101:4853e4862. https://doi.org/10.1007/s00253-017-8328-z

He, M.Q., R.L. Zhao, K.D. Hyde, D. Begerow, M. Kemler, A. Yurkove, E.H.C. McKenzie, O. Raspé, M. Kakishima, S. Sánchez-Ramírez, et al. 2019. Notes: Outline and divergence times of Basidiomycota. *Fungal Diversity* 99:105–367.

Hsieh, H.M. & Y.M. Ju. 2018. Medicinal components in *Termitomyces* mushrooms. *Applied Microbiology and Biotechnology* 102. https://doi.org/10.1007/s00253-018-8991-8

International Union for Conservation of Nature and Natural Resources. 2019. The IUCN red list of threatened species. Version 2019–3 (html). IUCN. ISSN 2307–8235. *Fungi Kingdom* 280.

Kark, S. 2007. Effects of ecotones on biodiversity. In S. Levin (ed.), *Encyclopedia of Biodiversity*, pp. 1–10. Academic Press, San Diego.

Kewessa, G., T. Dejene, W. Tadesse & P. Martín-Pinto. 2024. Fungal biodiversity in Ethiopia: Distribution, threats and prospects. In J. Putzke (ed.), *The Diversity of the Fungal World*. IntechOpen. https://doi.org/10.5772/intechopen.1005910

Kim, N.K., J.H. Lee, J.W. Jo & J.K. Lee. 2017. A checklist of mushrooms of Cambodia. *Journal of Forest and Environmental Science* 33:49–65.

Kost, G. 2002. Contributions to tropical fungi I. Ecology and distribution of fungi of Kenya (East Africa). *Feddes Repertorium* 113:132–151.

López-Quintero, C.A., G. Straatsma, A.E. France-Molanno & T. Broekhout. 2012. Macrofungal diversity in Colombian Amazon forests varies with regions and regimes of disturbance. *Biodiversity and Conservation* 21:2221–2243.

MEWNR. 2015. Kenya Biodiversity Atlas. *Kenya's Natural Capital, a Biodiversity Atlas*. Ministry of Environment Natural Resources and Regional Development Authorities, Nairobi, Kenya.

Molina, R. 1994. The role of mycorrhizal symbioses in the health of giant redwoods and other forest ecosystems 1. In *Symposium on Giant Sequoia*. Visalia, California.

Muchane, M.N. 2013. *Influences of Agricultural Management Practices on Arbuscular Mycorrhizal Fungal Symbioses in Kenyan Agro-Ecosystems*. Ph.D. Thesis. Wageningen University.

Muchane, M.N., S.K. Njuguini, K.M. Waithaka & S.M. Muchai. 2022. An annotated checklist of the Macrofungi Species in Kenya—Mycological Survey, Literature review, fungarium and iNaturalist online data base. In M.N Muchane, S.K. Njuguini & S.M. Muchai (eds.), *Inventory and Monitoring of Macrofungi Diversity using Citizen Science Programs and iNaturalist Data Collection Tools, Technical Report*. National Geographical Society (NGS).

Muchane, M.N. & K.M. Waithaka. 2023. Status of fungi resources in Lake Victoria Watershed project. In V. Mainga, T. Thenya, P. Kirika, M. Muchane et al. (eds.), *Assessment of Biodiversity Resources and Biodiversity Hotspot Areas. Technical Report*. Tree For The Future. Kisumu, Kenya.

Muchane, M.N. & K.M. Waithaka. 2024, March. Unlocking fungi diversity in Lake Olbolossat: Field and literature review report. In Theya et al. (eds.), *Unlocking Biodiversity Data to Facilitate Designation of Lake Olbolossat as The 7th Ramsar Site In Kenya East Africa*. Wildlife Society (EAWS).

Muchane, M.N. & K. Waithaka. 2025. *Status of Fungi Resources in Lolldaiga Valley Ecosystem (LVE)*. Narenda Development Company. Narenda Consultants, Kenya.

Muchane, M.N., K.M. Waithaka & T. Terer. 2021. Macro-fungi community of Lake Ol Bolossat Basin and associated riparian ecosystems in Central, Kenya. *International Journal of Natural Resource Ecology and Management* 6:6–15. https://doi.org/10.11648/j.ijnrem.20210601.13

Muchane, M.N., K.M. Waithaka, T. Thenya & J. Okumu. 2025. *Macrofungi Species Occurrences in Lake Olbolosat Ecosystems*. The East African Wildlife Society, Uganda. https://doi.org/10.15468/mq73zj

Muchane, M.N., S.K. Mwai, S.M. Muchai & H. Saado. 2016. Effects of land use on diversity of macrofungi in Kereita forest kikuyu escarpment, Kenya. *Technical Report*. National Museums of Kenya and Tropical Biology Association, Nairobi, Kenya.

Muchane, M.N., S.K. Njuguini, M. Obunga, N. Kinyua, K.M. Waithaka & T. Terer. 2019. Fungi diversity of Upper Tana River Catchment (UTRC) in Central Kenya. *Technical Report*. National Museums of Kenya, Nairobi.

Müller, G.M., J.P. Schmit, P.R. Leacock & B. Buyck. 2007. Global diversity and distribution of macrofungi. *Biodiversity Conservation* 16:37–48. https://doi.org/10.1007/s10531-006-9108-8

Niego, A.G., S.S. Rapior, N. Thonglang, O. Raspire, K.D. Hyde & P. Motimer. 2023. Reviewing the contributions of macrofungi to forest ecosystem processes and services. *Fungal Biology Reviews* 100294. https://doi.org/10.1016/j.fbr.2022.11.002

Njuguini, K.S. & M.N. Muchane. 2023. Domestication of wild edible mushrooms in Eastern Africa: A review of research advances and future prospects. *Mantar Dergisi* 14:22–50.

Njuguini, S.K., M.N. Muchane, P. Wachira, S. Okoth, M. Muchane & H. Saado. 2018. Effects of Land use on the diversity of macrofungi in Kereita Forest Kikuyu Escarpment, Kenya. *Current Research in Environmental & Applied Mycology* 8:254–281. https://doi.org/10.5943/cream/8/2/10

Nyaga, J.J., J. Jefwa & V. Raul. 2009. Macrofungi of Mt. Kenya. In W. Musila (ed.), *Mt. Kenya Forest Biodiversity Assessment. National Museums of Kenya and Kenya Wildlife Service*. Nairobi, Kenya.

Onsando, J.M., P. Wargo & S.W. Waudo. 1997. Distribution, severity, and spread of Armillaria root disease in Kenya tea plantations. *Plant Disease* 81:133–137.

Otieno, N.C. 1966. Contributions to a knowledge of termite fungi in East Africa. *Proceedings of the East African Academy* 2:109–121.

Otieno, N.C. 1968. Further contributions to a knowledge of termite fungi in East Africa. *Sydowia* 22:160–165.

Otieno, W. 2002. *Armillaria Root Rot of Tea in Kenya Characterization of the Pathogen and Approaches to Disease Management*. Ph.D. Thesis, Wageningen University, Netherlands.

Parihara, M., A. Rakshita, K. Ranab, G. Tiwaric & S. Jatav. 2020. The effect of arbuscular mycorrhizal fungi inoculation in mitigating salt stress of pea (*Pisum sativum* L.). *Communications in Soil Science and Plant Analysis* 51:1545–1559. https://doi.org/10.1080/00103624.2020.1784917

Pegler, D.N. 1977. *A Preliminary Agaric Flora of East Africa*. Lubrecht & Cramer Ltd, London, England.

Pegler, D.N. & R.W. Rayner. 1969. A contribution to the agaric flora of Kenya. *Kew Bulletin* 23:347–412. https://doi.org/10.2307/4117177

Pitman, N.C. & P.M. Jørgensen. 2002. Estimating the size of the world's threatened flora. *Science* 298:989.

Polhill, D. 1988. *Flora of Tropical East Africa: Index of Collection Localities*. Royal Botanic Gardens, Kew, London.

Rossman, A. 1994. A strategy for an all-taxa inventory of fungal biodiversity. In C.I. Peng & C.H. Chou (eds.), *Biodiversity and Terrestrial Ecosystems*, pp. 169–194. Academia Sinica Monograph Series No. 14. Taipei.

Roux, J., G. Meke, B. Kanyi, L. Mwangi, A. Mbaga, G.C. Hunter & M.J. Wingfield. 2005. Diseases of plantation forestry trees in eastern and southern Africa. *South African Journal of Science* 101:409–413.

Smith, S.E. & D.J. Read. 2007. *Mycorrhizal Symbiosis*, 3rd ed. Academic Press, London, Britain.

Srivastava, M.P. 2021. Bio-diversity of wild mushrooms and their future perspectives. *International Journal of Plant and Environment*. https://doi.org/10.18811/ijpen.v7i02.7

Tapwal A., R. Kumar & P. Shailesh. 2013. Diversity and frequency of macrofungi associated with wet ever green tropical forest in Assam, India. *Biodiversitas* 14:73–78.

Tibuhwa, D.D., M.N. Muchane, C.W. Masiga, C. Mugoya & S.M. Muchai. 2011. An inventory of macro-fungi and their diversity in the Serengeti-Masai mara ecosystem, Tanzania and Kenya. *Journal of Biological Sciences* 11:399–410.

Vasco-Palacios, A.M. & A.E. Franco-Molano. 2013. Diversity of Colombian macrofungi. (Ascomycota–Basidiomycota). *Mycotaxon* 121:1–58.

Vesala, R., T. Niskanen, K. Liimatainen, H. Boga, P. Pellikka & J. Rikkinen. 2017. Diversity of fungus-growing termites (*Macrotermes*) and their fungal symbionts (*Termitomyces*) in the semiarid Tsavo Ecosystem, Kenya. *Biotropica* 49:402–412.

Wijayawardene, N.N., K.D. Hyde, K.V. Mikhailov, et al. 2024. Classes and phyla of the kingdom Fungi. *Fungal Diversity* 128:1–165. https://doi.org/10.1007/s13225-024-00540-z

Zhou, Y., B. Liu, Y. Mbuni, X. Yan, G. Mwachala, G. Hu & Q. Wang. 2017. Vascular flora of Kenya, based on the Flora of Tropical East Africa. *PhytoKeys* 90:113–126. https://doi.org/10.3897/phytokeys.90.20531

12 Diversity, Ecological Patterns, and Biogeography of Myxomycetes in Island Tropical Forests

A Case Study From the Philippines

Thomas Edison E. dela Cruz, Sittie Aisha B. Macabago, Nikki Heherson A. Dagamac, and Melissa H. Pecundo

12.1 INTRODUCTION

The Philippines, located in the western Pacific, is divided into three major island groups—Luzon, Visayas, and Mindanao. All of these belong to a recognized global biodiversity hotspot (Heaney 1998). Geologically, the archipelago is a composite of multiple island arcs and crustal fragments assembled over tens of millions of years. Much of the country is underlain by Cretaceous to Eocene ophiolites, thick successions of Cretaceous to Eocene turbidites and carbonate rocks, Oligocene to Miocene granodioritic intrusions, Neogene andesitic intrusive and volcanic units, and extensive Tertiary sedimentary sequences (Yumul et al. 1997). These rock assemblages reflect a long history of subduction, volcanism, arc-arc collisions, and terrane accretion that collectively shaped the modern islands (Karig 1983; Hall 2002). Luzon and Mindanao formed from independent island-arc systems, while the Visayas developed from a mosaic of smaller crustal blocks with distinct geological histories (Holloway 1982; Mitchell et al. 1986).

Biogeographically, the Philippines lies adjacent to Wallacea, a transitional zone between the Asian and Australasian regions that is characterized by deep-water barriers and exceptional endemism. Although Palawan intermittently connected to Borneo during Pleistocene low sea levels, the remainder of the archipelago remained isolated, producing unique patterns of species diversification (Rangin et al. 1990; Aurelio 2000). Throughout the Cenozoic, continued subduction, terrane accretion, volcanic activity, and tectonic uplift gave rise to major mountain ranges, active volcanic systems, and heterogeneous habitats (Hall 2002; Yumul et al. 2008). Coastal uplift, subsidence, and rapid sedimentation reshaped ecosystems, creating lowlands, coral reefs, river valleys, and highland forests. Today, the Philippines remains geologically active, with frequent earthquakes, ongoing crustal deformation, and more than 20 active volcanoes—including Mayon, Taal, and Mount Pinatubo—part of the Pacific "Ring of Fire." These dynamic processes contribute not only to natural hazards but also to fertile soils, mineral resources, and highly productive habitats. Among the latter are forest habitats in the Philippines (Figure 12.1).

The tropical maritime climate of the Philippines, with average temperatures of 25–28°C, high humidity, and abundant rainfall, further interacts with its geologically diverse landscape to shape ecosystems. According to PAGASA (2025), two main seasons dominate. These are the wet season that occurs from June to November, driven by the southwest monsoon, and the dry season that occurs from December to May, influenced by the northeast monsoon. Regional variations exist,

DOI: 10.1201/9781003647140-12

FIGURE 12.1 Some forest habitats in the Philippines. (A) Island forest ("sleeping dinosaur island") in Mati City, Davao Oriental, (B) Island limestone (karst) forest in Hundred Islands National Park, Pangasinan, (C) Mountain grassland, (D) Pygmy forest in Mt. Hamiguitan, Davao Oriental, (E) Mossy forest at peak of Mt. Banahaw, Quezon, (F) Pine forest in Baguio City, Benguet, and (G) Secondary forest in Bohol Island.

Source: Photos by T.E.E. dela Cruz

with western islands experiencing distinct wet and dry periods, while eastern islands receive more consistent rainfall. Four climate types based on rainfall defined the many regions in the country. Type I has two distinct seasons, being dry from November to April and wet from June to October, with maximum rain between June to September as seen in the Western Visayas and some parts of Luzon. Type II has no dry season but a pronounced rainy period between December to February such as is found in some parts of the Bicol Region and in Eastern Visayas. Type III has a short or no pronounced dry season with rainfall that is more evenly distributed or a dry spell between

December and May. This can be observed in Mindoro Island and in the Cagayan Valley. Type IV has rainfall that is distributed throughout the year. This is the case in some central and southern areas as well as in mountainous regions. Also, frequent typhoons frequently enter the country and influence erosion, sediment deposition, and species distribution patterns.

Biodiversity in the Philippines is among the highest in the world, with over 52,000 described species and remarkable rates of endemism in plants, birds, mammals, and reptiles (Barcelona et al. 2013; Birdlife International 2025). Iconic species include the country's national bird, the Philippine eagle (*Pithecophaga jefferyi*), the tamaraw or the Mindoro dwarf buffalo (*Bubalus mindorensis*), and Philippine tarsier (*Carlito syrichta*), alongside numerous rare orchids and pitcher plants. Marine ecosystems are equally rich, as the archipelago lies within the Coral Triangle, the global center of marine biodiversity (Bemis et al. 2023). Beyond macroorganisms, recent studies highlight extensive microbial and fungal diversity, including macrofungi, endophytic fungi, soil microbes, and marine-associated taxa, many of which are unique to the Philippine archipelago (Notarte et al. 2018; Apurillo et al. 2019; Jacob et al. 2022). A recent checklist of novel microbes including fungi that were discovered in the Philippines support the country's rich microbial diversity (dela Cruz et al. 2024). These microorganisms are essential for nutrient cycling, symbiotic interactions, and overall ecosystem functioning.

12.2 EARLY STUDIES ON PHILIPPINE MYXOMYCETES

Among the earliest reports of myxomycetes in the Philippines were the 20 species mentioned in the seminal book on myxomycetes by Martin and Alexopoulos (1969). These early records included *Calomyxa metallica* (Berk.) Nieuwl., *Cribraria atrofusca* G.W. Martin & Lovejoy, *Craterium minutum* (Leers) Fr., *Didymium anellus* Morgan, *Didymium leoninum* Berk. & Broome, *Erionema aureum* Penz., *Lamproderma cribrariodes* [current valid name: *Meriderma cribrarioides* (Fr.) Mar. Mey. & Poulain], *Perichaena minor* [current valid name: *Hemitrichia minor* G. Lister], *Physarum bivalve* [current valid name: *Angioridium sinuosum* (Bull.) Grev.], *Physarum echinosporum* Lister, *Physarum flavicomum* Berk., *Physarum gyrosum* [current valid name: *Fuligo gyrosa* (Rostaf.) E. Jahn], *Physarum nicaraguense* T. Macbr., *Physarum pezizoideum* [current valid name: *Trichamphora pezizoidea* Jungh.], *Physarum psittacinum* Ditmar, *Physarum pulcherrimum* Berk. & Ravenel, *Physarum restisporum* G.W. Martin, K.S. Thind & Rehill, *Physarum roseum* [current valid name: *Craterium roseum* (Berk. & Broome) J.M. García-Martín & Lado], *Physarum stellatum* (Massee) G.W. Martin, and *Tubifera bombarda* [current valid name: *Alwisia bombarda* Berk. & Broome]. This was followed by the study of Uyenco (1973), who documented 18 species belonging to 10 genera. Dogma (1975) provided a longer list of 46 species in 20 genera for the country. However, it was Don Reynolds, a Peace Corp volunteer at the University of the Philippines Los Baños (UPLB) in the early 70s, who provided the most extensive list (107 species) in 1981. Reynolds (1981) reported myxomycete collections that had been recorded by American botanists E. B. Copeland, A. D. E. Elmer, and E. D. Merrill along with other species that he was able to document during his stint at UPLB. Of the 107 species included in his annotated list, 53 species were listed as new records for the country, reflecting the potential richness of myxomycetes in the Philippines. Many of these species are still deposited as voucher specimens at the Museum of Natural History Mycological Herbarium located at UPLB in Laguna (dela Cruz et al. 2009) and in other collections in the US.

Recent studies on Philippine myxomycetes have increased the number of recorded taxa in the country from 107 (Reynolds 1981) to 160 (Dagamac and dela Cruz 2015, 2019). The studies of Dagamac et al. (2010, 2011) added five new records of corticolous myxomycetes and another five new records from Mount Arayat National Park, Pampanga. dela Cruz et al. (2011) and Macabago et al. (2012) reported two and six newly recorded species of myxomycetes. The highest number of new records was that of dela Cruz et al. (2014) with 19 species added to the list of Philippine myxomycetes. Continuous documentation of myxomycetes in relatively underexplored areas such as the Bicol Peninsula (Dagamac et al. 2017c) and in Aurora and Quezon (Cabutaje et al. 2021) added additional taxa, bringing the total number of recorded species to 160, included two newly discovered species—*Craterium retisporum* G. Moreno, D.W. Mitch. & S.L. Stephenson (Moreno et al. 2009) and *Lycogala acinonychum* Leontyev, Schnittler, T.E. dela Cruz, M.F.B. Eloreta & T. van der Heul (Leontyev et al. 2023). The sampling localities where myxomycetes have been thus far recorded in the Philippines are shown in Figure 12.2. A one-stop-shop webpage on myxomycetes (https://philmyxos.wordpress.com/) has been created to consolidate all these findings. An interactive database

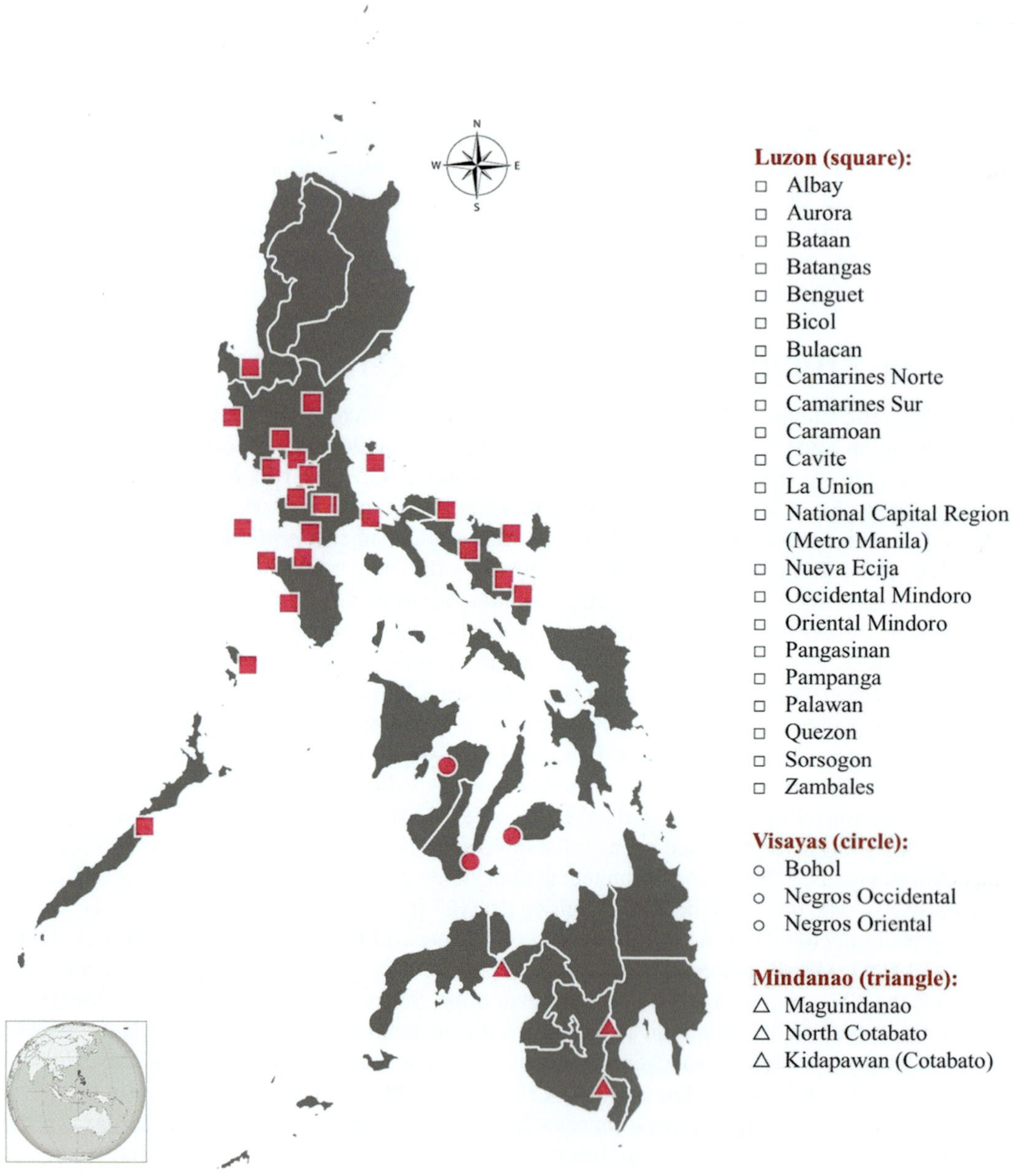

FIGURE 12.2 Philippine map showing sampling localities where myxomycetes were reported in the county.

Source: Map of the Philippines and sampling points were generated using SimpleMaps (https://simplemaps.com/world). Globe map: obtained from Wikimedia Commons (TUBS, CC BY-SA 3.0, https://creativecommons.org/licenses/by-sa/3.0).

and photoguides were also developed to aid in species identification (Dagamac et al. 2011; dela Cruz et al. 2012; Macabago and dela Cruz 2012). Moreover, there were some early efforts to study myxomycetes for their potential application in enzyme production and bioremediation (Macabago and dela Cruz 2014; Rea-Maminta et al. 2015).

12.3 DIVERSITY AND BIOGEOGRAPHY OF MYXOMYCETES IN PHILIPPINE ISLAND ECOSYSTEMS

The archipelagic landscape of the Philippines provides an ideal system for studying island biogeography of myxomycetes. However, most previous studies have focused on updating the myxobiota of the country (Macabago et al. 2017). In addition, despite the increase in myxomycete studies in the Philippines, the majority of these studies are focused on the Luzon islands, the largest island in the country, with a few studies on the Visayas islands and minimal studies on the Mindanao islands. In this section, we present the different studies of myxomycetes carried out on many of the major islands in the country.

12.3.1 Luzon Island

Luzon, the largest island in the Philippines, hosts an extraordinary range of vegetation types and forested landscapes, making it the center of myxomycete research in the country. Most Philippine surveys have been carried out here, forming much of what is currently known about species richness, substrate preference, and ecological patterns of myxomycetes (dela Cruz et al. 2014; Dagamac and dela Cruz 2019). Its dramatic variation in topography and climate creates a mosaic of habitats where this group of eukaryotic amoeboid protists flourish.

Over the past two decades, building upon Reynolds' annotated list (1981) and the efforts of dela Cruz et al. (2009), Luzon has remained a focal point for myxomycete research, with surveys covering nearly every major habitat type on the island. Studies have documented species in lowland forests (Dagamac et al. 2015a; Pecundo et al. 2020), mountains and protected landscapes (Dagamac et al. 2014, 2015a; dela Cruz et al. 2014), upland terrains (Balaoro-Banzuela et al. 2023), and pine forests (Kuhn et al. 2013a). Research has also addressed coastal and beach forests (Cabutaje et al. 2021; Balaoro-Banzuela et al. 2023.), limestone and karst landscapes (dela Cruz et al. 2011; Dagamac et al. 2015b; Pecundo et al. 2017, 2021), ecological sites and urban parks (Macabago et al. 2010; Accad et al. 2023), national parks (Atayde et al. 2012; Corpuz et al. 2012; Pecundo et al. 2023), volcanic environments (Isagan et al. 2020), and fragmented forest patches (Rea-Maminta et al. 2015; Bernardo et al. 2018). Several islands associated with Luzon, including Anda, Polillo, Lubang, the Hundred Islands, and Caramoan Island, have also been surveyed (Kuhn et al. 2013b; Viray et al. 2014; Macabago et al. 2012; dela Cruz et al. 2011; Macabago et al. 2020a). Together, these studies have expanded the known myxomycete diversity in Luzon and contributed to the description of new species such as *Craterium retisporum* from Pangasinan (Moreno et al. 2009) and *Lycogala acinonychum* from *Camarines Sur* (Leontyev et al. 2023).

Despite Luzon's myxomycete research prominence, its coverage is uneven, with Northern Luzon remaining one of the least explored regions (see also Figure 12.2). Only a handful of studies have been published from the north. In Benguet, Kuhn et al. (2013a) reported eight species representing genera such as *Arcyria*, *Comatricha*, *Didymium*, *Perichaena*, *Physarum*, and *Stemonitis*. For the Ilocos Region, dela Cruz et al. (2011) documented 30 species in 10 genera from the Hundred Islands National Park in Pangasinan, while Kuhn et al. (2013b) recorded 24 species across 11 genera on Anda Island, noting particularly abundant taxa including *Arcyria*, *Craterium*, *Diderma*, and *Physarum*. This study also added three new records [*Craterium microcarpum* H.Z. Li, Yu Li & Shuang L. Chen, *Physarum decipiens* M.A. Curtis, and *Elaeomyxa miyazakiensis* (Emoto) Hagelst.] for the Philippines. More recently, Balaoro-Banzuela et al. (2023) examined upland and coastal terrains in San Fernando City, La Union, using moist chambers prepared from leaf litter. They documented 27 species belonging to 10 genera and found richer assemblages in upland habitats, emphasizing that differences in terrain formed clearer ecological clusters than differences in substrate type.

Central Luzon is one of the best explored regions in the Philippines for myxomycetes, with at least eight published studies. Early work by dela Cruz et al. (2010) offered a first look at the Subic Watershed Forest Reserve in Zambales, where they documented 14 species representing seven genera and three orders. A few years later, surveys expanded to the lowland forests of Bataan National Park in Bataan and the Subic Bay Wildlife Forest Reserve in Zambales. These surveys were more productive, yielding 54 and 57 species, respectively, and adding 19 new records for the country (dela Cruz et al. 2014). As Philippine myxomycete research progressed, investigators began to look beyond simple species lists and focused on environmental factors shaping these communities. For example, Dagamac et al. (2014) examined Mt. Arayat National Park in Pampanga and demonstrated how elevation can structure myxomycete assemblages where the mid-elevation zone around 700–800 m had the highest taxonomic diversity, with peak species richness at 700 m. Moreover, Rea-Maminta et al. (2015) investigated forests with volcanic and ultramafic soils in Bataan, Pangasinan, and Zambales. Ultramafic sites supported higher diversity than volcanic sites, suggesting that many myxomycete species can tolerate and thrive even under these chemically challenging conditions. Their surveys recorded 40 species regardless of the sites, and consistently abundant taxa included *Arcyria cinerea* (Bull.) Pers., *Diachea leucopodia* (Bull.) Rostaf., *Diderma effusum* (Schwein.) Morgan, and *Perichaena corticalis* (Batsch) Rostaf. Karst landscape added another layer to the region's ecological diversity. At Minalungao National Park in Nueva Ecija, a survey of the karst forest yielded 28 myxomycete species, with moist chamber cultures successfully recovering 80% of these taxa (Pecundo et al. 2021). Most of the species belonged to the order Physarales, members of which produce lime in their fruiting bodies. This pattern suggests that the high-calcium environment of karst selectively favors lime-producing myxomycetes, thus potentially shaping the composition and structure of these communities. Interestingly, researchers also began exploring how natural disturbances influence species patterns. Studies conducted in Aurora and Quezon provinces took advantage of recent typhoon events, sampling aerial and ground litter months after typhoons. Despite sampling at different recovery intervals—seven months in Aurora and one month in Quezon—the surveys yielded 42 species and 48 species, respectively (Cabutaje et al. 2021). Inland forests consistently supported higher diversity, richness, and evenness than coastal sites, although beach forests maintained comparable species composition. The authors further suggested that rapid litter turnover and wind-driven spore movement during typhoons may help explain this resilience (Cabutaje et al. 2021). Figure 12.3 shows some of the species recorded from these surveys in Aurora, Nueva Ecija and Quezon.

In Southern Luzon, the provinces of Batangas and Quezon have also been surveyed for myxomycetes. Cheng et al. (2013) conducted the first assessment on the northern slope of Mt. Makulot in Batangas and documented 21 species in 10 genera. Most species were collected from aerial leaf litter and twigs, whereas only six were recovered from ground leaf litter, pointing to the importance of canopy-derived substrates at this site. Isagan et al. (2020) later broadened the geographic scope by examining both Mt. Makulot and the volcanic lake island of Napayong (now known as Laurel Island). Their survey recorded 35 species in 14 genera, exceeding the richness reported by Cheng et al. (2013) for Mt. Makulot alone. This expanded comparison revealed species unique to each locality and underscored ecological differences between the crater-rim forest of Mt. Makulot and the isolated island environment of Napayong within Taal Lake, also in Batangas. The surveys in Batangas have demonstrated the diversity of myxomycetes in Southern Luzon's upland and volcanic habitats. Attention has also turned east to Quezon Province, where contrasting forest types provide additional insights into regional community patterns. In Quezon National Park, karst forest surveys recorded 35 species in 16 genera (Dagamac et al. 2015a), reflecting the influence of alkaline limestone habitats and the presence of several rarely occurring taxa, similar to patterns observed in Minalungao National Park in Nueva Ecija (Pecundo et al. 2021). Nearby Polillo Island yielded 34 species in 16 genera (Viray et al. 2014), with ground litter supporting the highest richness and several abundant taxa, further emphasizing the differences between island and mainland forests in Quezon.

FIGURE 12.3 Fruiting bodies of some myxomycetes collected from Aurora, Nueva Ecija and Quezon in Luzon Island, Philippines. (A) *Arcyria denudata* (L.) Wettst., (B) *Cribraria violacea* Rex, (C) *Craterium leucocephalum* (Pers. ex J.F. Gmel.) Ditmar, (D) *Diderma hemisphaericum* (Bull.) Hornem., (E) *Diachea leucopodia* (Bull.) Rostaf., (F) *Hemitrichia calyculata* (Speg.) M.L. Farr, (G) *Nannengaella mellea* (Berk. & Broome) J.M. García-Martín, J.C. Zamora & Lado., (H) *Ophiotheca chrysosperma* Curr., and (I) *Physarum echinosporum* Lister.

Source: Photos by M. H. Pecundo and E. M. Cabutaje

The patterns observed in Quezon were shown to be mirrored in the Bicol Peninsula, where surveys across lowland tropical wet to montane forests have revealed additional contributions to Luzon's myxomycete diversity. Surveys across seven localities (175–435 m) recorded 57 species in 18 genera, including eight new Philippine records and *Stemonaria fuscoides* Nann.-Bremek. & Y. Yamam. as the first record for the Asian Paleotropics (Dagamac et al. 2017c). Forest-edge habitats along Mt. Isarog yielded 24 species in 10 genera (Eloreta et al. 2020), illustrating localized differences in species composition. Work around Bulusan Lake emphasized the role of substrate heterogeneity, identifying 25 species in 12 genera across ten substrate types (dela Cruz et al. 2023). Additional surveys of isolated islands in the Caramoan group further expanded regional diversity, contributing 16 new morphospecies, including *Lamproderma arcyrioides* (Sommerf.) Rostaf., to the Bicol Peninsula checklist (Macabago et al. 2020a).

Separated from Luzon, Mindoro Island hosts distinctive habitats shaped by its geographic isolation, diverse forest structures, and varying levels of human disturbance. On Lubang Island, an offshore island of Mindoro, Macabago et al. (2012) collected specimens from ground leaf litter and twigs, including several species new to the Philippines. The first survey on Mindoro proper, conducted in Puerto Galera, Oriental Mindoro, by Dagamac et al. (2015b), focused on the effects of forest type and human disturbance. They found that species richness was highest in inland lowland forests, while coastal forests had fewer specimens. The authors suggested that differences in forest structure and disturbance likely influenced species occurrence, with some sites exhibiting a higher number of unique species. Subsequent surveys by Pecundo et al. (2020) expanded the geographic scope to two lowland mountain forests—Mt. Malasimbo in Puerto Galera, Oriental Mindoro and Mt. Siburan in Sablayan, Occidental Mindoro—using moist chamber cultures to assess overall diversity. While richness varied between sites, both forests shared many species, indicating that Mindoro's lowland forests support largely overlapping myxomycete communities despite local differences in habitat and disturbance.

While regional surveys highlight diversity and ecological patterns of myxomycetes across landscapes, substrate-specific studies provide insight into the finer ecological relationships that shape its communities. Early research on corticolous myxomycetes—those inhabiting the bark surfaces of living trees—documented seven species on *Samanea saman*, five of which were new records for the Philippines (Dagamac et al. 2010). Since then, corticolous myxomycetes have gained increasing attention in the Philippines, particularly in Luzon. Follow-up surveys examining the bark of mature trees, including *Gmelina arborea*, *Mangifera indica*, *Parkia timoriana*, *Pinus* sp., *Pterocarpus indicus*, and *Shorea contorta*, yielded 17 species (Policina and dela Cruz 2020), emphasizing how tree characteristics may also influence species assemblages. Grass litter has also emerged as an important microhabitat. Carascal et al. (2017) reported 27 species in 13 genera from 300 grass litter samples, while Pecundo et al. (2023) recovered 27 species in 13 genera across multiple substrates (e.g., cow dung, grass litter, broadleaf litter, needle leaves of *Pinus insularis,* inflorescences, fern fronds, and lianas), with grass litter supporting the highest diversity. These substrates further demonstrated how substrate heterogeneity shapes myxomycete occurrence throughout Luzon.

12.3.2 Palawan Island

Palawan occupies a unique geological position in the Philippine archipelago, forming part of the northeastern margin of the Sunda Shelf. During Pleistocene glacial periods, lowered sea levels repeatedly exposed land bridges connecting Palawan with northern Borneo, resulting in flora and fauna with closer affinities to Borneo than to the predominantly oceanic islands of the Luzon–Mindanao region (Esselstyn et al. 2010; Piper et al. 2011). Early myxomycete surveys by Reynolds (1981) documented 26 species in Palawan, and this record was expanded by Pecundo et al. (2017) to 33 species in 15 genera with reports from Puerto Princesa Subterranean River National Park, including 22 species newly reported for the island. Their work highlighted the importance of aerial leaf litter and woody substrates, with specialized microhabitats such as grass litter and dried inflorescences supporting species like *Arcyria cinerea*, *Didymium squamulosum* (Alb. & Schwein.) Fr. & Palmquist, *Echinostelium minutum* de Bary, and *Lamproderma scintillans* (Berk. & Broome) Morgan. More recently, Macabago et al. (2020b) surveyed coastal and inland forests on Coron Island, recording 25 morphotaxa, including *Badhamia macrocarpa* H.Z. Li as a new Philippine record. Combined, these studies raise Palawan's known myxomycete diversity to 56 morphospecies, contributing substantially to the 160 species recorded for the country. Comparative analyses by

Macabago and Stephenson (2021) demonstrated that Palawan's species composition shares greater similarity with the rest of the Philippines (CC = 0.50) than with Borneo (CC = 0.39), indicating that, although Palawan is separated from the main Luzon landmass, its myxomycete communities remain closely aligned with the Philippine archipelago. Despite this richness, surveys remain concentrated in northern and central Palawan, leaving southern and transitional areas largely unexplored. Ongoing anthropogenic pressures, including deforestation, agricultural expansion, and limestone quarrying, further emphasize the need for continued research and conservation of myxomycete communities in Palawan Island.

12.3.3 Visayas Islands

To date, two major groups of islands in the Visayas region have been comprehensively assessed for myxomycetes. These are the Bohol group of Islands (Macabago et al. 2017) and the Negros Island (Alfaro et al. 2014; dela Cruz et al. 2021; Lim et al. 2021). Bohol is an island province located in the Central Visayas region of the Philippines. An ecological approach was employed by Macabago et al. (2017) in their study of myxomycetes on Bohol Island in 2014, examining the potential role of dispersal in the occurrence of myxomycetes along a disturbance gradient following an earthquake and typhoon. Utilizing the moist chamber culture technique, this study identified 54 morphotaxa of myxomycetes belonging to 20 genera, and seven of these were found to be abundant across variable habitats, including heterogeneous natural forests, monotypic man-made woodlands, grasslands, indigenous woodlands, mixed-vegetation islets, and mixed-vegetation coastal forested sites. Analyses of the results suggested that environmental alterations could be a factor affecting the abundance of myxomycetes in different habitats, and that dispersal plays a role in shaping the distribution patterns of myxomycete assemblages associated with particular plant communities. This study increased the number of species of myxomycetes known to the Philippines to 158 with the addition of eight new records—*Comatricha elegans* (Racib.) G. Lister, *Comatricha laxa* Rostaf., *Didymium bahiense* Gottsb., *Physarina echinospora* K.S. Thind & Manocha, *Physarum bitectum* G. Lister, *Physarum sulphureum* Alb. & Schwein., *Physarum serpula* Morgan, and *Physarum straminipes* Lister. This study also highlighted the need for an extensive survey to assess the impacts of conservation and restoration on areas susceptible to ecological shifts, such as those caused by the changing global climate.

Negros Island, located in the Western Visayas, is the fourth-largest island in the Philippines. Referred to as the Negros Island Region, it is an administrative region that includes three provinces (Negros Occidental, Negros Oriental, and Siquijor). Alfaro et al. (2014) collected forest floor litter and agricultural plantation substrates, providing the first reports of myxomycetes from Negros Occidental. In this rapid survey, the two habitat types considered were lowland forested areas in Mt. Kanlaon National Park and agricultural areas in a sugarcane (*Saccharum officinarum*) plantation. Morphological characterization of the myxomycetes from moist chambers identified 28 species belonging to the genera *Arcyria*, *Ceratiomyxa*, *Collaria*, *Comatricha*, *Craterium*, *Cribraria*, *Diderma*, *Didymium*, *Hemitrichia*, *Lamproderma*, *Physarum*, *Stemonitis*, *Trichia*, and *Tubifera*. In both the agricultural and forested areas, *Arcyria cinerea* was the single most abundant myxomycete species. There were more myxomycetes recorded in the mountainous forest (27) compared to agricultural sites (2), verifying that a habitat with more heterogeneous plant communities is characterized by more diverse assemblages of myxomycetes.

In 2017, field specimens of myxomycetes and plant substrates were collected in an expedition to explore the myxomycetes of Negros Oriental. The localities studied were along the forest habitats of Casaroro and Pulangbato Falls in Valencia (dela Cruz et al. 2021) and the habitats represented by successional and mangrove forests in the municipalities of Bindoy, Bacong, Ayungon, and the city of Tanjay (Lim et al. 2021). While both studies were conducted in the southern part of Negros Oriental, the former focused on the assemblages of myxomycetes in forested areas along two natural falls (Casaroro and Pulangbato) in the municipality of Valencia. This town, which is considered 65% mountainous, with elevations ranging from 200 to 500 m above sea level, has a land area of

14,749 ha. Leading up to the collection areas near the waterfalls, the terrain featured steep slopes and a moderate canopy cover, with both sites characterized by a Type III climate described as wet from June to November and relatively dry for the rest of the year. The average rainfall and annual temperature have been recorded as 900 mm and 27.8°C, respectively. Human activities, such as slash-and-burn for forest clearing and construction of infrastructures, were observed during the random collection of various plant-based substrates to prepare the moist chamber cultures. Seventy-nine percent of the moist chambers showed evidence of myxomycete growth and ultimately yielded 34 species belonging to 15 genera (*Arcyria, Ceratiomyxa, Clastoderma, Collaria, Comatricha, Cribraria, Diachea, Diderma, Didymium, Hemitrichia, Lamproderma, Perichaena, Physarum, Stemonitis* and *Stemonitopsis*). Twenty species were common between the two sampling locations; however, the forest areas leading to Casaroro Falls exhibited a higher taxonomic diversity than those of Pulangbato Falls, despite Pulangbato having a slightly higher number of recorded species.

The study by Lim et al. (2021) surveyed two successional forests and two coastal mangrove forests in various municipalities in Negros Oriental. These chosen sites were (1) two successional forests, namely Liptong Woodland, which is a forest patch and reservation in the municipality of Bacong and Ayungon Forest, which is a successional forest in the municipality of Ayungon, and (2) two mangrove forests, namely the Tanjay mangrove forest, which is a reforestation site for mangroves located along the coast in the city of Tanjay, and Tinaogan mangrove forest, a coastal mangrove forest in the municipality of Bindoy. A total of 19 species of myxomycetes in 12 genera were identified from moist chamber cultures, whereas 13 species in nine genera were identified from each successional forest site, and only 3–4 species from 3–4 genera of myxomycetes were recorded from the mangrove forests. Among these, *Didymium nigripes* (Link) Fr., *Stemonitis fusca* Roth, and *Diderma effusum* (Schwein.) Morgan were the most abundant species. In this study, five species—*Didymium nigripes, Lamproderma scintillans, Arcyria cinerea, Stemonitis fusca*, and *Physarum oblatum* T. Macbr.—were also reported for the first time for mangrove forests in the country.

12.3.4 Mindanao Islands

To date, at least three published studies on myxomycetes have been reported from the Mindanao Islands, despite being the second largest island in the Philippines. Buisan et al. (2019) reported myxomycetes from banana plantations in North Cotabato and Maguindanao. This was then followed by studies in rice farms in Cotabato Province, Southern Mindanao (Buisan et al. 2020), and an experimental forest in Kidapawan City, Cotabato Province (Buisan and Dagamac 2024).

The study by Buisan et al. (2019) recorded five species of myxomycetes, namely *Arcyria cinerea, Didymium nigripes, Didymium squamulosum, Perichaena depressa* Lib., and *Perichaena vermicularis* (now known as *Gulielmina vermicularis* [Schwein.] García-Cunch., J.C. Zamora & Lado) collected from banana leaves in two areas, which were in Pikit, North Cotabato (Barangay Balong and Barangay Gocotan) and in Pagalungan, Maguindanao (Barangay Layog). This study reported a 38% positive rate for moist chambers, either as fruiting bodies or plasmodia, with 31% observed from the pseudostem and only 7% found to be growing on actual ground leaf litter. Not long after that, Buisan et al. (2020) reported the first records of myxomycetes on rice litter from farms in Midsayap and Kabacan, Southern Mindanao. A comparable (38%) moist chamber positive rate was observed for rice litter substrates that yielded three cosmopolitan species (*Arcyria cinerea, Diderma hemisphaericum*, and *Perichaena depressa*). The moist chambers from the organic rice farm in Midsayap generally yielded more myxomycetes, except for *A. cinerea*, which was marginally more evident in the conventional rice farm in Kabacan. In a more recent paper on myxomycetes from Mindanao, Buisan and Dagamac (2024) studied myxomycete composition relative to forest zones (core, midline, or edge), and identified 15 species. These were *Arcyria cinerea, Didymium nigripes, Diderma effusum, D. hemisphaericum, Gulielmina vermicularis, Hemitrichia serpula* (Scop.) Rostaf. ex Lister, *Ophiotheca pedata* (Lister & G.Lister) García-Cunch., J.C. Zamora & Lado, *Perichaena chrysosperma* (now *Ophiotheca chrysosperma* Curr.), *P. depressa, Physarum album* (Bull.) Chevall., *P. cinereum* (Batsch) Pers., *P. decipiens*, *P. lakhanpalii* (now *Nannengaella lakhanpalii* [Nann.-Bremek. & Y. Yamam.] J.M. García-Martín, J.C. Zamora & Lado), *Stemonitis fusca*, and *S. pallida* Wingate. Among these, *Perichaena depressa* was found to be the most abundant, followed by *A. cinerea* and *S. fusca*. Fifty-one percent of the moist chambers prepared with aerial leaf litter, ground leaf litter, and woody substrates from the ARDEC Experimental Forest, a secondary-growth forest in Kidapawan City, produced myxomycetes, with plasmodia as the more evident form than fruiting bodies. There was no significant difference in alpha diversity values among forest zones, sites, or substrates; however, cluster analysis revealed that the forest edge and core were more similar to each other than to the forest midline.

12.3.5 Smaller Philippine Island Ecosystems

Hundred Islands and Anda Island, Pangasinan. Pangasinan is a coastal province situated on the western part of the Philippines, bordering the Lingayen Gulf and the West Philippine Sea. dela Cruz et al. (2011) and Kuhn et al. (2013b) published diversity assessment studies of myxomycetes from groups of islands in Pangasinan, specifically the Hundred Islands and Anda Island. Located on the northwestern portion of the Lingayen Gulf, Anda Island is positioned in close proximity north of the Hundred Islands. Climate data indicate a minimum and maximum temperature of 23.3°C and 32.4°C, respectively, and an average monthly rainfall from 6.1 mm in February to 608.6 mm in August in this area. In 2008, substrates from aerial and ground leaf litter, twigs, and dead barks of living trees were collected to prepare moist chambers from seven geographically isolated islands, characterized by limestone forests dominated by small trees, shrubs, grasslands, and some mangrove forests, from the Hundred Islands National Park in Alaminos, Pangasinan. The isolated islands were Camantiles Island, Monkey Island (0.19 sq. km), Children's Island (0.024 sq. km), Governor's Island (0.074 sq. km), Century Park Island (0.13 sq. km), Sison Island (0.058 sq. km), and Shell Island (0.0056 sq. km) (dela Cruz et al. 2011). The identification of the collected myxomycetes revealed a total of 30 species, belonging to ten genera—*Arcyria, Collaria, Calomyxa, Comatricha, Diderma, Didymium, Lepidoderma, Perichaena, Physarum*, and *Stemonitis*. Century Island showed the highest yield of myxomycetes, while the Governor's and Sison islands indicated high taxonomic diversity. Children's and Monkey islands revealed a high similarity in terms of species composition; however, low percentage similarity values were recorded, indicating that the study sites differed in the relative abundance of species. In addition, two species were reported as new records for the Philippines. These were *Lepidoderma tigrinum* (now *Diderma tigrinum* [Schrad.] Prikhodko, Shchepin, Novozh. López-Vill., G. Moreno & Schnittler) and *Perichaena pedata*.

Anda Island, comprising the municipality of Anda together with minor islets northwestern of the Hundred Islands, was the study site of Kuhn et al. (2013b), who delivered the first reports of myxomycetes on the island. With a land area of 83.8 square kilometers, the island is characterized by tropical secondary forest, interspersed with patches of rice fields and residential houses. An area in the secondary forest, located near the local highway and adjacent to a rice field, was selected as the sampling site to assess the diversity and abundance of myxomycetes found on Anda Island, Pangasinan. From the 180 moist chambers set up from ground litter and twigs, 55% yielded evidence of myxomycetes, and 24 species of myxomycetes belonging to 11 genera were collected and identified: *Arcyria* (2 species), *Collaria* (1), *Comatricha* (1), *Craterium* (2), *Diachea* (1), *Diderma* (2), *Didymium* (1), *Elaeomyxa* (1), *Perichaena* (1), *Physarum* (11), and *Stemonitis* (1). Ground leaf litter substrates produced more myxomycetes than twigs (29% versus 26%). Species diversity, richness, and evenness indices revealed a rich assemblage of myxomycetes, with three species—*Craterium microcarpum*, *Physarum decipiens*, and *Elaeomyxa miyazakiensis*—representing new records for the Philippines.

Polillo Island. The Polillo group of islands is located in the province of Quezon, Southern Luzon, Philippines, west of the Pacific Ocean. With over 600 km^2 of land, the main island of Polillo, which is the namesake of the Polillo islands, is subdivided across three municipalities—Burdeos (northeast), Panukulan (northwest), and Polillo (southern). The island itself is covered in forests, but the northern section has been affected by illegal logging activities. These forests serve as a refuge for many threatened and endemic species and are situated in watersheds that provide valuable sources of potable water, food, timber, and medicinal resources (Viray et al. 2014). The Southeast monsoon occurs from May to September, while the Northeast monsoon takes place from October to April, with almost no dry season throughout the year. The maximum rainfall occurs from November to January, when typhoons are also typically present. Previous studies have observed dominant forest trees within the island, including antipolo (*Artocarpus blancoi*), narra (*Pterocarpus indicus*), batino (*Alstonia macrophylla*), anubing (*Artocarpus rubrovenius*), lanite (*Kibatalia gitingensis*), layasin (*Maranthes corymbosa*), takip-asin (*Macaranga bicolor*), and tibig (*Ficus nota*) (Gonzalez 2007). In the study by Viray et al. (2014), 34 species of myxomycetes were collected from moist chambers prepared with ground leaf litter and twigs and these were identified as representing 16 genera (*Arcyria, Ceratiomyxa, Comatricha, Collaria, Cribraria, Diachea, Diderma, Didymium, Echinostelium, Hemitrichia, Lamproderma, Lycogala, Perichaena, Physarum, Stemonitis*, and *Trichia*). There were more species recorded from leaf litter than twigs (24 versus 20), where *Arcyria cinerea* and *Cribraria microcarpa* were found to be the most abundantly occurring in this first report of myxomycetes from Polillo Island.

Napayong Island. Napayong Island, now officially called "Laurel Island," under the municipality of Tanauan in the province of Batangas, is situated within Taal Lake, which encompasses Taal Volcano Island. The island is characterized by a Type 1 climate with two pronounced seasons, a dry season from November to April and a wet season from May to October. The average temperature is reported to be 25.7°C, with May having the highest average temperature of 28.8 °C. The highest precipitation is expected in July, with an average of 320 mm (https://en.climate-data.org). In 2014, Isagan et al. (2020) collected decayed aerial and ground leaf litter, twigs, and woody vines from study areas Napayong Island and Mt. Makulot, located in the adjacent mainland within the municipality of Cuenca in the same province, to assess and compare the occurrence and diversity of myxomycetes in the forest areas of Napayong Island with those of Taal Lake and the Taal Volcano crater rim in Mt. Makulot. In this study, 35 species of myxomycetes were collected and identified. These represented 14 genera—*Arcyria*, *Ceratiomyxa*, *Clastoderma*, *Collaria*, *Comatricha*, *Cribraria*, *Diachea*, *Diderma*, *Didymium*, *Hemitrichia*, *Lamproderma*, *Perichaena*, *Physarum*, and *Stemonitis*. Seven species were exclusive to Napayong Island, which also exhibited higher species diversity, while five species were uniquely present on Mt. Makulot. This study represented the first investigation of myxomycetes on Napayong Island, a small island uniquely situated within a lake that lies on another island.

Lubang Island. Under the jurisdiction of the province of Occidental Mindoro, the Lubang Group of Islands, situated at the northern tip of Mindoro and approximately 100 km southwest of Manila, constitutes an isolated geographical entity that is detached from any continental landform. Lubang Island is the largest of the seven islands in the group and is divided into two municipalities—Lubang in the northwest section and Looc in the southeastern portion of the island. Macabago et al. (2012, 2016) collected both plant substrates for use in setting up the moist chambers and myxomycete specimens that had naturally fruited in the field. Two distinct forest habitats (1) on the periphery of the island, referred to as coastal forests and (2) on two accessible trails leading up to the summit of Mt. Gonting near the center of the island, referred to as mountain forests, were selected as sampling sites on Lubang island, which has a rugged terrain, with strips of coastal lowlands, mountain ranges, valleys, plateaus, and rolling lands along the coastal region. In this first exploration of myxomycetes on Lubang Island, the total number of species known from the country was increased to 128, with six new records for the Philippines (Macabago et al. 2012). The new records were *Arcyria globosa* Schwein., *Collaria rubens* (Lister) Nann.-Bremek., *Comatricha robusta* (T.N. Lakh. & K.G. Mukerji) Nann.-Bremek. & Y. Yamam., *Craterium atrolucens* Flatau, *Lamproderma cacographicum* Bozonnet, Mar. Mey. & Poulain, and *Perichaena microspora*. Occurrence and abundance data from the moist chamber collections of myxomycetes in this study were analyzed using species and taxonomic diversity indices, as well as distribution patterns within the community (Macabago et al. 2016). The most abundant species were *Arcyria cinerea, Diachea leucopodia, Diderma effusum, Lamproderma scintillans*, and *Physarum cinereum*. The highest productivity of myxomycetes, either as plasmodia or fruiting bodies, and the highest taxonomic diversity were observed from ground litter substrates collected from the mountain forests. In general, the assemblages of myxomycetes on Lubang Island were found to be similar within a particular area or within the same forest type.

Caramoan Islands. The municipality of Caramoan in the province of Camarines Sur, which is located on the east coast of the Bicol Peninsula facing the Pacific Ocean, features a rugged land with forested areas, white beaches, lagoons, caves, and rocky cliffs. Off the coast of the Caramoan mainland are numerous islands and islets collectively called the "Caramoan Islands." Four islands of various sizes and distances from one another in this archipelago (Lahos, Matukad, Minalahos, and Busdak) were surveyed to characterize the assemblages of myxomycetes through a correlational study using geographical and ecological distance and to update the myxomycetes known from the Bicol Peninsula (Macabago et al. 2020a). Several species of trees and other plants were noted during this project, including narra (*Pterocarpus indicus*), molave (*Vitex* sp.), ipil (*Intsia bijuga*), and other dipterocarp trees, bamboos, some shrubs, grasses, and mangroves along the coast. Using the moist chamber culture technique, a total of 38 morphospecies of myxomycetes were recorded and identified to belong to 20 genera—*Arcyria, Badhamia, Ceratiomyxa, Clastoderma, Collaria, Comatricha, Craterium, Cribraria, Diachea, Dictydiaethalium, Diderma, Didymium, Fuligo, Hemitrichia, Lamproderma, Lycogala, Perichaena, Physarum, Stemonaria*, and *Stemonitis*. A new record for the country (*Lamproderma arcyrioides* [Sommerf.] Rostaf.) and 16 new morphospecies were obtained from the Bicol Peninsula, bringing the total to 73. Lahos Island was found to be the most species-diverse, followed by Matukad Island, which had the highest species richness and taxonomic diversity. Pairwise comparisons using community similarity indices and clustering analyses consistently revealed that Lahos and Matukad were the most similar to each other, being geographically close yet the most distant from the mainland. In contrast, the two smallest islands, located nearest to each other and to the mainland, clustered together in the analysis but exhibited the lowest pairwise percentage similarity value. Interestingly, the ecological patterns observed in this study seem to align more closely with the unified neutral theory of biodiversity and biogeography than with the insular biogeography theory.

12.4 ECOLOGICAL STUDIES ON PHILIPPINE MYXOMYCETES

12.4.1 General Distribution Patterns

Globally, more than 1,000 morphospecies of myxomycetes are recognized. The country of the Philippines has emerged as a biodiversity hotspot for these organisms, with more than 160 species recorded to date, which represents about 20% of recorded taxa worldwide. These numbers are mostly accounted for by records obtained from moist chambers cultures prepared with decaying organic substrates from the field rather than as fruiting bodies that have developed in nature. Most surveys conducted in Luzon (Dagamac and dela Cruz 2019) and the smaller islands in the Visayas region (Alfaro et al. 2014) and Mindanao (Buisan et al. 2020) have revealed both cosmopolitan species (e.g., *Arcyria cinerea* and *Didymium hemisphaericum*) and localized taxa associated with specific microhabitats.

While most species of myxomycetes are broadly distributed across continents (Dagamac et al. 2017a), studies in the Philippines show varying occurrences of species and the extent of their distributional ranges across different islands. For example, some taxa are reported to appear more frequently in moist chambers collected in montane mossy forests of Mindanao compared to the lowland forests of Luzon, the largest group of islands in the archipelago. Although degrees of regional endemism or at least strong island-specific occurrence patterns have not been highly investigated among the scattered studies conducted throughout the Philippines, these variations may result from differences in climate, substrate availability, and historical biogeographic processes linked to the archipelagic structure of the Philippines. In terms of vertical dispersion analysis of myxomycete composition, a simulation conducted in an urban setting by Siscar et al. (2024) has implicated cosmopolitan myxomycetes associated with high-rise skyscrapers.

12.4.2 Elevation Gradients

The fragmented geography of the Philippines creates natural barriers and corridors for spore dispersal. Myxomycete spores are lightweight and can be dispersed by wind, water, and animal vectors, enabling them to colonize across islands (Schnittler et al. 2022). However, despite this high dispersal potential, differences in island size, elevation, forest cover, and climatic regime influence community composition (Macabago et al. 2017).

Among these factors is elevation, which significantly influences myxomycete distribution. Initial studies examining the elevational distribution of myxomycetes in the Philippines were carried out in Mt. Arayat National Park (Dagamac et al. 2011). The study yielded a total of 30 species of myxomycetes from five different elevational transect points, with the highest richness and composition of myxomycete species recorded at the substrates collected at the highest elevation (Dagamac et al. 2014). Unlike the more established Neotropical counterparts, the results of this initial paper seem to contradict the known hypothesis of the "*increasing elevation, decreasing diversity*" pattern among myxomycetes. This can be attributed to the fact that the current elevational extent for study was only 700–800 meters above sea level, whereas the mountain peak was at 1,300 m. Recently, a paper by Racoma et al. (2025) investigating the different elevational transects of Mt. Calavite Wildlife Sanctuary in Occidental Mindoro has supported the claim for an elevational diversity gradient. This study has identified heterogeneous habitats that occupy the lowland dipterocarp forest, as well as the highland forest, which is dominated by grasses and tall bamboos (Komoda et al. 2025).

Additional investigations at higher elevations are therefore necessary to fully understand the elevational diversity patterns of myxomycetes in the Philippines and determine if this observed trend persists, similar to the observations based on molecular patterns of cryptic nivicolous myxomycetes (Dahl et al. 2018). In addition, variations in local habitat features, such as canopy cover and leaf litter density, coupled with microclimatic factors like temperature and precipitation, can significantly influence myxomycete abundance and species richness across different elevational zones. Furthermore, studies have demonstrated that geographic distance correlates with fruiting body assemblages in certain habitats, such as valleys and high-light environments, suggesting that limits on dispersal play a significant role in structuring myxomycete communities (Dagamac et al. 2012). This suggests that while myxomycetes may exhibit high dispersal potential, localized environmental conditions and geographic barriers can still result in distinct community assemblages. Indeed, similar patterns of species turnover have been observed in other ecological contexts, where sites at varying elevations frequently share less than 50% of their taxa, suggesting that elevation acts as a significant driver of community structure (Racoma et al. 2025). The influence of meso-climatic differences, shaped by elevation and slope aspect, further contributes to varied soil characteristics, including moisture and temperature, which in turn affect protist community composition. For instance, certain environmental variables such as understory density, temperature, leaf litter thickness, and tree height exhibit strong correlations with species distribution and abundance, influencing the availability of suitable microhabitats for myxomycetes (Dagamac et al. 2015a). Furthermore, changes in microbial communities of tree ferns along elevation gradients underscore the dynamic interplay between environmental heterogeneity and species distribution, suggesting that similar patterns may hold for myxomycetes (Vélez-Múnera et al. 2024).

The intricate relationship between diverse environmental factors and elevational gradients highlights the necessity of carrying out comprehensive, multi-year sampling efforts to accurately capture the full extent of myxomycete diversity and its drivers in mountainous regions of the Philippines. Such detailed investigations could employ quantitative molecular techniques (Almadrones-Reyes et al. 2019; Shchepin et al. 2019) such as qPCR to assess mechanisms through which temperature influences fruiting success across various myxomycete species (Buisan and Dagamac 2021; Schnittler et al. 2017). This approach would provide a more robust understanding of how soil temperature regimes, influenced by elevation, impact the phenology and distribution of myxomycetes, drawing parallels with studies on other microbial groups (Ocenar-Bautista et al. 2025). These detailed examinations are crucial for elucidating the complex ecological drivers that shape myxomycete diversity and distribution, particularly given the observed high turnover rates in other fungus-like protist communities in other situations such as along ridge-to-reef gradients (Claudio-Paragas et al. 2024).

12.4.3 Species-Substrate Relationship

Myxomycetes are often categorized based on the substrates from which they are associated. For instance, some species predominantly inhabit decaying wood, while others are more frequently found on leaf litter (Redeña-Santos et al. 2017), bark of living trees (Policina and dela Cruz 2020), or even dung (Abdel-Raheem 2002). This substrate specificity is not merely incidental but often reflects distinct physiological adaptations to variations in nutrient availability, moisture retention, and microbial communities associated with these diverse microhabitats (Redeña-Santos et al. 2018). In the context of the Philippines, understanding these intricate relationships is crucial for unraveling the ecological roles and biodiversity patterns of myxomycetes within its tropical ecosystems (Balaoro-Banzuela et al. 2023). Local studies investigating these relationships have begun to reveal unique associations, underscoring the influence of specific lignocellulosic matrices and associated microbial flora on myxomycete distribution and community structure (Novozhilov et al. 2020). For example, specific tree species may harbor distinct myxomycete assemblages due to variations in their wood chemistry, bark texture, or decomposition rates, influencing the availability of essential micronutrients and moisture (Mifsud 2020). Moreover, the decomposition stages of these substrates, ranging from freshly fallen logs to highly degraded detritus, also play a significant role in determining myxomycete succession and species composition (Bernardo et al. 2018). These microhabitat preferences are further modulated by prevailing climatic conditions such as humidity and temperature, which are particularly pronounced in the tropical climate of the Philippines (Dagamac and dela Cruz 2015). Listed below are some of the common types of substrates with which myxomycetes are associated and collected from in local studies in the Philippines.

- Corticolous species (growing on bark): Common in humid forests where tree bark provides suitable microhabitats. Species of *Arcyria* and *Hemitrichia* are representatives.
- Lignicolous species (on decaying wood): Decayed logs harbor diverse species such as *Stemonitis* and *Comatricha*, depending on the stage of wood decomposition.
- Litter-inhabiting species (from forest floor litter): Many species of *Physarum* and *Perichaena* are abundant in leaf litter substrates.
- Coprophilous species (on dung): Less studied in the Philippines but occasionally reported in grassland or disturbed habitats.

While the true diversity of myxomycetes is likely underestimated due to their cryptic nature and the limitations of morphological identification (Dagamac et al. 2017b), ongoing surveys are crucial for detecting and diagnosing new occurrences, particularly in less-explored regions of the Philippines with more varied ecosystems (e.g., peatlands, marshes, and caves). Indeed, the collection methods, often relying on moist chamber cultures prepared with substrate samples, highlight the dependence on specific microenvironmental conditions for development of fruiting bodies, which are often overlooked in general ecological surveys (Nguyen et al. 2019). Nevertheless, fruiting body-based surveys can provide valuable insights into the myxamoebal communities present in the soil, effectively indicating specific habitat features much like other indicator groups such as plants or amphibians. This suggests that even when myxomycetes are not overtly fruiting, their presence as amoeboid organisms within the soil is indicative of specific environmental parameters and resource availability (Ocenar-Bautista et al. 2024).

12.4.4 Phenology and Seasonal Dynamics

The seasonal shifts in precipitation and humidity characteristic of tropical climates profoundly impact the life cycle and distribution patterns of these ubiquitous, yet often overlooked, protists (Macabago et al. 2010). Therefore, understanding the phenology of myxomycetes in the Philippines necessitates a detailed examination of their responses to these distinct climatic periods, particularly the pronounced rainy and dry seasons (Rea-Maminta et al. 2015; Dagamac et al. 2012). The Philippines experiences distinct wet and dry seasons, and precipitation is the primary driver of myxomycete fruiting. This climatic dichotomy, characterized by significant variations in rainfall, directly influences substrate moisture and microbial activity, which are critical for myxomycete plasmodial development and sporulation (Buisan and Dagamac 2024). The rainy season is characterized by high humidity and moist substrates, which favor the fruiting of most species following a dry period. Field surveys show peak diversity during or immediately after continuous rains which is then followed by sunny period to induce fruiting. Conversely, species adapted to drier conditions might exhibit delayed fruiting or specialized resting stages during prolonged wet periods, awaiting more favorable conditions. On the other hand, during the dry season, some drought-tolerant myxomycete species persist, but overall fruiting frequency declines in the field. Our observations are that samples collected during the dry season have a higher success of fruiting when subjected to a moist chamber set-up in comparison to substrates that were collected after the onset of rain. During the dry season, the scarcity of moisture typically limits plasmodial growth and fruiting directly, perhaps in the field, though some desiccation-tolerant species may persist as sclerotia, reactivating with the onset of favorable conditions (Dahl et al. 2018).

The Coronas Climate Classification System is one of the most enduring frameworks for describing the climatic diversity of the Philippines. It was first proposed by Fr. José Algué, S.J., and later refined by Fr. José Coronas, S.J., in the early twentieth century to provide a more practical basis for understanding tropical rainfall distribution across the archipelago. Unlike temperate-based classifications such as Köppen-Geiger, which emphasize temperature gradients, the Coronas system focuses primarily on rainfall seasonality, which is the most ecologically and agriculturally significant variable in a tropical island setting. The system divides the Philippines into four climate types. Here we present the climate types and the myxomycete surveys so far conducted in areas affected or influenced by these climate types.

- Type I—Distinct dry season from November to April and a pronounced wet season during the rest of the year. Central Luzon and nearby provinces are typically Type I (distinct dry/wet season) although local topography (mountains, lakes) modulates microclimate. Mt. Arayat in Central Luzon has an interactive species list and surveys documenting many myxomycete taxa (Dagamac et al. 2011) while La Mesa Eco-park in Metro Manila (Macabago et al. 2010) and other Luzon urban/riparian sites also have published species lists.
- Type II—No dry season but with a maximum rainfall period from November to January. Many areas of Mindanao are Type II but still depends on local monsoon influence. Studies reported the first records of myxomycetes on rice litter from Cotabato (southern Mindanao), documenting cosmopolitan species (e.g., *Arcyria cinerea*, *Diderma hemisphaericum* and *Perichaena depressa*) and showing that agroecosystems can host myxomycetes (Buisan et al. 2020; Buisan and Dagamac 2024). Moist-chamber based study found higher yields in organic versus conventional fields.

- Type III—Without very pronounced wet or dry seasons, although conditions tend to be relatively dry from December to February and wet during the rest of the year. Mindoro is spatially variable but typically shows Type III seasonality depending on the exposure; overall western islands often have clearer dry seasons. First records from Lubang were published as island-specific notes of myxomycetes which was followed by studies from Puerto Galera, Oriental Mindoro and Sablayan, Occidental Mindoro (Macabago et al. 2012; Dagamac et al. 2015b; Pecundo et al. 2020).
- Type IV—Rainfall more or less evenly distributed throughout the year. A dedicated survey for Coron/Palawan (Pecundo et al. 2017) reported 56 morphospecies for the Palawan group, including some new country records. This is one of the better-documented island groups in the Philippines. Another Type IV climate is found in the peninsular landscapes of Bicol where a rapid assessment conducted by Dagamac et al. (2017c) reported the first records of *Stemonaria fuscoides*.

12.5 COMPUTER-BASED ECOLOGICAL MODELING OF MYXOMYCETES

Given the scattered nature of biodiversity records, SDMs such as MaxEnt (Maximum Entropy modeling) have been recently applied to predict the potential distribution of Philippine myxomycetes. These models utilize species occurrence data, along with environmental variables (temperature, precipitation, and vegetation indices), to identify suitable habitats. Results often show strong correlation with precipitation seasonality and forest cover (Dagamac et al. 2021; Almadrones-Reyes and Dagamac 2018). Thus far for the Philippines, the models reported by Limbo-Dizon et al. (2022) are based on the distributional ranges of three commonly occurring bright-spored myxomycetes (*Arcyria cinerea*, *Perichaena depressa*, and *Hemitrichia serpula*) in the Philippines. This study highlighted how studies in Mindanao often become a biological blind spot for myxomycetes due to the limitations of records for these organisms.

In a future context, the integration of satellite-derived data, such as the NDVI (Normalized Difference Vegetation Index) and land cover maps (Almadrones-Reyes and Dagamac 2023), provide spatial proxies for forest productivity and microclimate. GIS-based modeling enables the visualization of potential hotspots (Olfato-Parojinog et al. 2023), the overlap of species ranges, and the identification of conservation priority areas. By integrating climate time series (rainfall and humidity) with fruiting data, predictive models can forecast seasonal peaks in fruiting. These models have practical applications for planning field surveys and monitoring ecological responses to climate change.

12.6 CONCLUDING POINTS

Myxomycete studies in the Philippines has shown a tremendous increase in the number of publications from relatively few studies (only about three publications before 1981) to more than 50 papers from 2009 until present. This is also reflected in a spike in the number of species recorded in the country, with the addition of more than 50 new records and two new species from the earlier annotated list of 107 species. In addition, there is a changing landscape in Philippine myxomycete research, from earlier efforts of documenting species and reports of species lists to distribution and diversity studies and other assessment of ecological patterns. Myxomycete research has also expanded from substrate-specific surveys to comparative evaluation of diversity across various habitats and ecosystems. There also have been efforts to use myxomycetes as model organisms for impact studies (e.g. of natural and man-made disturbances) and on their possible application in enzyme production and bioremediation. With the advent of artificial intelligence and machine learning, studies on Philippine myxomycetes evolved from development of photo guides and interactive database into ecological modeling. However, many areas remained unexplored or little explored for myxomycetes. Ecological modeling has identified many island ecosystems in the Visayas and in Mindanao as blind spot on myxomycetes. The species associated with various microhabitats and the associations of myxomycetes with other organisms such as endemic plants and animals remained understudied. Phylogenetic studies on Philippine myxomycetes are absent. The need to bring myxomycetes to the limelight amidst its seemingly non-economic importance, the training of future myxomycete researchers and the involvement of Filipino enthusiasts through citizen-science initiatives can drive further progress to the field.

REFERENCES

Abdel-Raheem, A.M. 2002. Myxomycetes from upper Egypt. *Microbiological Research* 157:47–67.

Accad, N.S.B., D.A.C. Bernardo, V.D.A. Candelaria, L.M.L. Deleniana, J.C.B. Olivares, N.H.A. Dagamac & T.E.E. dela Cruz. 2023. Occurrence of myxomycetes in elevated urban residential gardens in Metro Manila, Philippines. *Philippine Journal of Systematic Biology* 17:16–22.

Alfaro, J.R.D., D.L.I.M. Alcayde, J.B. Agbulos, N.H.A. Dagamac & T.E.E. dela Cruz. 2014. The occurrence of myxomycetes from a lowland montane forest and agricultural plantations of Negros Occidental, Western Visayas, Philippines. *Fine Focus* 1:7–20.

Almadrones-Reyes, K.J. & N.H.A. Dagamac. 2018. Predicting local habitat suitability in changing climate scenarios: Applying species distribution modelling for *Diderma hemisphaericum*. *Current research in Environmental & Applied Mycology* 8:492–500.

Almadrones-Reyes, K.J. & N.H.A. Dagamac. 2023. Land-use/land cover change and land surface temperature in Metropolitan Manila, Philippines using lands at imagery. *GeoJournal* 88:1415–1426.

Apurillo, C.C.S., L. Cai & T.E.E. dela Cruz. 2019. Diversity and bioactivities of mangrove fungal endophytes from Leyte and Samar, Philippines. *Philippine Science Letters* 12:33–48.

Atayde, R.J.D., B.H. Sebastian, J.J.T. Bumatay, M.J.D.G. Magpayo, et al. 2012. Myxomycetes (plasmodial slime molds) from Biak-na-Bato National Park and Angat Dam Watershed, Bulacan, Philippines. *Philippine BIOTA Journal* 45:45–64.

Aurelio, M.A. 2000. Shear partitioning in the Philippines: Constraints from Philippine Fault and global plate motion data. *The Island Arc* 9:584–597.

Balaoro-Banzuela, R.C., C.E. Ocenar-Bautista, D.E. Buebos-Esteve, C.Y. Claudio-Paragas, K.J. Almadrones-Reyes, J.L.M. Bernardo, J.C. Redeña-Santos & N.H.A. Dagamac. 2019. Distribution of 18s rDNA variants of *Diderma hemisphaericum* (Myxomycota) specimens obtained from moist chambers of materials collected in the Philippines and Vietnam. *Philippine Journal of Systematic Biology* 13:19–28.

Balaoro-Banzuela, R.C., C.E. Ocenar-Bautista, D.E. Buebos-Esteve, C.Y. Claudio-Paragas, et al. 2023. Rapid diversity assessment of litter myxomycete assemblages in the upland and coastal terrains of San Fernando City, La Union, Philippines. *Biodiversitas* 24:2877–2886.

Barcelona, J., D. Nickrent, J. Lafrankie, J.R. Callado & P. Pelser. 2013. Co's Digital Flora of the Philippines: Plant identification and conservation through cybertaxonomy. *Philippine Journal of Science* 142:57–67.

Bemis, K.E., M.G. Girard, M.D. Santos, K.E. Carpenter, J.R. Deeds, D.E. Pitassy, N.A.L. Flores, E.S. Hunter, A.C. Driskell, K.S. Macdonald III, L.A. Weigt & J.T. Williams. 2023. Biodiversity of Philippine marine fishes: A DNA barcode reference library based on voucher specimens. *Scientific Data* 10:411.

Bernardo, J.L.M., L.J.Q. Arioder, K.J. Almadrones-Reyes & N.H.A. Dagamac. 2018. Myxomycete communities occurring in fragmented forest patches in two municipalities of Laguna, Philippines. *Community Ecology* 19:289–299.

Birdlife International. 2025. *Birdlife International Country Profile: Philippines*. http://www.birdlife.org/datazone/country/philippines

Buisan, P.N.H.N., D. Abu, J.P. Catipay, C.J. Dango, J. Supremo & N.H.A. Dagamac. 2020. Documenting the first records of myxomycetes on rice litter of Cotabato, Southern Mindanao, Philippines. *Karstenia* 58:250–259.

Buisan, P.N.H.N. & N.H.A. Dagamac. 2021. Elucidating hidden slime mold diversity in Southeast Asia: A review of potential methods. *Slime Molds* 1:V1A3.

Buisan, P.N.H.N. & N.H.A. Dagamac. 2024. Do myxomycete assemblages differ in the forest core vs. its edge? Rapid assessment of a tropical secondary-growth experimental forest in Southern Mindanao, Philippines. *Protistology* 18:325–343.

Buisan, P.N.H.N., Z.M. Pasandalan, A.D. Pandulo, N.A. Balabagan, T.O. Farid, T.A. Abubakar, A.M.M. Pantog, D.A. Labas & N.H.A. Dagamac. 2019. Records of cosmopolitan myxomycetes occurring on banana litter from Southern Mindanao, Philippines. *Philippine Journal of Systematic Biology* 13:55–57.

Cabutaje, E.M., M.H. Pecundo & T.E.E. dela Cruz. 2021. Diversity of myxomycetes in typhoon-prone areas: A case study in beach and inland forests of Aurora and Quezon Province, Philippines. *Sydowia* 73:113.

Carascal, M., M.A.D. Rea, N.H.A. Dagamac & T.E.E. dela Cruz. 2017. Myxomycetes associated with grassland litter in the Philippines. *Current Research in Environmental & Applied Mycology* 7:56–63.

Cheng, C.B.T., K.N.T. Yu, M.L. Campos, J.M.V. Adora, G.C.P. Pascua, M.V.B. Pangilinan, A.T. Buaya & T.E.E. dela Cruz. 2013. Myxomycetes (plasmodial slime molds) along the northern slope of Mt. Makulot, Cuenca, Batangas, Philippines. *Asian Journal of Biodiversity* 4:65–83.

Claudio-Paragas, C.Y., C.E. Ocenar-Bautista, R.C. Balaoro-Banzuela, L.A.D.A. Tan, D.E. Buebos-Esteve & N.H.A. Dagamac. 2024. Dictyostelids isolated from the soils of San Fernando City, La Union, Philippines. *Protistology* 18:52–59.

Corpuz, I.R., C.C. Martinez, K.A. Petilla, J.N. Baranda, A. Buaya & T.E.E. dela Cruz. 2012. Occurrence and diversity of myxomycetes (plasmodial slime molds) in Mt. Palay-palay National Park, Cavite, Philippines. *Acta Manilana* 60:57–65.

Dagamac, N.H.A., B. Bauer, J. Woyzichovski, O.N. Shchepin, Y.K. Novozhilov & M. Schnittler. 2021. Where do nivicolous myxomycetes occur?–Modeling the potential worldwide distribution of *Physarum albescens*. *Fungal Ecology* 53:101079.

Dagamac, N.H.A. & T.E.E. dela Cruz. 2015. Myxomycetes research in the Philippines: Updates and opportunities. *Mycosphere* 6:784–795.

Dagamac, N.H.A. & T.E.E. dela Cruz. 2019. The Philippine slime molds after Dogma's 1975 list–How far have we been. *Philippine Journal of Systematic Biology* 13:58–65.

Dagamac, N.H.A., T.E.E. dela Cruz, M.V.B. Pangilinan & S.L. Stephenson. 2011. List of species collected and interactive database of myxomycetes (plasmodial slime molds) for Mt. Arayat National Park, Pampanga, Philippines. *Mycosphere* 2:449–455.

Dagamac, N.H.A., T.E.E. dela Cruz, M.A.D. Rea-Maminta, J.A.D. Cruz & M. Schnittler. 2017c. Rapid assessment of myxomycete diversity in the Bicol Peninsula, Philippines. *Nova Hedwigia* 104:31–46.

Dagamac, N.H.A., D.V. Leontyev & T.E.E. dela Cruz. 2010. Corticolous myxomycetes associated with *Samanea samans* (Jacq.) Merr. collected from different sites in Luzon Island, Philippines. *The Philippine Biota* 20.

Dagamac, N.H.A., Y.K. Novozhilov, S.L. Stephenson, C. Lado, C. Rojas, T.E.E. dela Cruz, M. Unterseher & M. Schnittler. 2017a. Biogeographical assessment of myxomycete assemblages from Neotropical and Asian Palaeotropical forests. *Journal of Biogeography* 44:1524–1536.

Dagamac, N.H.A., M.A.D. Rea-Maminta & T.E.E. dela Cruz. 2015a. Plasmodial slime molds of a tropical karst forest, Quezon National Park, the Philippines. *Pacific Science* 69:411–422.

Dagamac, N.H.A., M.A.D. Rea-Maminta, N.S. Batungbacal, S.H. Jung, C.R.T. Bulang, A.G.R. Cayago & T.E.E. dela Cruz. 2015b. Diversity of plasmodial slime molds (myxomycetes) in coastal, mountain, and community forests of Puerto Galera, Oriental Mindoro, the Philippines. *Journal of Asia-Pacific Biodiversity* 8:322–329.

Dagamac, N.H.A., C. Rojas, Y.K. Novozhilov, G.H. Moreno, R. Schlueter & M. Schnittler. 2017b. Speciation in progress? A phylogeographic study among populations of *Hemitrichia serpula* (Myxomycetes). *PloS One* 12:e0174825.

Dagamac, N.H.A., S.L. Stephenson & T.E.E. dela Cruz. 2012. Occurrence, distribution and diversity of myxomycetes (plasmodial slime moulds) along two transects in Mt. Arayat National Park, Pampanga, Philippines. *Mycology* 3:119–126.

Dagamac, N.H.A., S.L. Stephenson & T.E.E. dela Cruz. 2014. The occurrence of litter myxomycetes at different elevations in Mt. Arayat National Park, Pampanga, Philippines. *Nova Hedwigia* 98:187–196.

Dahl, M.B., O. Shchepin, C. Schunk, A. Menzel, Y.K. Novozhilov & M. Schnittler. 2018. A four year survey reveals a coherent pattern between occurrence of fruit bodies and soil amoebae populations for nivicolous myxomycetes. *Scientific Reports* 8:11662.

dela Cruz, T.E.E., R.P.D. Fabul, F.P.R. Romero, M.C.B. Cruz & E.M. Cabutaje. 2021. Diversity of myxomycetes from forest habitats leading to Casaroro and Pulang Bato Falls in Negros Oriental, Philippines. *Slime Molds* 1:V1A9.

dela Cruz, T.E.E., R.V. Kuhn, A.O.M. Javier, C.M. Parra & T.H. Quimio. 2009. Status of the myxomycete collection at the UPLB—Museum of Natural History (UPLB-MNH) Mycological Herbarium. *Philippine Journal of Systematic Biology* 3:97–111.

dela Cruz, T.E.E., R.V. Kuhn, A.O.M. Javier, C.P. Rodillas, C.M. Parra, L.H.M. Corpuz & R.D. McHugh. 2011. Occurrence and distribution of plasmodial myxomycetes in Hundred Islands National Park, Pangasinan, Philippines. *Acta Manilana* 59:65–74.

dela Cruz, T.E.E., L.C.J. Llames, P.J.V. Glori, R.R. Sanvictores, J.E.V. Cabales, G.C.C. Aldover, J.H.M. Rejano, B.P. Akmad, S.B. Lopez, H.V.N. Esmundo, R.K.D. Arbes, M.K.T. Morato, A.T. Agustin, J.A.J. Nohay, B.T. Cortes, J.C. Lagman, J.J.T. Bellen, J.B. Sabado, K.OB. Martin, R.M. Bennett. 2024. A checklist of novel microbes discovered in the Philippines. *Philippine Journal of Science* 153:257–297.

dela Cruz, T.E.E., M.V.B. Pangilinan, R.J. Cruz, E.E. de Jesus, et al. 2010. A checklist of plasmodial myxomycetes (slime molds) from Subic Watershed Forest Reserve, Zambales, Philippines. *Acta Manilana* 58:41–45.

dela Cruz, T.E.E., M.V.B. Pangilinan & R.A. Litao. 2012. Printed identification key or web-based identification guide: An effective tool for species identification? *Journal of Microbiology and Biology Education* 13:180–182.

dela Cruz, T.E.E., M.A.D. Rea, H.T.M. Tran, T.W. Ko Ko & S.L. Stephenson. 2014. A comparative species listing of myxomycetes from tropical (Philippines) and temperate (United States) forests. *Mycosphere* 5:299–311.

dela Cruz, T.E.E., R.S. Taguchi, J.M.A. Dotig, R.J.P. Tolete, et al. 2023. Myxomycetes associated with plant-based substrata collected around Bulusan Lake in Sorsogon, Philippines. *Slime Mold.* 3:V3A1. https://doi.org/10.5281/zenodo.6795776

Eloreta, M.F.B.M., M.S. Policina & T.E.E. dela Cruz. 2020. Occurrence and diversity of myxomycetes along the forest edges of Mt. Isarog National Park, Camarines Sur, Philippines. *Current Research in Environmental and Applied Mycology* 10:377–385. https://doi.org/10.5943/cream/10/1/30

Esselstyn, J.A., C.H. Oliveros, R.G. Moyle, A.T. Peterson, J.A. McGuire & R.M. Brown. 2010. Integrating phylogenetic and taxonomic evidence illuminates complex biogeographic patterns along Huxley's modification of Wallace's Line. *Journal of Biogeography* 37:2054–2066.

Gonzalez, J.C.T. 2007. *Report on the Fauna and Flora Inventory of Priority Sites in the Polillo Islands Anibawan, Lipata and Carlagan. Pioneering Community-Based Conservation Sites in the Polillo Islands.* Polillo Islands Biodiversity Conservation Foundation Inc., Fauna & Flora International and the Darwin Initiative for Conservation, United Kingdom.

Hall, R. 2002. Cenozoic geological and plate tectonic evolution of SE Asia and the SW Pacific. *Journal of Asian Earth Sciences* 20:353–431.

Heaney, L.R. 1998. Philippine biodiversity: Insights and prospects for conservation. *Environment* 40:12–20.

Holloway, N.H. 1982. North Palawan Block, Philip-pines—its relation to Asian mainland and role in evolution of South China Sea. *Bulletin of the American Association of Petroleum Geologists* 66:1355–83.

Isagan, M.D.E., M.K.C. Carbonell, Y.A. Dalangin, A.J.L. Lapira, M.H. Pecundo & T.E.E. dela Cruz. 2020. Species listing and diversity of myxomycetes from Mt. Makulot and Napayong Island in Taal Lake, Batangas, Philippines. *MycoAsia* 2020/05:1–14.

Jacob, J.K., E.S. Romorosa, E.D.L.R. Arguelles & J.A.G. Paguirigan. 2022. Checklist of reported macrofungi in the Philippines. *Philippine Journal of Science* 151:1657–1676.

Karig, D.E. 1983. Accreted terranes in the northern part of the Philippine archipelago. *Tectonics* 2:211–236.

Komoda, A.T., L.A.D.A. Tan, A.M.G. Tomimbang, K.A.R. Tan, J.C. Redeña-Santos, N.H.A. Dagamac & C.E. Ocenar-Bautista. 2025. Protosteloid amoebal assemblages as microbial models for elevational diversity gradient in tropical montane landscape. *Sydowia* 77:129–139.

Kuhn, R.V., A.O.M. Javier, C.P. Rodillas, C.M. Parra, L.H.M. Corpuz, L.S. Moron & T.E.E. dela Cruz. 2013a. Occurrence and distribution of myxomycetes (plasmodial slime molds) in three provinces of Luzon Island, Philippines. *Philippine Science Letters* 6:1–7.

Kuhn, R.V., A.O.M. Javier, C.P. Rodillas, C.M. Parra, L.H.M. Corpuz, A.T. Buaya & T.E.E. dela Cruz. 2013b. Diversity of plasmodial myxomycetes from Anda Island, Pangasinan, Philippines. *Biotropia* 20:1–9.

Leontyev, D., Y. Ishchenko & M. Schnittler. 2023. Fifteen new species from the myxomycete genus Lycogala. *Mycologia* 115:524–560.

Lim, A.R.U., R.M.N. Silva, G.R.E. Lesaca, V.J.C. Mapalo, M.H. Pecundo & T.E.E. dela Cruz. 2021. First survey of myxomycetes in successional and mangrove forests of Negros Oriental, Philippines. *Slime Molds* 1:V1A7. https://doi.org/10.5281/zenodo.5098397

Limbo-Dizon, J.E., K.J. Almadrones-Reyes, S.A.B. Macabago & N.H.A. Dagamac. 2022. Bioclimatic modeling for the prediction of the suitable regional geographical distribution of selected bright-spored myxomycetes in the Philippine archipelago. *Biodiversitas: Journal of Biological Diversity* 23:2285–2294.

Macabago, S.A.B., N.H.A. Dagamac & T.E.E. Dela Cruz. 2010. Diversity and distribution of plasmodial myxomycetes (slime molds) from La Mesa Ecopark, Quezon City, Philippines. *Biotropia* 17:51–61.

Macabago, S.A.B., N.H.A. Dagamac, T.E.E. dela Cruz & S.L. Stephenson. 2017. Implications of the role of dispersal on the occurrence of litter-inhabiting myxomycetes in different vegetation types after a disturbance: A case study in Bohol Islands, Philippines. *Nova Hedwigia* 104:221–236.

Macabago, S.A.B. & T.E.E. dela Cruz. 2012. Development of a myxomycete photoguide as a teaching tool for microbial taxonomy. *Journal of Microbiology and Biology Education* 13:67–69.

Macabago, S.A.B. & T.E.E. dela Cruz. 2014. Preservation and enzyme production of myxomycetes from Lubang Island, Occidental Mindoro, Philippines. *Philippine Science Letters* 7:331–336.
Macabago, S.A.B., T.E.E. dela Cruz & S.L. Stephenson. 2012. First records of myxomycetes from Lubang Island, Occidental Mindoro, Philippines. *Sydowia* 64:109–118.
Macabago, S.A.B., T.E.E. dela Cruz & S.L. Stephenson. 2020a. Myxomycetes of the Caramoan Islands and an update on the species found in the Bicol Peninsula, Philippines. *Philippine Journal of Systematic Biology* 14:1–9.
Macabago, S.A.B., T.E.E. dela Cruz & S.L. Stephenson. 2020b. Myxomycetes of Coron Island and additions to the Myxomycetes of Palawan group of islands in the Philippines. *Karstenia* 58:157–167.
Macabago, S.A.B. & S.L. Stephenson. 2021. Does Huxley's line apply to myxomycetes? *Pacific Science* 75:531–541.
Macabago, S.A.B., S.L. Stephenson & T.E.E. dela Cruz. 2016. Diversity and distribution of myxomycetes in coastal and mountain forests of Lubang Island, Occidental Mindoro, Philippines. *Mycosphere* 7:18–29.
Martin, G.W. & C.J. Alexopoulos. 1969. *The Myxomycetes*. University of Iowa Press, Iowa City.
Mifsud, S. 2020. An updated checklist of the myxomycetes in the Maltese islands—An overview of an ongoing research. *Microbial Biosystems* 5:9–19.
Mitchell, A.H.G., F. Hernandez & A.P. dela Cruz. 1986. Cenozoic evolution of the Philippine archipelago. *Journal of Southeast Asian Earth Sciences* 1:3–22.
Moreno, G., D.W. Mitchell, S.L. Stephenson & T.E. dela Cruz. 2009. A new species of *Craterium* (Myxomycetes) with reticulate spores. *Boletín de la Sociedad Micológica de Madrid Micol. Madrid* 33:175–180.
Nguyen, L.T.T., O. Sanchez-Mahecha, K.J. Almadrones-Reyes, J.C. Redeña-Santos & N.H.A. Dagamac. 2019. Occurrence of leaf litter inhabiting myxomycetes from lowland forest patches of Northern and Central Vietnam. *Tropical Ecology* 60:495–506.
Notarte, K.I., T. Yaguchi, K. Suganuma & T.E. dela Cruz. 2018. Antibacterial, cytotoxic and trypanocidal activities of marine-derived fungi isolated from Philippine macroalgae and seagrasses. *Acta Botanica Croatica* 77:141–151.
Novozhilov, Y.K., O.N. Shchepin, E.S. Popov, A.V. Alexandrova, M. Schnittler, N.H.A. Dagamac & A.N. Kuznetsov. 2020. Myxomycetes associated with mountain tropical forests of Bidoup Nui Ba and Chu Yang Sin national parks (Dalat Plateau, southern Vietnam). *Nova Hedwigia* 110:185–224.
Ocenar-Bautista, C.E., R.C. Balaoro-Banzuela, C.Y. Claudio-Paragas, D.E. Buebos-Esteve & N.H.A. Dagamac. 2024. First records of protosteloid amoebae isolated from coastal litter in the Philippines. *Check List* 20:249.
Ocenar-Bautista, C.E., R.C. Balaoro-Banzuela, N.H.A. Dagamac & A.T. Komoda. 2025. Diversity and distribution of protosteloid amoebae in the ridge-to-reef ecosystem of San Fernando City, Philippines. *Acta Protozoologica* 64:11–20.
Olfato-Parojinog, A., P.A. Sobremonte-Maglipon, J.E. Limbo-Dizon, K.J. Almadrones-Reyes & N.H.A. Dagamac. 2023. Land use/land cover changes (LULCC) using remote sensing analyses in Rizal, Philippines. *GeoJournal* 88:6105–6118.
PAGASA. 2025. *Climatological Normals. Philippine Atmospheric, Geophysical and Astronomical Services Administration*. https://www.pagasa.dost.gov.ph/climate/climatological-normals
Pecundo, M.H., A.T. Buaya & T.E.E. dela Cruz. 2023. A rapid survey of myxomycetes associated with different substrates on Luzon Island, Philippines. *Slime Molds* 3:V3A2.
Pecundo, M.H., L.M. Espiritu, R.L.G. Carillaga, A.F.E. Samaniego & E.M. Cabutaje. 2021. First report of myxomycetes in the karst forest of Minalungao National Park, Nueva Ecija, Philippines with updates on the limestone-inhabiting myxomycetes of the Philippines. *Current Research in Environmental & Applied Mycology* 11:230–247.
Pecundo, M.H., N.H.A. Dagamac & T.E.E. dela Cruz. 2020. A comparative diversity study of myxomycetes in the lowland forests of Mt. Malasimbo and Mt. Siburan, Mindoro Island, Philippines. *Karstenia* 58:275–291.
Pecundo, M.H., N.H.A. Dagamac, S.L. Stephenson & T.E.E. dela Cruz. 2017. First myxomycete survey in the limestone forest of Puerto Princesa Subterranean River National Park, Palawan, Philippines. *Nova Hedwigia* 104:129–141.
Piper, P.J., J. Ochoa, E.C. Robles, H. Lewis & V. Paz. 2011. Palaeozoology of Palawan Island, Philippines. *Quaternary International* 233:142–158.
Policina, M.S. & T.E.E. dela Cruz. 2020. Records of corticolous myxomycetes from selected trees in Angat Watershed Forest Reserve, Bulacan, Philippines. *Philippine Journal of Systematic Biology* 14:1–6. https://doi.org/10.26757/pjsb2020c14007
Racoma, C.J.M., J.C. Redeña-Santos & N.H.A. Dagamac. 2025. Distribution and diversity of myxomycetes along the elevational belt of Mt. Calavite Wildlife Sanctuary (MCWS), Occidental Mindoro, Philippines. *Ecologies* 6:7.

Rangin, C., L. Jolivet, M. Pubellier & The Tethys Pacific Working Group. 1990. A simple model for the tectonic evolution of Southeast Asia and Indonesia region for the last 43 m.y. *Bulletin de la Société Géologique de France* 8:889–905.

Rea-Maminta, M.A.D., N.H.A. Dagamac, F.Z. Huyop, R.A. Wahab & T.E.E. dela Cruz. 2015. Comparative diversity and heavy metal biosorption of myxomycetes from forest patches on ultramafic and volcanic soils. *Chemistry and Ecology* 31:741–753.

Redeña-Santos, J.C., J.A.U. Dunca, D.V. Thao & N.H.A. Dagamac. 2017. Myxomycetes occurring on selected agricultural leaf litters. *Studies in Fungi* 2:171–177.

Redeña-Santos, J.C., D. Van Thao, M. Schnittler & N.H.A. Dagamac. 2018. The first report of composition and occurrence of myxomycete assemblages in protected and unprotected plantation forests: A comparative study in Thai Nguyen City, Northern Vietnam. *Plant Ecology and Evolution* 151:231–240.

Reynolds, D.R. 1981. Southeast Asian myxomycetes II. Philippines. *Philippine Journal of Biology* 10:127–150.

Schnittler, M., N.H.A. Dagamac, J. Woyzichovski & Y.K. Novozhilov. 2022. Biogeographical patterns in myxomycetes. In *Myxomycetes*, pp. 377–416. Academic Press.

Schnittler, M., O.N. Shchepin, N.H.A. Dagamac, M. Borg Dahl & Y.K. Novozhilov. 2017. Barcoding myxomycetes with molecular markers: Challenges and opportunities; Barcoding myxomycetes with molecular markers: Challenges and opportunities. *Nova Hedwigia* 104:323–341.

Shchepin, O.N., M. Schnittler, N.H.A. Dagamac, D.V. Leontyev & Y.K. Novozhilov. 2019. Unexplored diversity of microscopic myxomycetes. *Plant Ecology and Evolution* 152:499–506.

Siscar, A.H., D.R.C. Sombilon, L.E.I. Octaviano, R.M. Bennett & T.E.E. dela Cruz. 2024. Vertical spore dispersal of myxomycetes in an urban high-rise residential building. *Philippine Journal of Systematic Biology Online ISSN* 2508:0342.

Uyenco, F.R. 1973. Myxomycetes of the Philippines. *University of the Philippines Natural Science Research Center Technical Report* 12:1–23.

Vélez-Múnera, M., M. Avitia, E.A. Rebollar & S. Ramírez-Barahona. 2024. Changes in the phyllosphere and rhizosphere of a cloud forest tree fern along an elevation gradient. *Research Square* (preprint). https://doi.org/10.21203/rs.3.rs-5374836/v1

Viray, A.T., D.D.S. Rotap, L.L. Migraso, N.C.I. Sibbaluca, E.C.V. Escobar, A.T. Buaya & T.E.E. dela Cruz. 2014. Occurrence and diversity of myxomycetes (slime molds) in Polillo Island, Quezon Province, Philippines. *Acta Manilana* 62:9–17.

Yumul, G.P. Jr., G.R. Alce, C.B. Imalanta & R.T. Datuin, R.T. 1997. Distribution, geochemistry and mineralization potentials of Philippine ophiolite and ophiolitic sequences. *Ofioliti* 22:47–56.

Yumul, G.P., C.B. Dimalanta, V.B. Maglambayan & E.J. Marquez. 2008. Tectonic setting of a composite terrane: A review of the Philippine island arc system. *Geosciences Journal* 12:7–17.

13 Fungi in Brazilian Forests
Diversity, Ecology and Conservation

Solange Xavier-Santos, Diogo Henrique Costa de Rezende, Elisandro Ricardo Drechsler-Santos, Laise de Holanda Cavalcanti, and Lucas Leonardo-Silva

13.1 INTRODUCTION

According to the Global Forest Resources Assessment (FRA) 2020 published by the Food and Agriculture Organization of the United Nations (FAO 2020), the world has a total forest area of 4.06 billion hectares (ha), which corresponds to 31% of the total land area. The tropical domain accounts for the largest proportion of the world's forests (45%), followed by the boreal, temperate, and subtropical domains. Five countries—the Russian Federation, Brazil, Canada, the United States of America, and China—contain more than half (54%) of the world's forests.

Brazil holds the largest area of tropical forests on Earth, sheltering approximately 13% of the global forest cover within its territory (FAO 2020). These forests represent a complex of ecosystems that sustains environmental services vital for climate balance and the maintenance of life. However, beneath the dense arboreal biomass described in national forest inventories lies a vast and often underestimated network. This network is made up mainly of fungi. This Brazilian Funga is the invisible architect of these forests, acting in everything from root symbiosis to carbon cycling.

This chapter seeks to overcome the barrier of historical invisibility, tracing back to the beginnings of mycological research in the country, unveiling diversity, ecological patterns, and the necessary strategies for fungal conservation. In doing so, it contributes to integrating Funga into the legal frameworks of national biodiversity protection, aiming to make it a priority in environmental policies and social engagement.

13.2 THE BRAZILIAN FORESTS: TYPOLOGY AND CHARACTERIZATION

By concentrating about 13% of the world's forest cover, as previously indicated, Brazil establishes itself as the largest reservoir of tropical biodiversity on the planet (FAO 2020). This immense forest cover is not homogeneous but rather is composed of a mosaic of biomes that present distinct phytophysiognomic and climatic characteristics. According to the mapping by the National Register of Public Forests of the Brazilian Forest Service (Brazil 2024) and the Brazilian Institute of Geography and Statistics (IBGE 2012), Brazil possesses different forest types (Figures 13.1 and 13.2) distributed among its six main biomes (Figure 13.3). These forest types are listed below.

- Amazon. This is the biome that occupies the largest extent of the territory of Brazil, which also contains the highest concentration of forests and the greatest diversity of typologies, where the following predominate:
  - *Dense Ombrophilous Forest (or Terra Firme Forest) and Open Ombrophilous Forest*: These forest types are characterized by high rainfall and constant temperatures. The

DOI: 10.1201/9781003647140-13

former is the predominant type in the biome, corresponding to the classic Amazon rainforest, with tall trees (up to 50 m) and closed canopies, creating a humid and shaded internal environment. The latter features more widely spaced trees, often with the presence of palms or bamboos (*tabocais*).

  - *Floodplain Forests, such as Várzea and Igapó*: These are adapted to the seasonal flooding of rivers. The *várzea* forest type occurs in areas flooded by "white water" rivers, which are rich in sediments and nutrients, and where soil fertility is periodically renewed by floods. The *igapó* forest type is located in areas flooded by "black water" or "muddy" rivers, which are acidic and nutrient-poor. The vegetation is adapted to waterlogged and acidic soils for long periods.
  - *Seasonal Forests*: Although less extensive than the ombrophilous forests, seasonal forests occur at the edges of the biome, in transition areas between the Amazon and other biomes, such as the Cerrado. Unlike ombrophilous forests, which is consistently humid, the seasonal forests are defined by the rhythm of rainfall. They can be semideciduous, where 20 to 50% of the trees lose their leaves during the dry season, or deciduous, when this number exceeds 50%.
  - *Campinarana*: Also known as "Amazonian *caatinga*" due to its more open appearance and smaller trees (5 to 15 m) compared to the dense forest. It is one of the most unique and fragile forest types in Brazil. It develops over sandy, whitish, highly acidic, and nutrient-poor soils with deficient drainage.
- Cerrado. Although considered the most biodiverse savannah in the world, this biome also includes dense forest formations, such as those listed below.
  - *Cerradão (or Savannah Forest)*: This is a dense forest formation featuring typical savannah vegetation species but with an arboreal stature (8 to 15 m), resembling a dry forest. It occurs in interfluvial areas (far from rivers) in poor and acidic (dystrophic) soils. The trees are mostly evergreen or semideciduous (losing few leaves during the dry season) because their very deep roots allow them to reach water in the water table.
  - *Dry Forest or Seasonal Deciduous Forest*: This type is found on more fertile soils, rich in calcium and magnesium (eutrophic soils). At the height of the dry season, the forest appears to die, as almost all trees lose their leaves to prevent water loss, since their roots are usually not as deep as those in the *cerradão.*
  - *Gallery Forest* (or *Alluvial Ombrophilous Forest*): These are evergreen forests (they do not lose their leaves during the dry season) that line the rivers, forming tunnels over them. They are humid ecosystems "hidden" within a dry biome.
  - *Riparian Forest (or Alluvial Seasonal Semideciduous Forest)*: These also are forests that occur along riverbanks, but on intermediate to large rivers; therefore, they do not form a tunnel over rivers as the rivers are too wide for the canopies to meet. Unlike the gallery forest, during the dry season, some trees lose their leaves, increasing the light penetration to the forest floor.
- Atlantic Forest. Although fragmented, this type of forest possesses one of the highest rates of endemism in the world. Due to variations in elevation and proximity to the sea, the biome is represented by three different types of forests.
  - *Dense Ombrophilous Forest (or Tropical Rainforest) and Open Ombrophilous Forest*: These forests are humid, with rainfall well-distributed throughout the year. The former is the classic image of the Atlantic Forest, found on the slopes of the Serra do Mar; it is composed of tall trees, a closed canopy, and many epiphytes (bromeliads and orchids). The latter features a more spaced canopy, lower density, and greater light penetration. Consequently, there are fewer epiphytes; they represent the transition between dense forests and drier formations, which are more common in the Northeast.
  - *Mixed Ombrophilous Forest*: This name refers to the *Araucaria* forest, typical of the South of the country and high-elevation areas of the Southeast. These forests experience a colder climate, with occasional frosts. They are of high ecological and cultural value.

   - *Seasonal Semideciduous Forest (or Sub-caducifolious Tropical Forest) and Deciduous Forest (or Caducifolious Tropical Forest)*: These forests occupy large areas of the interior, where there is a well-defined dry season. They are heavily impacted by agribusiness. In the former type, only a portion of the trees lose their leaves during the dry season. In the latter, this occurs with most of the trees; they are rarer and form the transition with the Cerrado and Caatinga.
- Caatinga. This is an exclusively Brazilian biome. Despite having predominantly sparse vegetation due to the semi-arid climate, it features some formations classified as forest, such as the examples listed below.
   - *Seasonal Deciduous Forest (or Arboreal Caatinga)*: Composed of trees up to 20 m tall, which lose their leaves entirely during the dry season, but form a thick and shaded canopy in their interior during the rainy season.
   - *Dry Forest*: Composed of forests situated near slopes and mountain tops, with leaves remaining during the dry season.
   - *High-Elevation Humid Forests (Brejos de Altitude)*: These correspond to "islands" of moist forest on mountaintops within the semi-arid region. They feature evergreen vegetation with tree species typical of the Atlantic Forest, the Amazon, and the *Caatinga*.
- Pantanal. This is the largest continuous floodplain in the world, where forest cover occurs in a fragmented manner and presents a unique ecological dynamic, highly dependent on the flood pulse regime. The subtypes are listed below.
   - *Riparian Forests (Matas Ciliares) and Forested Savannah*: These are forests that follow the river courses or are adapted to constant soil moisture and seasonal floods, respectively.
   - *Capões and Cordilheiras*: These are on sandy elevated areas that do not flood. While *capões* are small, circular arboreal "islands"—possibly originating from termite mounds or indigenous earthworks (*aterros*)—*cordilheiras* are elongated terraces that support extensive stretches of *cerradão* forests.
- Pampa. Located exclusively in the state of Rio Grande do Sul, in the far south of the country, it is defined by a temperate climate, with well-distributed rainfall and temperatures that can drop below zero in winter. Although dominated by grassy fields, this biome contains important forest fragments as indicated below.
   - *Seasonal Deciduous Forest (or Matas de Encosta)*: This occurs especially on hill slopes and transition areas. The trees lose more than 50% of their leaves during the harsh winter or dry periods.
   - *Dense Ombrophilous Forest (Mata de Galeria or Mata de Sanga)*: These forests occur along river courses amidst open grasslands, functioning as biological corridors. They feature constant humidity, and the trees do not lose their leaves.

Restinga and *Mangroves* are other vegetation types that may present arboreal formations, but whose classification as forests is controversial, as their structure and distribution are primarily governed by edaphic and hydrodynamic factors rather than by climate, and they encompass a wide range of physiognomies that do not fit the classical concept of forests. Forest restinga formations occur exclusively within the Atlantic Forest biome and develop on sandy substrates under strong marine influence. Mangroves, in turn, occur in both Atlantic Forest and Amazon biomes and consist of coastal arboreal or shrubby halophytic formations subjected to direct fluviomarine influence (transition zones between terrestrial and marine environments), developing on muddy, poorly drained, anoxic soils rich in organic matter, and harboring highly specialized species with aerial roots (pneumatophores and rhizophores) and salt-excreting glands.

It is worth noting that the Amazon contains approximately 90% of all recognized forest areas. The Seasonal Climate Forests, which respond to the rhythm of rainfall, are crucial for understanding the resilience of the *Cerrado* and *Caatinga*. Their primary ecological feature is deciduousness—the loss of leaves (*caducifolia*) that allows them to conserve water. Unlike the Amazon, many of these forests occur in higher fertility soils, which creates a very different organic matter decomposition dynamic in the soil. On the other hand, in humid climate or ombrophilous forests, found in the

FIGURE 13.1 Some of the different Brazilian forest types. Amazon Biome: (a) Dense Ombrophilous Forest, (b) Open Ombrophilous Forest, (c) *Campinarana*, (d) Floodplain Forest; Cerrado Biome: (e) *Cerradão*, (f) Dry Forest, (g) Riparian Forest (*Mata Ciliar*), (h) Gallery Forest.

Amazon and the slopes of the Atlantic Forest, there is no severe dry period. They are defined by rapid nutrient cycling and exhibit high biomass density with vertical stratification (multiple vegetation layers). Finally, transition zones or ecotones, where biomes meet, deserve special attention, as they are areas of high biodiversity, harboring species from two or more biomes.

The heterogeneity of forest typologies and the dynamics of these forests both shape and are shaped by local fungal communities, establishing an intrinsic correlation between forest types and

FIGURE 13.2 Some of the different Brazilian forest types. Atlantic Forest Biome: (a) Dense Ombrophilous Forest, (b) Mixed Ombrophilous Forest, (c) Seasonal Semideciduous Forest, (d) Seasonal Deciduous Forest; Pantanal Biome: (e) Forested Savannah, (f) *Capão*; Caatinga: (g) Seasonal Deciduous Forest, (h) High-Elevation Humid Forests (*Brejos de Altitude*); Pampa Biome: (i) *Sanga* Forest, (j) Slope Forest (*Mata de Encosta*).

the ecological niches of the fungi present. Understanding this mutual dependence is the first step toward including fungi in national conservation plans, transforming historical invisibility into scientific and legal recognition.

13.3 ECOSYSTEM SERVICES OF FUNGI IN BRAZILIAN FORESTS

Despite their divergent evolutionary trajectories, fungi act synergistically in forest maintenance. They are the "workers" that are not always visible, of the ecosystem services that sustain all these environments; in other words, they are a structural component of the health of the ecosystems. Their role can be categorized into supporting services, such as nutrient cycling and soil formation; regulatory and water resistance services; provisioning services, such as genetic and food resources; and cultural services, including ethnomycology and bioindication, among others.

In ombrophilous forests (in the Amazon and the Atlantic Forest), saprotrophic fungi are responsible for the accelerated decomposition of leaf litter, converting complex organic matter into inorganic nutrients readily available to plants. This process allows the forest to "feed on itself" and is vital in tropical soils, which are often poor in chemically active nutrients, where the forest depends on immediate biological recycling to maintain its biomass (Stark and Jordan 1978; Soares and Froufe 2015; Cunha et al. 2020).

Especially in Seasonal Forests and Savannahs (both in the Cerrado and the Caatinga), fungi that form symbiotic associations with tree roots (mycorrhizae) play a crucial role in the hydric resistance of these plants. In this way, the fungi expand the root's capacity to absorb water and phosphorus, which is vital in ecosystems with severe dry periods. The resilience of these forests in the face of climate change directly depends on the effectiveness of these soil communication networks, which provide them with greater tolerance to prolonged periods of drought and thermal stress. It is estimated that approximately 80 to 90% of tropical plants depend on this relationship to survive water stress (Moreira and Siqueira 2006; Smith and Read 2008; Siqueira et al. 2002, 2010; Winagraski et al. 2019).

Brazilian forests host a vast diversity of macroscopic fungi with food, medicinal and technological potential, many of which are historically used by traditional communities and Indigenous Peoples. The Funga provides resources ranging from native edible mushrooms to biochemical compounds used in the pharmaceutical industry, representing an economic asset that remains underexplored within the National Register of Public Forests (Mendoza et al. 2022; Drewinski et al. 2024).

Fungi also act as indicators of ecological integrity and forest health. The presence of specific fungal communities, especially poroid fungi and lichenized fungi (lichens), signals the stage of forest succession—whether the forest is old-growth and preserved or in the early stages of regeneration—and thus habitat quality, serving as essential tools for environmental monitoring and scientific education (Lepinay et al. 2021; Pires et al. 2025).

It must also be considered that certain species of fungi in specific ecosystems, such as the *Araucaria* Forest in Southern Brazil, occur only in association with the pine *Araucaria angustifolia*. In these areas, ectomycorrhizae are common, forming "sheaths" (mantles) that envelop the roots—unlike the mycorrhizae in other Brazilian forests, which are predominantly of the arbuscular type (AMF), where fungal hyphae penetrate the root's cortical cells without forming external structures visible to the naked eye. Because they are in colder regions, decomposition in *Araucaria* forests is slower, and ectomycorrhizal fungi help the tree capture organic nitrogen (Breuninger et al. 2000). Thus, the preservation of these mixed ombrophilous forests is essential for the conservation of this unique Funga.

The same attention must be given to the High-elevation Humid Forests *(brejos de altitude)* of the Caatinga. As islands of moisture, they would function as refuges for fungi that would not survive the inhospitable conditions of the surrounding semi-arid region, representing hotspots of fungal biodiversity that remain poorly studied. Fungi occurring in other Caatinga ecosystems operate in pulsating cycles, with rapid sporulation during rare rainfall events, and exhibit highly efficient dormancy strategies in the arid soil. In this biome, AMF are important allies in the recovery of degraded areas, mitigating impacts and minimizing the consequences of climate change (Garcia-Pichel and Sala 2022).

In addition, endophytic fungi play a fundamental role in plant protection, especially in the forest ecosystems of the Atlantic Forest. In this environment, high humidity and tree density favor the Funga that inhabits the interior of foliar tissues without causing apparent damage. This symbiosis acts as a vital chemical barrier: Through the production of bioactive secondary metabolites, it inhibits colonization by pathogens and reduces herbivory, thus contributing to the survival of trees in areas of high biological pressure (Coertjens et al. 2024).

As the holder of the largest tropical forest heritage, Brazil is expected to be the world's largest potential reservoir of tropical fungal diversity. Although approximately 8,400 species of fungi *sensu lato* are currently known and cataloged in the country, distributed across its various biomes (Flora e Funga do Brasil 2026).

13.4 FUNGI IN BRAZILIAN FORESTS

13.4.1 A Brief History of Brazilian Mycology

The history of knowledge on fungi in Brazil closely parallels the development of biological collections and botanical studies in the country and is deeply linked to the high environmental heterogeneity of the Brazilian territory, characterized by a broad mosaic of vegetation types and climatic conditions. Despite this singular potential for fungal diversity, the earliest advances in fungal studies were largely exogenous, based on naturalist expeditions and the shipment of specimens to Europe, where they were described and incorporated into foreign collections (Fidalgo 1968; Capelari et al. 1998).

The transition to a more structured Mycology occurred in the first half of the twentieth century, with particular emphasis on the work of Father Johannes Rick, widely regarded as a central figure in the history of Brazilian Mycology. Rick organized one of the largest fungal collections in the country at the Anchieta Herbarium (PACA), comprising thousands of specimens, and published extensively on macrofungi, especially the Basidiomycota (Maia et al. 2015). From the mid-twentieth century, mycological research in Brazil became institutionalized. In the Northeast, Augusto Chaves Batista, a disciple of Father Camille Torrend, founded the Institute of Mycology, the country's only institution exclusively devoted to fungal studies. Between 1954 and 1967, Batista and his team produced over 700 publications covering taxonomy, systematics (describing numerous new genera and species), ecology, industrial applications, and medically relevant fungi (Bezerra et al. 2017).

In the 1960s, a second research hub emerged in southeastern Brazil at the former Instituto de Botânica de São Paulo (now Instituto de Pesquisas Ambientais). Researchers including Alcides Ribeiro Teixeira, João Salvador Furtado, Maria Eneyda Pacheco Kauffmann Fidalgo, Oswaldo Fidalgo, and Vera Lucia Ramos Bononi advanced macrofungal studies, particularly of lignicolous species, through inventories, taxonomic revisions, and specialist training (Maia et al. 2015; Leonardo-Silva and Xavier-Santos 2023).

As a direct consequence of this historical process, knowledge of Brazilian fungal diversity remained fragmented for decades and dispersed across publications primarily focused on taxonomy, as well as regional or localized species lists. This scenario began to change more consistently only in the early 2010s, with the intensified efforts toward cataloging, data curation, and the integration of information into collaborative platforms and national databases (Maia et al. 2015; Leonardo-Silva and Xavier-Santos 2023). The consolidation of these initiatives led to a rapid increase in the documented fungal diversity of Brazil, as reflected in widely cited syntheses reporting a marked rise in the number of recognized species within a relatively short period, with the Basidiomycota and Ascomycota emerging as the most representative groups.

13.4.2 Fungal Diversity and Ecological Patterns

As previously presented, Brazil harbors one of the most complex and extensive vegetation mosaics on the planet, resulting from the interaction of climatic, edaphic, geomorphological, and historical gradients. This environmental heterogeneity supports a wide range of forest formations, from humid and seasonally flooded tropical forests to dry forests and savanna woodlands (Brasil 2024), creating a particularly favorable setting for fungal diversification. The structural complexity of vegetation, combined with high primary productivity and the continuous renewal of plant biomass, generates a broad array of ecological niches and substrates that are essential for the establishment and coexistence of different fungal groups (Swift et al. 1979; Medeiros et al. 2015; Peay et al. 2016).

From a quantitative perspective, the currently documented diversity of fungi *sensu stricto* in Brazil comprises approximately 7,918 species (Flora e Funga do Brasil 2026). This figure represents an updated benchmark of recorded richness and reflects substantial advances in taxonomic curation, the digitization of biological collections, and the integration of data derived from regional inventories and systematic revisions. Nevertheless, there is broad consensus that this documented richness accounts for only a fraction of the country's actual fungal diversity, being strongly influenced by historical inequalities in sampling effort, persistent geographic gaps, and taxonomic limitations, particularly in forested environments (Maia et al. 2015; Pagin-Cláudio et al. 2022; Leonardo-Silva and Xavier-Santos 2023).

The diversity and distribution of fungal communities in Brazilian forests are shaped by biome-specific climatic, hydrological, and structural conditions (Peay et al. 2016; Brasil 2024; Pires et al. 2025). Consequently, fungal diversity in Brazil reflects both ecological responses specific to each forest context and the historically uneven development of mycological research across the country.

The spatial distribution of recorded fungal richness reveals marked contrasts among Brazilian biomes. The Atlantic Forest has the highest number of known species, followed by the Amazon, whereas the Cerrado, Caatinga, Pantanal, and Pampa exhibit progressively lower values (Flora e Funga do Brasil 2026) (Figure 13.3). This pattern reflects, in part, ecological differences such as

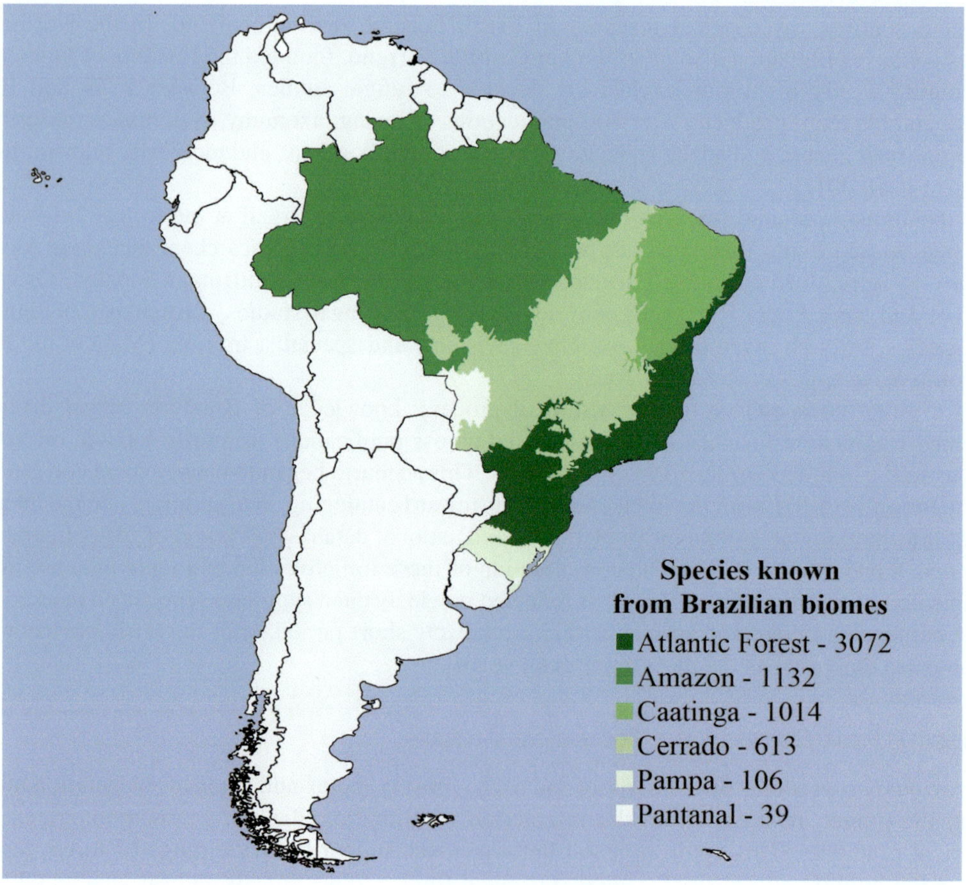

FIGURE 13.3 Known species of fungi *sensu stricto* across Brazilian biomes.

greater moisture availability, forest continuity, and the volume of dead wood, but it is also strongly shaped by sampling effort historically concentrated in regions with a higher density of research institutions, well-established biological collections, and a longer tradition of mycological studies. Consequently, the currently documented richness should not be interpreted as a faithful representation of the actual fungal diversity, particularly in biomes that have been historically underexplored.

In the humid tropical forests of the Atlantic Forest and the Amazon, high humidity throughout the year, combined with elevated plant productivity and the continuous supply of leaf litter and dead wood, provides favorable conditions for the establishment of diverse fungal communities (Gibertoni et al. 2007; Farias et al. 2025). In the Atlantic Forest-a complex of formations that includes dense ombrophilous forests, restingas, and mangroves, currently reduced to a small fraction of its original extent and recognized as a global biodiversity hotspot (Brasil 2024)-the high number of known fungal species reflects not only favorable environmental conditions but, above all, a long history of systematic sampling carried out over decades by multiple research groups across the country. These efforts have focused predominantly on wood-inhabiting fungi, particularly poroid species, which explains the strong representation of this group in the biota recorded for the biome (Baltazar and Gibertoni 2009; Pagin-Cláudio et al. 2022; Pagin-Cláudio and Gugliotta 2025). Among the most frequently recorded families are the Polyporaceae and Hymenochaetaceae, followed by the Meruliaceae, Phanerochaetaceae, and Schizoporaceae, all composed predominantly of lignocellulolytic species.

In contrast, the Brazilian Amazon, despite its continental scale and global ecological relevance, remains characterized by significant gaps in knowledge regarding fungal diversity (Cavalcante et al. 2020; Farias et al. 2025). Available studies indicate a Funga dominated by lignocellulolytic species, particularly Agaricomycetes associated with the decomposition of dead wood (Xavier et al. 2018). Even at local scales, surveys reveal high spatial heterogeneity, with community composition strongly influenced by substrate type and the stage of wood decomposition (Medeiros et al. 2015; Gibertoni et al. 2016; Komura et al. 2017).

Beyond formal taxonomic knowledge, studies conducted in collaboration with Indigenous Peoples of the Amazon demonstrate that regional fungal diversity includes culturally significant species that are recognized, named, and used in diverse traditional contexts. Ethnomycological research among Yanomami communities documents the use of macrofungi both as food and in artisanal practices, such as *Marasmius yanomami* J.S. Oliveira & N.K. Ishik., employed in female basketry, revealing a functional and symbolic dimension of fungal diversity that is largely absent from conventional scientific collections (Sanuma et al. 2016; Yanomami et al. 2019). Together, these findings underscore the ecological and sociocultural relevance of Amazonian fungi and indicate that the forest harbors a rich functional diversity that is central to understanding ecological and biogeographic processes in humid tropical forests.

The Brazilian Cerrado, the second largest biome in South America, comprises a complex mosaic of phytophysiognomies ranging from open grassland formations to dense forests, including *cerradões*, gallery forests, and seasonal forests (Ribeiro and Walter 2008). This structural heterogeneity, combined with a climatic regime marked by strong hydrological seasonality, creates a unique set of ecological conditions that shape the diversity, distribution, and dynamics of fungal communities. Although historically regarded as poor in fungal diversity, the Cerrado has revealed an unexpectedly rich Funga as systematic research efforts have expanded over recent decades (Alvarenga et al. 2015; Leonardo-Silva et al. 2020; Dianese et al. 2022; Leonardo-Silva and Xavier-Santos 2023; Regio et al. 2024).

Knowledge of fungi in the biome has undergone a substantial change since the 2010s. A comprehensive survey of corticioid and poroid fungi compiled 47 studies published between 1876 and 2021, more than half of which were produced in the last decade, indicating a rapid expansion of mycological research in the Cerrado (Leonardo-Silva and Xavier-Santos 2023). This effort resulted in documentation of 223 species and 387 occurrence records, with a predominance of genera such as *Hymenochaete*, *Ganoderma*, *Amauroderma*, *Polyporus*, *Trametes*, and *Fuscoporia*. The dataset also includes Cerrado-endemic species and taxa that, within the Brazilian context, are known exclusively from this biome despite having broader global distributions. Together, these findings underscore the singular character of the Cerrado Funga and its potential to harbor rare and geographically restricted species.

From an ecological perspective, the Cerrado Funga is strongly structured by the availability and continuity of woody substrates, with a predominance of wood-inhabiting fungi (Leonardo-Silva and Xavier-Santos 2023), reflecting the central role of plant biomass decomposition in nutrient cycling within seasonally dry environments. Gallery forests and denser forest patches function as microclimatic refugia, buffering the effects of the dry season and favoring fungal development. Fire, one of the main ecological and evolutionary drivers of the biome (Coutinho 1990), directly influences this dynamic by temporarily reducing substrate availability and suppressing lignicolous fungi, while simultaneously increasing environmental heterogeneity through the creation of charred trunks and newly exposed substrates. These conditions are exploited by opportunistic, disturbance-tolerant, and generalist species, including representatives of the genera *Auricularia*, *Coriolopsis*, *Fomes*, *Lentinus*, and *Pycnoporus* (Figure 13.4). Although evidence suggests resilience in part of the Cerrado fungal community, with rapid recolonization following the reestablishment of moisture and vegetation cover, the cumulative effects of frequent and intense fires on fungal diversity remain poorly understood.

A similar situation is observed in the Caatinga, a biome exclusively found in Brazil and characterized by a semi-arid climate and pronounced hydrological seasonality (Brasil 2024). Revisions of historical collections combined with recent field surveys have substantially increased the number of recorded

FIGURE 13.4 Species of fungi colonizing substrates in areas of Cerrado after fire events.

species (Flora e Funga do Brasil 2026), revealing a Funga adapted to conditions of water stress and strongly associated with woody substrates and more stable microhabitats, such as riparian forests and highland humid enclaves (*brejos de altitude*) (Drechsler-Santos et al. 2009, 2013; Neves et al. 2013).

The Pantanal exhibits an ecological dynamic governed by the annual flood pulse (Brasil 2024), which structures the availability of substrates for fungal development. The limited number of available studies indicate communities dominated by fungi associated with the decomposition of wood and leaf litter (Bononi et al. 2008), substrates that may become exposed after flood events, with community composition varying according to the intensity and duration of inundation. In contrast, the Brazilian Pampa, characterized by natural grasslands and a subtropical climate, remains among the least studied biomes, indicating a high and still largely unexplored potential for fungal diversity (Flora e Funga do Brasil 2026).

Beyond its ecological relevance, the Brazilian Funga includes a substantial number of edible species. A recent synthesis on wild edible mushrooms in Brazil compiled bibliographic records, molecular data, and ethnomycological information, documenting 409 species distributed across different biomes, with prominence in the Amazon and the Atlantic Forest (Drewinski et al. 2024). Genera such as *Auricularia*, *Lentinus*, *Panus*, *Pleurotus*, and *Favolus* are among the most frequently recorded, being recognized both for their species richness and for their nutritional and cultural importance, especially among Indigenous Peoples and traditional communities.

An analysis of species distribution across biomes also reveals the presence of widely distributed taxa within Brazilian territory. Species such as *Trametes villosa* (Sw.) Kreisel occur across all biomes, whereas *Favolus brasiliensis* (Fr.) Fr., *Hexagonia hydnoides* (Sw.) M. Fidalgo, *H. variegata* Berk., *Lentinus crinitus* (L.) Fr., *L. velutinus* Fr., *Fuscoporia gilva* (Schwein.) T. Wagner & M. Fisch., and *Pycnoporus sanguineus* (L.) Murrill are recorded from multiple biomes (Leonardo-Silva and Xavier-Santos 2023). These species, mostly wood-decomposers, are characterized by broad ecological tolerance and play a central role in decomposition processes and nutrient cycling across different Brazilian forest types.

Overall, the available data indicates that although wood-decomposing fungi ranks among the most frequently recorded groups in all biomes, this pattern largely reflects the historical focus of mycological research. Other groups—including agaricoid, gasteroid, gelatinous, mycorrhizal, endophytic, and soil-associated fungi—have also been investigated and exhibit high diversity, although still unevenly documented among regions and biomes.

Therefore, understanding fungal diversity in Brazilian forests requires an integrative perspective that simultaneously considers environmental heterogeneity, functional diversity, and the historical development of mycological research. Brazilian forests not only sustain a highly diverse and functionally essential Funga but also represent one of the major scientific challenges in terms of documentation, synthesis, and conservation of this key component of biodiversity.

13.5 CONSERVATION OF THE FUNGI IN BRAZILIAN FORESTS

Despite hosting one of the largest reservoirs of biological diversity on Earth, Brazil has historically overlooked fungi in biodiversity conservation policies. Until very recently, conservation initiatives, legal frameworks, and monitoring programs were almost exclusively designed around plants and animals, leaving the Funga largely invisible in environmental decision-making. This omission is particularly paradoxical given the central ecological role of fungi in Brazilian forests, where they underpin nutrient cycling, carbon dynamics, soil formation, symbiotic networks, and ecosystem resilience. Over the last decade, however, Brazil has experienced a rapid and significant shift, from neglect to action, placing fungal conservation onto the national biodiversity agenda (Drechsler-Santos et al. 2025).

13.5.1 From Invisibility to Recognition: Historical Context and Conceptual Advances

Historically, fungi have lagged behind other organisms in conservation. A clear example is that the first species of fungi (two lichens) were included in the IUCN Red List only in 2003, representing a delay of 50 years compared to other organisms (Mueller et al. 2022; IUCN 2025). This scenario has been attributed to several factors, including a paucity of data and the early misconception that a rigorous assessment of the conservation status of fungi was not feasible (Dahlberg and Mueller 2011).

The lack of knowledge of the conservation status of Brazilian fungal species persisted for even longer. Consequently, for much of Brazil's conservation history, fungi were excluded from the legal and conceptual understanding of biodiversity. Environmental impact assessments, protected area management plans, and threatened species lists were restricted to *Flora* and *Fauna*, reinforcing a false dichotomy that ignored fungi as independent conservation targets. Although this historical gap reflects a broader global pattern, it has been especially consequential in megadiverse tropical countries where fungal diversity is exceptionally high yet poorly documented.

Several actions have contributed to changing this pattern. One of them was the publication of operational guidelines for assessing the conservation status of fungi, as proposed by Dahlberg and Mueller (2011). Also, a critical conceptual advance was the formal recognition of Funga as a term equivalent to *Fauna* and *Flora*, acknowledging fungi as a fundamental and autonomous component of biodiversity (Kuhar et al. 2018). In Brazil, adoption of this framework catalyzed a broader cultural and institutional rethinking of fungal diversity, making it possible to integrate fungi into national biodiversity platforms, conservation discourse, and public policy (Drechsler-Santos et al. 2025).

13.5.2 Diversity Shortfalls and Conservation Challenges in Brazilian Forests

Brazilian forests, including the Amazon, Atlantic Forest, Cerrado forest formations, Caatinga woodlands, Pantanal, and Pampa-associated forests, harbor an immense yet unevenly documented fungal diversity. Recent syntheses estimate that only about 14–16% of the species of fungi predicted to occur in Brazil are currently represented by accessible digital records, with a much smaller fraction assessed for conservation status (Drechsler-Santos et al. 2025). This combination of Linnean, Wallacean, and Scottian shortfalls profoundly limits conservation planning.

Spatial patterns of known fungal diversity largely reflect historical sampling biases rather than true richness. The Atlantic Forest stands out as the best-documented domain, largely due to its long history of scientific infrastructure and intense research activity. In contrast, vast forested regions of the Amazon, Pantanal, and ecotonal zones remain severely under-sampled, hindering robust assessments of threat, endemism, and population trends. These gaps are particularly problematic in a context of accelerating land-use change and climate-driven disturbances affecting Brazilian forests.

The threats faced by fungi in Brazilian forests mirror, but do not fully overlap with, those affecting plants and animals. Habitat loss and degradation due to agriculture, silviculture, urban expansion, infrastructure development, and mining are the most frequently cited pressures in Red List assessments of Brazilian fungal species (Mueller et al. 2022; Drechsler-Santos et al. 2025). Forest fragmentation disrupts fungal dispersal, reduces host availability for symbiotic species, and alters microclimatic conditions essential for reproduction and persistence.

Climate change represents an additional and increasingly relevant threat. Shifts in temperature and precipitation regimes affect fungal phenology, fruiting patterns, host associations, and the stability of mycorrhizal networks (Gange et al. 2011; Bennett and Classen 2020; Andrew 2025). Although climate change is still underrepresented in fungal assessments, available evidence suggests that many forest-associated fungi, particularly endemics with narrow ecological niches, are likely to be highly sensitive to these changes.

13.5.3 Red Lists and the Assessment of Fungal Extinction Risk

Despite recent progress, fungi remain underrepresented in the IUCN Red List globally (Mueller et al. 2022). Brazil, however, has emerged as a global leader in reversing this trend. Since the first Brazilian fungal species were added to the IUCN Red List in 2019, the number of assessed species has increased rapidly, with over 120 species currently evaluated, approximately half of which are threatened or near threatened (IUCN 2025).

These assessments reveal a strong bias toward conspicuous macrofungi, particularly the Basidiomycota, reflecting both methodological challenges and taxonomic expertise gaps. Lichens and microfungi, despite their ecological importance, remain comparatively underassessed. Nevertheless, Red List data already provide critical insights into threat patterns and priority actions, clearly demonstrating the urgency of integrating fungi into national conservation frameworks.

A landmark recent development is the imminent publication of the first official National Red List of Brazilian Funga, made possible by new legal instruments and coordinated efforts among researchers, governmental agencies, and the IUCN Species Survival Commission (SSC). This milestone represents a fundamental shift in Brazilian conservation policy, enabling fungi to be formally recognized as targets of protection, management, and restoration actions.

13.5.4 Legal Frameworks and Public Policies for Fungal Conservation

The most significant breakthrough for fungal conservation in Brazil came with Decree No. 12.137 (2024), which officially incorporated fungi into national environmental legislation. This decree expanded the mandate of the Centro Nacional de Conservação da Flora (CNCFlora) to include the Funga, establishing a legal foundation for assessing extinction risk, implementing conservation actions, and integrating fungi into environmental impact assessments (Drechsler-Santos et al. 2025).

This legal recognition aligns fungal conservation with broader national and international commitments, including Brazil's National Biodiversity Strategy and Action Plan and obligations under the Convention on Biological Diversity. Importantly, it enables threatened fungal species to be incorporated into existing conservation instruments, such as National and Territorial Action Plans for Threatened Species, protected area planning, and mitigation hierarchies applied to development projects.

13.5.5 Institutional Leadership and Collaborative Networks

Institutional advances have been central to Brazil's recent progress. The creation of the Brazil Fungal Specialist Group (BrazFunSG) within the IUCN SSC marked a pivotal moment, providing an organized platform for training, assessments, policy dialogue, and international collaboration (Drechsler-Santos et al. 2023). Through workshops, capacity-building programs, and coordinated assessments, BrazFunSG has helped overcome historical barriers to fungal conservation expertise in the country.

At the same time, initiatives such as the Brazilian Threatened Fungi Collection (CFAB–MIND.Funga) represent important steps toward *ex situ* conservation by preserving, among others, isolates of *Bondarzewia loguerciae* Salvador-Mont., Rajchenb., Kossmann, Bittencourt & Drechsler-Santos (Figure 13.5) and *Fomitiporia nubicola* Alves-Silva, Bittencourt & Drechsler-Santos (Figure 13.6), two critically endangered species. In addition, this initiative promotes data integration and public accessibility of information on threatened fungal species. These efforts highlight the increasing institutional maturity of fungal conservation in Brazil and its integration with cutting-edge research, citizen science, and technological innovation (Mind.Funga 2026 [continuously updated]).

13.5.6 Integrative Approaches to Fungal Conservation in Forests

Effective fungal conservation in Brazilian forests requires integrative strategies that go beyond species-level protection. Habitat conservation remains the most efficient short-term approach, particularly for poorly known groups. Protecting forest integrity, structural complexity, and ecological processes inherently safeguards fungal diversity, including species that cannot yet be individually assessed.

FIGURE 13.5 Basidiome of the critically endangered *Bondarzewia loguerciae* in the field.

FIGURE 13.6 Basidiome of the critically endangered *Fomitiporia nubicola* in the field.

Case studies from southern Brazil demonstrate the potential of integrative approaches that combine taxonomic research, ecological data, spatial prioritization, Red List assessments, and stakeholder engagement (Costa-Rezende et al. 2022). Such frameworks allow conservation actions to be implemented despite incomplete knowledge, while progressively refining priorities as new data become available.

Long-term success will also depend on addressing chronic knowledge gaps, especially for microfungi and symbiotic guilds such as arbuscular mycorrhizal fungi. Although methodological constraints currently limit formal extinction-risk assessments for some groups, emerging tools, such as environmental DNA, predictive modeling, and trait-based approaches, offer promising avenues for future integration into conservation planning.

13.5.7 Communication, Education, and Societal Engagement

Brazilian society has traditionally exhibited negative cultural perceptions of fungi. Overcoming this barrier is essential for the long-term success of fungal conservation. Strategic communication, environmental education, and outreach initiatives linking fungi to ecosystem services, food security, and climate resilience are increasingly recognized as core components of conservation strategies.

Citizen science platforms, training programs, and collaboration with local communities hold significant potential to expand data collection, monitoring, and public awareness at scales unattainable by professional scientists alone. These approaches can play a decisive role in consolidating fungal conservation as a socially supported and politically viable agenda.

13.5.8 Perspectives and Future Directions

Brazil now possesses an unprecedented combination of legal recognition, institutional capacity, and scientific expertise to advance fungal conservation in its forests. The challenge ahead lies in translating this emerging framework into sustained action. Continued political commitment, stable funding for mycological research, and integration of fungi into cross-sectoral conservation policies are critical to ensuring long-term impact.

By shifting from species neglect to active inclusion, Brazil not only addresses a historical imbalance in biodiversity conservation but also offers a globally relevant model for integrating fungi into ecosystem-based conservation strategies. In a country defined by its forests, conserving the Funga is inseparable from conserving the ecological integrity, resilience, and future of Brazilian forest ecosystems.

13.6 ACKNOWLEDGMENTS

Appreciation is extended to the National Council for Scientific and Technological Development (CNPq) for providing the productivity grants to SX-S (process 313260/2022-2) and ERD-S (process 310150/2022-1). Thanks also go to the Coordination for the Improvement of Higher Education Personnel (CAPES) for the PDJ scholarship awarded to LL-S (process 88887.189344/2025-00).

REFERENCES

Alvarenga, R.L.M., L.R.R. Naves & S. Xavier-Santos. 2015. The genus *Auricularia* Bull. ex Juss. (Basidiomycota) in Cerrado (Brazilian Savanna) areas of Goiás state and the Federal District, Brazil. *Mycosphere* 6:532–541.

Andrew, C. 2025. Not always optimal: Fungal fruiting triggers indicate climate sensitivity in cooler regions. *Fungal Ecology* 75:101416.

Baltazar, J.M. & T.B. Gibertoni. 2009. A checklist of the aphyllophoroid fungi (Basidiomycota) recorded from the Brazilian Atlantic Forest. *Mycotaxon* 109:439–442.

Bennett, A.E. & A.T. Classen. 2020. Climate change influences mycorrhizal fungal-plant interactions, but conclusions are limited by geographical study bias. *Ecology* 101:e02978.

Bezerra, J.D.P., M.H.C. Maciel, J.L. Bezerra, O.M.C. Magalhães & M.C. Souza-Mota. 2017. The contribution of Augusto Chaves Batista (1916–1967) to Mycology in Brazil. *Gaia Scientia* 11(2): 250–273.

Bononi, V.L.R., A.K.M. Oliveira, J.R. Quevedo & A.M. Gugliotta. 2008. Fungos macroscópicos do Pantanal do Rio Negro, Mato Grosso do Sul, Brasil. *Hoehnea* 35:489–511.

Brasil. 2024. Ministério do Meio Ambiente e Mudança do Clima. *Florestas do Brasil: dados e estatísticas 2024*. MMA, Brasília, DF. https://www.gov.br/florestal/pt-br/centrais-de-conteudo/publicacoes/publicacoes-diversas/flobr_2024_v0-24___online.pdf

Breuninger, M., W. Einig, E. Magel, S. Cardoso & R. Hampp. 2000. Mycorrhiza of Brazil pine (*Araucaria angustifolia* [Bert. O. Ktze.]). *Plant Biology* 2:4–10.

Capelari, M., A.M. Gugliotta & M.B. Figueiredo. 1998. O estudo de fungos macroscópicos no estado de São Paulo. In C.A. Joly & C.E.M. Bicudo (eds.), *Biodiversidade do estado de São Paulo, Brasil: síntese do conhecimento ao final do século XX*, pp. 9–23. São Paulo Research Foundation, São Paulo, Brazil.

Cavalcante, F.S.A., M.C.C. Campos & J.P.S. Lima. 2020. Studies of fungi in the state of Amazonas, Brazil in the last 10 years. *Ciência e Natura* 42:e38.

Coertjens, N.C., M.S. Mascarenhas & M. Batistote. 2024. Brazilian endophytic fungi: Biotechnological potential in evidence. *Diversitas Journal* 9:2835–2851.

Costa-Rezende, D., E.R. Drechsler-Santos, L. Gumbinger, F. Bittencourt & V. Ferreira-Lopes. 2022. An integrative approach for fungal conservation in southern Brazil. *Oryx* 56:13.

Coutinho, L.M. 1990. Fire in the tropical biota: Ecosystem processes and global challenges. In J.G. Goldammer (ed.), *Fire in the Ecology of the Brazilian Cerrado*, pp. 82–105. Springer-Verlag, Berlin, Germany.

Cunha, H.F.V., K.M. Andersen, L.F. Lugli, F.D. Santana, I.F. Aleixo, A.M. Moraes & C.A. Quesada. 2020. Direct and indirect effects of nutrient limitation on Amazonian forest productivity. *Nature Communications* 11:1–11.

Dahlberg, A. & G.M. Mueller. 2011. Applying IUCN red-listing criteria for assessing and reporting on the conservation status of fungal species. *Fungal Ecology* 4:147–162.

Dianese, J.C.F., C.A. Inácio, A.A. de Carvalho Junior, M.D.M. dos Santos, T. Cantillo-Pérez & D.B. Pinho. 2022. Exploring the overlooked diversity of plant-associated Cerrado microfungi. *Revista Annual de Patologia de Plantas* 28:69–101

Drechsler-Santos, E.R., M.A.Q. Cavalcanti & T.B. Gibertoni. 2009. A re-evaluation of the lignocellulolytic Agaricomycetes from the Brazilian semi-arid region. *Mycotaxon* 108:241–244.

Drechsler-Santos, E.R., T.B. Gibertoni & M.A.Q. Cavalcanti. 2013. New records of Auriculariales, Hymenochaetales and Polyporales (Fungi: Agaricomycetes) for the Caatinga biome. *Check List* 9:800–805.

Drechsler-Santos, E.R., T. Kossmann, F. Bittencourt, V. Ferreira-Lopes, K.M. da Cunha, G.L. Robledo & G.M. Mueller. 2023. A new specialist group for Brazilian fungi. *Oryx* 57:421–422.

Drechsler-Santos, E.R., T. Kossmann, F. Bittencourt, V. Ferreira-Lopes, G.D.S. Seger & G.M. Mueller. 2025. Brazil as a global player in fungal conservation: A rapid shift from neglect to action. *Perspectives in Ecology and Conservation* 23:246–254.

Drewinski, M.P., E.R. Drechsler-Santos, G.D.S. Seger, V. Ferreira-Lopes, L. Gumbinger, T. Kossmann, et al. 2024. Over 400 food resources from Brazil: Evidence-based records of wild edible mushrooms. *IMA Fungus* 15:40.

Farias, V.P., O.S. Medeiros, L. Gumbinger & E.R. Drechsler-Santos. 2025. Distribution patterns of wood-decay macrofungi (Agaricomycetes) in floodplain forest islands of the eastern Amazon. *Journal of Fungi* 11:288.

Fidalgo, O. 1968. Introdução à história da micologia brasileira. *Rickia* 3:1–44.

Flora e Funga do Brasil. 2026. *Jardim Botânico do Rio de Janeiro*. http://floradobrasil.jbrj.gov.br/.

Food and Agriculture Organization of the United Nations (FAO). 2020. *Global Forest Resources Assessment 2020: Main Report*. United Nations, New York.

Gange, A.C., E.G. Gange, A.B. Mohammad & L. Boddy. 2011. Host shifts in fungi caused by climate change? *Fungal Ecology* 4:184–190.

Garcia-Pichel, F. & O. Sala. 2022. Expanding the pulse–reserve paradigm to microorganisms on the basis of differential reserve management strategies. *BioScience* 72:638–650.

Gibertoni, T.B., P.J.P. Santos & M.A.Q. Cavalcanti. 2007. Ecological aspects of Aphyllophorales in the Atlantic rain forest in northeast Brazil. *Fungal Diversity* 25:49–67.

Gibertoni, T.B., P.S. Medeiros, A.M.S. Soares, A.C. Gomes-Silva, P.J.P. Santos, H.M.P. Sotão, L.V. Ferreira & E. Savino. 2016. The distribution of polypore fungi in endemism centres in Brazilian Amazonia. *Fungal Ecology* 20:1–6.

Instituto Brasileiro de Geografia e Estatística (IBGE). 2012. *Manual técnico da vegetação brasileira*, 2nd ed.

IUCN. 2025. *The IUCN Red List of Threatened Species* (Version 2025-2). https://www.iucnredlist.org.

Komura, D.L., C.E. Zartman, A.F. Cunha & W.E. Magnusson. 2017. How do seasonality, substrate, and management history influence macrofungal fruiting assemblages in a central Amazonian forest? *Biotropica* 49:643–652.

Kuhar, F., G. Furci, E.R. Drechsler-Santos & D.H. Pfister. 2018. Delimitation of Funga as a valid term for the diversity of fungal communities: The Fauna, Flora & Funga proposal (FF&F). *IMA Fungus* 9:A71–A74.

Leonardo-Silva, L., L.B. Silva & S. Xavier-Santos. 2020. Poroid fungi (Agaricomycetes, Basidiomycota) from Floresta Nacional de Silvânia—a conservation unit of Brazilian Savanna. *Microbial Biosystems* 5:100–107.

Leonardo-Silva, L. & S. Xavier-Santos. 2023. Corticioid and poroid fungi from Brazilian Cerrado: A history of research and a checklist of species. *Anais da Academia Brasileira de Ciências* 95:e20220165.

Lepinay, C., E. Heegaard, K. Eldegard, B. Schulz & R. Halvorsen. 2021. Successional development of fungal communities associated with decomposing deadwood in a natural mixed temperate forest. *Journal of Fungi* 7:412.

Maia, L.C., A.A. de Carvalho Jr., L.H. Cavalcanti, A.M. Gugliotta, E.R. Drechsler-Santos, A.L.M.A. Santiago, M.C. Abrahão et al. 2015. Diversity of Brazilian fungi. *Rodriguésia* 66:1033–1045.

Medeiros, O.S., J.H. Cattanio & H.M.P. Sotão. 2015. Riqueza e relação dos fungos poroides lignolíticos (Agaricomycetes) com o substrato em floresta da Amazônia. *Boletim do Museu Paraense Emílio Goeldi* 10:423–436.

Mendoza, A.Y.G., F.G. Pinheiro & J.G. Barros. 2022. Potencialidades biotecnológicas dos fungos da Amazônia brasileira: uma revisão sistemática. *Diversitas Journal* 7:2381–2400.

Mind.Funga. 2026. *Brazilian Threatened Fungi Collection (CFAB–MIND.Funga)*. Universidade Federal de Santa Catarina, Brazil. https://mindfunga.ufsc.br/

Moreira, F.M.S. & J.O. Siqueira. 2006. *Microbiologia e bioquímica do solo*, 2nd ed. Editora Federal University of Lavras, Brazil.

Mueller, G.M., A. Dahlberg, B. Walker, E.N. Lughadha & C. Ituarte. 2022. What do the first 597 global fungal Red List assessments tell us about the threat status of fungi? *Diversity* 14:736.

Neves, M.A., I.G. Baseia, E.R. Drechsler-Santos & A. Góes-Neto. 2013. *Guide to the common fungi of the semiarid region of Brazil*. Editora TECC, Florianópolis, Brazil.

Pagin-Cláudio, F. & A.M. Gugliotta. 2025. Six new records of polypores (Agaricomycetes, Basidiomycota) from Southeast Brazil (Atlantic Forest) and perspectives in basic mycological research. *Rodriguesia* 76:e00672024.

Pagin-Cláudio, F., A.M. Gugliotta & R.M. Pires. 2022. Checklist of polypores (Agaricomycetes, Basidiomycota) reveals seventeen new records in Brazilian Atlantic Forest fragments at the Poços de Caldas Plateau, Minas Gerais. *Rodriguésia* 73:e00832021.

Peay, K.G., P.G. Kennedy & J.M. Talbot. 2016. Dimensions of biodiversity in the Earth mycobiome. *Nature Reviews Microbiology* 14:434–447.

Pires, R.M., A.M. Gugliotta, F. Pagin-Cláudio & M.C. Abrahão. 2025. Drivers of fungal succession in three different woody species in a tropical forest fragment. *Cryptogamie, Mycologie* 46:1–20.

Regio, N.C., M.C. Westphalen & S. Xavier-Santos. 2024. Contributions to the fungal diversity of the Cerrado: New records of lignocellulolytic Agaricomycetes. *Iheringia, Série Botânica* 79:e20241102.

Ribeiro, J.F. & B.M.T. Walter. 2008. As principais fitofisionomias do bioma Cerrado. In S.M. Sano, S.P. Almeida & J.F. Ribeiro (eds.), *Cerrado: Ecologia e Flora*, pp. 151–212. Embrapa Cerrados, Brazil.

Sanuma, O.I., F. Yanomami & N.K. Ishikawa. 2016. *Cogumelos: enciclopédia dos alimentos Yanomami (Sanöma)*. Instituto Socioambiental, Brazil.

Siqueira, J.O., M.R. Lambais & S.L. Stürmer. 2002. *Fungos micorrízicos arbusculares: biologia e aplicação*. Editora Federal University of Lavras, Brazil.

Siqueira, J.O., F.A. de Souza, E.J.B.N. Cardoso & S.M. Tsai (eds.). 2010. *Micorrizas: 30 anos de pesquisas no Brasil*. Editora Federal University of Lavras, Brazil.

Smith, S.E. & D.J. Read. 2008. *Mycorrhizal Symbiosis*, 3rd ed. Academic Press, New York.

Soares, M.T.S. & L.C.M. Froufe. 2015. Estimativa de ciclagem de nutrientes em ecossistemas florestais por meio da produção e decomposição de serapilheira. In L.M. Parron et al. (eds.), *Serviços ambientais em sistemas agrícolas e florestais do Bioma Mata Atlântica*, pp. 155–170. Brazilian Agricultural Research Corporation, Brasília, Brazil.

Stark, N.M. & C.F. Jordan. 1978. Nutrient retention by the root mat of an Amazonian rain forest. *Ecology* 59:434–437.

Swift, M.J., O.W. Heal & J.M. Anderson. 1979. *Decomposition in Terrestrial Ecosystems*. University of California Press, Oakland, California.

Winagraski, E., G. Kaschuk, P.H.R. Monteiro, C.G. Auer &. A.R. Higa. 2019. Diversity of arbuscular mycorrhizal fungi in forest ecosystems of Brazil: A review. *Cerne* 5:25–35.

Xavier, W.K.S., O.S. Medeiros & H.M.P. Sotão. 2018. Riqueza de Agaricomycetes poroides da Serra do Navio, Amazônia oriental, com novo registro de *Oxyporus lacera* para o Brasil. *Boletim do Museu Paraense Emílio Goeldi* 13:303–315.

Yanomami, F.D.C.P., N.K. Ishikawa & V. Py-Daniel. 2019. *Përisi: përisiyoma pë wãha oni = Marasmius yanomami: o fungo que as mulheres Yanomami usam na cestaria*. s.n. BMF Gráfica, Editora Federal University of Lavras, Brazil.

14 Plant Fungal Pathogens in the Greater Mekong Subregion (GMS)

Tikka Dewage Chamarika Priyadarshani, Garumuni Dilrukshi Nadeeshani Menike, Mudugama Vidanage Sachini Sithma Mudugama, and Samantha Chandranath Karunarathna

14.1 INTRODUCTION

The Greater Mekong Subregion (GMS) is a verdant tapestry spanning approximately 2.5 million km^2 and encompassing Cambodia, the Lao People's Democratic Republic, Myanmar, Yunnan Province of the People's Republic of China, Thailand, and Vietnam (Figure 14.1). Encompassing one of the longest rivers in the world, the area is characterized by a varied climate, diverse vegetation types and habitats, and falls within both temperate and tropical regions. It is a biodiversity hotspot, harbouring some of the world's best examples of tropical and evergreen forests and is home to a high diversity of endemic and endangered organisms (Chaiwan et al. 2021).

The climate in GMS ranges from continental and temperate to tropical monsoonal, creating favorable conditions for a wide variety of fungal plant pathogens. Specifically, most of the higher elevation is snow-covered, and sub-zero temperatures are prevalent during the wet season. Even in the dry season, temperatures remain around 13 °C, causing it to have a temperate climate. The Yunnan province of China, in particular, has a relatively temperate climate year-round. In contrast, the lower part of the Mekong River Basin has a tropical monsoonal climate, with seasons not distinctly demarcated. The temperature ranges from a maximum of 32 °C to a minimum of 15–23 °C throughout the year. The average annual rainfall is around 1,000–4,000 mm (Mekong River Commission). Due to these conditions, the GMS region is considered a hotspot for fungal pathogens, with a diverse array of fungal inhabitants.

Agriculture is predominant in this area, as most of the land receives rainfall during the tropical monsoon season, providing a stable water supply for the cultivation of crops. While rice is the most widely cultivated food crop throughout the region, there are other crops such as banana, mango, coconut, cassava, jackfruit, cashew, lime/lemon, lemon grass in Cambodia (National Institute of Statistics 2023); maize, cassava, banana, and coffee in Laos, rubber, sugarcane, cotton, oil palm, coffee, tea, oilseeds and vegetables in Myanmar (EuroCham Myanmar 2020); corn, palm oil, soybean, sorghum, peanut, cotton in Thailand (Foreign Agricultural Service, USDA 2025); cassava, corn, sweet potatoes, nuts, banana, coconut, citrus, coffee and tea in Vietnam; and corn, barley, wheat, rapeseed, sweet potatoes, soybeans, tea, sugarcane, tobacco, and cotton in Yunan, China. Consequently, the abundant host diversity across the region's agricultural landscape is a key factor in the proliferation of numerous plant pathogenic fungi.

Since agriculture is the primary source of income in the GMS region, rural-level farmers tend to engage in cross-border exchanges of seeds, seedlings/planting materials across countries without strict phytosanitary protocols, thereby facilitating the rapid dispersal of pathogens (Chaithra et al.

DOI: 10.1201/9781003647140-14

FIGURE 14.1 The Greater Mekong Subregion (GMS).

2024). Also, agricultural trade routes, such as banana exports from Laos to China or cassava trade between Cambodia and Vietnam, play an important role in disseminating fungal pathogens across countries (Chittarath et al. 2018).

In addition, the interconnected agroecological landscapes of rice, maize, and cassava cultivation across national borders facilitate the free movement of fungal pathogens to new locations, as there are few natural geographical barriers among these countries. As these pathogens can disperse through the air (e.g., *Pyricularia oryzae*), water (e.g., *Rhizoctonia solani*), or as soilborne propagules (e.g., *Fusarium* spp.), regionwide disease epidemics can arise. This presents a significant challenge for managing them at a national level (Mwangi et al. 2023). Moreover, when pathogen populations from different locations mix across borders, there is a high likelihood of recombination and the exchange of genetic material, thereby increasing overall diversity (Duan et al. 2024). This is problematic not only in breeding programs for resistance cultivars but also undermines the uniform disease control strategies of a region. Ultimately, these localized outbreaks can lead to regional agricultural crises, as they are rarely confined to a single region, resulting in severe yield losses across multiple regions. Eventually, this will be a threat to food security. Moreover, the region is also rich in fungal diversity and homes a vast array of fungal species, many of which are yet to be discovered (Hapuarachchi et al. 2019; Luangharn et al. 2021).

In the Greater Mekong Subregion, numerous studies have shown that China (specifically Yunnan Province) and Thailand are characterized by a staggering diversity of macro- and microfungi (Ren et al. 2022). Fungi in this region serve important roles as decomposers, parasites, and symbionts, while also providing medicine, food, and income for local communities (Li et al. 2018). For example, nearly 9,000 documented edible species of mushrooms, accounting for more than 40% of the world's edible mushrooms, are found in the GMS, providing a significant source of income for rural economies. Thriving on diverse substrates, fungi play other pivotal roles, including the regulation of nutrients and as a key component of the carbon cycle. They function as primary decomposers that break down decaying organic matter into simpler substances for their own consumption as well as for other organisms, and they are among the few organisms that can effectively break down woody material that is highly resistant to degradation due to its lignin composition. Moreover, ectomycorrhizal fungi form symbiotic relationships with plant species, enhancing plant growth and survival by assisting plants in nutrient uptake via mycorrhizae. These fungi also influence soil structure and produce specific secondary metabolites, which are crucial for human health and well-being (Carris et al. 2012; Chaiwan et al. 2021; Phonemany et al. 2022). Although most fungi are non-pathogenic to plants and animals, a subset of taxa can cause various diseases in these organisms (Agrios 2005). Among these are plant pathogenic fungi that are parasites and can cause diseases characterized by various plant symptoms (Carris et al. 2012).

In general, fungi that colonize plants represent fewer than 10% of all known species. Although not all plant-parasitic fungi are pathogenic, phytopathogenic fungi are the primary cause of most plant diseases (Carris et al. 2012). These fungi are predominantly members of the phyla Ascomycota and Basidiomycota, and have evolved to obtain their food requirements and reproduce within living or dead plant hosts while causing various plant diseases that impair the plant productivity (Doehlemann et al. 2017). Based on their economic and scientific importance, Dean et al. (2012), reported that some of the most important global plant pathogens include *Blumeria graminis* (Figure 14.2A, B), *Botrytis cinerea* (Figure 14.2C), *Colletotrichum* spp., *Fusarium graminearum* (Figure 14.2D), *Fusarium oxysporum* (Figure 14.2E), *Puccinia* spp. (Figure 14.2G), *Pyricularia oryzae* (Figure 14.2H, I), *Melampsora lini, Zymoseptoria tritici* (Figure 14.2M), and *Ustilago maydis* (Figure 14.2L). Being a region of rich fungal diversity, the flora of GMS is no exception with respect to susceptibility to some of these plant pathogenic fungi.

Some of the most significant plant fungal pathogens in this region include pathogenic members of the genus *Ganoderma* and members of the family *Didymosphaeriaceae*. These fungi cause several diseases in plants such as white rot, stem rot and wood decay, inflicting severe losses to both forestry and agriculture (Luangharn et al. 2021; Ren et al. 2022; Karunarathna et al. 2024). The highly virulent tropical race 4 (TR4) strain of *Fusarium oxysporum* f. sp. *cubense* (Foc), with a wide host range, severely impedes banana cultivations in the region (Chittarath et al. 2022) along with other numerous disease-causing fungi which will be discussed later in this chapter.

This chapter has the objective of providing an overview of the most prevalent species of fungi species in the region and the control methods used for these fungi. The various species of fungi are listed, and their host range and the optimal control strategies currently practiced in the region are described. In addition, recent advances in molecular approaches and highlight the future directions of mycology-based research in the region are discussed.

14.2 PLANT FUNGAL DISEASES REPORTED IN GMS

Among fungal pathogens infecting plants in the GMS, the genus *Ganoderma*, a wood-decaying fungus that plays a critical role in nutrient cycling and carbon sequestration, holds an important role. This is because the region is a hotspot for species of *Ganoderma*, with approximately 22 of the 80 species identified worldwide. Members of this genus serve dual purposes. They are beneficial fungi because decompose wood in forest ecosystems, are mycorrhizal partners with trees, and they interact with other organisms, including humans, providing valuable medicine and an income source. Another function is their potential to function as phytopathogens, causing diseases such as white rot, stem rot, and wood decay in economically important trees in the GMS. Some of these species of

FIGURE 14.2 Some major fungal pathogens reported from GMS. (A, B) *Blumeria graminis* on *Elymus repens*, (C) *Botrytis cinerea* on grapes, (D) *Fusarium graminearum* on *Zea mays* corn, (E) *Fusarium oxysporum* on pumpkin, (F) *Ganoderma boninense* on a tree base, (G) *Puccinia sorghi* on *Zea mays* leaves, (H, I) *Pyricularia oryzae* on rice plants, (J, K) *Rhizoctonia solani* on crabgrass and potato, respectively, (L) *Ustilago maydis* on *Zea mays* corn, (M) *Zymoseptoria tritici* on *Triticum aestivum* leaves.

Source: Photos used under the CC BY-NC 4.0 [Attribution-NonCommercial 4.0 International].

Ganoderma present paradoxical cases: They are simultaneously cultivated for their high economic value as medicinal mushrooms (e.g., *G. australe* and *G. lucidum*) while also having the potential to cause plant diseases. With growth facilitated by warm and humid conditions, upon spore germination fungal hyphae invade host tissues through root contact, wounds, or natural openings like stomata for nutrient acquisition by decomposing lignin and cellulose, while simultaneously causing disease symptoms such as wilting, yellowing, decay, and, ultimately, the death of the plant. They also produce fruiting bodies (basidiocarps) under favorable conditions (Luangharn et al. 2021; Karunarathna et al. 2024; Karunarathna et al. 2025).

Fusarium wilt, also known as Panama disease, is another growing fungal menace to the flora, particularly those intended for human consumption in the GMS. Of particular concern is a highly virulent strain of *Fusarium oxysporum* f. sp. *cubense* (Foc), going by the name of Foc R4, causing significant yield losses on Cavendish banana cultivations in GMS. In general, *Fusarium oxysporum* is a soil-borne, asexually reproducing plant pathogenic fungus known for its remarkable persistence and adaptability. The pathogen can remain dormant in soil through chlamydospores for up to 30 years or survive asymptotically in non-host plants as endophytes. When this fungus encounters a suitable host, it can penetrate plant roots and reproduce in xylem vessels, thereby causing vessel occlusions that impede the transport of essential nutrients necessary for plant growth and reproduction, ultimately leading to the death of the host. As for Foc R4, the strain is subdivided into subtropical race 4 (SR4), which infects Cavendish bananas in the subtropics, and the more damaging tropical race 4 (TR4), which has a wider host range and affects Cavendish in both the tropics and subtropics. Over recent years, the fungus has been rapidly spreading across Laos, Vietnam, Myanmar, Cambodia, and Thailand, and several of these countries are major Cavendish producers. This spread has coincided with efforts by private Chinese companies to expand banana cultivation in these regions, using possibly infected planting material. In addition to this strain, Foc R1 and R2-like strains are also found to infect other banana cultivars in GMS (Zheng et al. 2018; Chittarath et al. 2022).

Theologized to have originated in China itself, rice blast disease, caused by *Pyricularia oryzae* (synonymously called *Pyricularia oryzae*), is another significant biotic constraint in the GMS (Le et al. 2010; Zewdu 2021). The causal organism of this disease is a filamentous ascomycete that primarily spreads via asexual spores (three-celled conidia) and exhibits a polycyclic disease cycle. It infects all above-ground parts of rice plants at any growth stage through a specialized penetration structure known as the "appressorium." Upon entering the host tissue, the fungal hyphae rapidly spread into adjacent tissues while suppressing the defense system of the host. Initially appearing as small, whitish-grey, diamond-shaped lesions with dark borders, the latter later coalesce, spreading to almost every part of the plant and gradually killing it. In the meantime, the fungus also forms spores and rapidly spreads to neighboring crop fields, up to roughly 230 m from the original fungus through wind and water splash, with these favored by dark, humid conditions. The ability of the pathogen to complete its life cycle within a very short time and the ability of a single lesion to produce thousands of spores render the disease highly virulent and challenging to manage (Asibi et al. 2019). Several studies have also shown that the GMS exhibits a diversity of blast pathotypes with regional differences, some of which are highly virulent toward specific resistant genes (Le et al. 2010; Fukuta et al. 2014).

14.3 THE PLANTS AFFECTED BY FUNGAL DISEASES IN GMS

Within the genus *Ganoderma*, *G. boninense* (Figure 14.2F) is the most devastating disease-causing agent for oil palm in the GMS, leading to yield losses of up to 80% in plantations across Thailand due to basal stem rot. *Ganoderma pseudoferreum* and *G. philippii* are responsible for red root disease in rubber cultivations of Thailand. Other pathogenic species of *Ganoderma* within this region include *G. applanatum, G. australe, G. colossum,* and *G. lucidum.* These fungi cause white rot disease in various forest tree species, including oaks, pines, mangroves, and black pepper, as well as various horticultural plants, including *Acacia* spp., *Cassia* spp., *Pinus* spp., and fruit trees such as avocado (Karunarathna et al. 2024; Karunarathna et al. 2025).

The primary plant affected by *Fusarium oxysporum* f. sp. *cubense* (Foc) in GMS is the banana. Given the region's significance as a major banana-producing area, the majority of cultivars remain highly susceptible to Fusarium wilt. The highly virulent Foc tropical race 4 (TR4) strain, which is prevalent in this region, currently affects exclusively Cavendish banana cultivation. However, considering that more than 80% of banana cultivation in the GMS, particularly in Laos, Myanmar and Vietnam, consists of Cavendish varieties and also given the fact that more than 50% of global banana production is Cavendish, there is high potential risk of this strain developing into an epidemic as in the case of Foc race 1 that completely wiped out "Gros michel" banana cultivations in America. There is also the potential for the race to infect other banana cultivars in the future. Other banana varieties, such as Pisang Awak and Pisang Mas, cultivated in the region are affected by FocR1/R2 (Zheng et al. 2018; Chittarath et al. 2022).

As a relatively flat area traversed by rivers and plains favorable for rice growth, the GMS has vast rice production. With countries such as Vietnam and Thailand taking prominence, the region accounts for nearly 44% of the world's rice output (World Bank). Rice blast poses a critical threat to rice cultivation in this region, as the prevailing environmental conditions, including 12-hour periods of moderate temperatures (25–30 °C), high relative humidity (90–92%), and elevated moisture levels, are optimal for pathogen development (Zewdu 2021). With the replacement of wild rice varieties by intensive cultivation of high-yielding, short-duration IRRI-recommended varieties that require high inputs of nitrogen, combined with the pathogen's capacity to rapidly overcome the host resistance of blast-resistant rice varieties, this disease has been causing outbreaks that are difficult to control. During recent years, the pathogen has evolved into several virulent blast races in the region (Fukuta et al. 2020). Yield losses due to blast in rice cultivation range from 1 to 50% depending on the cultivar and environmental conditions. There is also the potential for complete crop failure in severe cases (Zewdu 2021). While the pathogen principally survives on rice, it is also capable of completing its life cycle by living on weed species such *Eleusine coracana*, *E. indica*, *Panicum* spp., and *Setaria* spp. (Khadka et al. 2025).

14.4 COUNTRY-WISE OVERVIEW (CAMBODIA, LAOS, MYANMAR, THAILAND, VIETNAM, AND YUNNAN-CHINA)

Detailed summaries of the major crops cultivated in each country, as well as economically important fungal diseases and the pathogens that cause them are provided in Table 14.1.

From disease populations at different locations mixing across borders, there is a high chance of recombination and the exchange of genetic material within a particular population, thus increasing overall diversity (Duan et al. 2024). This will be problematic not only in breeding programs for resistance cultivars but also in undermining the uniform disease control strategies of a region. Ultimately, these localized outbreaks can lead to regional agricultural crises, as they are rarely confined to a single region, resulting in severe yield losses across multiple regions. Eventually, this will be a threat to food security.

14.4.1 CAMBODIA

The primary agricultural crop is rice; however, cassava, maize, mung bean, and soybean are also considered major crops. As the staple food and main source of income, rice Blast disease, caused by *Pyricularia oryzae*, is the most important fungal plant pathogen in Cambodia (Chou et al. 2020). *Rhizoctonia solani* (Figure 14.2J, K) is the pathogen responsible for the rice sheath blight, which also results in destructive effects on rice yields. These diseases also play a vital role, contributing to Cambodia's comparatively low rice productivity relative to other rice-growing countries in Southeast Asia (Kea et al. 2016).

Apart from these main pathogens, several fungal species also affect major crop cultivation in Cambodia (Table 14.1). However, accurate and reliable information on the impacts and yield losses caused by these pathogens is lacking, as the disease diagnostic infrastructure in other countries in the region is underdeveloped. Disease reports are primarily through collaborations with research institutes from other countries. Therefore, many outbreaks are likely underreported, especially from remote provinces.

TABLE 14.1
Fungal Pathogens in the Great Mekong Subregion (GMS).

Country	Crop	Disease	Pathogen/s	References
Cambodia	Rice	Rice Blast	*Pyricularia oryzae*	
		Sheath Blight	*Rhizoctonia solani*	
		Brown leaf spot	*Curvularia lunata*	Tann and Soytong (2017)
		Brown spot	*Cochliobolus miyabeanus*	Saito et al. (2025)
		Leaf scald of rice	*Monographella albescens*	Saito et al. (2025)
		Narrow brown spot	*Sphaerulina oryzina*	Saito et al. (2025)
	Cassava	Cassava witches' broom disease	*Moniliophthora perniciosa* and/ or *Ceratobasidium theobromae*/ phytoplasma	Alvarez et al. (2014); Leiva et al. (2023)
	Maize	Downy Mildew	*Peronosclerospora neglecta*	Sudsanguan et al. (2024)
		Southern maize leaf blight	*Bipolaris maydis*	Belfield and Brown (2009)
		Stalk rot and ear rot	*Fuusarium* spp.	Belfield and Brown (2009)
	Mung bean	Cercospora leaf spot	*Cercospora canescens*	Martin et al. (2020)
		Powdery mildew	*Erysiphe polygoni*	Martin et al. (2020)
	Soybean	Downy mildew	*Peronospora manshurica*	Belfield et al. (2011)
		Rust	*Phakopsora pachyrhizi*	Belfield et al. (2011)
Laos	Rice	Rice Blast	*Pyricularia oryzae*	Xangsayasane et al. (2020)
	Cassava	Cassava witches' broom disease	*Ceratobasidium* spp.	Leiva et al. (2023)
	Maize	Downy Mildew	*Peronosclerospora neglecta*	Lim et al. (2023); Sudsanguan et al. (2024)
		Grey leaf spot	*Cercospora sorghi*	Lao PDR Trade Portal (2023)
	Banana	Fusarium wilt	*Fusarium oxysporum*	Chittarath et al. (2018)
		Freckle disease	*Phyllosticta* spp.	Anderson et al. (2021)
	Coffee	Anthracnose	*Colletotrichum gloeosporioides*	Vilavong and Soytong (2017)
Myanmar	Rice	Rice blast	*Pyricularia oryzae*	FAO (2022)
		Sheath rot	*Sarocladium oryzae*	FAO (2022)
		Sheath Blight	*Rhizoctonia solani*	FAO (2022)
		Brown leaf spot	*Drechslera oryzae*	FAO (2022)
		Narro brow leaf spot	*Cercospora janceana*	FAO (2022)
		False smut	*Ustilaginoidea virens*	FAO (2022)
		Stem rot	*Nakataea oryzae*	FAO (2022)
		Bakanae disease	*Fusarium moniliforme*	FAO (2022)
	Green gram/Black gram	Anthracnose	*Colletotrichum lindemuthianum*	FAO (2022)

(Continued)

TABLE 14.1 *(Continued)*
Fungal Pathogens in the Great Mekong Subregion (GMS).

Country	Crop	Disease	Pathogen/s	References
		Rust	*Uromyces phaseoli*	FAO (2022)
		Powdery mildew	*Erysiphe polygoni*	FAO (2022)
		Root Rot	*Thanatephorus cucumeris*	FAO (2022)
		Leaf Spot	*Cercospora canescens*	FAO (2022)
		Leaf Spot	*Corynespora cassiicola*	FAO (2022)
		Collar Rot	*Corticium rolfsii*	FAO (2022)
	Green gram/Black gram seeds	Fusarium Wilt	*Fusarium udum*	FAO (2022)
		Leaf spot	*Cercospora* spp.	FAO (2022)
	Chickpea	Aspergillus ear rot	*Aspergillus flavus*	FAO (2022)
		Fusarium wilt of chickpea	*Fusarium oxysporum*	FAO (2022)
		Collar rot	*Aspergillus niger*	FAO (2022)
		Basal stem rot	*Rhizoctonia solani*	FAO (2022)
	Maize	Corn Blight	*Helminthosporium turcicum/ Helminthosporium maydis*	IPPC (2015)
		Corn Rust	*Puccinia sorghi*	IPPC (2015)
		Sheath Blight	*Rhizoctonia solani*	IPPC (2015)
		Ear Rot	*Cladosporium herbarum*	IPPC (2015)
		Stalk rot	*Fusarium moniliforme*	IPPC (2015)
		collar rot	*Aspergillus niger*	IPPC (2015)
		Cercosporiosis/Gray leaf spot	*Cercospora sorghi*	IPPC (2015)
		Root and foot rot	*Cochliobolus sativus*	IPPC (2015)
		Sclerotium rot	*Corticium rolfsii*	IPPC (2015)
		Charcoal rot	*Macrophomina phaseolina*	IPPC (2015)
		Rust	*Puccinia purpurea*	IPPC (2015)
		Brown stripe downy mildew of maize	*Sclerophthora rayssiae*	IPPC (2015)
		Maize leaf blight	*Setosphaeria turcica*	IPPC (2015)
		Head smut of maize	*Sphacelotheca reiliana*	IPPC (2015)
		False smut	*Ustilaginoidea virens*	IPPC (2015
Thailand	Rice	Sheath rot	*Sarocladium oryzae*	Bigirimana et al. (2015)
			Sarocladium attenuatum	Unartngam et al. (2024)
		Rice blast	*Pyricularia oryzae*	Phanon et al. (2024)
		Sheath blight	*Rhizoctonia solani*	Joomdok et al. (2022)
		Brown leaf spot	*Bipolaris oryzae*	Dethoupa et al. (2022)
		False smut	*Ustilaginoidea virens*	Akphumkitanun et al. (2025)
		Bakanae diseases	*Fusarium fujikuroi*	Pathom-aree et al. (2019)
		Riceberry Rice Leaf Spot	*Curvularia hominis*	Chaisarn et al. (2023)

TABLE 14.1 ***(Continued)***
Fungal Pathogens in the Great Mekong Subregion (GMS).

Country	Crop	Disease	Pathogen/s	References
		Leaf spot	*Curvularia asianensis*	Cheewangkoon et al. (2021)
			Curvularia lunata	Cheewangkoon et al. (2021)
			Curvularia verruculosa	Cheewangkoon et al. (2021)
		Seedling rot	*Curvularia lunata and Helminthosporium oryzae*	Limtong et al. (2020)
	Sugarcane	Red rot disease	*Colletotrichum falcatum*	Srijudanu et al. (2023)
			Fusarium moniliforme	
		Pokkah Boeng disease	*F. moniliforme*	Vishwakarma et al. (2013)
	Cassava	Black stem and root rot	*Lasiodiplodia theobromae, Neoscytalidium dimidiatum,* and *L. euphorbiaceicola*	Sangpueak et al. (2023)
		Fusarium root rot	*Fusarium solani*	Saengchan et al. (2022)
		Tuber rot	*Fusarium incarnatum*	
		Anthracnose disease	*Colletotrichum* sp.	Sangpueak et al. (2018)
	Maize	Downy Mildew	*Peronosclerospora neglecta*	Sudsanguan et al. (2024)
		Northern corn leaf blight	*Helminthosporium turcicum*	Mongkol et al. (2025)
		Leaf Spot	*Bipolaris maydis* and *Bipolaris turcica.*	Sudsanguan et al. (2021
		Blight	*Curvularia akaii, Curvularia clavata, Curvularia geniculata* and *Curvularia lunat*	Sudsanguan et al. (2021)
		Southern Corn Rust	*Puccinia polysora*	Oo et al. (2024)
	Oil Palm	basal stem rot (BSR)	*Ganoderma boninense*	Malinee et al. (2021)
		leaf spot disease	*Pestalotiopsis theae*	Suwannarach et al. (2013)
			Curvularia oryzae	Sunpapao et al. (2014)
		Leaf blight	*Alternaria longipes*	Suwannarach et al. (2015)
	Rubber	Leaf fall disease	*Neopestalotiopsis cubana* and *N. formicarum*	Pornsuriya et al. (2020)
			Colletotrichum siamense	Kaewchai et al. (2024)
		Powdery mildew	*Erysiphe quercicola*	Liyanage et al. (2017)
		Wilt	*Chalaropsis thielavioides*	Li et al. (2020)
Vietnam	Rice	Blast	*Pyricularia oryzae*	Nguyet et al. (2020)
		Sheath rot	*Sarocladium oryzae fungus* and *Fusarium fujikuroi*	Lim et al. (2023)
		Sheath blight	*Rhizoctonia solani*	Matsumoto and Cuong (2014)
	Cassava	Cassava witches' broom disease	*Candidatus Phytoplasma asteris*	Alvarez et al. (2013)
		Root rot	*Cunninghamella elegans*	Pham and Tran (2021)
	Coffee	Leaf Rust	*Hemileia vastatrix*	Le et al. (2022)
	Rubber	Die back of seeding	*Botryodiplodia theobromae*	Tran et al. (2011)
		Oidium leaf fall	*Oidium heveae*	Nguyen et al. (2015)
		Powdery mildew	*Erysiphe quercicola*	Tam et al. (2016)
	Sugarcane	Eye pot	*Helminthosporium sacchari*	Hoat et al. (2012)
		Fusarium sett and stem rot	*Fusarium moniliforme*	
		Leaf blight	*Nigrospora sphaerica*	
		Leaf spot	*Septoria demidovae*	
		brown stripe disease	*Helminthosporium stenospilum*	

(Continued)

TABLE 14.1 *(Continued)*
Fungal Pathogens in the Great Mekong Subregion (GMS).

Country	Crop	Disease	Pathogen/s	References
China-Yunan	Sugarcane	Brown stripe disease	*Bipolaris secalis* and *B. setariae*	Li et al. (2025)
		Eye spot	*Bipolaris sacchari*	Huang (2004)
		Pokkah boeng	*Fusarium monillforme*	
		Smut	*Ustilago scitaminea*	
		Red rot	*Colletotrichum alcatum*	
		Ring spot	*Phyllosticta saccharicola*	
		Yellow spot	*Mycovellosiella koepkei*	
		Pineapple rot	*Thielaviopsis paradoxa*	
	Rubber	Leaf spot	*Lasiodiplodia pseudotheobromae*	Wu et al. (2019)
			Bipolaris bicolor	Liang et al. (2019)
		Corynespora leaf fall	*Corynespora cassiicola*	Pu et al. (2007)
		Powery mildew	*Erysiphe quercicola*	Cao et al. (2024)
	Maize	White spot	*Diaporthe eres*	Yang et al. (2024)

14.4.2 Laos (Lao People's Democratic Republic)

Although rice is the staple food in Laos and the country is considered to have one of the highest levels of rice biodiversity, the status and impact of rice diseases are poorly documented in the peer-reviewed literature. After rice, maize is the second most important agricultural product in Laos, based on the area harvested (UNCTAD 2020). Continuous monocropping of bananas in the Laos has led to soil degradation, and recovery will take decades. If unaddressed, this will also contribute to the low production of the banana (Mekong eye). The banana is a staple food and fruit crop in Laos with cultural, nutritional, and economic importance. However, information on banana pathogens is lacking (Vansilalom 2016).

Rice blast (*P. oryzae*) and maize downy mildew (*Peronosclerospora neglecta*) are notable, with maize downy mildew causing severe yield losses of 40%-60%, even reaching 100% under favorable conditions. Cassava witches' broom disease severity has also been highlighted. Coffee anthracnose and banana Fusarium wilt have been reported, but economic data are unclear.

14.4.3 Myanmar

Rice, maize, beans, and pulses (especially chickpeas, green gram, and black gram) are the economically important crops produced in Myanmar. As the largest bean and pulse exporter in Asia and the second-largest in the world, beans and pulses can be considered a major agricultural product of the country (EuroCham Myanmar Agriculture Guide 2020). In addition to these major cash crops, there are other relatively less significant crops such as groundnuts, sesame, and sunflowers, as well as fruits like watermelons, oranges, and chilies (EuroCham Myanmar Agriculture Guide 2020).

Rice blast disease, caused by *Pyricularia oryzae*, is the most prevalent rice disease in Myanmar. This disease is prevalent across all major rice-growing ecosystems—irrigated, rainfed upland, and lowland—but is most severe in rainfed upland. Maize diseases include rust and blights, with limited impact on specifics.

14.4.4 Thailand

As one of the world's largest rice producers, Thailand cultivates rice for both local consumption and export (Seekham et al. 2024). Numerous studies have been conducted in the rice sector to enhance productivity. In addition to rice, cassava, corn, sugarcane, oil crops, and perennial trees such as para rubber are considered economically important cultivated crops in Thailand (Chainuvati and Athipanan 2001). Additionally, the disease records are well-documented. Several novel or first reports such as leaf blight in rice were noted from Thailand. Apart from other common major rice diseases, false smut disease, caused by *Ustilaginoidea virens*, has increased in severity over the past decade and is therefore attracting attention from the research community (Akphumkitanun et al. 2025).

Some diseases, such as cassava witches' broom and sugarcane red rot, have substantial impacts. Although sugarcane is a significant crop in Thailand's agricultural sector, accounting for 8% of the country's total agricultural area (Yingkamhaeng and Vanichsriratana 2024), there is limited literature available on the fungal diseases affecting sugarcane cultivation in the country.

14.4.5 Vietnam

Approximately 40% of the total land area is used for agriculture, and under favorable conditions, rice dominates throughout, followed by maize and rubber (FAO 2018). Rice diseases are similar to those found in other countries, including blast, sheath rot, and sheath blight, as well as high-impact cassava witches' broom disease, which has an incidence of up to 80% and substantial yield reduction. Prevalent throughout Vietnam, with incidence rates sometimes exceeding 75–80%, this disease causes severe reductions in root yield and starch content by up to 30% (Pardo et al. 2023).

14.4.6 Yunnan, China

With a rich biodiversity and diverse ecosystems, the agricultural industry in Yunnan produces a wide range of crops, including rice, corn, barley, wheat, rapeseed, sweet potatoes, soybeans, tea, sugarcane, tobacco, and cotton. Topography is a major constraint to agriculture since it determines which crops can be cultivated. In general, *Pyricularia oryzae* is the pathogen responsible for rice blast disease, which is common in all GMS nations, including Cambodia, Laos, Myanmar, Thailand, Vietnam, and Yunnan, China. Its ability for rapid adaptation makes *P. oryzae* a widespread pathogen across the region. In addition, common agricultural practices in the region facilitate the spread of the pathogen. Similarly, *Rhizoctonia solani*, which causes sheath blight, is frequently observed in GMS rice crops and thrives in the area's warm, humid climate, posing a persistent endemic threat. Cassava witches' broom disease, common in Cambodia, Laos, Vietnam, and Thailand, is associated with various species of fungi and phytoplasmas, causing significant yield losses and raising questions about its spread through planting materials.

More regional, yet economically significant, patterns can be observed in other important fungal infections. Although it is sometimes overlooked until outbreaks happen, reports of false smut disease of rice (*Ustilaginoidea virens*) are rapidly occurring in Myanmar, Thailand, and Yunnan. This disease is linked to humid settings that are favorable to its life cycle. Infected seedling mobility likely facilitates regional transmission within rubber plantations, as rubber leaf fall and seedling dieback are primarily observed in Thailand and Yunnan. Red rot and sugarcane brown stripe are also prevalent in Thailand and Yunnan, indicating a relationship between disease pressure and clusters of commercial sugarcane agriculture. The majority of reports of coffee leaf rust (*Hemileia vastatrix*) originate from Vietnam, a region in close proximity to high coffee-producing areas, posing a threat to nearby regions. Finally, reports of *Fusarium oxysporum* f. sp. *cubense*, commonly known as banana wilt, have been made in Laos and nearby regions, reflecting its gradual regional expansion and threatening banana cultivation across the GMS.

14.5 METHODS USED IN CONTROLLING AND PREVENTING PLANT FUNGAL DISEASES

Regardless of their taxonomic classification, the successful control of different phytopathogenic fungi requires meticulous identification, a thorough understanding of the fungi involved, and knowledge of the susceptible growth stages and favorable environmental factors for pathogen progression. The current norm is to use chemicals to control plant fungal diseases. However, given the long-term negative impacts these synthetic chemicals have on ecosystems and human health, other measures such as following proper management and prevention strategies, developing resistant varieties, utilizing biological control measures and botanicals, and adopting modern approaches that integrate technology, are increasingly being emphasized (Degani 2025).

Management strategies for fungal pathogens such as species of *Ganoderma* encompass multiple approaches. Agricultural practices such as the intentional introduction of different pathogenic strains of *Ganoderma* can reduce disease severity through competitive interactions. Cultural management practices include land management through proper plant spacing, crop rotation with non-host species, intercropping, trenching, sanitation, and poisoning infected trees. Biological control agents, including certain fungi (*Trichoderma* spp. and *Gliocladium viride*), bacterial strains (*Pseudomonas fluorescens*, *Bacillus* spp., *Streptomyces* spp., and *Burkholderia* spp.), and some actinomycetes, are also used for disease management. These biocontrol agents not only suppress pathogens but also enhance plant defense responses by priming the innate immune system and improving nutrient acquisition through the solubilization of soil nutrients, thereby helping build a stronger plant. Synthetic fungicides, as well as naturally occurring plant extracts and compounds, are also often used. Some other approaches to disease control include the use of resistant varieties and relying on integrated approaches rather than relying solely on one management strategy (Karunarathna et al. 2024).

Given the tenacious survival potential of the pathogen under adverse conditions for prolonged periods, no effective control measures are available to date for controlling Fusarium wilt in banana cultivation in the GMS. Prevention is the key to mitigating new incursions of the Foc TR4 strain. Recent studies have revealed significant knowledge gaps among farmers regarding this disease, despite its widespread occurrence. Hence, farmers need to be made aware of the importance of early identification and control of the pathogen to prevent further spread through the movement of infected banana suckers, contaminated nursery soils, farming equipment, and irrigation waterways. It is also essential to implement proper biosecurity and quarantine measures when introducing new planting material or equipment into fields. Despite a lack of commercial-level control methods, different management strategies such as the use of biological control agents like *Trichoderma* spp., *Streptomyces* spp., and *Bacillus* spp. have displayed the potential to effectively control Foc TR4 by significantly reducing the disease (Zheng et al. 2018; Damodaran et al. 2020; Chittarath et al. 2022; Wang et al. 2022; Liu et al. 2024).

Due to their rapid proliferating nature, crop diseases such as rice blast (caused by *Pyricularia oryzae*) are worrisome and directly threaten the food security and income sources of farmers in the GMS, as rice is a staple food and an essential source of calories. In developing nations like Vietnam, deployment of resistant cultivars is considered the most straightforward and cost-effective way to prevent the disease. However, the efficacy of this method is short-lived because the pathogen can mutate and produce pathotypes that overcome host resistance (Le et al. 2010). Hence, it is beneficial to shift from conventional breeding techniques to more advanced genomic and molecular-based approaches. Crop management practices, including split nitrogen fertilizer applications and incorporating FYM to optimize soil parameters, play a critical role in mitigating the disease by minimizing conditions favorable for its development. Cultivation practices such as early planting, field sanitation, ploughing, maintaining optimal field water conditions, and crop rotation or varietal rotation can also help reduce inoculum levels and mitigate the disease's effects. Despite the availability of chemicals to manage the pathogen, a major issue farmers face when using chemicals is the development of resistance. This can be avoided by rotating chemicals of different groups with different modes of action. However, biocontrol agents such as *Bacillus* spp. and disease forecasting are viable alternatives for pathogen management (Le et al. 2010; Khadka et al. 2025).

14.6 CURRENT STATUS AND FUTURE DIRECTIONS

The identification of the most prevalent species of fungi is the primary research focus in the region at present. Scientists have successfully identified the sources of inoculum and modes of transmission of fungal pathogens in the region. Suitable control strategies are being used up to the level needed. Further improving mycology-based research is badly needed. Such research requires funding in order to be carried out.

14.7 MOLECULAR TAXONOMY AND THE DEVELOPMENT OF A FUNGAL TAXONOMIC DATA REPOSITORY

Most of the identification methods have been conducted by the mycology research teams in the region, and this is still ongoing. The vast number of species of fungi, both beneficial and harmful, continues to be diagnosed and confirmed. The diversity of fungal organisms is significantly higher in the region than most other places in the world, and a purely taxonomic-based classification is necessary, since most of the taxonomic data in the data repository can be generated from the GMS.

14.8 LAND-BASED INNOVATIONS OF FUNGAL PLANT PATHOGEN CONTROL

Most of the plant pathogenic fungal genera were described have been properly diagnosed. Agronomists and plant protectionists are also learning about the causes of the rapid spread of diseases and effective strategies for disease transmission. Conventional and chemical-based disease control strategies are commonly used for fungal disease management. However, this is the ideal time to shift control methods toward sustainable land-based innovation. In that case, the region might be rich in some antagonistic species of fungi and these can be exploited for their control potential.

As the scientific community understands and harnesses the fungal diversity in the GMS, the collective efforts of researchers and stakeholders are instrumental in realizing the multitude of benefits that mycorrhizal symbiosis offers in various fields of science, agriculture, and environmental conservation.

14.9 CONCLUSIONS

In this chapter, we have identified the most commonly occurring fungal species in the GMS. We also identified some plant pathogenic fungal species. We paid our greater attention to the currently used control strategies. Thus, it creates a clear link between existing knowledge and future research. For future directions, greater attention should be given to implementing sustainable solutions for managing common plant pathogenic diseases in the region.

REFERENCES

Agrios, G.N. 2005. *Plant Pathology*. Elsevier, Amsterdam, Netherlands.

Akphumkitanun, R., C. Kaewkrajay & T. Dethoup. 2025. Fungicidal effects of fungicides in controlling rice seedling rot, brown spot, and false smut under field conditions in Thailand. *European Journal of Plant Pathology* 73:723–730.

Alvarez, E., J.M. Pardo & M.J. Truke. 2014. Detection and identification of '*Candidatus Phytoplasma asteris*'-related phytoplasma associated with a witches' broom disease of cassava in Cambodia. *Phytopathology* 104:S3.7.

Alvarez, E., J.M. Pardo, J.F. Mejia, A. Bertaccini, N.D. Thanh & T.X. Hoat. 2013. Detection and identification of '*Candidatus Phytoplasma asteris*'-related phytoplasmas associated with a witches' broom disease of cassava in Vietnam. *Phytopathogenic Mollicutes* 3:77–81.

Anderson, J.M., C. O'Dwyer, S. Phanthavong, K. Sengsoulichan, A. Drenth, S. Vilavong & L.W. Burgess. 2021. First report of *Phyllosticta* spp. associated with banana freckle disease in southern Lao PDR. *Australasian Plant Disease Notes* 16(1).

Asibi, A.E., Q. Chai & J.A. Coulter. 2019. Rice blast: A disease with implications for global food security. *Agronomy* 9:451.

Belfield, S. & C. Brown. 2009. A guide to upland cropping in Cambodia: Maize. *Australian Centre for International Agricultural Research (ACIAR) Monograph* 140:43.

Belfield, S., C. Brown & M. Martin. 2011. A guide to upland cropping in Cambodia: Soybean. *ACIAR Monograph No. 146. Australian Centre for International Agricultural Research: Canberra.* 72.

Bigirimana, V.P., G.K.H. Hua, O.I. Nyamangyoku & M. Höfte. 2015. Rice sheath rot: An emerging ubiquitous destructive disease complex. *Frontiers in Plant Science* 6:1066.

Cao, X., Q. Han, Y. Xiao, J. He, X. Chuan, G. Jiang, J.S. West & X. Xu. 2024. Population genetic structure of the rubber tree powdery mildew pathogen (*Erysiphe quercicola*) from China. *Plant Disease* 10:62–70.

Carris, L.M., C.R. Little & C.M. Stiles. 2012. Introduction to fungi. *The Plant Health Instructor* 48.

Chainuvati, C. & W. Athipanan. 2001. Crop diversification in Thailand. *FAO Website*.

Chaisarn, D., N. Suwannarach, S. Lumyong, S. Youpensuk & W. Pathom-aree. 2023. First report of *Curvularia hominis* as a causal agent of rice berry rice leaf spot in northern Thailand. *Chiang Mai Journal of Science* 50:1–10. e2023066.

Chaithra, M., S. Shreedevasena, H.M. Akshay Kumar, C. Vimal Kumar, R. Gangaraj, Pankhuri Singhal, Kodaru Anitha & R.D. Prasad. 2024. *Emerging and Re-Emerging Plant Diseases and Their Management Strategies. https://www.researchgate.net/publication/382253961_Emerging_and_Re-emerging_Plant_Diseases_and_their_Management_Strategies*

Chaiwan, N., D. Gomdola, S. Wang, J. Monkai, S. Tibpromma, M. Doilom, D.N. Wanasinghe, P.E. Mortimer, S. Lumyong & K.D. Hyde. 2021. An online database providing updated information of microfungi in the Greater Mekong Subregion. *Mycosphere* 12:1513–1526. https://doi.org/10.5943/mycosphere/12/1/19

Cheewangkoon, R., R. Kodsueb, S. Monkhung, K.V.A. Karunarathna, T. Wongwan, P. Withee, H. Sukanya & T. Nisachon. 2021. Phytopathogenic fungi identification in Thailand: Current trends and future perspectives. *Chiang Mai Journal of Science* 48:726–759.

Chittarath, K., C.H. Nguyen, W.C. Bailey, S.J. Zheng, D. Mostert, A. Viljoen, A.F. Tazuba, W. Ocimati, E. Kearsley, T.Y. Chi & N.T. Tho. 2022. Geographical distribution and genetic diversity of the banana *Fusarium* wilt fungus in Laos and Vietnam. *Journal of Fungi* 8:46.

Chittarath, K., D. Mostert, K.S. Crew, A. Viljoen, G. Kong, A.B. Molina & J.E. Thomas. 2018. First report of *Fusarium oxysporum* f. sp. *cubense* Tropical Race 4 (VCG 01213/16) associated with Cavendish bananas in Laos. *Plant Disease* 102:449.

Chou, C., N. Castilla, B. Hadi, T. Tanaka, S. Chiba & I. Sato. 2020. Rice blast management in Cambodian rice fields using *Trichoderma harzianum* and a resistant variety. *Crop Protection* 135:104864.

Damodaran, T., S. Rajan, M. Muthukumar, R. Gopal, K. Yadav, S. Kumar, I. Ahmad, N. Kumari, V.K. Mishra & S.K. Jha. 2020. Biological management of banana *Fusarium* wilt caused by *Fusarium oxysporum* f. sp. *cubense* tropical race 4 using antagonistic fungal isolate CSR-T-3 (*Trichoderma reesei*). *Frontiers in microbiology* 11:595845.

Dean, R., J.A. Van Kan, Z.A. Pretorius, K.E. Hammond-Kosack, A. Di Pietro, P.D. Spanu, J.J. Rudd, M. Dickman, R. Kahmann, J. Ellis & G.D. Foster. 2012. The Top 10 fungal pathogens in molecular plant pathology. *Molecular Plant Pathology* 13:414–430.

Degani, O. 2025. Plant fungal diseases and crop protection. *Journal of Fungi* 11:274.

Dethoupa, T., A. Jantasornb, N. Kaewsalongc. 2022. Efficacy of the antagonistic fungus *Talaromyces tratensis* KUFA 0091 in controlling rice blast and brown leaf spot diseases in field trials. *Agriculture and natural resources* 56:697–704.

Doehlemann, G., B. Ökmen, W. Zhu & A. Sharon. 2017. Plant pathogenic fungi. *Microbiology Spectrum* 5. https://doi.org/10.1128/microbiolspec.funk-0023-2016.

Duan, G., Y. Liu, C. Zheng, K. Yu, J. Xie, B. Wang, H. Zheng, W. Tang, J. Bao, Z. Wang & M. Chen. 2024. Chinese populations of *Magnaporthe oryzae* serving as a source of human-mediated gene flow to Asian countries: A population genomic analysis. *Journal of Fungi* 10:739.

EuroCham Myanmar Agriculture Guide. 2020. *European Chamber of Commerce in Myanmar*. Yangon, Myanmar.

FAO. 2018. Country Gender Assessment of Agriculture and the Rural Sector in Viet Nam. Nay Pyi Taw. Hanoi, Viet Nam.

FAO. 2022. Handbook on the Integrated Crop Management of Green Gram and Chickpea for Farmer Field Schools in Central Dry Zone of Myanmar. Nay Pyi Taw. Nay Pyi Taw, Myanmar.

Foreign Agricultural Service, US Department of Agriculture. 2025. Report on the Grain and feed Monthly, Thailand. U.S. department of Agriculture, Bangkok.

Fukuta, Y., I. Koga, T. Ung, K. Sathya, A. Kawasaki-Tanaka, Y. Koide, K. Nobuya, M. Obara, H. Yadana & N. Hayashi. 2014. Pathogenicity of rice blast (*Pyricularia oryzae* Cavara) isolates from Cambodia. *Japan Agricultural Research Quarterly: JARQ* 48:155–166.

Fukuta, Y., T.L. Nguyen, T.T.H. Pham, T.L. Trinh, V. du Pham, N. Hayashi, A. Kawasaki-Tanaka, M. Obara, B.T.T. Tran & C.B. Bui. 2020. Pathogenicity of rice blast (*Pyricularia oryzae* Cavara) isolates from Mekong River Delta, Vietnam. *Japan Agricultural Research Quarterly: JARQ* 54:239–252.

Hapuarachchi, K.K., S.C. Karunarathna, P. Phengsintham, H.D. Yang, P. Kakumyan, K.D. Hyde & T.C. Wen. 2019. Ganodermataceae (Polyporales): Diversity in Greater Mekong Subregion countries (China, Laos, Myanmar, Thailand and Vietnam). *Mycosphere* 10:221–309.

Hoat, T.X., D.V.T. Thanh, M. Dickinson, N.G. Bon, M.V. Quan, V.D. Hien, N.D. Thanh, L.T. Thuy & N.N. Vien. 2012. Disease problems of sugarcane in Vietnam, with special reference to phytoplasma. *Functional plant science and biotechnology* 6:117–123.

Huang, S.H. 2004. Progress of sugarcane disease research in China: Recent developments. *Sugar Tech* 6:261–265.

Joomdok, J., S. Saepaisan, A. Sunpapao, R. Pongpisutta, T. Monkham, J. Sanitchon & S. Chankaew. 2022. Identification of *Rhizoctonia solani*, as the cause of rice sheath blight and the source of its resistance, from Thai indigenous lowland rice germplasm. *Euphytica* 218:6.

Kaewchai, S., P. Paopradit, C. Pornsuriya, R.S. Jayawarrdena & K.D. Hyde. 2024. Pathogenicity and molecular phylogenetic analysis reveal *Colletotrichum siamense* as causal agent of leaf fall disease in rubber trees (*Hevea brasiliensis*) in southern Thailand. *Agriculture and Natural Resources* 58:49–58.

Karunarathna, S.C., N.M. Patabendige, W. Lu, S. Asad & K.K. Hapuarachchi. 2024. An in-depth study of phytopathogenic Ganoderma: Pathogenicity, advanced detection techniques, control strategies, and sustainable management. *Journal of Fungi* 10:414.

Karunarathna, S.C., N.M. Patabendige, T. Luangharn & K.K. Hapuarachchi. 2025. Ganodermataceae—current status, research, and development in Lower Mekong Basin. *Frontiers in Cellular and Infection Microbiology* 15:1545135.

Kea, S., H. Li & L. Pich. 2016. Technical efficiency and its determinants of rice production in Cambodia. *Economies* 4:22.

Khadka, R.B., H.K. Manandhar, S. Shrestha, B. Acharya, P. Sharma, S. Baidya, V.S. Luu & K.D. Joshi. 2025. Defending rice crop from blast disease in the context of climate change for food security in Nepal: A review. *Frontiers in Plant Science* 16:1511945.

Lao PDR Trade Portal, Department of Agriculture. 2023. *Instruction on the Phytosanitary Conditions for the Export of Maize from Lao PDR to China*. Accessed 10 September 2025. https://www.laotradeportal.gov.la/upload/files/legal_20241227041155.pdf

Le, M.T., T. Arie & T. Teraoka. 2010. Population dynamics and pathogenic races of rice blast fungus, *Magnaporthe oryzae* in the Mekong Delta in Vietnam. *Journal of General Plant Pathology* 76:177–182.

Le, C.T.M., I. Okane, Y. Ono, Y. Tsuda & Y. Yamaoka. 2022. Incidence of coffee leaf rust in Vietnam, possible original sources and subsequent pathways of migration. *Frontiers in Plant Science* 13:872877.

Leiva, A.M., J.M. Pardo, W. Arinaitwe, J. Newby, P. Vongphachanh, K. Chittarath, S. Oeurn, L.T. Hang, A. Gil-Ordóñez, R. Rodriguez & W.J. Cuellar. 2023. *Ceratobasidium* sp. is associated with cassava witches' broom disease, a re-emerging threat to cassava cultivation in Southeast Asia. *Scientific Reports* 13:22500.

Li, H., J. Guo, S.C. Karunarathna, L. Ye, J. Xu, K.D. Hyde & P.E. Mortimer. 2018. Native forests have a higher diversity of macrofungi than comparable plantation forests in the Greater Mekong Subregion. *Forests* 9:402.

Li, X., J. Li, H.Y. Bai, K. Xu, R. Zhang & Q. Huang. 2020. First report of wilt of rubber tree caused by *Chalaropsis thielavioides* in China. *Plant Disease* 105:1221.

Li, J., X. Wang, R. Zhang, H. Shan, Y. Li, C. Wang & Y. Huang. 2025. Identification and characterization of *Bipolaris secalis* and *B. setariae* associated with brown stripe disease on *Saccharum officinarum* in Yunnan, China. *Plant Disease* 109:562–569.

Liang, X., Y. Peng, Y. Liu, M. Wang, Y. Yang & Y. Zhang. 2019. First report of *Bipolaris bicolor* causing a leaf spot disease on rubber tree. *Journal of Phytopathology* 167:553–557.

Lim, J.A., J.S. Yaacob, S.R.A. Mohd Rasli, J.E. Eyahmalay, H.A. El Enshasy & M.R.S. Zakaria. 2023. Mitigating the repercussions of climate change on diseases affecting important crop commodities in Southeast Asia, for food security and environmental sustainability—a review. *Frontiers in Sustainable Food Systems* 6.

Limtong, S., P. Into & P. Attarat. 2020. Biocontrol of rice seedling rot disease caused by *Curvularia lunata* and *Helminthosporium oryzae* by epiphytic yeasts from plant leaves. *Microorganisms* 8:647.

Liu, S., W. Yang, X. Yang, R. Gong, D. Xiang & C. Li. 2024. Integrated control of *Fusarium* wilt in banana by *Bacillus velezensis* EB1 and potassium sorbate. *BMC microbiology* 24:457.

Liyanage, K.K., S. Khan, S. Brooks, P.E. Mortimer, S.C. Karunarathna, J. Xu & K.D. Hyde. 2017. Taxonomic revision and phylogenetic analyses of rubber powdery mildew fungi. *Microbial Pathogenesis* 105:185–195.

Luangharn, T., S.C. Karunarathna, A.K. Dutta, S. Paloi, I. Promputtha, K.D. Hyde, J. Xu & P.E. Mortimer. 2021. Ganoderma (Ganodermataceae, Basidiomycota) species from the greater Mekong subregion. *Journal of Fungi* 7:819.

Malinee, R., D. Stratoulias & N. Nuthammachot. 2021. Detection of oil palm disease in plantations in Krabi province, Thailand with high spatial resolution satellite imagery. *Agriculture* 11:251.

Martin, R., S. Montgomery, S. Yous & R. Rien. 2020. Mungbean Production Guide for Cambodian Conditions, ACIAR *Monograph No. 214, Australian Centre for International Agricultural Research, Canberra*. 76.

Matsumoto, M. & H.V. Cuong. 2014. Genetic characterization of the rice sheath blight pathogen *Rhizoctonia solani* AG1-IA in North Vietnam by rep-pcr and sequence analysis. *Journal of Plant Pathology* 96:377–380.

Mongkol, W., C. Jompuk, P. Jompuk, K. Laosatit, W. Bunkoed, C. Phruetthithep & C. Kerdsri. 2025. Improvement of sweet corn resistance to northern corn leaf blight (NCLB) and downy mildew (DM) through breeding program in Thailand. *Journal of Applied Biology & Biotechnology*. *Journal of Applied Biology & Biotechnology* 13:38–50.

Mwangi, R.W., M. Mustafa, K. Charles, I.W. Wagara & N. Kappel. 2023. Selected emerging and reemerging plant pathogens affecting the food basket: A threat to food security. *Journal of Agriculture and Food Research* 14:100827.

National Institute of Statistics, Ministry of Planning, Cambodia. 2023. Report on the Cambodia Agriculture Survey 2023: Crop Production.

Nguyen, D.H., P.V. Nguyen, A.N. Nguyen, T.D. Phan, T.Q.C. Vu, T.H.T. Nguyen, M.K. Nguyen & V.T. Truong. 2015. *International Rubber Conference Proceedings*. Ho Chi Minh City, Vietnam.

Nguyet, N.T.M., H.H. Long, N.B. Ngoc, N.T. Nhai, N.T.T. Thuy, N. Hayashi & Y. Fukuta. 2020. Diversity and distribution of rice blast (*Pyricularia oryzae* Cavara) races in Vietnam. *Plant Disease* 104:381–387.

Oo, N.N., V. Ruanjaichon, K. Laosatit, T. Toojinda & J. Unartngam. 2024. Genome Wide Association Study (GWAS) for southern corn rust (SCR) disease resistance in maize. *Thai Agricultural Research Journal* 42:71–85.

Pardo, J.M., K. Chittarath, P. Vongphachanh, L.T. Hang, S. Oeurn, W. Arinaitwe, R. Rodriguez, S. Sophearith, A.I. Malik & W.J. Cuellar. 2023. Cassava Witches' Broom Disease in Southeast Asia: A review of its distribution and associated symptoms. *Plants* 12:2217.

Pathom-aree W., S. Kreawsa, M. Kamjam, S. Tokuyama, S. Yoosathaporn & S. Lumyong. 2019. Potential of selected mangrove *streptomyces* as plant growth promoter and rice bakanae disease control agent. *Chiang Mai Journal of Science* 46:261–276.

Pests of Corn in Myanmar, Report No. MMR-05/1. 2015. International Plant Protection Convention (IPPC), Food and Agriculture Organization of the United Nations.

Pham, C.V. & H.T. Tran. 2021. *Cunninghamella elegans* causing cassava root rot in Vietnam. *Australasian Plant Disease Notes* 16:14.

Phanon, O., W. Intanoo, S. Korinsak & J. Unartngam. 2024. Variability of *Pyricularia oryzae* populations causing agents of rice blast in Thailand. *Thai Agricultural Research Journal* 42:14–23.

Phonemany, M., A.G.T. Niego, S.C. Karunarathna, Y. Hu, T. Luangharn, P. Sysouphanthong, S. Lumyong, R.L. Zhao, M.Q. He, N. Thongklang & K.D. Hyde. 2022. A comprehensive online database of mushrooms in the Greater Mekong SubRegion. *Chiang Mai Journal of Science* 49:538–550.

Pornsuriya, C., T. Chairin, N. Thaochan & A. Sunpapao. 2020. Identification and characterization of *Neopestalotiopsis* fungi associated with a novel leaf fall disease of rubber trees (*Hevea brasiliensis*) in Thailand. *Journal of Phytopathology* 168:416–427.

Pu, J., X. Zhang, Y. Qi, X.Y. Xian, H.Q. Zhang & H. Zhang. 2007. First record of *Corynespora* leaf fall disease of *Hevea* rubber tree in China. *Australasian Plant Disease Notes* 2:35–36.

Ren, G., D.N. Wanasinghe, A.R.G. De Farias, K.D. Hyde, E Yasanthika, J. Xu, A. Balasuriya, K.W.T. Chethana & H. Gui. 2022. Taxonomic novelties of woody litter fungi (Didymosphaeriaceae, Pleosporales) from the Greater Mekong Subregion. *Biology* 11:1660.

Saengchan, C., P. Phansak, K. Thumanu, S. Siriwong, T.L. Thanh, R. Sangpueak, W. Thepbandit, N.K. Papathoti & N. Buensanteai. 2022. Resistance induction by salicylic acid formulation in cassava plant against *Fusarium solani*. *The Plant Pathology Journal* 38:212–219.

Saito, K., N. Pangesti, R.J. Flor, C. Uch, A. Laborte, S. Khay, C. Orn, K. Kong & V. Seng. 2025. History and progress in rice research and its future perspective in Cambodia. *Plant Production Science* 28:122–138.

Sangpueak, R., S. Duchanee, C. Saengchan, N.K. Papathoti, N.H. Hoang, T.L. Thanh, P. Phansak & N. Buensanteai. 2023. Identification of cassava black stem and root rot agents in Thailand. *Chilean Journal of Agricultural Research* 83:70–82.

Sangpueak, R., P. Phansak & N. Buensanteai. 2018. Morphological and molecular identification of *Colletotrichum* species associated with cassava anthracnose in Thailand. *Journal of Phytopathology* 166:129–142.

Seekham, N., N. Kaewsalong & T. Dethoup. 2024. Efficacy of *Trichoderma* obtained from healthy rice seeds in promoting seedling growth and controlling rice seed rot and false smut diseases under field conditions. *European Journal of Plant Pathology* 169:657–668.

Srijudanu, A., O. Piasai, W. Leesutthiphonchai, M. Wanitch, C. Chaisiri & N. Khewkhom. 2023. Biocontrol of *Colletotrichum falcatum* causing red rot disease of sugarcane using non-toxigenic *Aspergillus* sp., *Chiang Mai Journal of Science* 50:e2023032.

Sudsanguan, M., C. Doungsa-ard, P. Athipunyakom, S. Seemadua, A. Kitjaideaw & S. Valyasevi. 2021. *Khon Kaen Agriculture Journal Supplement* 1:856–863.

Sudsanguan, M., R.G. Shivas, Y.P. Tan, C. Doungsa-ard, K. Chittarath, C.H. Heng & S. Nalumpang. 2024. Maize downy mildew (*Peronosclerospora neglecta*) in Cambodia, Lao PDR, and Thailand. *Chiang Mai Journal of Science* 51:e2024062.

Sunpapao, A., J. Kittimorakul & C. Pornsuriya. 2014. Disease Note: Identification of *Curvularia oryzae* as cause of leaf spot disease on oil palm seedlings in nurseries of Thailand. *Phytoparasitica* 42:529–533.

Suwannarach, N., J. Kumla & S. Lumyong. 2015. First report of Alternaria leaf blight disease on oil palm caused by *Alternaria longipes* in Thailand. *Phytoparasitica* 43:57–59.

Suwannarach, N., K. Sujarit, J. Kumla, B. Bussaban & S. Lumyong. 2013. First report of leaf spot disease on oil palm caused by *Pestalotiopsis theae* in Thailand. *Journal of General Plant Pathology* 79:277–279.

Tam, L.T.T., H.V. Cuong, N.M. Khue, M.V. Tri, H.M. Thanh, P.N. Dung, T.X. Hoat & N.V. Liem. 2016. First report of powdery mildew caused by *Erysiphe quercicola* on *Hevea brasiliensis* in Vietnam. *Plant Disease* 100:1239–1239.

Tann, H. & K. Soytong. 2017. Biological control of brown leaf spot disease caused by *Curvularia Lunata and* field application method on rice variety IR66 in Cambodia. *AGRIVITA Journal of Agricultural Science* 39:111–117.

Tran, A.P., T.D. Phan, D.H. Nguyen, N.M. Nguyen & A.N. Nguyen. 2011. Disease caused by *Botryodiplodia theobromea* pat on rubber tree in Vietnam: Current status and recent studies In *IRRDB International Rubber Conference*, 15–16 December 2011. Chiang Mai, Thailand.

Unartngam, J, N. Kopmoo, U. Pinruan, C. Kosawang & H.J.L. Jørgensen. 2024. Molecular and morphological identification of *Sarocladium* species causing sheath rot of rice in Thailand and their division into physiological races. *Journal of Fungi* 10:535.

United Nations Conference on Trade and Development (UNCTAD). 2020. Analyzing the maize value chain for export in Lao People's Democratic Republic. Geneva, Switzerland.

USDA Foreign Agricultural Service, US Department of Agriculture. *Thailand Production*.

Vansilalom, V. 2016. Bananas, Laos and Economic Development: Farming Practice and Farmers' Perceptions. Ph.D. Thesis, James Cook University, Canberra, Australia.

Vilavong, S. & K. Soytong. 2017. Application of a new bio-formulation of *Chaetomium cupreum* for biocontrol of *Colletotrichum gloeosporioides* causing coffee anthracnose on arabica variety in Laos. *AGRIVITA Journal of Agricultural Science* 39:303–310.

Vishwakarma, S.K., P. Kumar, A. Nigam, A. Singh & A. Kumar. 2013. Pokkah boeng: An emerging disease of sugarcane. *Journal of Plant Pathology & Microbiology* 4:2.

Wang, J., B. Cai, K. Li, Y. Zhao, C. Li, S. Liu, D. Xiang, L. Zhang, J. Xie & W. Wang. 2022. Biological control of *Fusarium oxysporum* f. sp. *cubense* tropical race 4 in banana plantlets using newly isolated Streptomyces sp. WHL7 from marine soft coral. *Plant Disease* 106:254–259.

Wu, R.H., Y. Zhang & Z.P. Li. 2019. First report of leaf spot on rubber tree caused by *Lasiodiplodia pseudotheobromae* in China. *Plant Disease* 103:766–766.

Xangsayasane, P., C. Boualaphanh, C. Bounphanousay, V. Bounphanousay, P. Manivong, S. Voradeth, P. Inthapanya, T. Sato, M. Obara, N. Hayashi & Y. Fukuta. 2020. Genetic variation of rice blast (*Pyricularia oryzae*) isolates in Laos. *Plant Health Progress* 21:248–255.

Yang, L., R. Lei, X.L. Yang, Y.J. Zhang & P.S. Wu. 2024. First report of *Diaporthe eres* causing white spot of maize in China. *Crop Protection* 184:106829.

Yingkamhaeng, N. & W. Vanichsriratana. 2024. Current situation and trends in Thailand's sugarcane sector. *Sugar Tech* 26:1088–1095.

Zewdu, Z. 2021. Rice blast biology and reaction of host to the disease. *World News of Natural Sciences* 39:11–21.

Zheng, S.J., F.A. García-Bastidas, X. Li, L. Zeng, T. Bai, S. Xu, K. Yin, H. Li, G. Fu, Y. Yu & L. Yang. 2018. New geographical insights of the latest expansion of *Fusarium oxysporum* f. sp. *cubense* tropical race 4 into the greater Mekong subregion. *Frontiers in Plant Science* 9:457.

15 Ectomycorrhizal Fungi of the Central Guiana Shield

Terry W. Henkel

15.1 INTRODUCTION

Neotropical forests are well known for their rich biodiversity. Fungal diversity in neotropical forests is also high, but plant-mutualistic ectomycorrhizal (ECM) fungi were traditionally thought to be underrepresented (e.g., Redhead 1968; Thomazini 1974; Bereau et al. 1997). In contrast to temperate and boreal forests, where ECM fungi may represent up to 40% of total macrofungal diversity, lowland neotropical ECM fungi are associated with a narrow range of ECM woody host plants characterized by regionally restricted, high-density distributions or dispersed, low-density distributions in rainforests otherwise composed of non-ECM trees (Corrales et al. 2018). These factors, combined with limited regional sampling, have resulted in a paucity of information concerning the distribution and diversity of ECM fungi and plants in the lowland Neotropics (Corrales et al. 2022).

Here I report on the diversity of ECM fungi associated with a range of host plants at a variety of spatial scales in the Central Guiana Shield (Guyana) of northeastern South America, a neotropical region with numerous ECM associations (ter Steege et al. 2019). Since 1997 this project has combined macrofungal sporocarp surveys with molecular detection of ECM fungi based on DNA sequences from roots of host plant taxa in monodominant and tree-diverse rainforests. The project has conducted sporocarp inventories of all known groups of ECM fungi, including epigeous, resupinate, and hypogeous species, and employed molecular methods to investigate ECM fungal diversity on roots of host plants in the Fabaceae, Cistaceae/Dipterocarpaceae, and Polygonaceae. This chapter updates work by Henkel et al. (2012) and Smith et al. (2011, 2013, 2017) regarding ECM fungal diversity of the Central Guiana Shield. Since publication of those studies, additional field work has expanded the number and taxonomic range of ECM fungi known from the region.

15.2 THE GUIANA SHIELD

The Guiana Shield is a geomorphological province in northeastern South America encompassing the countries of Guyana, Suriname, French Guiana, and portions of Venezuela, Brazil, and Colombia (Figure 15.1). The Precambrian basement rock of the Shield is overlain in its central portion by sandstone and related sedimentary rocks of the Roraima Formation, collectively forming the Guayana Highlands concentrated in central and eastern Venezuela, west central Guyana, and extreme northern Brazil, with outliers in eastern Colombia and central Suriname (Maguire 1945; Gibbs and Barron 1993; Rull et al. 2019). The Roraima Formation is of great antiquity, with a minimum age of 1.6 Byr, with subsequent periods of step-faulting and uplift (Gibbs and Barron 1993; Hammond 2005). The resulting terrain of this eroded landscape is of high relief and characterized by deeply cut creek and river valleys amidst remnant plateaus and pinnacled ridgelines (Henkel 2003). Soils derived from Guiana Shield erosionates are oligotrophic throughout the region, with a pronounced effect on contemporary canopy tree community composition relative to other lowland tropical forests of South America (Hammond 2005). In particular, Guiana Shield forests have a lower canopy tree species diversity and a preponderance of Fabaceae, Lecythidaceae, and Chrysobalanaceae relative to forests of central and upper Amazonia (ter Steege et al. 2006, 2023; Luize et al. 2024).

DOI: 10.1201/9781003647140-15

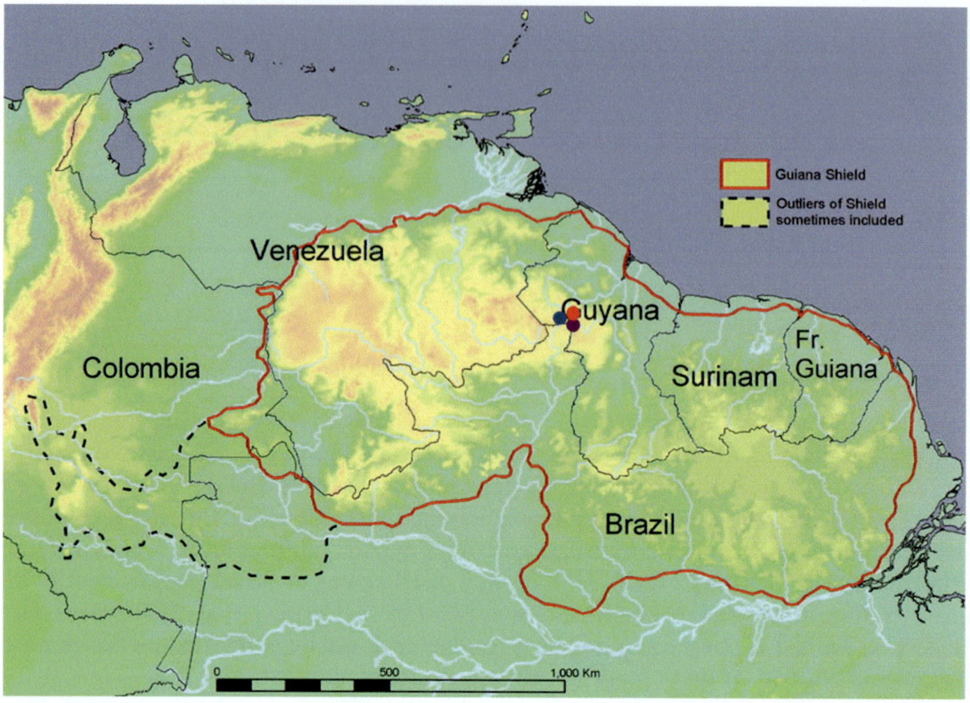

FIGURE 15.1 The Guiana Shield region. Field sites for this study are indicated by circular symbols: red = Upper Potaro River Basin; blue = Upper Mazaruni River Basin; purple = Upper Ireng River Basin.

Source: (Map source: Tom Hollowell, NMNH Informatics–Biological Diversity of the Guiana Shield Map, Public Domain, https://commons.wikimedia.org/w/index.php?curid=13259772).

15.3 ECTOMYCORRHIZAL PLANTS OF THE GUIANA SHIELD

Several lines of evidence indicate that, relative to other parts of greater Amazonia, the Guiana Shield region has a concentration of woody plant genera that host ECM fungi (Moyersoen 1993; Corrales et al. 2018; Vasco-Palacios et al. 2018; ter Steege et al. 2019, 2023). This is especially true for Guyana, in the central part of the Shield (Figure 15.2; Henkel et al. 2002; Henkel 2003; McGuire et al. 2008; DeGagne et al. 2009; Henkel and Mayor 2019). For the current study, sites for exploration and sampling were selected within Guyana's Pakaraima Mountains and the surrounding lowlands and were described in detail in Henkel et al. (2002, 2012), Henkel (2003) and Smith et al. (2011, 2013). These included (1) sites with established permanent plots with detailed woody plant and macrofungal sporocarp inventories and confirmed ECM host plants occurring at high (*Dicymbe corymbosa* Spruce ex Benth., *Dicymbe altsonii* Sandw.) and low (*Dicymbe jenmanii* Sandw., *Aldina insignis* (Benth.) Endl.) conspecific densities (Upper Potaro and Upper Ireng River Basins, west-central Pakaraimas), and (2) areas of the Pakaraimas and outwash drainages off the Pakaraima plateau with confirmed or putative ECM host plants ranging from low (*D. altsonii*, *Coccoloba excelsa* Benth., *Coccoloba* cf. *marginata* Benth.; Upper Demerara River Basin) to high densities (*Pakaraimaea dipterocarpacea* Maguire & P.S. Ashton, *D. jenmanii*, Upper Mazaruni River Basin).

15.3.1 *Dicymbe corymbosa* Monodominant Forests

Ectomycorrhizal *D. corymbosa* (Fabaceae subfam. Detarioideae) monodominant forests contain large, mast seeding canopy trees, high-density recruitment banks of seedlings and saplings, and diverse communities of ECM fungi (e.g., Henkel 2003; Henkel et al. 2005a, 2012). Large *D. corymbosa* trees

FIGURE 15.2 Ectomycorrhizal host plants of the Central Guiana Shield. (A, B) *Dicymbe corymbosa* Spruce ex Benth., (C) *Dicymbe altsonii* Sandw., (D) *Aldina insignis* (Benth.) Endl., (E) *Pakaraimaea dipterocarpacea* Maguire & P.S. Ashton, (F) *Dicymbe jenmanii* Sandw.

Source: Photos by Terry Henkel

constitute 50–90% of stand basal area and exhibit reiterative shoot and root growth that result in long-lived, multi-stemmed individuals with root mounds (Figure 15.2A–B; Woolley et al. 2008). Thick layers of litterfall are trapped within stem interstices and root mounds of these trees and are colonized by ECM roots, and are putative sites for nutrient recycling (Mayor and Henkel 2005; Woolley et al. 2008). *Dicymbe corymbosa* monodominant forests cover substantial areas of the Upper Potaro, Mazaruni, and Ireng River Basins in Guyana (Fanshawe 1952; Richards 1996; Degagne et al. 2009).

15.3.2 *Dicymbe corymbosa*-*Dicymbe altsonii*-*Aldina insignis* Co-Dominant Forests

In the Upper Potaro Basin of Guyana forests co-dominated by the ECM canopy trees *D. corymbosa* and *D. altsonii* (Fabaceae subfam. Detarioideae) with scattered canopy trees of ECM *Aldina insignis* (Benth.) Endl. (Fabaceae subfam. Papilionoideae) were noted by Fanshawe (1952), identified by Degagne et al. (2009), and described by Smith et al. (2011). *Dicymbe altsonii* and *A. insignis* have typical tree architecture with a single bole and horizontal crown, although *A. insignis* is more strongly buttressed (Figure 15.2C–D). As noted above, *D. corymbosa* has a reiterative physiognomy, including a vertically layered crown, multiple stems that vary in size, and adventitious roots, resulting in a large litter-trapping capacity (Woolley et al. 2008).

15.3.3 *Pakaraimaea dipterocarpacea-Dicymbe jenmanii* Co-Dominant Forests

In the Upper Mazaruni Basin immediately west of Mt. Ayanganna (2041 m), the highest sedimentary mountain in Guyana, are savanna communities intermixed with patches of closed-canopy fringing forest. As detailed by Smith et al. (2013), the savanna-fringing forests are dominated by ECM *P. dipterocarpacea* with *D. jenmanii* as a common subdominant, with a upper canopy height of 20 m (Figure 15.2E–F). This area, known as the Pegaima savanna, is ca. 25 km from the rainforest sites on the eastern side of Mt. Ayanganna in the Upper Potaro Basin, where multi-year sampling of ECM fungi in *Dicymbe*-dominated forests has occurred (Henkel et al. 2012). The Potaro and Pegaima sites, while geographically close, vary dramatically in annual precipitation (ca. 4000 mm and ca. 2000 mm, respectively); the Pegaima site is drier due to its position within the rain shadow of Mt. Ayanganna (Fanshawe 1952; Hijmans et al. 2005). Pegaima soils are highly oligotrophic white sands derived from the Roraima Formation sandstone (Gibbs and Barron 1993), whereas the soils from the Potaro rainforest study sites are laterites derived from igneous intrusives (Henkel 2003). At Pegaima forest edge trees exhibit burn scars, indicating periodic anthropogenic or natural fires, whereas fires are rare in the Potaro rainforests (Hammond and ter Steege 1998).

15.4 ECTOMYCORRHIZAL FUNGI OF THE CENTRAL GUIANA SHIELD

Below I describe collection and identification protocols for ECM fungi documented from the Central Guiana Shield from 1997 to 2025, summarize total taxa known as of 2026, and profile taxa within the specific orders. Note that taxonomic authors for species mentioned in the text can be found in Table 15.1.

TABLE 15.1
Ectomycorrhizal Fungi Recovered as Sporocarps from the Central Guiana Shield of Guyana, 1997–2025.

Order	Family	Species/morphospecies[1]	Representative vouchers[2]
Agaricales	Amanitaceae	*Amanita* cf. *crebresulcata* Bas complex #1	TH 8342, 7911, 7513, 7470
		Amanita cf. *crebresulcata* Bas complex #2	TH 10095, 10041, 9792
		Amanita albirosea ined.	TH 9128
		Amanita andrewi ined.	TH 8739, CL 118
		Amanita aurantiobrunnea Simmons, T.W. Henkel & Bas	TH 11383, 11382, *6431*
		Amanita calochroa C.M. Simmons, T.W. Henkel & Bas	TH 10048, *6246*, NS 5283
		Amanita cf. *campinaranae* Bas	TH 11212, 11065, CL 100
		Amanita craseoderma Bas	TH 11134, 10070, CL 110
		Amanita cyanochlorinosma Mighell & T.W. Henkel	TH 11174, 9737, *9172*
		Amanita cyanopus C.M. Simmons, T.W. Henkel & Bas	TH 11335, *7083*, NS 5292
		Amanita edmundi ined.	MCA 3910
		Amanita floccosa ined.	TH 11161, 11051, CL 75
		Amanita fulvoalba Mighell & T.W. Henkel	TH 11137, *10395*, 10394
		Amanita furfuraceostipitata ined.	TH 8224, CL 121, CL 108

TABLE 15.1 ***(Continued)***

Ectomycorrhizal Fungi Recovered as Sporocarps from the Central Guiana Shield of Guyana, 1997–2025.

Order	Family	Species/morphospecies[1]	Representative vouchers[2]
		Amanita glutinosa ined.	TH 11222, 11195
		Amanita griseopulveracea ined.	TH 11351, 10389, 10351
		Amanita guyanavaginata ined.	TH 8920, 8165
		Amanita guyanensis Mighell & T.W. Henkel	TH 10081, 9772, *9767*
		Amanita lanivolva Bas	TH 9151, 8123, 7514
		Amanita pakaraimensis ined.	TH 8247
		Amanita parvulicana ined.	TH *11217*, 10127, CL *120*
		Amanita parvulirosea ined.	TH 11216, *11172*, CL 113
		Amanita patamonarum ined.	TH 8854
		Amanita perphaea Simmons, T.W. Henkel & Bas	TH 11152, 7535, *6229*
		Amanita planosquamosa ined.	TH 11209, 9251
		Amanita tuberculivolva ined.	TH 10126, 9837, *9834*
		Amanita vaginochlorinosma ined.	TH 8201
		Amanita validorobusta ined.	TH 8956
		Amanita xerocybe Bas	TH 11161, 11063, NS 5282
		Amanita sp. 1	CL 5
		Amanita sp. 2	CL 29, CL 45, CL 68
		Amanita sp. 3	CL 35, CL 69
		Amanita sp. 4	CL 36
		Amanita sp. 5	CL 71
		Amanita sp. 6	CL 76
		Amanita sp. 7	CL 119, 112, 28
		Amanita sp. 8	CL 85
	Callistosporiaceae	*Guyanagarika anomala* Sanchez-Garcia, T.W. Henkel & Aime	TH *7419*
		Guyanagarika aurantia Sanchez-Garcia, T.W. Henkel & Aime	TH 11413, 11137, *9693*
		Guyanagarika pakaraimensis Sanchez-Garcia, T.W. Henkel & Aime	TH *8941*, 8512, 6618
	Cortinariaceae	*Cortinarius altissimus* Harrower & T.W. Henkel	TH 11054, *8211*, NS 5348
		Cortinarius andrewii T.W Henkel, Siegel & Aime	TH 11246, *11081*, 11060
		Cortinarius magnibrunneus ined.	TH 11425, 11356, 11069
		Cortinarius magniviscidus T.W. Henkel, Siegel & Aime	TH 11078, 11061, *9178*
		Cortinarius parviviscidus T.W. Henkel, Siegel & Aime	TH 11357, 11328, *9113*
		Cortinarius purpureobrunneus ined.	TH 11417, 11208, 11059
		Cortinarius purpureoscabrosus ined.	TH 11424, 11422, 11412
		Phlegmacium purpureofirmum T.W. Henkel, Siegel & Aime	TH 11079, *11068*, 11067
		Phlegmacium purpureosolidum T.W. Henkel, Siegel & Aime	TH 11058, *11052*, 9124

(Continued)

TABLE 15.1 ***(Continued)***

Ectomycorrhizal Fungi Recovered as Sporocarps from the Central Guiana Shield of Guyana, 1997–2025.

Order	Family	Species/morphospecies[1]	Representative vouchers[2]
	Entolomataceae	*Entoloma fragilum* Largent & Aime	MCA *2415*
		Entoloma illinitum Largent & Aime	MCA *2488*
		Entoloma olivaceocoloratum Largent & T.W. Henkel	TH 9136, 9041, *8855*
		Entoloma rugosostriatum Largent & T.W. Henkel	TH 11381, 11371, *6766*
		Entoloma grandicaeruleum ined.	TH 9118
		Entoloma tadangensis ined.	TH 9137, 9123
		Entoloma dicymbophilum Largent & T.W. Henkel	TH 11407, 11327, *9740*
		Entoloma aldinophilum Largent & T.W. Henkel	TH 11416, 11406, *11390*
	Hymenogastraceae	*Phaeocollybia* cf. *amazonica* Singer	TH 11434, 9153
	Inocybaceae	*Inocybe ayangannae* Matheny, Aime & T.W. Henkel	TH 9825, 8160, MCA *1232*
		Inocybe pulchella Matheny, Aime & T.W. Henkel	TH 9666, MCA *1879*, NS 5330
		Inocybe epidendron Matheny, Aime & T.W. Henkel	TH 9186, 8297, MCA *1473*
		Inocybe marginata Matheny, Aime & T.W. Henkel	TH 8290, MCA *3190*, 2166
		Inocybe lasseri Dennis	MCA 1971
		Inocybe liliacinasquamosa Matheny, Aime & T.W. Henkel	TH 9838, MCA *976*, NS 5293
		Inocybe amazoniensis Singer & I.J.A. Aguiar	MCA 3142
		Inocybe enigmata Matheny & Aime	MCA *1868*, 1490
		Inocybe lepidotella Matheny & Aime	MCA *1881*
		Inocybe magnifolia Matheny, Aime, & T.W. Henkel	TH 11220, 9132, MCA *2441*
		Inocybe rhodella Matheny, Aime & M.E. Sm.	MCA *3917*
Boletales	Boletaceae	*Austroboletus festivus* (Singer) Wolfe	TH 11221, 9250, 9058
		Austroboletus rostrupii (Syd. & P. Syd.) E. Horak	TH 11373, 10071, 7065
		Binderoboletus segoi T.W. Henkel & Husbands	TH 11338, *8035*, 5334
		Boletellus ananas var. *ananas* (M.A. Curtis) Murrill	TH 10075, 9782, 9668
		Boletellus dicymbophilus Fulgenzi & T.W. Henkel	TH 11346, 11131, *8616*
		Boletellus exiguus T.W. Henkel & Fulgenzi	TH 11185, 10085, *8696*
		Boletellus piakaii T.W. Henkel & Fulgenzi	TH 11175, 10391, *8728*
		Brasilioporus rufonigricans (T.W. Henkel) A.C. Magnago & T.W Henkel	TH 11199, 11084, *6366*

TABLE 15.1 *(Continued)*
Ectomycorrhizal Fungi Recovered as Sporocarps from the Central Guiana Shield of Guyana, 1997–2025.

Order	Family	Species/morphospecies[1]	Representative vouchers[2]
		Castellanea pakaraimophila T.W. Henkel & M.E. Sm.	TH 9670, *9514*
		Chalciporus cf. *trinitensis* Singer	TH 11153, 8012
		Chalciporus guyanensis ined.	TH *10390*
		Costatisporus cyanescens T.W. Henkel & M.E. Sm.	TH 11164, *9061*, NS 5325
		Fistulinella cinereoalba Fulgenzi & T.W. Henkel	TH 9233, *8471*, 8108
		Guyanaporus albipodus T.W. Henkel & Husbands	TH 10088, 9701, *8488*
		Jimtrappea guyanensis T.W. Henkel & M.E. Sm.	TH 10077, 9689, *9163*
		Neotropicomus parvogracilis (T.W. Henkel & Husbands) A.C. Magnago & T.W. Henkel	TH 10118, 9614, *9209*
		Nevesoporus exiguus (T.W. Henkel) A.C. Magnago & T.W. Henkel	TH 10119, 9658, *6283*
		Phylloporus alveolatus ined.	TH 11428, 9173
		Phylloporus colligatus Neves & T.W. Henkel	TH 11379, 11339, *8026*
		Phylloporus irengensis ined.	TH 11429, 11353, 11329
		Pulveroboletus cf. *rosaemariae* Singer	TH 10065, 9571, 8232
		Pulveroboletus squamosus ined.	TH 11336, 11168, 11074
		Singerocomus inundabilis (Singer) T.W. Henkel	TH 11163, 10111, 9199
		Singerocomus rubriflavus T.W. Henkel & Husbands	TH 11427, 11196, *9585*
		Sutorius cf. *eximius* (Peck) Halling, M. Nuhn & Osmundson	TH 11349, 11166, 9770
		Tylopilus balloui (Peck) E. Horak complex #3	TH 11140, 10101, 10026
		Tylopilus balloui (Peck) E. Horak complex #1	TH 11347, 11129, 10049
		Tylopilus balloui (Peck) E. Horak complex #2	TH 11340, 11177, 8218
		Tylopilus cf. *aquarius* (Singer) Wartchow, Barb.-Silva, B. Ortiz & Ovrebo	TH 11348, 11176, NS 5270
		Tylopilus cf. *nigripes* Barb.-Silva & Wartchow	TH 11197, NS 5268
		Tylopilus orsonianus Fulgenzi & T.W. Henkel	TH 11337, 11200, *8106*
		Tylopilus pakaraimensis T.W. Henkel	TH 11354, 11050, *6426*
		Tylopilus potamogeton var. *irengensis* T.W. Henkel	TH 11355, 11124, *6266*
		Tylopilus vinaceipallidus (Corner) T.W. Henkel	TH 10094, 10027, 9613

(Continued)

TABLE 15.1 ***(Continued)***

Ectomycorrhizal Fungi Recovered as Sporocarps from the Central Guiana Shield of Guyana, 1997–2025.

Order	Family	Species/morphospecies[1]	Representative vouchers[2]
		Xerocomus cf. *amazonicus* Singer	TH 11333, 10071, 10052
		Xerocomus cyaneibrunnescens T.W. Henkel & Husbands	TH 11123, 9601, *9155*
		Xerocomus potaroensis T.W. Henkel & Husbands	TH 11133, 10086, *9260*
	Diplocystidiaceae	*Tremellogaster surinamensis* E. Fisch.	TH 9829, 9195
	Sclerodermataceae	*Scleroderma sinnamariense* Mont.	TH 8281
		Scleroderma viridostellatum ined.	TH 9787
		Scleroderma sp. 1	MES 350
		Scleroderma sp. 2	TH 8493
		Scleroderma sp. 3	MES 346
		Scleroderma sp. 4	TH 9596
		Scleroderma sp. 5	TH 9597
		Scleroderma sp. 6	MES 351
		Scleroderma sp. 7	TH 11436, 11359, 11358
Cantha-rellales	Hydnaceae	*Cantharellus guyanensis* Mont.	TH 11400, 10046, 9201
		Craterellus atratoides T.W. Henkel, Aime et A.W. Wilson	TH 10069, 9655, *9232*
		Craterellus atratus (Corner) Yomyart et al.	TH 10093, 9777, 9679
		Craterellus cinereofimbriatus T.W. Henkel, Aime & A.W. Wilson	TH 11414, 10113, *9211*
		Craterellus excelsus T.W. Henkel & Aime	TH 10042, *8235*, NS 5299
		Craterellus olivaceoluteus T.W. Henkel, Aime & A.W. Wilson	TH 10130, 10103, *9205*
		Craterellus pleurotoides (T.W. Henkel, Aime & S.L. Mill.) A.W. Wilson	TH 10045, *8741*, NS 5287
		Craterellus strigosus T.W. Henkel, Aime & A.W. Wilson	TH 10129, *9204*
	Clavulinaceae	*Clavulina albofragilis* ined.	TH 10812,10104, 9678
		Clavulina amazonensis Corner	TH 10090, 9815, 9223
		Clavulina caespitosa T.W. Henkel, Meszaros & Aime	TH 11385, 10034, *8709*
		Clavulina cerebriformis Uehling, T.W. Henkel & Aime	TH 10132, MCA *4022*
		Clavulina cinereoglebosa Uehling, T.W. Henkel & Aime	TH 9652, 8561, MCA *4023*
		Clavulina cirrhata (Berk.) Corner	TH 11179, 10116, 10035
		Clavulina connata (Berk.) Corner	TH 9586
		Clavulina craterelloides Thacker & T.W. Henkel	TH 11415, 10043, *8243*
		Clavulina dicymbetorum T.W. Henkel, Meszaros & Aime	TH 10066, 9776, *8730*
		Clavulina effusa Uehling, T.W. Henkel & Aime	TH 10099, *9193*, 8511

TABLE 15.1 *(Continued)*
Ectomycorrhizal Fungi Recovered as Sporocarps from the Central Guiana Shield of Guyana, 1997–2025.

Order	Family	Species/morphospecies[1]	Representative vouchers[2]
		Clavulina griseohumicola T.W. Henkel, Meszaros & Aime	TH 11150, 9759, *8729*
		Clavulina guyanensis Uehling & T.W. Henkel	TH 10813, 9257, *9245*
		Clavulina humicola T.W. Henkel, Meszaros & Aime	TH 10815, 10037, *8737*
		Clavulina kunmudlutsa T.W. Henkel & Aime	TH 10047, 9575, *8932*
		Clavulina monodiminutiva T.W. Henkel, Meszaros & Aime	TH 10121, 10029, *8738*
		Clavulina nigricans Thacker & T.W. Henkel	TH 11396, 11186, *8284*
		Clavulina pakaraimensis Uehling, T.W. Henkel & Aime	TH 10062, 9224, *9194*
		Clavulina rosiramea Uehling, T.W. Henkel & Aime	TH 9775, *8954*, 7652
		Clavulina sprucei (Berk.) Corner complex #1	TH 10814, 8221, NS 5324
		Clavulina sprucei (Berk.) Corner complex #2	MCA 3989
		Clavulina sprucei (Berk.) Corner complex #3	TH 9122
		Clavulina tepurumenga T.W. Henkel & Aime	TH 10036, 9761, *8217*
		Clavulina sp. 1	TH 8265
		Clavulina sp. 2	NS 5286
Eurotiales	Elaphomycetaceae	*Elaphomyces adamantinus* Castellano & T.W. Henkel	TH *9660*
		Elaphomyces compleximuris Castellano, T.W. Henkel & S.L. Mill.	TH 11404, 11159, *8880*
		Elaphomyces digitatus Castellano, T.W. Henkel & S.L. Mill.	TH 11170, *8887*, NS 5321
		Pseudotulostoma volvatum O.K. Mill. & T.W. Henkel	TH 11180, *7022*, NS 5274
Gomphales	Gomphaceae	*Phyllobolites* aff. *miniatus* (Rick) Singer	TH 8525, 8016
Hymenochaetales	Hymenochaetaceae	*Coltricia astipitata* ined.	MCA 2018
		Coltricia cinnamomea (Jacq.) Murrill	MCA 1601
		Coltricia cinnamomeoides ined.	TH 9534
		Coltricia fibrosa Aime & Ryvarden	MCA *2054*
		Coltricia montagnei (Fr.) Murrill	TH 8558, 8344
		Coltricia cf. *montagnei* (Fr.) Murrill complex #1	TH 11362, 9176, 9108
		Coltricia cf. *montagnei* (Fr.) Murrill complex #2	TH 11401

(Continued)

TABLE 15.1 *(Continued)*

Ectomycorrhizal Fungi Recovered as Sporocarps from the Central Guiana Shield of Guyana, 1997–2025.

Order	Family	Species/morphospecies[1]	Representative vouchers[2]
		Coltricia verrucata Aime, T.W. Henkel & Ryvarden	MCA 2118, *962*
		Coltricia spicata ined.	MCA 1339, 1327
		Coltriciella navispora T.W. Henkel, Aime & Ryvarden	TH 8559, *7576*
		*Coltriciella oblectabilis (*Lloyd) Kotl., Pouzar & Ryvarden	TH 9187, 8560, NS 5328
Hysterangiales	Hysterangiaceae	*Hysterangium brunneoglebum* ined.	TH 10054, 9832, 9806
		Hysterangium olivaceorubescens ined.	TH 10079, 10055, NS 5342
Russulales	Russulaceae	*Lactarius* cf. *brasiliensis* (Singer) Silva-Filho & Wartchow	TH 9631, 7677
		Lactarius anomalous ined.	TH 9671, 9522
		Lactarius aurantiolamellatus T.W. Henkel, Delgat & Aime	TH 11184, 11154, 10800
		Lactarius dicymbophilus T.W. Henkel, Nuytinck & Verbeken	TH 11203, *10810*, 9801
		Lactarius edmundi ined.	TH 11085
		Lactarius guyanensis T.W. Henkel, Nuytinck & Verbeken	TH 11126, 9672, *9630*
		Lactarius humiphilus T.W. Henkel, Nuytinck & Verbeken	TH 11206, *10784*, 9809
		Lactarius lignyotoides ined.	TH 10793, 9239
		Lactarius mycenoides T.W. Henkel, Nuytinck & Verbeken	TH 10798, *9234*, 8673
		*Lactifluus brunellus (*S.L. Mill., Aime & T.W. Henkel) DeCrop	TH 11183, 9130, *7641*
		Lactifluus cf. *annulifer* (Singer) Nuytinck	TH 11162, 10122, 9014
		Lactifluus cf. *panuoides* (Singer) DeCrop	TH 10785, 10098, 9764
		Lactifluus multiceps (S.L. Mill., Aime & T.W. Henkel) DeCrop	TH 9154, 8632, *7656*
		Lactifluus pruinolamellatus ined.	TH 11178, 11057, 9217
		Lactifluus subiculatus S.L. Mill., Aime & T.W. Henkel	TH 10802, 9763, SLM *10047*
		Russula cf. *pluvialis* Singer complex #1	TH 11189, 9018
		Russula cf. *pluvialis* Singer complex #2	TH 9568, 9230
		Russula cf. *puiggarii* (Speg.) Singer complex #1	TH 11331, 9702, 9222
		Russula cf. *puiggarii* (Speg.) Singer complex #2	TH 11045, 8308, SLM 10049
		Russula cf. *puiggarii* (Speg.) Singer complex #3	TH 11165, MCA 3954
		Russula aurantioceps ined.	TH 9842, 9572, 8468
		Russula batistae Singer	TH 10028, 8236, 8227
		Russula caeruleovirida ined.	TH 9568
		Russula campinensis (Singer) T.W. Henkel, Aime & S.L. Mill.	TH 11409, 11182, NS 5288
		Russula caulofructis ined.	TH 8299

TABLE 15.1 ***(Continued)***

Ectomycorrhizal Fungi Recovered as Sporocarps from the Central Guiana Shield of Guyana, 1997–2025.

Order	Family	Species/morphospecies[1]	Representative vouchers[2]
		Russula cf. *amnicola* Singer "NPBD"	TH 9242, 8228, 7446
		Russula cf. *leguminosarum* Singer	TH 9547, 9227, 9017
		Russula concentrica ined.	TH 9157
		Russula gelatinivelata S.L. Mill., Aime & T.W. Henkel	TH 11375, 11076, *6410*
		Russula grandicrocea ined.	TH 8320, 7909, 7880
		Russula granularoceps ined.	MCA 1856
		Russula lilliputia S.L. Mill, T.W. Henkel & Aime	SLM *10046*
		Russula magnirubra ined. complex #1	MCA 3957, SLM 10163
		Russula magnirubra ined. complex #2	MCA 3912, SLM 10090
		Russula metachromatica Singer	TH 11411, 11188, 9564
		Russula metachromatica subsp. *metachromatica* Singer	TH 9218, 8215, 7678
		Russula myrmecobroma S.L. Mill., Aime & T.W. Henkel	TH 11332, 11194, *10109*
		Russula neobrunnea ined.	TH 11190, 7879, NS 5323
		Russula neorubra ined.	TH 9529
		Russula obtusopunctata Buyck	TH 9606, 9228, 9019
		Russula pakaraimaeae S.L. Mill. & T.W. Henkel	TH 9515, *9503*
		Russula paxilliformis S.L. Mill., Aime & T.W. Henkel	TH 11077, 9006, SLM *10097*
		Russula purpureonigricans ined.	TH 11191, 9140, SLM 10111
		Russula rubroglutinata S.L. Mill, T.W. Henkel & Aime	TH 11430, 9784, SLM *10095*
		Russula translucidostipitata ined.	MCA 4008
		Russula venezueliana Singer	TH 7874, 7534, SLM 10014
		Russula sp. 1	NS 5320
		Russula sp. 2	TH 11423, CL 41
Sebacinales	Sebacinaceae	*Sebacina azurea* ined.	TH 8996, 8588, 8537
		Sebacina incrustans (Pers.) Tul. & C. Tul.	TH 9654, 9164, 8974
		Sebacina ocreata (Berk.) Oberw., Garnica & K. Riess	TH 9817, 9192, 7552
Thelephorales	Bankeraceae	*Neosarcodon pakaraimensis* (A. Grupe & T.W. Henkel) L. Xiao et al.	TH 9554, *9513*
	Thelephoraceae	*Thelephora ascendens* T.W. Henkel, Alvarez-Manjarrez, Aime & M.E. Sm.	TH 11256, 11255, *10400*
		Thelephora bicoloricystidia T.W. Henkel & Alvarez-Manjarrez	TH *11244*, 10414, 8978
		Thelephora compacta T.W. Henkel & Alvarez-Manjarrez	TH 11230, *10402*, 9167
		Thelephora ianthinus T.W. Henkel & Alvarez-Manjarrez	TH *11251*, 9569
		Thelephora irengensis T.W. Henkel & Alvarez-Manjarrez	TH 11361, *11360*

(Continued)

TABLE 15.1 ***(Continued)***

Ectomycorrhizal Fungi Recovered as Sporocarps from the Central Guiana Shield of Guyana, 1997–2025.

Order	Family	Species/morphospecies[1]	Representative vouchers[2]
		Thelephora singularis T.W. Henkel, Alvarez-Manjarrez & Aime	TH 11426, *11139*, 10092
		Thelephora tenebrispurpurea T.W. Henkel & Alvarez-Manjarrez	TH 11254, *11228*, 11223
		Thelephora ventricobasidia T.W. Henkel, Alvarez-Manjarrez & M.E. Smith	TH 11257, 11233, *9653*
		Thelephora vidaturasexualis T.W. Henkel & Alvarez-Manjarrez	TH *9167*
		Thelephora sp. 1	TH 8568, 8544
		Thelephora sp.2	TH 11252
Trechisporales	Hydnodontaceae	*Hydnodon thelephorus* (Lév.) Banker complex #1	TH 8550
		Hydnodon thelephorus (Lév.) Banker complex #2	TH 9247
		Hydnodon thelephorus (Lév.) Banker complex #3	TH 11397, 8204

[1] Taxa lacking epithets are morphologically distinct but yet to be identified to species; taxa with epithets followed by "ined." have been tentatively determined as new to science but are yet to be formally described.

[2] TH numbers are in Terry Henkel's collection series; MCA numbers are in M. Catherine Aime's collection series; SLM numbers are in Steven L. Miller's collection series (University of Wyoming); MES numbers are in Matthew E. Smith's collection series; CL numbers are in Casey Ledford's number series; NS numbers are in Noah Siegel's collection series; numbers in *italic* are type specimens.

15.4.1 Collections and Identifications

Among the various collecting sites noted in Henkel et al. (2002, 2012), Smith et al. (2011, 2013, 2017), and Henkel and Largent (2023, 2026), macrofungal species were categorized as ECM if they were in genera or lineages for which the ECM symbiosis has been reported or demonstrated (Miller 1983; Singer 1986; Tedersoo et al. 2010a). These included species within ECM genera of the Agaricomycetes (Basidiomycota) orders and families Agaricales (Amanitaceae, Callistosporiaceae, Cortinariaceae, Entolomataceae, Hymenogastraceae, Inocybaceae), Boletales (Boletaceae, Diplocystidiaceae, Sclerodermataceae), Cantharellales (Hydnaceae, Clavulinaceae), Gomphales (Gomphaceae), Hymenochaetales (Hymenochaetaceae), Hysterangiales (Hysterangiaceae), Russulales (Russulaceae), Sebacinales (Sebacinaceae), Thelephorales (Bankeraceae, Thelephoraceae), Trechisporales (Hydnodontaceae), and the Ascomycota order Eurotiales (Elaphomycetaceae). Fungi were placed in preliminary genera based on morphology for agaricoid and boletoid fungi following Singer (1986), Corner and Bas (1962) and Kalichman et al. (2020); for cantharelloid and clavarioid fungi, Corner (1950, 1966, 1970); for Bankeraceae, Maas Geesteranus (1964) and Douch et al. (2024); for Hymenochaetaceae, Gilbertson and Ryvarden (1986) and Ryvarden (2024); for Hysterangiaceae, Castellano et al. (1989) and Davoodian et al. (2021); for Thelephoraceae, Corner (1968), Larsen (1968) and Kõljalg et al. (2024); for Sebacinaceae, Oberwinkler (1964), Wells and Bandoni (2001), and Oberwinkler et al. (2014); and for Elaphomycetaceae, Trappe (1979), Miller et al. (2001), and Castellano et al. (2012). Determinations of previously described species were made in consultation with Bas (1978), Corner (1950, 1966, 1970, 1972), Dennis (1970), Pegler (1983), Singer et al. (1983), Singer et al. (1991) and other primary sources. Fungi were identified at several levels of certainty: (1) at the species level for taxa formally described between 1999 and 2026 by Terry Henkel and colleagues; (2) at the species level as previously described taxa; (3) at the species level for species new to science but not yet formally published (designated here with provisional binomial followed by "ined."); (4) at the morphospecies level (i.e., morphologically distinct at the species level but not yet determined; designated here with genus name followed by "sp. 1", "sp. 2," etc.); and (5) as species complexes; these taxa were identified at the species or morphospecies level and were morphologically identical among different collections, but molecular data indicated that cryptic sympatric species exist within the taxon. Representative voucher collections and holotypes from all fungal taxa are deposited at the University of Guyana Herbarium, with duplicates and isotypes at Cal Poly Humboldt Fungarium.

15.4.2 Total Taxa Collected as Sporocarps

A total of 231 species/morphospecies of ECM fungi have been discovered from 1997 to 2025 (Table 15.1). These fungi occur in 11 orders, 20 families, and 43 genera. Below are synopses of the major and minor orders and families of these ECM fungi. This study constitutes the most complete regional documentation of ECM fungal diversity in lowland neotropical forests.

15.4.3 Agaricales

The Agaricales assemblage consists of 69 species/seven genera/six families (Table 15.1). The most speciose family by far is Amanitaceae, with 37 species/morphospecies known for *Amanita*. Among these *Amanita* species, 13 are currently described; of these, seven were described from Guyana by Simmons et al. (2002) and Mighell et al. (2019), with six proving conspecific with species described by Bas (1978) from collections by Rolf Singer in the Central Amazon (Simmons et al. 2002). Thirteen of the remaining *Amanita* species are known to be new and have received provisional names, and eight of these are currently undergoing formal description (T.W. Henkel & C.W. Ledford, unpubl. data). The remaining eight *Amanita* are distinct morphospecies that have yet to receive formal taxonomic attention. While many morphological highlights occur among the *Amanita* species, *A. xerocybe*, *A. calochroa* and *A. cyanopus* stand out due to their brilliantly and unusually colored, pulverulent universal veils (Figure 15.3A–C; Simmons et al. 2002)

Inocybaceae is the second-most speciose family in the Agaricales, with 11 *Inocybe* species (Table 15.1). Nine of these were new to science and have been formally described (Matheny et al. 2003, 2012), with two new distribution records for species previously described from the lowland Neotropics (Dennis 1953; Singer et al. 1983). Morphological highlights include *I. liliacinasquamosa* with its purple pileus and stipe squamules (Figure 15.3D), and *I. pulchella* with its extremely small basidiomata (< 6 mm pileus width) occurring in bryophyte mats on trunks of *Dicymbe* trees (Figure 15.3E; Matheny et al. 2003).

Cortinariaceae consists of nine species (seven *Cortinarius* and two *Phlegmacium*), all new to science (Table 15.1). Seven of these species have been formally described (Harrower et al. 2015; Siegel et al. 2024), and two await formal description. The significance of the Guyana Cortinariaceae assemblage lies in the fact that records for the family are scarce from the lowland tropics worldwide (Corrales et al. 2022), despite the large diversity of Cortinariaceae at higher latitudes of the northern and southern hemispheres (Bhunjun et al. 2022). Morphological highlights include the world's tallest *Cortinarius* species, *C. altissimus*, which combines its dark violet basidiomata with a squamulose pileus and a stipe regularly 200–300 mm in length (Figure 15.3F; Harrower et al. 2015).

Entolomataceae consists of eight species of *Entoloma* s.str. (=*Entoloma* subgen. *Entoloma*). *Entoloma* s.str. species are recognized among other genera/subgenera of Entolomataceae by their tricholomatoid basidiomata, sinuate lamellae, and isodiametric basidiospores, among other features, and are considered to be ECM (see Henkel and Largent 2026 and references contained therein). All of the Guyana species are new to science and occur in forests dominated by *Dicymbe* species or *A. insignis*; six of them have been formally described (Largent et al. 2008; Henkel and

FIGURE 15.3 Ectomycorrhizal fungi of the Central Guiana Shield, (A) *Amanita xerocybe* (NS 5282), (B) *Amanita calochroa* (NS 5283), (C) *Amanita cyanopus* (NS 5292), (D) *Inocybe lilacinasquamosa* (NS 5293), (E) *Inocybe pulchella* (NS 5330), (F) *Cortinarius altissimus* (NS 5348), (G) *Entoloma aldinophilum* (TH 11406), (H) *Guyanagarika aurantia* (TH 11137), (I) *Phaeocollybia* cf. *amazonica* (TH 11434), (J) *Austroboletus rostrupii* (TH 9038), (K) *Xerocomus cyaneibrunnescens* (TH 11130), (L) *Costatisporus cyanescens* (NS 5325), (M) *Tremellogaster surinamensis*.

Source: Photos A–F, H, K, L, M by Noah Siegel; Photos G, I, J by Terry Henkel

Largent 2023, 2026), and two with provisional names await description (Table 15.1). Morphological highlights include *E. aldinophilum* with its robust basidiomata, olivaceous yellow pileus, bright yellow lamellae, stipe, and exposed trama, and association with *A. insignis* in the Upper Ireng Basin (Figure 15.3G; Henkel and Largent 2026).

Callistosporiaceae consists of three species of *Guyanagarika* (Table 15.1). The genus was erected and the species described by Sanchez-Garcia et al. (2016) from material collected in Guyana's Upper Potaro, Mazaruni, and Ireng Basins in association with *D. corymbosa* or *P. dipterocarpacea*. While the three *Guyanagarika* species occur sympatrically and are indistinguishable in their large, tricholomatoid, entirely orange basidiomata and have no consistent micromorphological differences, molecular data identified three cryptic species within the group, and they were described accordingly (Figure 15.3H; Sanchez-Garcia et al. 2016). Ectomycorrhizas formed by *G. aurantia* with *D. jenmanii* were detected with molecular methods by Smith et al. (2013).

Hymenogastraceae is represented by a single species, *Phaeocollybia* cf. *amazonica*, known from three collections from *Dicymbe* forests of the Upper Potaro and Upper Ireng Basins (Table 15.1). Temperate and neotropical montane *Phaeocollybia* species are presumed ECM (Norvell and Exeter 2008). The Guyana material is morphologically consistent with *P. amazonica* originally described from Bolivian Amazonia (Figure 15.3I; Singer 1962) and is not conspecific with any of the numerous neotropical, oak-associated *Phaeocollybia* species known from montane Costa Rica and Colombia (Singer 1987; Horak and Halling 1991).

15.4.4 Boletales

The Boletales assemblage consists of 47 species/20 genera/three families (Table 15.1). The most speciose family is Boletaceae, with 37 species/morphospecies known across 20 genera. Among these species 19 have been formally described from Guyana (e.g., Henkel 1999, 2001; Fulgenzi et al. 2007, 2008; Neves et al. 2010; Husbands et al. 2013; Smith et al. 2015; Henkel et al. 2016), 14 were conspecific with previously described species from the Central Amazon (e.g., Singer et al. 1983), with the remaining four new and provisionally named but yet to be described. Many of the Guyana Boletaceae species were described within the traditional genera *Austroboletus*, *Boletellus*, *Chalciporus*, *Fistulinella*, *Phylloporus*, *Pulveroboletus*, *Tylopilus*, or *Xerocomus* using north temperate morphogeneric concepts. While numerous species fit well within a traditional morphogenus, some combine diagnostic features of multiple genera (e.g., *Tylopilus exiguus*; Henkel 1999). Subsequent molecular work has led to transfer of these "transgeneric" species to new genera based on Brazilian material (e.g., *Brasilioporus*, *Neotropicomus*, *Nevesoporus*; Magnago et al. 2022) as well as the erection of genera nova to accommodate other new epigeous species (e.g., *Binderoboletus*, *Guyanaporus*, *Singerocomus*; Henkel et al. 2016). *Singerocomus* has since been discovered in Brazil (Magnago et al. 2018). Whole-genome sequencing has supported the validity of these new genera (Tremble et al. 2024). Boletaceae morphological highlights include *Austroboletus rostrupii* with its tall, slender basidiomata with tawny brown, glutinous pileus, lacunose-reticulate stipe, and pink mature tubes (Figure 15.3J) and *Xerocomus cyaneibrunnescens* with its brown, dry, rugulose pileus, pale greenish cream, brown-staining tubes, and tan, brown-staining stipe with a blue-green apical band (Figure 15.3K; Husbands et al. 2013).

Three hypogeous sequestrate Boletaceae have been discovered in Guyana, each a new genus and species supported by morphological and molecular data (Smith et al. 2015). Of note among these putatively mycophagy-dispersed species is *Costatisporus cyanescens* with its deeply buried, ovoid basidiomata with a white, smooth peridium that slowly stains dark blue, and brown, loculate, gelatinizing gleba with a strong chocolate-nutty odor (Figure 15.3L).

Sclerodermataceae consists of nine species/morphospecies of *Scleroderma* (Table 15.1). These include the pantropical *S. sinnamariense* with its brilliant yellow peridium, the provisionally named *S. viridostellatum* ined. with its green, floccose peridium that recurves into stellate rays to expose the gleba, and seven distinct morphospecies awaiting formal description. The remaining Boletales family is Diplocystidiaceae, represented by a single species, *Tremellogaster surinamensis*. This fungus was originally described by Fischer (1924) from Suriname and features an epigeous, entirely sequestrate basidioma that in Guyana only occurs in *D. corymbosa* monodominant stands on lateritic slope soils and is considered ECM (T.W. Henkel pers. obs.). Basidiomata of *T. surinamensis* form gregariously at the soil surface, are globose and up to 70 mm broad, are subtended by a dense mass of thick white hyphal cords, and have a three-layered peridium with a thick, gelatinized middle layer, all surrounding a dark grey mature gleba composed of hydrophobic basidiospores (Figure 15.3M). The peridium does not open at maturity, but instead weathers away with heavy raindrop battering, eventually exposing the gleba and aerially dispersing some basidiospores while others ooze away in the disintegrating gelatinous peridium (T.W. Henkel pers. obs.). *Tremellogaster surinamensis* has also been reported from the Colombian Amazon (Vasco-Palacios et al. 2005).

15.4.5 Russulales

The Russulales assemblage currently consists of 47 spp/three genera/one family (Table 15.1). Among the three Russulaceae genera *Russula* is the most diverse with 33 species, of which six have been formally described from Guyana (Miller at al. 2012, 2024, 2025), 13 appear conspecific with previously described Amazonian taxa (e.g., Singer et al. 1983), 12 are new species provisionally named but not yet formally described, with two additional unnamed morphospecies. Morphological highlights include *R. campinensis* with its extremely small (2–7 mm wide), white, laterally stipitate, strongly acrid, gregarious basidiomata forming on humic mats on trunks of living *Dicymbe* trees, these always subtended by conspecific ectomycorrhizas (Figure 15.4A; Henkel et al. 2000). *Russula lilliputia* is likely the world's smallest centrally stipitate *Russula* with its 3–5 mm wide, brownish red, sulcate pileus, white lamellae, and white, densely pruinose stipe (Miller et al. 2024). *Russula pakaraimaeae* has the distinction of being the most abundant ECM fungal species recovered from root tips of *P. dipterocarpacea* (Smith et al. 2013; Miller et al. 2024).

Lactarius s.str. has nine species, with five formally described from Guyana (Nuytinck et al. 2023), one conspecific with a Brazilian species, and three provisionally named (Table 15.1). The five new species from Guyana formed a distinct, ancient lineage within *Lactarius* subgen. *Plinthogalus* (Hackel et al. 2022; Nuytinck et al. 2023). The lineage includes the morphological standouts *L. mycenoides* with its small (12–14 mm pileus width), off-white, *Mycena*-like basidiomata forming on bryophyte mats on the trunks of living *Dicymbe* trees, and *L. aurantiolamellatus* with its large basidiomata with grey-brown, rugulose pileus, light orange lamellae, and copious white latex that slowly turns pinkish grey (Figure 15.4B–C; Nuytinck et al. 2023).

Lactifluus contains six species, three of which have been described from Guyana (Miller et al. 2002), two are conspecific with extraregional neotropical species, and one provisionally named (Table 15.1). *Lactifluus* cf. *annulifer* is distinguished as one of the few known annulate *Lactifluus* species in the Neotropics (e.g., Duque Barbosa et al. 2020; Silva-Filho et al. 2023). Four of the *Lactifluus* species occurring in Guyana form small basidiomata arising gregariously from subicular mats that develop apogeotropically on sapling stems or lower trunks of *D. corymbosa* (Henkel et al. 2000; Miller et al. 2002). Among this subiculate assemblage *L. brunellus* is particularly striking in its pleurotoid basidiomata with a brown pileus, white to cream, lactescent lamellae, and formation of troops on a densely fasciculate subiculum enshrouding multiple sapling stems, lower tree trunks and exposed crown roots of *D. corymbosa* and covering up to 15 m diam areas of ground, the subicula enervated throughout with equally fasciculate, conspecific ectomycorrhizas (Figure 15.4D; Miller et al. 2002; Miller and Henkel 2004).

15.4.6 Cantharellales

The Cantharellales assemblage consists of 32 spp./three genera/two families (Table 15.1). The most speciose family is Clavulinaceae, with 24 known *Clavulina* species/morphospecies. Among these

FIGURE 15.4 Ectomycorrhizal fungi of the Central Guiana Shield. (A) *Russula campinensis* (NS 5288), (B, C) *Lactarius aurantiolamellatus* (TH 11154), (D) *Lactifluus brunellus* (TH 11183), (E) *Clavulina effusa* (TH 9193), (F) *Clavulina craterelloides* (TH 10043), (G) *Craterellus excelsus* (NS 5299), (H) *Craterellus pleurotoides* (NS 5287), (I) *Cantharellus guyanensis* (TH 9201), (J) *Thelephora ascendens* (TH 10400), (K) *Neosarcodon pakaraimensis* (TH 9513), (L) *Coltricia oblectabilis* (NS 5328).

Source: Photos A–D, H–I, M by Noah Siegel; Photos E–G, J–L by Terry Henkel

species 14 have been formally described from Guyana (Thacker and Henkel 2004; Henkel et al. 2005; Uehling et al. 2012), six were previously described from elsewhere in greater Amazonia (Berkeley 1856; Corner 1970), two are new with provisional names, with two remaining unidentified morphospecies. The *Dicymbe* forests of the Upper Potaro Basin are a globally significant diversity hotspot for *Clavulina*, with far more species occurring sympatrically than any other known site, and the genus appears to be of tropical origin (Kennedy et al. 2012). A number of extreme morphological diversions from the standard *Clavulina* branching coralloid basidioma occur in these species, most graphically illustrated in *C. effusa* which is resupinate with short branching coralloid basidiomata (Figure 15.4E), *C. craterelloides* which has large infundibuliform basidiomata borne in caespitose clusters (Figure 15.4F), and *C. cerebriformis* and *C. cinereoglebosa*, both of which are entirely resupinate (Figure 15.4G; Thacker and Henkel 2004; Uehling et al. 2012). A number of *Clavulina* species conspecific with Guyana taxa have been found in the Colombian Amazon in association with the ECM dipterocarp *Pseudomonotes tropenbosii* A.C. Londoño, E. Alvarez & Forero (Vasco-Palacios and Boekhout 2022).

The Hydnaceae contains eight species, seven of which are *Craterellus*, with six formally described from Guyana and one previously described from the Amazon (Corner 1966; Henkel et al. 2006, 2009; Wilson et al. 2012; Henkel et al. 2014). Morphological highlights among *Craterellus* include the edible *C. excelsus* with its large caespitose clusters of grey-brown, broadly infundibuliform basidiomata regularly exceeding 100 mm in height (Figure 15.4H; Henkel et al. 2004a, 2009). *Craterellus pleurotoides* has tiny spathulate and sessile basidiomata forming on bryophyte and humus mats on seedlings and trunks of *Dicymbe* trees (Figure 15.4I; Henkel et al. 2006a). The lone *Cantharellus* species, *C. guyanensis*, was originally described from French Guiana (Montagne 1854), has bright orange basidiomata, fruits gregariously in Guyana's *Dicymbe* forests on sandy soils, and is a superb edible mushroom (Figure 15.4J; Henkel et al. 2004a; Henkel et al. 2012). The species has the broadest distribution currently known for an ECM fungus in the South American lowland tropics, with conspecific collections from eastern Venezuela, Guyana, French Guiana, and central, northeastern, and southeastern Brazil (Henkel et al. 2014).

15.4.7 Thelephorales

The Thelephorales assemblage consists of 12 spp./two genera/two families (Table 15.1). The most speciose family is Thelephoraceae, with 11 known *Thelephora* species/morphospecies. All but one of these species have resupinate basidiomata and would have been placed until recently in the genus *Tomentella* (Kõljalg et al. 2024). Among these species nine have been formally described from Guyana, including the lone pileate-stipitate species *T. singularis*, and two remain as unidentified morphospecies (Alvarez-Manjarrez et al. 2025, 2026). In root-tip sequencing studies *Thelephora* species were the dominant ectomycorrhiza formers on both seedlings and adults of *Dicymbe* species as well as *P. dipterocarpaceae* and *A. insignis* (Smith et al. 2011, 2013, 2017; Delevich et al. 2021). Of note among the *Thelephora* species are the apogeotropic rhizomorphs of *T. ascendens* which climb *D. corymbosa* seedling stems to form resupinate basidiomata on the lower sides of living leaves (Figure 15.4K), and *T. ventricobasidia* and *T. bicoloricystidia* which form aerial ectomycorrhizas on adventitious roots descending gravitropically from trunks of *D. corymbosa* trees (Alvarez-Manjarrez et al. 2025, 2026).

Bankeraceae consists of a single species, *Neosarcodon pakaraimensis*, which was originally described in *Sarcodon* and is known only in association with *P. dipterocarpaceae* from the type locality in Guyana's Upper Mazaruni Basin (Figure 15.4L; Grupe et al. 2015). *Neosarcodon* was recognized as a monophyletic genus-level clade relative to *Sarcodon* and *Hydnellum* in the Bankeraceae by Wang et al. (2024) and is composed almost entirely of neotropical species (Grupe et al. 2016; Douch et al. 2024).

15.4.8 Hymenochaetales

While the globally distributed order Hymenochaetales consists primarily of wood-decay fungi (Larsson et al. 2006) the genera *Coltricia* and *Coltriciella* are considered ECM (e.g., Thoen and Ba 1989; Aime et al. 2003; Tedersoo et al. 2007, 2010a). In Guyana the coltricioid assemblage consists of 11 species/two genera/one family (Table 15.1). Of these, two new species of *Coltricia* and one of *Coltriciella* have been described (Aime et al. 2003, 2007), five were previously described, and three are new with provisional names. While terrestrially fruiting coltricioid species occur in the Guyana assemblage (e.g., *Coltricia montagnei*) the majority of *Coltricia* and *Coltriciella* species fruit in elevated positions on humus or bryophyte mats on the trunks of

Dicymbe trees (Figure 15.4M; Aime et al. 2003, 2007) and all are restricted to *Dicymbe* forests (Henkel et al. 2012).

15.4.9 Eurotiales

The Eurotiales (Eurotiomycetes, Ascomycota) assemblage consists of four spp./two genera in the Elaphomycetaceae (Table 15.1). While the vast majority of species in the Eurotiales are saprotrophic and few make macroscopic sporulation structures (Houbraken et al. 2020), the Elaphomycetaceae make mostly hypogeous, sequestrate and relatively large ascomata, are primarily mycophagy-dispersed, and are ECM (Tedersoo et al. 2010a). Three species of hypogeous, sequestrate *Elaphomyces* have been described from Guyana and were the first *Elaphomyces* records in South America (Figure 15.5A; Castellano et al. 2012, 2016). The fourth species, *Pseudotulostoma volvatum*, was originally described by Miller et al. (2001) and at the time was a monotypic genus closely related to *Elaphomyces* in the Elaphomycetaceae. *Pseudotulotoma volvatum* is an extraordinary species in its ascoma form and ascospore dispersal strategy. Unlike *Elaphomyces* species which are fully sequestrate, hypogeous and mycophagy-dispersed, *P. volvatum* develops initially as a closed ascoma in the upper litter layer and at maturity produces an epigeous stalk that pushes the hydrophobic gleba of ascospores and pseudocapillitum upward through the peridium to expose it to rain drop battering of ascospores (Figure 15.5B). *Pseudotulostoma volvatum* has been confirmed as ECM by morphological and molecular examination of ectomycorrhizas subtending the ascomata (Henkel et al. 2006b) and a second species in the genus was discovered in Japan (Asai et al. 2004). A broader regional distribution of *P. volvatum* is inferred from its recent documentation from the southern limit of the Guiana Shield in north-central Brazil (Komura et al. 2021) and the western limit in southeastern Colombia (Vasco-Palacios and Boekhout 2022).

FIGURE 15.5 Ectomycorrhizal fungi of the Central Guiana Shield. (A) *Elaphomyces digitatus* (NS 5321), (B) *Pseudotulostoma volvatum* (NS 5274), (C) *Sebacina ocreata* (TH 9192), (D) *Hydnodon thelephorus* (TH 11397), (E) *Hysterangium olivaceorubescens* ined. (NS 5342), (F) *Phyllobolites* aff. *miniatus* (TH 8016).

Source: Photos A–B, E by Noah Siegel; photos C–D, F by Terry Henkel

15.4.10 Sebacinales

The Sebacinales assemblage consists of three species of *Sebacina* (Sebacinaceae), two of which were previously described and one new with a provisional name (Table 15.1). Noteworthy among the *Sebacina* species is the ascent of *Dicymbe* seedlings via rhizomorphs by *S. incrustans* s.l. to form resupinate, gelatinous basidiomata on living stems and leaves. *Sebacina ocreata* was originally described by Berkeley (1856) from a Richard Spruce collection from the Upper Amazon and features terrestrially fruiting, erect, highly branched coralloid basidiomata (Figure 15.5C; Henkel et al. 2004b; Oberwinkler et al. 2014).

15.4.11 Trechisporales

Evidence suggests that in the Neotropics species of *Hydnodon* (Hydnodontaceae, Trechisporales) are ECM (Vanegas-León et al. 2019). In Guyana there are three morphospecies consistent morphologically with *Hydnodon thelephorus* (Coker and Beers 1951) which form epigeous, pileate-stipitate basidiomata with a dentate hymenophore but vary in overall basidioma color from white to pink to orange (Table 15.1; Figure 15.5D). These are considered unidentified ECM fungal morphospecies awaiting description, and all fruit in *D. corymbosa* monodominant stands in the Upper Potaro Basin (Henkel et al. 2012)

15.4.12 Hysterangiales

The Hysterangiales assemblage consists of two species of *Hysterangium* (Hysterangiaceae), both of which are new with provisional names and await formal description (Table 15.1). In Guyana these *Hysterangium* species along with *Elaphomyces* and several Boletaceae compose the ecological guild of hypogeous, sequestrate fungi with putative mycophagy-based spore dispersal (Figure 15.5E; Castellano et al. 2012, 2016; Smith et al. 2015).

15.4.13 Gomphales

The Gomphales consists of a single species, *Phyllobolites* aff. *miniatus*. This species, originally described in *Paxillus* by Rick (1906) from southeastern Brazil and later combined into *Phyllobolites* by Singer et al. (1983) is now considered a member of the Gomphaceae (Farid et al. 2018). The Guiana Shield species is macromorphologically consistent with *P. miniatus*, but the dissertation work of Linhares (2018) suggested that it is a new species of *Phyllobolites* awaiting formal description. Regardless, it is a striking agaricoid species with red pileus, white stipe, an off-white lamellate hymenophore producing a brown spore print, and is considered ECM as it fruits exclusively in *D. corymbosa* monodominant stands (Figure 15.5F; Henkel et al. 2012).

15.5 SUMMARY AND FUTURE DIRECTIONS

Many more ECM fungi remain to be discovered in the Central Guiana Shield. This is evidenced in Guyana by the steady accrual of new sporocarp taxa with successive multi-year sampling, and the large number of ECM fungal species OTUs recovered from root tips but yet to be found as sporocarps. The recent discoveries in Brazilian and Colombian peripheral areas of the Guiana Shield of ECM fungi that are conspecific with Guyanese taxa or are new species point to a broad regional distribution of ECM associations (e.g., Komura et al. 2021; Silva-Filho et al. 2021; Vasco-Palacios and Boekhout 2022). As noted by Corrales et al. (2022) a fertile ground for exploring ECM associations would be Venezuela, where few studies of ECM fungi and plants have been done. Situated between Guyana and Colombia and encompassing much of the Guiana Shield, Venezuela contains a wide array of putatively ECM host plants, including at least 15 species of *Dicymbe*, several species of *Aldina*, and one of *Pakaraimaea* (Moyersoen 1993, 2006; Hokche et al. 2008; Stergios and Aymard 2008; Ramos et al. 2016). Outside of closed canopy rainforests dominated by ECM trees, plant species more broadly distributed at low densities in the lowland South American tropics of *Neea*, *Gaupira*, *Pisonia* (Nyctaginaceae), and *Coccoloba* (Polygonaceae) are regularly ECM and may host diverse and unique ECM fungi (Tedersoo et al. 2010b; Hackel et al. 2022; Magnago et al. 2022; Furtado et al. 2023; T.W. Henkel unpubl. data). Efforts are currently underway to expand sampling into unexplored watersheds in the Central Guiana Shield to discover more ECM associations and further emphasize the region as a hotspot for tropical ECM fungi.

15.6 ACKNOWLEDGMENTS

Studies on ECM fungi in the Central Guiana Shield were supported by grants DEB-0918591 and DEB-1556338 from the National Science Foundation of the United States and by grants 6679–99, 7435–03 and 8481–08 from National Geographic Society's Committee for Research and Exploration to T.W.H. Research permits were granted by the Guyana Environmental Protection Agency. Field assistance was provided by members of the Patamona indigenous community including Chris Andrew, Justin Edmund, Francino Edmund and Luciano Edmund. Additional field assistants included M. Catherine Aime, Mei Lin Chin, Piakai Henkel, Dillon Husbands and others too numerous to mention. Noah Siegel is profusely thanked for providing many splendid photographs. A special thanks goes to numerous taxonomic collaborators, in particular M. C. Aime, M.A. Castellano, D.L. Largent, P.B. Matheny, S.L. Miller, M.E. Smith, and J.K. Uehling.

REFERENCES

Aime, M.C., T.W. Henkel & L. Ryvarden. 2003. Studies in neotropical polypores 15: New and interesting species from Guyana. *Mycologia* 95:614–619. doi:10.1080/15572536.2004.11833065

Aime, M.C., T.W. Henkel & L. Ryvarden. 2007. Studies in Neotropical polypores 22. Additional new and rare species from Guyana. *Synopsis Fungorum* 23:15–31.

Alvarez-Manjarrez, J., T.W. Henkel, A. Duncan, M.C. Aime, K.G. Peay & M.E. Smith. 2025. New species of *Thelephora* (Thelephorales, Basidiomycota) associated with *Dicymbe* and *Aldina* in Guyana. *Mycologia* 117:480–498. https://doi.org/10.1080/00275514.2025.2472593

Alvarez-Manjarrez, J., T.W. Henkel, A. Duncan & K.G. Peay. 2026. Additional new species of resupinate *Thelephora* (Thelephorales, Basidiomycota) associated with *Dicymbe* in Guyana, with confirmation of their aerial ectomycorrhizas. *Mycologia* (in review)

Asai, I., H. Sato & T. Nara. 2004. *Pseudotulostoma japonicum*, comb. nov. (= *Battarrea japonica*), a species of Eurotiales, Ascomycota. *Bulletin of the National Science Museum Series B* 30:1–7.

Bas, C. 1978. Studies in *Amanita*. I. Some species from Amazonia. *Persoonia* 10:1–22.

Bereau, M., M. Gazel & J. Garbaye. 1997. Mycorrhizal symbioses in the trees of tropical rainforest of French Guiana. *Canadian Journal of Botany* 75:711–716. https://doi.org/10.1139/b97-080

Berkeley, M.J. 1856. Decades LIX—LX. Rio Negro fungi. *Hooker's Journal of Botany* 8:233–241.

Bhunjun, C.S., T. Niskanen, N. Suwannarach, N. Wannathes, Y.J. Chen, E.H.C. McKenzie, S.N. Maharachchikumbura, B. Buyck, C.L. Zhao, Y.G. Fang, et al. 2022. The numbers of fungi: Are the most speciose genera truly diverse? *Fungal Diversity* 114:387–462. https://doi.org/10.1007/s13225-022-00501-4

Castellano, M.A., J.M. Trappe, Z. Maser & C. Maser. 1989. Keys to spores of the genera of hypogeous fungi of North Temperate forests with special reference to animal mycophagy. Mad River Press, Eureka.

Castellano, M.A., B.T.M. Dentinger, O. Séné, T.F. Elliott, C. Truong & T.W. Henkel. 2016. New *Elaphomyces* species (Elaphomycetaceae, Eurotiales, Ascomycota) from tropical rainforests of Cameroon and Guyana. *IMA Fungus* 7:59–73. https://doi.org/0.3852/12-061

Castellano, M.A., T.W. Henkel, S.L. Miller, M.E. Smith & M.C. Aime. 2012. New *Elaphomyces* species (Elaphomycetaceae, Eurotiales, Ascomycota) from Guyana. *Mycologia* 104:1244–1249. https://doi.org/10.3852/12-061

Coker, W.C. & A.H. Beers. 1951. *The Stipitate Hydnums of the Eastern United States*. University of North Carolina Press, Chapel Hill.

Corner, E.J.H. 1950. *A Monograph of Clavaria and Allied Genera*. Oxford University Press, London.

Corner, E.J.H. 1966. *A Monograph of Cantharelloid Fungi*. Annals of Botany Memoir No. 2. Oxford University Press, Oxford.

Corner, E.J.H. 1968. A monograph of *Thelephora* (Basidiomycetes). *Beihefte Nova Hedwigia* 27:1–97.

Corner, E.J.H. 1970. Supplement to "a monograph of *Clavaria* and allied genera". *Beihefte Nova Hedwigia* 33:1–299.

Corner, E.J.H. 1972. *Boletus in Malaysia*. Botanic Gardens, Singapore.

Corner, E.J.H. & C. Bas. 1962. The genus *Amanita* in Singapore and Malaya. *Persoonia* 2:241–304.

Corrales, A., T.W. Henkel & M.E. Smith. 2018. Ectomycorrhizal associations in the tropics–biogeography, diversity patterns, and ecosystem roles. *New Phytologist* 220:1076–1091. https://doi.org/10.1111/nph.15151

Corrales, A., R.A. Koch, A.M. Vasco-Palacios, M.E. Smith, Z.W. Ge & Henkel TW. 2022. Diversity and distribution of tropical ectomycorrhizal fungi. *Mycologia* 114:919–933. https://doi.org/10.1080/00275514.2022.2115284

Davoodian, N., T. Lebel, M.A. Castellano & K. Hosaka. 2021. Hysterangiales revisited: Expanded phylogeny reveals new genera and two new suborders. *Fungal Evolution and Systematics* 8:65–80. https://doi.org/10.3114/fuse.2021.08.06

Delevich, C., R.A. Koch, M.C. Aime & T.W. Henkel. 2021. Ectomycorrhizal fungal community assembly on seedlings of a Neotropical monodominant tree. *Biotropica* 53:1486–1497. https://doi.org/10.1111/btp.12989

Degagne, R.S., T.W. Henkel, S.J. Steinberg, L. Fox. 2009. Identifying *Dicymbe corymbosa* monodominant forests in Guyana using satellite imagery. *Biotropica* 41:7–15. https://doi.org/10.1111/j.1744-7429.2008.00446.x

Dennis, R.W.G. 1953. Les Agaricales de l'ile de la Trinité: rhodosporae-ochrosporae. *Bulletin Societe Mycologique France* 69:145–198.

Dennis, R.W.G. 1970. *The Fungus Flora of Venezuela and Adjacent Countries. Kew Bulletin Additional series III*. HMSO, London.

Douch, J.K., L.J. Vaughan, J.A. Cooper, G.D. Holmes, R. Robinson, F. Stefani, A. Idnurm & T.W. May. 2024. Taxonomic revision of fleshy species of *Hydnellum*, *Neosarcodon*, and *Sarcodon* (Thelephorales) from Australasia. *Mycologia* 116:965–992. https://doi.org/10.1080/00275514.2024.2363211

Duque Barbosa, J.A., L. Delgat, S. Galváo Elias, A. Verbeken, M.A. Neves & A. Carvalho, alves de. 2020. A new section, *Lactifluus* section *Neotropicus* (Russulaceae), and two new *Lactifluus* species from the Atlantic Forest, Brazil. *Systematics and Biodiversity* 18:347–361. https://doi.org/10.1080/14772000.2020.1769221

Fanshawe, D.B. 1952. *The Vegetation of British Guiana: A Preliminary Review*. Institute Paper No. 39, Imperial Forestry Institute, Oxford.

Farid, A., M. Gelardi, C. Angelini, A.R. Franck, F. Costanzo, L. Kaminsky, E. Ercole, T.J. Baroni, A.L. White, J.R. Garey, M.E. Smith & A. Vizzini. 2018. *Phylloporus* and *Phylloboletellus* are no longer alone: *Phylloporopsis* gen. nov. (Boletaceae), a new smooth-spored lamellate genus to accommodate the American species *Phylloporus boletinoides*. *Fungal Systematics and Evolution* 2:341–359. https://doi.org/10.3114/fuse.2018.02.10.

Fischer, E. 1924. *Mitteilungen der Naturforschenden Gesellschaft in Bern*. Wyss Erben, Bern.

Fulgenzi, T.D., T.W. Henkel & R.E. Halling. 2007. *Tylopilus orsonianus* sp. nov. and *Tylopilus eximius* from Guyana. *Mycologia* 99:622–627. https://doi.org/10.1080/15572536.2007.11832556

Fulgenzi, T.D., J.R. Mayor, T.W. Henkel & R.E. Halling. 2008. New species of *Boletellus* from Guyana. *Mycologia* 100:490–495. https://doi.org/10.3852/07-170R

Furtado, A.N.M, M. Leonardi, O. Comandini, A.C. Rinaldi & M.A. Neves. 2023. Guapirioid ectomycorrhiza: A novel fungus-plant subtype is described associated to *Guapira opposita* (Nyctaginaceae) in the Brazilian restinga. *Forest Systems* 32:e009. https://doi.org/10.5424/fs/2023322-19998

Gibbs, A.K. & C.N. Barron. 1993. *The Geology of the Guiana Shield*. Oxford University Press, Oxford, UK.

Gilbertson, R.L. & L Ryvarden 1986. *North American Polypores*. Vol. 1 Abortiporus Lindtneria. Fungiflora, Oslo.

Grupe, A.C., A.D. Baker, J.K. Uehling, T.J. Baroni, D.J. Lodge & T.W. Henkel. 2015. *Sarcodon* in the Neotropics I. New species from Guyana, Puerto Rico, and Belize. *Mycologia* 107:591–606. https://doi.org/10.3852/14-185

Grupe, A.C., A.M. Vasco-Palacios, M.E. Smith, T. Boekhout & T.W. Henkel. 2016. *Sarcodon* in the Neotropics II: Four new species from Colombia, and a key to the regional species. *Mycologia* 108:791–805. https://doi.org/10.3852/15-254

Hackel, J., T.W. Henkel, P.A. Moreau, E. De Crop, A. Verbeken, M. Sà, B. Buyck, M.A. Neves, A. Vasco-Palacios, F. Wartchow, H. Schimann, F. Carriconde, S. Garnica, R. Courtecuisse, M. Gardes, S. Manzi, E. Lousianna & M. Roy. 2022. Biogeographic history of a large clade of ectomycorrhizal fungi, the Russulaceae, in the Neotropics and adjacent regions. *New Phytologist* 236:698–713. https://doi.org/10.1111/nph.18365

Hammond, D.S. 2005. Biophysical features of the Guiana Shield. In D.S. Hammond (ed.), *Tropical Forests of the Guiana Shield. Ancient Forests in a Modern World*, pp. 15–194. CABI Publishing, Oxfordshire, United Kingdom.

Hammond, D.S. & H. ter Steege. 1998. Propensity for fire in Guianan rainforests. *Conservation Biology* 12:944–947. https://doi.org/10.1046/j.1523-1739.1998.012005944.x

Harrower, E., N. Bougher, C. Winterbottom, T.W. Henkel, E. Horak & P.B. Matheny. 2015. New species in *Cortinarius* section *Cortinarius* (Agaricales) from the Americas and Australasia. *MycoKeys* 11:1–21. https://doi.org/10.3897/mycokeys.11.5409

Henkel, T.W. 1999. New taxa and distribution records for *Tylopilus* from *Dicymbe* forests of Guyana. *Mycologia* 91:655–665. https://doi.org/10.1080/00275514.1999.12061065

Henkel, T.W. 2001. *Tylopilus pakaraimensis* sp. nov., a new bolete in *Tylopilus* section *Potamogetones* from Guyana. *Mycotaxon* 78:105–114.

Henkel, T.W. 2003. Monodominance in the ectomycorrhizal *Dicymbe corymbosa* (Caesalpiniaceae) from Guyana. *Journal of Tropical Ecology* 19:417–437. https://doi.org/10.1017/S0266467403003468

Henkel, T.W., M.C. Aime, M. Chin & C. Andrew. 2004a. Edible mushrooms from Guyana. *The Mycologist* 18:104–111. https://doi.org/10.1017/S0269915XO4003027

Henkel, T.W., M.C. Aime, M. Chin, S.L. Miller, R. Vilgalys & M.E. Smith. 2012. Ectomycorrhizal fungal sporocarp diversity and discovery of new taxa in *Dicymbe* monodominant forests of the Guiana Shield. *Biodiversity and Conservation* 21:2195–2220. https://doi.org/10.1007/s10531-011-0166-1

Henkel, T.W., M.C. Aime & H.K. Mehl. 2009. *Craterellus excelsus* sp. nov. from Guyana. *Mycotaxon* 107:201–208.

Henkel, T.W., M.C. Aime, H.K. Mehl & S.L. Miller. 2006a. *Cantharellus pleurotoides*, a new and unusual basidiomycete from Guyana. *Mycological Research* 110:1409–1412. https://doi.org/10.1016/j.mycres.2006.09.010

Henkel, T.W., M.C. Aime & S.L. Miller. 2000. Systematics of pleurotoid Russulaceae from Guyana and Japan, with notes on their ectomycorrhizal status. *Mycologia* 92:1119–1132. https://doi.org/10.1080/00275514.2000.12061259

Henkel, T.W., T.Y. James, S.L. Miller, M.C. Aime & O.K. Miller Jr. 2006b. The mycorrhizal status of *Pseudotulostoma volvata* (Elaphomycetaceae, Eurotiales, Ascomycota). *Mycorrhiza* 16:241–244. https://doi.org/10.1007/s00572-006-0040-2

Henkel, T.W. & J.R. Mayor. 2019. Implications of a long-term mast seeding cycle for climatic entrainment, seedling establishment, and persistent monodominance in a Neotropical, ectomycorrhizal canopy tree. *Ecological Research* 34:472–484. https://doi.org/10.1111/1440-1703.12014

Henkel, T.W., J.R. Mayor & L.P. Woolley. 2005a. Mast fruiting and seedling survival of the ectomycorrhizal, monodominant *Dicymbe corymbosa* (Caesalpiniaceae) in Guyana. *New Phytologist* 167:543–556. https://doi.org/10.1111/j.1469-8137.2005.01431.x

Henkel, T.W., R. Meszaros, M.C. Aime & A. Kennedy. 2005b. New species of *Clavulina* from the Pakaraima Mountains of Guyana. *Mycological Progress* 4:342–350. https://doi.org/10.1007/s11557-006-0140-6

Henkel, T.W. & D.L. Largent. 2023. The Entolomataceae of the Pakaraima Mountains of Guyana VII: *Entoloma dicymbophilum* sp. nov. and a key to Guyanese species of *Entoloma* s.str. *Edinburgh Journal of Botany* 80:1–9. https://doi.org/10.24823/EJB.2023.1943

Henkel, T.W. & D.L. Largent. 2026. The Entolomataceae of the Pakaraima Mountains of Guyana IX. *Entoloma aldinophilum* sp. nov., a striking species associated with *Aldina insignis*. *Edinburgh Journal of Botany* (in press)

Henkel, T.W., K. Obase, D. Husbands, J.K. Uehling, G. Bonito, M.C. Aime & M.E. Smith. 2016. New boletes from Guyana: *Binderoboletus segoi* gen. et sp. nov., *Guyanaporus albipodus* gen. et sp. nov., *Singerocomus rubriflavus* gen. et sp. nov., and a new combination for *Xerocomus inundabilis*. *Mycologia* 108:157–173. https://doi.org/10.3852/15-075

Henkel, T.W., P. Roberts & M.C. Aime. 2004b. Sebacinoid species from the Pakaraima Mountains of Guyana. *Mycotaxon* 89:433–439.

Henkel, T.W., J. Terborgh & R. Vilgalys. 2002. Ectomycorrhizal fungi and their leguminous hosts in the Pakaraima Mountains of Guyana. *Mycological Research* 106:516–532. https://doi.org/10.1017/S0953756202005919

Henkel, T.W., A.W. Wilson, M.C. Aime, J. Dierks, J.K. Uehling, M. Roy, H. Schimann, F. Wartchow & G.M. Mueller. 2014. Cantharellaceae of Guyana II. New species of *Craterellus*, new South American records for *Cantharellus guyanensis* and *Craterellus excelsus*, and a key to the Neotropical taxa. *Mycologia* 106:307–322. https://doi.org/10.3852/106.2.307

Hijmans, R.J., S.E. Cameron, J.L. Parra, P.G. Jones & A. Jarvis. 2005. Very high resolution interpolated climate surfaces for global land areas. *International Journal of Climatology* 25:1965–1978. https://doi.org/10.1002/joc.1276

Hokche, O., P.E. Berry & O. Huber. 2008. Nuevo catálogo de la flora vascular de Venezuela. *Fundación Instituto Botánico de Venezuela*. Caracas (Venezuela). p. 859.

Horak, E. & R.E. Halling. 1991. New records of *Phaeocollybia* from Colombia. *Mycologia* 83:464–472. https://doi.org/10.1080/00275514.1991.12026036

Houbraken, J., S. Kocsube, C.M. Visagie, N. Yilmaz, X.C. Wang, M. Meijer, B. Kraak, V. Hubka, K. Bensch, R.A. Samson & J.C. Frisvad. 2020. Classification of *Aspergillus*, *Penicillium*, *Talaromyces* and related genera (Eurotiales): An overview of families, genera, subgenera, sections, series and species. *Studies in Mycology* 95:5–169. https://doi.org/10.1016/j.simyco.2020.05.002

Husbands, D.R., T.W. Henkel, G. Bonito & M.E. Smith. 2013. New species of *Xerocomus* (Boletales) from the Guiana Shield, with notes on their ectomycorrhizal status and fruiting occurrence. *Mycologia* 105:422–435. https://doi.org/10.3852/12-146

Kalichman, J., P.M. Kirk & P.B. Matheny. 2020. A compendium of generic names of agarics and Agaricales. *Taxon* 69:1–23. https://doi.org/10.1002/tax.12240

Kennedy, P., P.B. Matheny, M. Ryberg, T.W. Henkel, J.K. Uehling & M.E. Smith. 2012. Scaling up: Examining the macroecology of ectomycorrhizal fungi. *Molecular Ecology* 21:4151–4154. https://doi.org/10.1111/j.1365-294X.2012.05703.x

Kõljalg, U., I. Saar & S. Svantesson. 2024. Merging the genus *Tomentella* with *Thelephora* (Fungi, Thelephorales). *Folia Cryptogamica Estonica* 61:67–86. https://doi.org/10.12697/fce.2024. 61.09

Komura, D.L., F. Wartchow & T.S. Cabral. 2021. *Pseudotulostoma volvatum* (Elaphomycetaceae, Ascomycota) from an Amazonian white-sand forest in Brazil: Odd and rare in the Neotropics? *Plant Systematics and Evolution* 307:65. https://doi.org/10.1007/s00606-021-01787-8

Largent, D.L., T.W. Henkel, M.C. Aime & T.J. Baroni. 2008. The Entolomataceae of the Pakaraima Mountains of Guyana. I. four new species of *Entoloma* s.str. *Mycologia* 100:132–140. https://doi.org/10.1080/15572536.2008.11832505

Larsen, M.J. 1968. *Tomentelloid fungi of North America*. Tech. Publ. 93. State University College of Forestry, Syracuse.

Larsson, K-H., E. Parmasto, M. Fischer, E. Langer, K.K. Nakasone & S.R. Redhead. 2006. Hymenochaetales: A molecular phylogeny for the hymenochaetoid clade. *Mycologia* 98:926–936. https://doi.org/10.1080/15572536.2006.11832622

Linhares, F.T.F. 2018. *Taxonomia e filogenia de Phyllobolites* Singer (Boletaceae, Basidiomycota). PhD dissertation, Universidade Federal de Santa Catarina. https://repositorio.ufsc.br/handle/123456789/193078

Luize, B.G., L. Ekelshot, K. Dexter, et al. 2024. The biogeography of the Amazonian tree flora. *Communications Biology*. https://doi.org/10.1038/s42003-024-06937-5

Maas Geesteranus, R.A. 1964. Notes on hydnums II. *Persoonia* 3:155–192.

Magnago, A.C., G. Alves-Silva, T.W. Henkel, M.A. Neves & R.B. Silveira. 2022. New genera, species, and combinations of Boletaceae from Brazil and Guyana. *Mycologia* 114:607–625. https://doi.org/10.1080/00275514.2022.2037307

Magnago, A.C., T.W. Henkel, M.A. Neves & R.M. Borges da Silveira. 2018. *Singerocomus atlanticus* sp. nov., and a first record for *Singerocomus rubriflavus* for Brazil. *Acta Brasilica Botanica* 32:222–231. https://doi.org/10.1590/0102-33062017abb0320

Maguire, B. 1945. Notes on the geology and geography of Tafelberg, Suriname. *Geographical Review* 35:563–579. https://doi.org/10.2307/210796

Mayor, J.R. & T.W. Henkel. 2005. Do ectomycorrhizas alter leaf-litter decomposition in tropical monodominant forests of Guyana? *New Phytologist* 169:579–588. https://doi.org/10.1111/j.1469-8137.2005.01607.x

McGuire, K.L., T.W. Henkel & I. de la Cerda. 2008. Dual mycorrhizal colonization of forest dominating tropical trees and the mycorrhizal status of non-dominant tree and liana species. *Mycorrhiza* 18:217–222. https://doi.org/10.1007/s00572-008-0170-9

Montagne, C. 1854. Cryptogamia guyanensis. *Annales des Science Naturelles, Botanique sér*. 4:91–144.

Matheny, P.B., M.C. Aime & T.W. Henkel. 2003. New species of *Inocybe* from *Dicymbe* forests of Guyana. *Mycological Research* 107:495–505. https://doi.org/10.1017/S0953756203007627

Matheny, P.B., M.C. Aime, M.E. Smith & T.W. Henkel. 2012. New species and records for *Inocybe* (Agaricales) from Guyana. *Kurtziana* 37:23–39. https://doi.org/10.1017/S0953756203007627

Mighell, K.S., T.W. Henkel, R.A. Koch, A. Goss & M.C. Aime. 2019. New species of *Amanita* subgen. *Lepidella* from Guyana. *Fungal Evolution and Systematics* 3:1–12. https://doi.org/10.3114/fuse.2019.03.01

Miller, O.K. Jr. 1983. Ectomycorrhizas in the Agaricales and gasteromycetes. *Canadian Journal of Botany* 61:909–916. https://doi.org/10.1139/b83-103

Miller, S.L. & T.W. Henkel. 2004. Biology and molecular ecology of subiculate *Lactarius* species from Guyana. *Memoirs of the New York Botanical Garden* 89:297–312.

Miller, S.L., M.C. Aime & T.W. Henkel. 2002. The Russulaceae of the Pakaraima Mountains of Guyana. I. New species of pleurotoid *Lactarius. Mycologia* 94:545–553. https://doi.org/10.1080/15572536.2003.11833220

Miller, S.L., M.C. Aime & T.W. Henkel. 2012. Russulaceae of the Pakaraima Mountains of Guyana. III. New species of *Russula* and *Lactifluus. Mycotaxon* 121:233–253.

Miller, S.L., M.C. Aime & T.W. Henkel. 2024. Russulaceae of the Pakaraima Mountains of Guyana 5. Two newly described diminutive species in a novel lineage of the crown clade of *Russula* (Russulaceae). *Phytotaxa* 668:117-129. https://doi.org/10.11646/phytotaxa.668.2.1

Miller, S.L., T.W. Henkel & M.C. Aime. 2025. *Russula rubroglutinata*, a new species from the Guiana Shield. In W. Dong et al. (eds) *2020–2025: Taxonomic and phylogenetic contributions to freshwater fungi and other fungal taxa. Fungal Diversity Notes*. pp. 229-233. https://doi.org/10.1007/s13225-025-00560-3

Miller, O.K. Jr., T.W. Henkel, T.Y. James & S.L. Miller 2001. *Pseudotulostoma*, a new genus in the Elaphomycetaceae from Guyana. *Mycological Research* 105:1268–1272. https://doi.org/10.1017/S095375620100466X

Moyersoen, B. 1993. Ectomicorrizas y micorrizas vesiculo-arbusculares en Caatinga Amazonica del Sur de Venezuela. *Scientia Guianae* 3:1–82.

Moyersoen, B. 2006. *Pakaraimaea dipterocarpacea* is ectomycorrhizal, indicating an ancient Gondwanaland origin for the ectomycorrhizal habit in Dipterocarpaceae. *New Phytologist* 172:753–762. https://doi.org/10.1111/j.1469-8137.2006.01860.x

Neves, M.A., T.W. Henkel & R.E. Halling. 2010. *Phylloporus colligatus*, a new species of gilled bolete from Guyana. *Mycotaxon* 111:143–148.

Norvell, L.L. & R.L Exeter. 2008. *Phaeocollybia of Pacific Northwest North America*. US Bureau of Land Management, Salem OR.

Nuytinck, J., T.W. Henkel, K. Milislav, C. Noordermeer, A. Verbeken & M.C. Aime. 2023. Russulaceae of the Pakaraima Mountains of Guyana. IV. New species forming a distinct lineage of *Lactarius* subg. *Plinthogalus. Mycologia* 115:69–86. https://doi.org/10.1080/00275514.2022.2125712

Oberwinkler, F. 1964. Intrahymeniale heterobasidiomyceten. *Nova Hedwigia* 1:489–499.

Oberwinkler, F., K. Riess, R. Bauer & S. Garnica. 2014. Morphology and molecules: The Sebacinales, a case study. *Mycological Progress* 13:445–470. https://doi.org/10.1007/s11557-014-0983-1

Pegler, D.N. 1983. *Agaric Flora of the Lesser Antilles*. Kew bulletin additional series IX. HMSO, London.

Ramos, G., H.C. de Lima, G. Prenner, L.P. de Queiroz, C.E. Zartman & D. Cardoso. 2016. Molecular systematics of the Amazonian genus *Aldina*, a phylogenetically enigmatic ectomycorrhizal lineage of papilionoid legumes. *Molecular Phylogenetics and Evolution* 97:11–18. https://doi.org/10.1016/j.ympev.2015.12.017

Redhead, J.F. 1968. Mycorrhizal associations in some Nigerian forest trees. *Transactions of the British Mycological Society* 51:377–387.

Richards, P.W. (ed.) 1996. *The Tropical Rainforest. An Ecological Study*. 2nd ed. Cambridge University Press, Cambridge, United Kingdom.

Rick, J.E. 1906. Pilze aus Rio Grande do Sul (Brazilien). *Brotéria, série Botânica* 5:4–53.

Rull, V., O. Huber, T. Vegas-Vilarrúbia & C. Señaris. 2019. Definition and characterization of the Pantepui biogeographical province. In V. Rull, T. Vegas-Vilarrúbia, O. Huber & C. Señaris (eds.), *Biodiversity of Pantepui*, pp. 3–32. Academic Press, Cambridge MA. https://doi.org/10.1016/B978-0-12-815591-2.00001-X

Ryvarden, L. 2024. *Poroid Fungi of North America*. Fungiflora 49. APS Press, St. Paul, MN.

Sanchez-Garcia, M., M.C. Aime, M.E. Smith & P.B. Matheny. 2016. *Guyanagarika*, a new ectomycorrhizal genus of Agaricales from the Neotropics. *Fungal Biology* 120:1540–1553. https://doi.org/10.1016/j.funbio.2016.08.005

Simmons, C., T.W. Henkel & C. Bas. 2002. The genus *Amanita* in the Pakaraima Mountains of Guyana. *Persoonia* 17:563–582.

Siegel, N., T.W. Henkel, S. Adams, J. Cooper & M.C. Aime. 2024. New Cortinariaceae species associated with *Dicymbe*, *Aldina*, and *Pakaraimaea* in Guyana. *Mycologia* 116:848–864. https://doi.org/10.1080/00275514.2024.2367399

Silva-Filho, A.G.S, M.C.A. Sá, D.L. Komura, R. Braga-Neto, I.G. Baseia, P. Marinho, J.M. Moncalvo, M. Roy & F. Wartchow. 2021. *Lactifluus* subg. *Lactariopsis* (Russulales/Basidiomycota) from Brazilian tropical forests. *Botany* 101:570–581. https://doi.org/10.1139/cjb-2023-0077

Singer, R. 1962. Diagnoses Fungorum novorum Agaricalium II. *Sydowia* 15:45–83.

Singer, R. 1986. *The Agaricales in Modern Taxonomy*. Koeltz, Koenigstein.

Singer, R. 1987. *Phaeocollybia* in the oak woods of Costa Rica, with notes on extralimital taxa. *Mycologia Helvetica* 2:247–266.

Singer, R., I. Araujo & M.H. Ivory. 1983. The ectotrophically mycorrhizal fungi of the neotropical lowlands, especially Central Amazonia. *Beihefte Nova Hedwigia* 77:1–352.

Singer, R., J. Garcia & L.D. Gomez. 1991. The boletineae of Mexico and Central America III. *Beihefte Nova Hedwigia* 102:1–99.

Smith, M.E., T.W. Henkel, M.C. Aime, A.K. Fremier & R. Vilgalys. 2011. Ectomycorrhizal fungal diversity and community structure on three co-occurring leguminous canopy tree species in a Neotropical rainforest. *New Phytologist* 192:699–712. https://doi.org/10.1111/j.1469-8137.2011.03844.x

Smith, M.E., T.W. Henkel, J.K. Uehling, A.K. Fremier, H.D. Clarke & R. Vilgalys. 2013. The ectomycorrhizal fungal community in a neotropical forest dominated by the endemic dipterocarp *Pakaraimaea dipterocarpacea*. *PLoS ONE* 8:e55160.

Smith, M.E., K. Amses, T.F. Elliott, M.C. Aime & T.W. Henkel. 2015. New sequestrate fungi from Guyana: *Jimtrappea guyanensis* gen. sp. nov., *Castellanea pakaraimophila* gen. sp. nov., and *Costatisporus cyanescens* gen. sp. nov. (Boletaceae, Boletales). *IMA Fungus* 6:263–283. https://doi.org/10.5598/imafungus.2015.06.02.03

Smith, M.E., T.W. Henkel, G. Williams, M.C. Aime & R. Vilgalys. 2017. Investigating niche partitioning of ectomycorrhizal fungi in specialized rooting zones of the monodominant leguminous tree *Dicymbe corymbosa*. *New Phytologist* 215:443–453. https://doi.org/10.1111/nph.14570

Stergios, B.D. & C.G.A. Aymard. 2008. A striking new species of *Aldina* (Fabaceae-Swartzieae Aldininae) from the Venezuelan Guayana highlands. *Harvard Papers in Botany* 13:29–33. https://doi.org/10.3100/1043-4534(2008)13[29:ASNSOA]2.0.CO;2

Tedersoo, L., T.W. May & M.E. Smith. 2010a. Ectomycorrhizal lifestyle in fungi: Global diversity, distribution, and evolution of phylogenetic lineages. *Mycorrhiza* 20:217–263. https://doi.org/10.1007/s00572-009-0274-x

Tedersoo, L., A. Sadam, M. Zambrano, R. Valencia & M. Bahram. 2010b. Low diversity and high host preference of ectomycorrhizal fungi in Western Amazonia, a neotropical diversity hotspot. *ISME Journal* 4:465–471. https://doi.org/10.1038/ismej.2009.131

Tedersoo, L., T. Suvi, K. Beaver & I. Sarr. 2007. Ectomycorrhizas of *Coltricia* and *Coltriciella* (Hymenochaetales, Basidiomycota) on Caesalpiniaceae, Dipterocarpaceae and Myrtaceae in Seychelles. *Mycological Progress* 6:101–107. https://doi.org/10.1007/s11557-007-0530-4

ter Steege, H., T.W. Henkel, N. Helal, B.S. Marimon, B.H. Marimon, et al. 2019. Rarity of monodominance in hyperdiverse Amazonian forests. *Nature Scientific Reports* 9:13822.

ter Steege, H., N.C.A. Pitman, I.L. Amaral, L. Coelho, F.D. Matos, D.L.L. Fihlo, et al. 2023. Mapping density, diversity and species-richness of the Amazon tree flora. *Communications Biology* 6. https://doi.org/10.1038/s42003-023-05514-6

ter Steege, H., N.C.A. Pitman, O.L. Phillips, et al. 2006. Continental-scale patterns of canopy tree composition and function across Amazonia. *Nature* 443:444–447. https://doi.org/10.1038/nature05134

Thacker, J. & T.W. Henkel. 2004. New species of *Clavulina* from Guyana. *Mycologia* 96:650–657. https://doi.org/10.1080/15572536.2005.11832961

Thoen, D. & A.M. Ba. 1989. Ectomycorrhizas and putative ectomycorrhizal fungi of *Afzelia africana* and *Uapaca guineensis* in southern Senegal. *New Phytologist* 113:549–559. https://doi.org/10.1111/j.1469-8137.1989.tb00368.x

Thomazini, L.I. 1974. Mycorrhiza in plants of the Cerrado. *Plant and Soil* 41:707–711. https://doi.org/10.1007/BF02185833

Trappe, J.M. 1979. The orders, families, and genera of hypogeous Ascomycotina (truffles and their relatives). *Mycotaxon* 9:297–340.

Tremble, K., T.W. Henkel, B. Bradshaw, C. Domnauer, L.M. Brown, L.X. Tham, G. Furci, M.C. Aime, J-M. Moncalvo & B.D.M. Dentinger. 2024. A revised phylogeny of the Boletaceae using whole genome sequences. *Mycologia* 116:392–408. https://doi.org/10.1080/00275514.2024.2314963

Uehling, J.K., T.W. Henkel, M.C. Aime & M.E. Smith. 2012. New species of *Clavulina* with effused or resupinate basidiomata from the Guiana Shield. *Mycologia* 147:547–556. https://doi.org/10.3852/11-130

Vanegas-León, M.L., M.A. Sulzbacher, A.C. Rinaldi, M. Roy, M.A. Selosse & M.A. Neves. 2019. Are Trechisporales ectomycorrhizal or non-mycorrhizal root endophytes? *Mycological Progress* 1231–1240. https://doi.org/10.1007/s11557-019-01519-w

Vasco-Palacios, A.M., A.E. Franco-Molano, C.A. López-Quintero & T. Boekhout. 2005. Macromicetes (Ascomycota, Basidiomycota) de la región del médio Caquetá, departamentos de Caquetá y Amazonas (Colombia). *Biota Colombiana* 6:127–140.

Vasco-Palacios, A.M., J. Hernandez, M.C. Peñuela-Mora, A.E. Franco-Molano & T. Boekhout. 2018. Ectomycorrhizal fungi diversity in a white sand forest in western Amazonia. *Fungal Ecology* 31:9–18. https://doi.org/10.1016/j.funeco.2017.10.003

Vasco-Palacios, A.M. & T. Boekhout. 2022. *Pseudomonotes tropenbosii*, an endemic dipterocarp tree from a neotropical terra-firme forest in Colombian Amazonia that hosts ectomycorrhizal fungi. In M.A. Lugo & M.C. Pagano (eds.), *Mycorrhizal fungi in South America. Fungal Biology,* pp. 47–78. Springer, Cham. https://doi.org/10.1007/978-3-031-12994-0_3

Wang, D., H. Feng, J. Zhou. et al. 2024. New insights into the stipitate hydnoid fungi *Sarcodon*, *Hydnellum*, and the formerly informally defined *Neosarcodon*, with emphasis on the edible species marketed in Southwest China. *IMA Fungus* 15:8. https://doi.org/10.1186/s43008-023-00138-1

Wells, K. & R. Bandoni. 2001. Heterobasidiomycetes. In D.J. McLaughlin, E.G. McLaughlin & P.A. Lemke (eds.), *The Mycota, vol. VII Part B: Systematics and Evolution*. Springer, Heidelberg, pp. 85–120.

Wilson, A.W., M.C. Aime, J. Dierks, G.M. Mueller & T.W. Henkel. 2012. Cantharellaceae of Guyana I. New species, distribution records, and a synopsis of known taxa. *Mycologia* 104:1466–1477. https://doi.org/10.3852/11-412

Woolley, L.P., T.W. Henkel & S.C. Sillett. 2008. Reiteration in the tropical monodominant tree *Dicymbe corymbosa* (Caesalpiniaceae) and its potential adaptive significance. *Biotropica* 40:32–43. https://doi.org/10.1111/j.1744-7429.2007.00348.x

16 Spegazzini's Legacy Revisited
A Century of Fungal Rediscovery in Paraguay

Michelle G. Campi, Yanine E. Maubet, and Juliana Moura-Mendes

16.1 INTRODUCTION

To learn about the history of mycology in Paraguay, it is necessary to travel back more than a century to the pioneering expeditions of the Italian-Argentine mycologist Carlos Luigi Spegazzini (Figure 16.1). Born in Italy in 1858, Spegazzini traveled to South America in 1879 and went on to become one of the most influential figures in the early study of fungi in the Neotropics. Over the course of his career, he published nearly 200 scientific papers and two books documenting the biodiversity of fungi in Argentina, Brazil, Chile, and Paraguay. In 1883, he joined an expedition to the Paraguayan Chaco, where he met his wife María de la Cruz Rodríguez and began a journey that marked the beginnings of his contributions to Paraguayan mycology. His marriage to María de la Cruz Rodríguez established a lasting personal bond with the country, complementing his scientific interest and ensuring that Paraguay would remain a recurring focus throughout his career. Later that same year, he initiated a series of manuscripts titled *Fungi Guaraniciti*, dedicated to the description of Paraguay's fungal diversity. This series, published in 14 parts in the *Anales de la Sociedad Científica Argentina* and the *Revista Argentina de Historia Natural* in Buenos Aires, spanned a period of nine years.

Although most of Spegazzini's work focused on the Argentinian mycobiota, he managed to describe approximately 2,000 species of fungi for Paraguay. Of these, about 250 were macroscopic forms, including mushrooms and polypores, thus providing the first comprehensive account of Paraguay's funga. Among these records, he described several new species, such as *Bresadolia paradoxa* Speg. and *Rickiella edulis* Speg. (as *Peziza edulis*) that to this day continue to be found by young or amateur mycologists of the country. The Swedish mycologist Lars Romell, a contemporary of Spegazzini, contributed to the study of Paraguayan fungi in 1901. His work, while not as extensive, provided additional records of the Paraguayan mycobiota.

16.2 SPEGAZZINI AND ROMELL

After the works of both Spegazzini and Romell, the study of the Paraguayan funga went into an almost century-long hiatus. It was not until the 1990s that new systematic efforts were undertaken to reassess Paraguay's fungal diversity. The most notable modern contributions came from Dr. Orlando F. Popoff and Dr. Jorge E. Wright (1994, 1998), who not only revisited and revised the previous work done by their predecessors but also described new species for the country with a special focus on corticioid wood-rotting fungi. Their preliminary checklist of the fungi of Paraguay documented approximately 90 species of polypores, offering the first comprehensive update since the earlier publication by Spegazzini in 1922.

Another 15 years passed before, for the first time in history, a group of Paraguayan biology students published a booklet on the fungal biota of Laguna Blanca, San Pedro, Paraguay (Campi et al. 2013).

DOI: 10.1201/9781003647140-16

FIGURE 16.1 Dr. Carolo Spegazzini, father of mycology in Argentina and Paraguay.

This modest volume presented descriptions of 35 species of macrofungi recorded for the country, with specimens deposited in a local herbarium. Although simple in scope, the publication marked an important milestone, since it set in motion a series of events that ultimately led to the creation of the first laboratory in Paraguay dedicated to the study of macrofungi at the Universidad Nacional de Asunción, Paraguay. From that group of students emerged new leaders who continue to guide research and education in Paraguayan mycology, ensuring that the legacy of Dr. Carlos Spegazzini remains alive and inspiring to this day.

16.3 THE CURRENT STATE OF THE STUDY OF MACROFUNGI IN PARAGUAY

For the Republic of Paraguay, the body of literature dedicated to the study of fungi remains remarkably sparse when compared with that of neighboring countries in South America. This discontinuous history has left large gaps in our understanding of fungal diversity in the region, despite Paraguay's geographic position at the intersection of major South American biomes such as the Chaco, the Atlantic Forest, and the Cerrado—all of which are known to harbor high biodiversity and endemism. On the other hand, this limited literature highlights the tremendous opportunities that exist for future research. Paraguay remains, in many respects, a "blank map" for mycological exploration. As such, the great task of cataloging the macroscopic funga of Paraguay began in 2014 with the creation of the Laboratory of Mycology at the Facultad de Ciencias Exactas y Naturales UNA-Paraguay. The numerous researchers who have contributed to our knowledge of the funga of Paraguay have worked hard throughout the past ten years to compile a list of approximately 200 records for the country, of which 86 species have not been previously described (Table 16.1).

TABLE 16.1

Species New to Science Described from Paraguay between 2013 and 2023.

Agaricus trisulphuratus	Campi et al. (2022a)	*Gymnopilus luteofolius*	Campi et al. (2021b)
Amylosporus guaraniticus	Campi et al. (2017b)	*Gymnopilus peliolepis*	Campi et al. (2021b)
Auricularia delicata	Campi et al. (2022a)	*Gymnopilus purpureosquamulosus*	Campi et al. (2021b)
Battarrea phalloides	Campi et al. (2017e)	*Hymenopellis radicata*	Campi et al. (2022a)
Blumenavia rhacodes	Campi et al. (2021a)	*Inonotus rickii*	Maubet et al. (2020)
Calvatia cyathiformis	Campi et al. (2021a)	*Inonotus splitgerberi*	Maubet et al. (2020)
Calvatia rugosa	Campi and Maubet (2016)	*Itajahya galericulata*	Campi et al. (2017a)
Chlorophyllum hortense	Maubet et al. (2022)	*Itajahya rosea*	Campi et al. (2017a)
Chlorophyllum sapukai	Maubet et al. (2022)	*Kusaghiporia talpae*	Campi et al. (2022b)
Clathrus argentinus	Campi et al. (2021a)	*Laccaria fraterna*	Campi et al. (2017f)
Clathrus chrysomycelinus	Campi et al. (2018b)	*Laetiporus gilbertsonii*	Campi et al. (2022b)
Clathrus columnatus	Campi et al. (2017a)	*Lentinus crinitus*	Campi et al. (2022a)
Clathrus crispus	Campi et al. (2018b)	*Leucocoprinus birnbaumii*	Campi et al. (2015b)
Coltricia stuckertiana	Maubet et al. (2020)	*Leucocoprinus cepistipes*	Campi et al. (2015b)
Cookeina colensoi	Campi et al. (2022b)	*Leucocoprinus cretaceous*	Campi et al. (2015b)
Coprinus comatus	Campi et al. (2016)	*Leucocoprinus fragilissimus*	Campi et al. (2021c)
Cotylidia diaphana	Campi et al. (2022b)	*Leucocoprinus straminellus*	Campi et al. (2015b)
Cyathus earlei	Campi et al. (2021a)	*Leucopaxillus gracillimus*	Campi et al. (2022)
Cyathus limbatus	Maubet et al. (2017)	*Lysurus sphaerocephalum*	Campi et al. (2021a)
Cyathus poeppigii	Campi and Maubet (2015b)	*Marasmius ferrugineus*	Madrignac et al. (2013)
Cyathus setosus	Maubet et al. (2017)	*Microporellus dealbatus*	Campi et al. (2022)
Cyathus stercoreus	Maubet et al. (2017)	*Mutinus argentinus*	Campi et al. (2017a)
Daedaela modesta	Cristaldo et al. (2022)	*Myriostoma calongei*	Campi et al. (2021a)
Daedalea aethalodes	Cristaldo et al. (2022)	*Myrmecopterula moniliformis*	Campi et al. (2022)
Donkia pulcherrima	Campi et al. (2017g)	*Phallus campanulatus*	Campi et al. (2021a)
Favolaschia calocera	Campi et al. (2022)	*Phallus indusiatus*	Campi et al. (2018b)
Filoboletus gracilis	Campi et al. (2017c)	*Phillipsia olivacea*	Campi et al. (2022a)
Fomitiporia neotropica	Maubet et al. (2020)	*Phylloporia chrysites*	Maubet et al. (2020)
Fulvifomes rhytiphloeus	Maubet et al. (2020)	*Phylloporia nodostipitata*	Maubet et al. (2020)
Fulvifomes wrightii	Martínez et al. (2023)	*Pisolithus arhizus*	Campi et al. (2015a)
Fuscoporia rhabarbarina	Maubet et al. (2020)	*Podaxis pistillaris*	Campi et al. (2015b)
Geastrum argentinum	Campi et al. (2018a)	Pleorutus djamour var. djamor	Flecha et al. (2014)

TABLE 16.1 (*Continued*)
Species New to Science Described from Paraguay between 2013 and 2023.

Geastrum coronatum	Campi and Maubet (2015a)	Pleurotus cystidiosus	Flecha et al. (2014)
Geastrum hariotii	Campi et al. (2018a)	*Pluteus fibulatus*	Campi et al. (2019)
Geastrum minimum	Campi et al. (2015a)	*Pluteus longistriatus*	Campi et al. (2019)
Geastrum pusillipilosum	Campi et al. (2021a)	*Pluteus petasatus*	Campi et al. (2019)
Geastrum sacatum	Campi and Maubet (2015a)	*Pluteus triplocystis*	Campi et al. (2019)
Geastrum triplex	Campi and Maubet (2015a)	*Tremella fuciformis*	Campi et al. (2022a)
Geastrum violaceum	Campi et al. (2013)	*Trogia cantharelloides*	Campi et al. (2022)
Geastrum schweinitizii	Campi and Maubet (2015a)	*Tropicoporus drechsleri*	Maubet et al. (2020)
Gymnopilus imperialis	Campi et al. (2021b)	*Tulostoma pygmaeum*	Campi et al. (2021a)
Gymnopilus lepidotus	Campi et al. (2021b)	*Xeromphalina tenuipes*	Campi et al. (2017c)

Among these new records, a remarkable diversity of forms, colors, morphologies, and ecological strategies can be observed. They range from large wood-decaying pathogens such as *Laetiporus gilbertsonii* Burds., *Kusaghiporia talpae* (Cooke) Campi, Drechsler-Santos & Robledo, and *Inonotus rickii* (Pat.) D.A. Reid to comparatively small saprotrophs, including *Cookeina colensoi* (Berk.) Seaver, *Cotylidia diaphana* (Cooke) Lentz, *Leucopaxillus gracillimus* Singer & A.H. Sm., and *Trogia cantharelloides* (Berk. ex Sacc.) Singer (Campi et al. 2022a, 2022b; Maubet et al. 2020). Particularly notable is the predominance of gasteroid fungi among these new records, ranging from the brightly colored rare stinkhorns such as *Lysurus sphaerocephalum* (Schltdl.) Hern. Caff., Urcelay, Hosaka & L.S. Domínguez or *Clathrus argentinus* L.S. Domínguez to the diverse earthstars.

One of the most remarkable findings of this recent wave of studies was the rediscovery of a species once thought to be extinct in Paraguay. *Rickiella edulis* (Speg.) Pfister, originally described by Spegazzini in 1891, had not been reported in the country for more than a century. Its absence from subsequent surveys, combined with extensive habitat loss in parts of its known range, led to the assumption that it had disappeared locally (Calle et al. 2020). In collaboration with Brazilian researcher Dr. Larissa Trierveiler-Pereira, researchers from Paraguay started the search for *Rickiella edulis*, thanks to the project titled "Population Survey, Molecular Characterization and Ex Situ Conservation of the Endangered South American Edible ||Fungal Species Rickiella Edulis," financed by the Mohamed bin Zayed Species Conservation Fund. After six months of looking in conservation sites occurring between ecosystems, such as the humid Chaco and the Atlantic Forest, the species was finally found. This rediscovery serves as a reminder that Paraguay's fungal diversity holds many surprises and that even species presumed lost may persist in isolated habitats awaiting careful documentation (Figure 16.2).

In addition to rediscoveries, recent research has also led to the description of species new to science. In 2017, the research team, in collaboration with Argentinian colleagues, described its first new species, a polypore discovered growing on a sewer structure on the university campus, most likely arising from buried decayed wood. *Amylosporus guaraniticus* Campi & Robledo was named in honor of the Guaraní, the first inhabitants of the region, and represents the sixteenth member of the genus and only the third species described for South America (Campi et al. 2017b).

This milestone was soon followed by additional discoveries in natural ecosystems. In 2022, *Chlorophyllum sapukai* Maubet, Campi & Robledo was described from collections made in the Mbaracayú Forest Nature Reserve, one of the last well-preserved remnants of the Atlantic Forest in Alto Paraguay. This discovery is particularly significant, as it highlights the role of Paraguay's remaining forest refuges in harboring unique and previously undocumented fungal diversity. Lastly, a lignolithic species from the dry and arid Paraguayan Chaco was described: *Fulvifomes wrightii* M. Martínez, Salvador-Montoya & Rajchenb. Together, these findings demonstrate both the potential

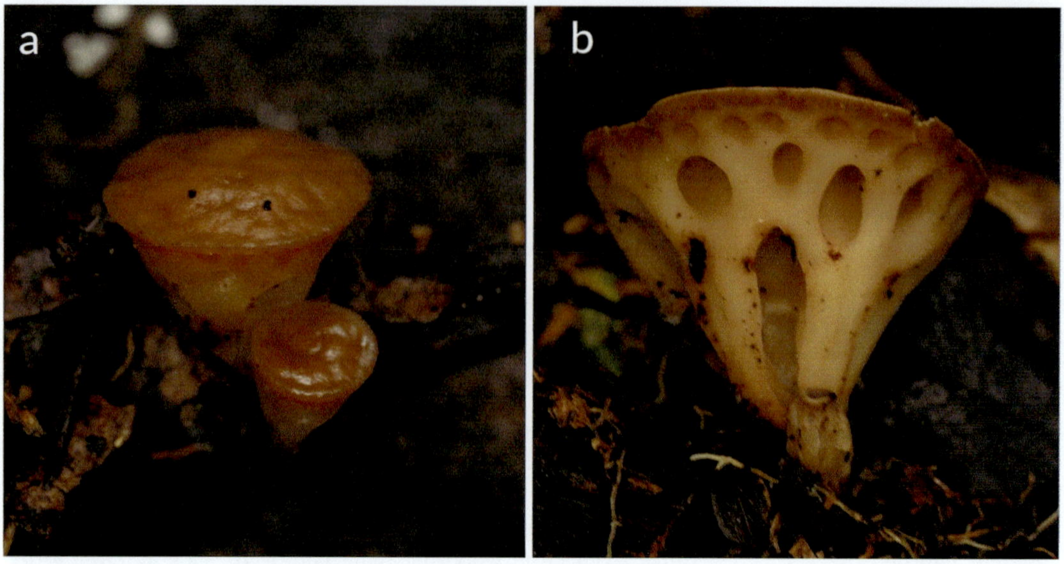

FIGURE 16.2 Macroscopic characters of fresh specimens of *Rickiella edulis*. (a) upper view and (b) lateral view with details of the lacunose excipular tissue.

for novel taxa to arise in unexpected urban contexts and the critical importance of conserving natural ecosystems, where much of Paraguay's hidden fungal richness still awaits discovery.

16.4 REDLISTING FUNGI WITHOUT A LIST: CHALLENGES AND PROSPECTS

The historical scarcity of bibliographic resources has important implications in other areas such as conservation or ecology. For one, it limits the inclusion of fungi in national biodiversity strategies, conservation planning, and ecological monitoring. Paraguay is recognized as a diverse country in terms of its flora and fauna, yet fungi remain virtually invisible in official reports, conservation programs, and environmental legislation. This invisibility is largely a reflection of the gaps in scientific knowledge. Without a solid taxonomic and ecological baseline, it is difficult to assess the conservation status of species of fungi, to identify threatened habitats, or to determine the impacts of deforestation, agriculture, and climate change on fungal communities.

Nevertheless, this challenge has also created an opportunity to align taxonomic surveys with international conservation initiatives. Since approximately 2020, fungal red-listing efforts have expanded in South America, initially through workshops in Brazil and later across the entire region, under the framework of the Global Fungal Red List Initiative (Mueller et al. 2014; Mueller et al. 2022). Paraguayan researchers have actively collaborated in these activities, contributing data and expertise gathered during fieldwork in the country. For example, *Phellinotus xerophyticus* Robledo, Urcelay & Drechsler-Santos, a species restricted to the Gran Chaco, where it grows exclusively on *Neltuma* spp., has been evaluated within this framework and is now considered a vulnerable species (Maubet et al. 2023a).

Among the newly recorded species, three taxa—*Clathrus argentinus* L.S. Domínguez, *Xylobolus gongylodes* Popoff & J.E. Wright, and *Geastrum violaceum* Rick (Figure 16.3)—have already been evaluated and included in the IUCN Global Fungal Red List (Maubet 2023; Maubet et al. 2023b, 2023c). Given that species of fungi are not restricted by national boundaries, these assessments were carried out in collaboration with researchers from neighboring countries, particularly Argentina and Brazil. These efforts not only ensure the representation of Paraguayan fungi in global conservation frameworks but also underscore the importance of integrating local field surveys with international red-listing initiatives. In doing so, they contribute to bridging existing knowledge gaps and advancing the recognition of fungi as a fundamental component of biodiversity and conservation strategies.

FIGURE 16.3 Sporoma of (a) *Clathrus argentinus*, (b) *Geastrum violaceum*, and (c) *Xylobolus gongylodes*, species that were evaluated and included in the IUCN Global Fungal Red List.

16.5 EDIBLE AND MEDICINAL MUSHROOMS FROM PARAGUAY, THE POTENTIAL OF NATIVE MUSHROOMS FROM THE NEOTROPICS

Paraguay, like several South American countries, does not have a mushroom-eating culture like some other American countries, such as the United States, Mexico, and Guatemala. Furthermore, the relationship of humans with mushrooms is often linked to myths and legends, and the Paraguayan society could be considered mycophobic (Campi et al. 2025). Although it is believed that the Indigenous People consumed mushrooms, this tradition was not transferred to subsequent generations, leaving its origins unclear. Recent research has revealed that mushrooms in Paraguay are primarily consumed by foreign colonies, primarily those composed of people from Germany, Switzerland, Russia, Japan, and Spain. Their ancestors brought this tradition from their native lands, but they have mostly not passed it down to the new generations living in Paraguay.

In Paraguay, the kingdom Fungi is stigmatized by a series of popular myths and legends. For the general population, talking about mushrooms is like talking about magical beings that frighten people or that allow them to travel astrally, and this perception about mushrooms has a long history. In Paraguayan culture, the Guaraní term *Pombero rekaká* (Pombero's excrement) is used to refer to mushrooms in general.

16.6 MYTHS AND LEGENDS: THE STORY OF THE "LORD OF THE NIGHT," *KARAI PYHARE*—THE *POMBERO*

The *Pombero* is one of the most iconic myths in Paraguay, a bilingual country with a rich Guaraní cultural heritage. This myth is associated with the protection of nature and wild animals. This mythical figure also forms part of the broader cultural landscape of Paraguay, northeastern Argentina, southern

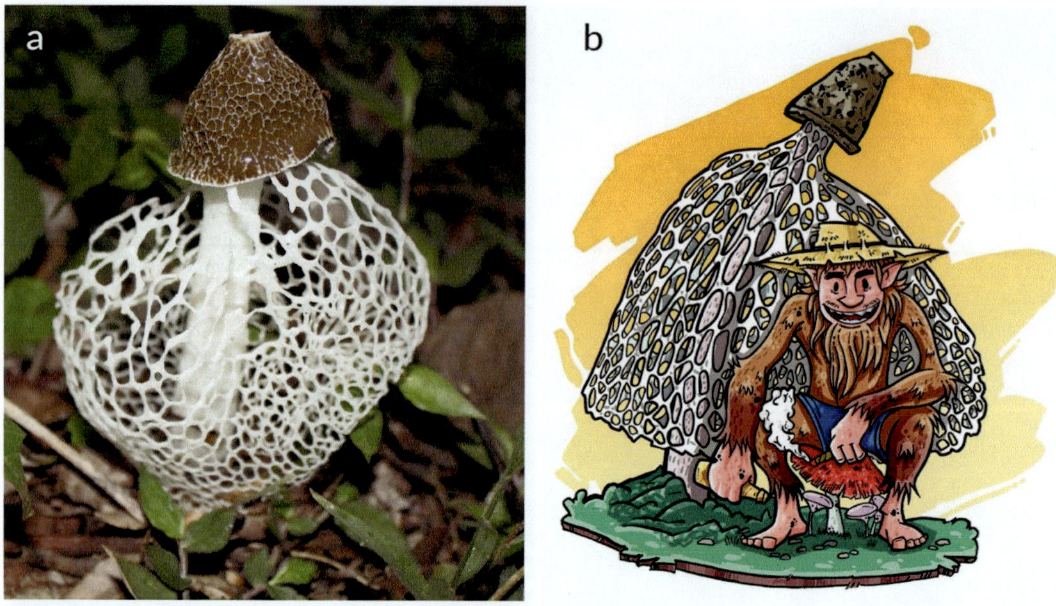

FIGURE 16.4 (a) Basidiome of *Phallus indusiatus*, (b) *Karaí Pyhare-Pombero*, and (c) FungiFest 2023 logo.

Brazil, and Uruguay (de Battini and Kornfeld 1980). The *Pombero* is typically described as a short, hairy man endowed with various supernatural powers, such as invisibility, ubiquity, camouflage, intangibility, and levitation (Jiménez 2023) (Figure 16.4b). Feared and respected by many, the *Pombero* is said to announce his presence with high-pitched whistles. He is believed to protect those who offer him gifts such as tobacco, honey, or sugarcane but to take revenge on anyone who disrespects him. His appearances are nocturnal, earning him the title *Señor de la noche, Karaí Pyhare.*

How did this mythical figure become associated with mushrooms? In Paraguay, mushrooms are colloquially referred to as *Pombero rekaka*—which translates to "the excrement of the *Pombero*." This is a uniquely Paraguayan expression (Campi et al. 2025). The phrase is believed to be inspired by *Phallus indusiatus* Vent. (Figure 16.4a), a curious mushroom with a phallic shape that appears suddenly at sunset, releases a foul smell to attract insects, and vanishes by morning—much like a fleeting trace of the *Pombero* himself. Though originally referring to this specific species, the expression is now used more broadly in Paraguayan culture to describe mushrooms in general.

16.7 BEFORE THE MYTHS: IN SEARCH OF THE ORIGIN OF THE LOST FUNGAL CULTURE

In Paraguay, five linguistic families and 19 Indigenous Peoples have been identified. These are the *Guaraní* (Aché, Avá Guaraní, Mbya, Pai Tavytera, Guaraní Ñandeva, Guaraní Occidental), *Maskoy* (Toba Maskoy, Enlhet Norte, Enxet Sur, Sanapaná, Angaité, Guaná), *Mataco-Mataguayo* (Nivaclé, Maká, Manjui), *Zamuco* (Ayoreo, Yvytoso, Tomáraho), and *Guaicurú* (Qom) (DGEEC 2013). Among these, the Guaraní are the most widely recognized Indigenous group in Paraguay (Lehner 2005).

Recent ethnomycological research suggests that Indigenous Peoples in Paraguay may have traditionally consumed mushrooms for both nutritional and medicinal purposes. However, this knowledge appears to have been largely lost in contemporary generations. It is believed that the process of Christianization contributed to the abandonment of the use of mushrooms since fungi were often associated with witchcraft or pagan rituals (Campi and Veloso 2025).

The *Tupí-Guaraní* peoples had specific terms in their language for edible mushrooms, some of which even described their taste. The word *Urupé* served as a general prefix for mushrooms.

FIGURE 16.5 Elders and children of the Aché people holding mushrooms collected from the forest.

For instance, *urupé-tinga* referred to species within the genus *Cantharellus*, *urupé-nambí* (ear mushroom) referred to species of *Polyporus*, and *urupé-ro* (bitter mushroom) was used for bitter species of the genus *Agaricus* (Servián and Campi 2020). Another graphic example of the use of mushrooms by Indigenous Peoples is the rock carving made by the Guaraní people of the *Misión Jesuítica de la Santísima Trinidad*, founded in 1706.

Another group that remembers tradition is the Aché's ancestral knowledge regarding fungi, but this knowledge may have been eroded over time. The forced displacement from their traditional territories, together with the drastic reduction of primary forest, likely contributed to this discontinuity in the transmission of ethnomycological knowledge. Yet, the fragments that persist—such as the use of *Lentaria* powder for calming bees or the existence of specific terms like Nambu puchi for *Geastrum*—hint at a deeper mycological awareness that may once have been more extensive. Even more intriguing is the symbolic dimension embedded in the language: Expressions such as *krando bua*, meaning "to pray," reveal that the concept of "krandó" transcends the material world of mushrooms and enters the sphere of spirituality. This association suggests that fungi may have held ritual, metaphoric, or cosmological significance for the Aché, even if their practical uses are no longer widely remembered. In this sense, the surviving words and isolated practices act as cultural residues, offering a glimpse into a broader, though now partially obscured, relationship between the Aché people and the fungal diversity of their ancestral forest (Campi et al. 2025) (Figure 16.5).

16.8 EDIBLE AND MEDICINAL MUSHROOMS FROM PARAGUAY, THE POTENTIAL OF NATIVE MUSHROOMS FROM THE NEOTROPICS

Paraguay is home to biodiversity hotspots such as the Alto Paraná Atlantic Forest. Across the globe, new species continue to be discovered each year, and many ecosystems in Paraguay have yet to be formally surveyed. The study of fungi in the country experienced a hiatus that lasted over a century. However, recent efforts have led to the discovery of new species and the rediscovery of species once believed to be extinct (Campi et al. 2017; Maubet et al. 2022).

TABLE 16.2
Bioprospected Species of Native Fungi from Paraguay.

Species	Biological properties	Nutritional properties	Chemical characterization	Toxicological profile	Antimicrobial properties	Bibliographic source
Amylosporus guaraniticus	✓					Campi et al. (2019)
Gloeophyllum striatum	✓					Campi et al. (2019)
Hydnopolyporus fimbriatus	✓					Campi et al. (2019)
Inonotus splitgerberi	✓					Campi et al. (2019)
Inonotus rickii	✓					Campi et al. (2019)
Lentinus lindquistii	✓					Campi et al. (2019)
Laccaria fraterna	✓					Campi et al. (2019)
Pisolithus arhizus	✓					Campi et al. (2019)
Trametes cubensis	✓					Campi et al. (2019)
Oudemansiella canarii	✓					Veloso et al. (2021)
Oudemansiella cubensis	✓	✓	✓	✓	✓	Mancuello et al. (2024)
Phlebopus beniensis	✓	✓	✓	✓	✓	Campi et al. (2023a)
Ganoderma tuberculosum	✓		✓			Campi et al. (2023b)
Stiptophyllum erubescens	✓		✓		✓	Mancuello et al. (2022)
Hornodermoporus martius	✓		✓		✓	Campi el al., (2023c)
Cristataspora flavipora	✓					Campi et al. (2021)
Ganoderma australe	✓					Campi et al. (2021)
Ganoderma martinicense	✓					Campi et al. (2021)
Ganoderma multicornum,	✓					Campi et al. (2021)

With the establishment of the Paraguay Fungal Study research team, initial reports on biological activity and chemical characterization have begun to emerge, underscoring the significant biotechnological potential of fungi. Over the past seven years, several species of mushrooms have been evaluated for their biological, chemical, and nutritional profiles (Table 16.2). Among the evaluated species, *Phlebopus beniensis* (Campi et al. 2023a) and *Oudemansiella cubensis*—the latter noted for its antigenotoxic properties (Mancuello et al. 2024)—stand out for their nutritional and bioactive potential.

Among the most captivating groups of fungi are the polypores, globally recognized for their medicinal properties. These fungi are integrated into the pharmacopeia and traditional medicine of various cultures around the world. Although not a monophyletic group, polypore fungi are distinguished by their poroid hymenophore and characteristic growth forms, and they are widely studied and appreciated (Robledo and Urcelay 2009). The genera best known for their biological potential are *Ganoderma, Inonotus, Hornodermo*porus, and *Stiptophyllum*. The potential of the cultivated mycelium and fruiting bodies has also been demonstrated. These initial findings highlight the remarkable potential of mushrooms in biotechnology, supporting the growing development of fungus-based products across diverse fields such as food, pharmaceuticals, cosmetics, and others.

In relation to edible mushrooms, the first formal studies on their nutritional properties were conducted by our research team, and several species classified as edible are still under evaluation. An initial outreach to the general public was achieved through the publication of *Wild and Edible Mushrooms of Paraguay and Some Recipes* (Servián and Campi 2020), which documented 19 edible species for the first time in Paraguay.

Building upon that foundation, the book *Edible and Medicinal Mushrooms of Paraguay* was published in 2025, presenting over 40 native fungal species with documented nutritional and medicinal properties. The following species are among those included in the book: *Agaricus brasiliensis* Fr., *Agaricus campestris* L., *Agaricus globocystidiatus* Drewinski & M.A. Neves, *Auricularia delicata* (Mont. ex Fr.) Henn., *Auricularia nigricans* (Sw.) Birkebak, Looney & Sánchez-García, *Agrocybe aegerita* (V. Brig.) Singer *Auricularia fuscosuccinea* (Mont.) Henn., *Calvatia cyathiformis* (Bosc) Morgan, *Calvatia rugosa* (Berk. & M.A. Curtis) D.A. Reid, *Clitocybe nuda* (Bull.) H.E. Bigelow & A.H. Sm., *Cookeina speciosa* (Fr.) Dennis, *Cookeina tricholoma* (Mont.) Kuntze, *Coprinus comatus* (O.F. Müll.) Pers., *Favolus rugulosus* Palacio & R.M. Silveira, *Fomes fasciatus* (Sw.) Cooke, *Ganoderma tuberculosum* Murrill, *Inonotus rickii* (Pat.) D.A. Reid, *Irpex rosettiformis* C.C. Chen & Sheng H. Wu, *Kusaghiporia talpae* (Cooke) Campi, Drechsler-Santos & Robledo, *Lactarius deliciosus* (L.) Gray, *Laetiporus gilbertsonii Burds.*, *Lentinus lindquist* (Singer) B.E. Lechner & Albertó, *Lentinus concavus* (Berk.) Corner, *Lentinus crinitus* (L.) Fr., *Macrocybe titans* (H.E. Bigelow & Kimbr.) Pegler, Lodge & Nakasone, *Macrolepiota bonaerensis* (Speg.), *Macrolepiota gracilenta* (Krombh.) Wasser, *Morchella esculenta* (L.) Pers., *Oudemansiella cubensis* (Berk. & M.A. Curtis) R.H. Petersen, *Panus neostrigosus* Drechsler-Santos & Wartchow, *Phlebopus beniensis* (Singer & Digilio) Heinem. & Rammeloo, *Pisolithus arrhizus* (Scop.) Rauschert, *Pleurotus cystidiosus* O.K. Mill., *Pleurotus albidus* (Berk.) Pegler, *Pleurotus djamor* (Rumph. Ex Fr.) Boedijn, *Pleurotus djamor* var. *roseus* Corner, *Pleurotus ostreatus* (Jacq.) P. Kumm, *Podaxis pistillaris,* (L.) Fr., *Polyporus tricholoma* Mont., *Pycnoporus sanguineus* (L.) Murrill, *Schizophyllum commune* Fr., *Suillus granulatus* (L.) Roussel, *Suillus salmonicolor* (Frost) Halling., *Tremella fuciformis* Berk., and *Tremella mesenterica* Retz. (Figure 16.6).

Of particular note is the first report of the edibility of species of *Kusaghiporia* in the Neotropics. While *K. usambarensis* had previously been identified as edible in Tanzania, East Africa (Hussein et al. 2018), Campi et al. (2025) were the first to report *K. talpae* as an edible species in the Neotropics. Research into the edibility of native Paraguayan mushrooms is accelerating, and it is anticipated that this preliminary list will expand as further studies are conducted.

16.9 SPEGAZZINI'S LEGACY LIVES ON: THE POWER OF SCIENTIFIC DISSEMINATION

Undoubtedly, one of the greatest challenges for our research team is conveying not only the passion but also the crucial importance of understanding native fungi. In response to this need, we founded

FIGURE 16.6 Edible mushrooms from Paraguay. (a) *Suillus granulatus*, (b) *Oudemansiella cubensis*, (c) *Macrolepiota bonaerensis*, (d) *Kusaghiporia talpae*, (e) *Laetiporus gilbertsonii*, (f) *Favolus rugulosus*.

FungiParaguay—a multidisciplinary organization dedicated to the study of Paraguayan fungi and the dissemination of scientific knowledge to the broader public.

Our commitment to strengthening collaborative networks also led us to take an active role in the Red Iberoamericana de Micología (RIIMICO). This regional platform has significantly increased our scientific impact. Through RIIMICO, we have advanced the exchange of methodologies, training opportunities, and joint research initiatives, aligning our local efforts with broader Ibero-American strategies for fungal conservation and mycological development (www.riimico.cemit.una.py).

Importantly, RIIMICO has become a space that promotes regional scientific production and visibility, fostering collaborative publications, supporting multicenter projects funded across Ibero-America, and strengthening opportunities for early-career researchers. The network also prioritizes scientific dissemination, offering open-access seminars and talks—many of them recorded and freely available on its YouTube channel—thus ensuring that knowledge reaches both academic and non-academic audiences. In addition, RIIMICO actively disseminates information about courses, workshops, scholarships, and training programs in mycology, contributing to capacity-building across the region and promoting equitable access to professional development.

Social media has played a pivotal role in fostering engagement, significantly enhancing our connection with the public at large. The public response has been overwhelmingly positive, with a growing demand for knowledge as well as for fungal-based food and medicinal products—far surpassing our initial expectations. To further promote this fascinating world, we launched *FungiFest*: a festival for both social recreation and scientific outreach that bridges academia and society. Researchers, producers, artists, and the general public can come together to celebrate and explore the fungal kingdom. After two successful editions, the festival has already become a beloved fungal tradition.

We firmly believe that humanity owes a debt to the fungal kingdom—one we can begin to repay by fostering awareness and appreciation. As Campi et al. (2025) rightly stated, "We cannot protect what we ignore, much less value what we do not understand." This is why we strive to cultivate a fungal culture from within academia, ensuring that future generations can both understand and protect this essential part of our natural world.

16.10 TRACING A CENTURY OF NEGLECT: THE RE-EMERGENCE OF CLINICALLY RELEVANT FUNGI IN PARAGUAY

Opportunistic fungal infections constitute a major and growing public health concern in Paraguay, particularly affecting immunocompromised populations such as people living with HIV (PLHIV). The most prevalent fungal pathogens include *Histoplasma capsulatum*, *Aspergillus* spp., *Candida* spp., and *Sporothrix* spp., among others. These are responsible for a wide range of clinical manifestations, from superficial infections to severe and potentially life-threatening systemic diseases (Nucci et al. 2010; Aguilar et al. 2020; Staicu et al. 2020; Aguilar et al. 2024; Silvero et al. 2025). Recent epidemiological studies have reported high levels of prevalence and mortality rates associated with these fungal infections, particularly histoplasmosis, cryptococcosis, and tuberculosis in PLHIV (Paul et al. 2022; Aguilar et al. 2024). For example, a cohort study involving 491 PLHIV in Paraguay found that 38% had at least one opportunistic infection, with histoplasmosis and cryptococcosis contributing substantially to mortality (Aguilar et al. 2024). Paracoccidioidomycosis, another systemic mycosis endemic to the region, predominantly affects rural male workers, characterized mainly by oral and cutaneous lesions, with an estimated mortality rate of approximately 9% (Staicu et al. 2020; Violeta Araújo López et al. 2024). Most of the patients diagnosed were from the eastern region of the country (Violeta Araújo López et al. 2024). Fungal keratitis, caused mainly by species of *Acremonium*, *Fusarium*, and *Curvularia*, represents an important cause of infectious corneal ulcers, particularly among agricultural workers exposed to plant debris and soil (Laspina et al. 2004; Sonego-Krone et al. 2006).

Environmental and food safety investigations have documented the presence of toxigenic fungi such as *Aspergillus* section Flavi, the *Fusarium graminearum* species complex, and *Penicillium* spp. as common contaminants of staple crops, including maize, rice, and wheat (Tonon et al. 1997; Alvarenga et al. 2022; Moura-Mendes et al. 2023). In response to this growing threat, the project "Investigation of New Antifungals and Associations against Emerging Pathogenic Fungi: A One Health approach," supported by Women in Science for the Developing World (OWSD)–UNESCO, is currently being implemented in Paraguay. This initiative adopts a comprehensive One Health framework to address sporotrichosis, bringing together key national actors across sectors: the academic institutions Universidad Nacional de Asunción (CEMIT/FCV) and the Universidad Nacional del Este (FACISA); the public health system through CED–MSPyBS; and the private sector represented by Brunelli Laboratorios. The project integrates human, animal, and environmental health approaches, with a strong emphasis on prevention, early detection, and community-based awareness to curb the spread of sporotrichosis at the national level. The presence of mycotoxin-producing strains in these agricultural products raises concerns about chronic exposure risks and their broader implications for food safety and public health.

Zoonotic fungal pathogens have recently gained attention due to the emergence of *Sporothrix brasiliensis* in Paraguay, marking a new chapter in the country's fungal epidemiology (Xavier et al. 2023; Silvero et al. 2025). The transmission of *S. brasiliensis* is primarily via zoonotic or environmental routes, with cats serving as the principal reservoir and vector responsible for the majority of human and animal infections (do Prado et al. 2023; Silvero et al. 2025). The first autochthonous feline sporotrichosis cases were confirmed in Ciudad del Este in 2022, followed by the first local human cases, highlighting the risk of cat-transmitted sporotrichosis and underscoring the urgent need for integrated One Health surveillance and control strategies (García Duarte et al. 2017; Rossow et al. 2020; do Prado et al. 2023). Environmental isolation of *S. brasiliensis* is rare but possible, typically from soil contaminated by infected cat feces (Rossow et al. 2020). In addition, Paraguay has recently reported its first case of *Lomentospora prolificans* (previously *Scedosporium prolificans*), an opportunistic and highly drug-resistant fungal pathogen. This case involved a pediatric patient who sustained a knee injury in a garden setting, illustrating the wide spectrum of environmental fungal pathogens and the potential for accidental inoculation during routine activities (Ramirez-Garcia et al. 2018; Amarilla et al. 2024). These findings underscore the importance of a One Health approach integrating human, animal, and environmental health perspectives to improve understanding and mitigate fungal risks of disease (Figure 16.7).

Despite advances in clinical and environmental research on fungal infections in Paraguay, these diseases remain widely underdiagnosed and neglected. Diagnostic resources, including rapid assays for fungal antigens such as *Histoplasma* and *Cryptococcus*, are limited, restricting early detection and timely treatment (Rodrigues and Nosanchuk 2020; Medina et al. 2022). Moreover, surveillance systems for fungal diseases and antifungal resistance monitoring remain seriously insufficient, impeding public health responses and epidemiological understanding (Paul et al. 2022; Aguilar et al. 2024). Research efforts are often fragmented, with minimal integration across clinical, environmental, and

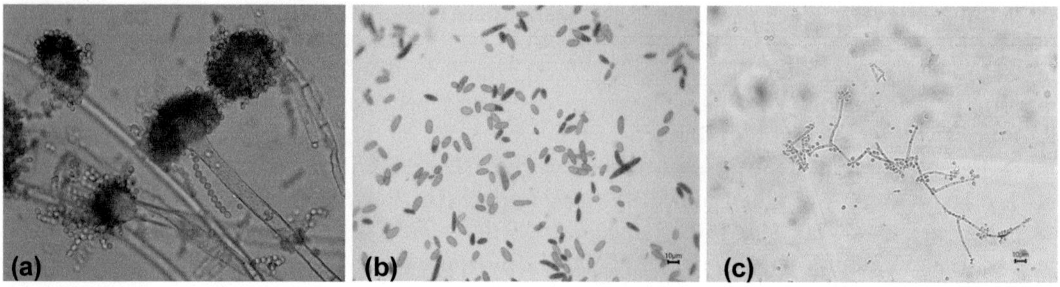

FIGURE 16.7 Relevant fungi under the One Health approach. Microscope observation of (a) *Aspergillus* sp., (b) *Fusarium* sp., and (c) *Sporothrix* sp. Scale bars = 10 μm.

epidemiological data streams, resulting in persistent gaps in knowledge about incidence, transmission dynamics, and resistance patterns.

To mitigate the growing burden of fungal diseases, Paraguay urgently needs a strengthened diagnostic capacity, including expanded access to rapid fungal antigen detection, molecular identification tools, and enhanced laboratory infrastructure. Comprehensive surveillance programs that integrate clinical and veterinary health data, coupled with environmental monitoring, are essential. Furthermore, promoting multidisciplinary collaboration across human health, veterinary, agricultural, and environmental sectors will improve understanding of fungal pathogen ecology and epidemiology. Such integrated efforts are critical for the development of targeted public health interventions, the improvement of patient outcomes, and the control of zoonotic and environmental fungal threats.

16.11 BEYOND REDISCOVERY: WHY CONSERVING PARAGUAY'S FUNGAL HERITAGE MATTERS TODAY

Fungi are among the most metabolically versatile and biotechnologically valuable organisms, providing essential products and services for pharmaceuticals, agriculture, industry, and environmental management. The full realization of their potential—particularly in biodiverse regions such as Paraguay—depends on the conservation of their rich and largely untapped biodiversity.

Despite their importance, fungi are often overlooked in conservation efforts, even though they play key roles in ecosystem functioning, soil health, and plant productivity (Lofgren and Stajich 2021; Attili-Angelis et al. 2022; Barron 2023). Paraguay, located within a region of high fungal diversity, holds significant promise for the discovery of novel species and biotechnological applications. However, the lack of targeted research and conservation initiatives threatens the loss of unique fungal resources before their potential can be realized (Lofgren and Stajich 2021; Attili-Angelis et al. 2022; Barron 2023; Campi et al. 2025).

Conserving fungal biodiversity is essential not only for maintaining ecosystem services but also for ensuring a continuous supply of new biotechnological tools and products. This requires integrating fungi into conservation policies, supporting dedicated research, and establishing collections and databases to preserve genetic resources (Lofgren and Stajich 2021; Attili-Angelis et al. 2022; Barron 2023).

Biological collections are the backbone of fungal biodiversity conservation, serving as repositories for genetic resources, reference materials for taxonomic and ecological studies, and platforms for biotechnological innovation. In Paraguay, no official national collection of fungal biological resources currently exists. Existing resources are maintained through the efforts of research groups such as the macrofungal collection of the FUNGI Paraguay group (FACEN-UNA) and the Colección de Cultivos de Microorganismos (CCM-UNA) housed at CEMIT-UNA. However, the CCM-UNA currently operates primarily on a demand-driven basis for research projects and lacks the consistent financial support required for its role as a national reference collection. The CCM-UNA is actively consolidating its infrastructure and pursuing sustainable funding to ensure the long-term preservation and distribution of fungal cultures. As a member of the Latin American Federation of Culture Collections (FELACC), it contributes to capacity-building initiatives and collaborative networks across the region. In addition, through a partnership with NAPI Taxonline, the CCM-UNA is authenticating several fungal isolates with broad biotechnological potential, which are already widely used in plant breeding, biotechnology, and human and animal health research.

Strengthening CCM-UNA as a national reference collection is imperative for conserving Paraguay's fungal heritage and positioning the country as a regional leader in fungal biotechnology and conservation.

REFERENCES

Aguilar, G., P. Araujo, G. Lird, S. Insaurralde, A. Kawabata, E. Ayala, J. Irala & R. Argüello. 2020. Identificación y perfil de sensibilidad de *Candida* spp. aisladas de hemocultivos en hospitales de Paraguay. *Review of Panamanian Salud Publica* 44. https://doi.org/10.26633/RPSP.2020.34

Amarilla, S., Z. Morel, M. Feijoo, W. Cabañas & H. Jiménez. 2024. Primer caso de infección por Lomentospora prolificans en Paraguay. Reporte de Caso. In *Congresos Regionales de Pediatría*, p. 71. https://revistaspp.org/suplementos/SUPLEMENTO_TEMAS_LIBRES_SPP_2024.pdf

Aguilar, G., P. Araujo, G. Lird, S. Insaurralde, A. Kawabata, E. Ayala, J. Irala, R. Arguello, G. Aguilar, G. Lopez, O. Sued, N. Medina, D.H. Caceres, J. Pereira, A. Jordan, V. Lezcano, C. Vicenti & G. Benitez. 2024. Implementation of a rapid diagnostic assay package for cryptococcosis, histoplasmosis and tuberculosis in people living with HIV in Paraguay. *BMC Infectious Diseases* 24:1–9. https://doi.org/10.1186/S12879-024-09257-5/FIGURES/2

Alvarenga, A.AA., J.C.M.I. Ouchi, C.C.C. Martínez, J.M. Mendes, A.A. Colmán, D.F. Ríos, P.D. Arrua, C.A.B. Guerreño, M.M. Kohli, M.L. Ramírez, et al. 2022. Trichothecene genotype profiling of wheat *Fusarium graminearum* species complex in Paraguay. *Toxins* 14:257. https://doi.org/10.3390/TOXINS14040257

Attili-Angelis, D., M.H. Taniwaki, N. da Silva, V.M. Oliveira & N. Lima. 2022. Microbial *ex situ* preservation supporting science and bioeconomy in Brazil. *Biota Neotropica* 22:e20221438. https://doi.org/10.1590/1676-0611-BN-2022-1438

Barron, E. 2023. Conservation of abundance: How fungi can contribute to rethinking conservation. *Conservation Society* 21:99–109. https://doi.org/10.4103/CS.CS_23_22

Calle, A., E.R. Drechsler-Santos, L. Funez, T. Kossmann, K.M. da Cunha, D. Pfister, G. Robledo, P. Sandoval-Leiva, T. Sanjuan & A.M. Vasco-Palacios. 2020. *Rickiella edulis. The IUCN Red List of Threatened Species* 2020:e.T172818281A172861317. https://dx.doi.org/10.2305/IUCN.UK.2020-2.RLTS.T172818281A172861317.en

Campi, M., B. de Madrignac, A. Flecha, A. Ortellado, M. Martínez & L. Rojas. 2013. *Hongos de la Reserva Natural Laguna Blanca*. San Lorenzo, Paraguay.

Campi, M. & Y. Maubet. 2015a. Especies de Geastrum (Geastraceae, Basidiomycota) nuevos registros para Paraguay. *Steviana* 7:79–88. https://doi.org/10.56152/StevianaFacenV7A6_2015

Campi, M. & Y. Maubet. 2015b. *Cyathus poeppigii* (Agaricales, Basidiomycetes): nuevo registro para Paraguay. *Steviana* 7:74–78.

Campi, M. & Y. Maubet. 2016. Nuevos registros de *Calvatia rugosa* (Berk. & MA Curtis) DA Reid y *Vascellum pampeanum* (Speg.) Homrich (Agaricaceae-Basidiomycota) en Paraguay. *Steviana* 8(1):43–49.

Campi, M., Y. Maubet, J. Armoa & P. Sandoval. 2017g. *Donkia pulcherrima* (Polyporales, Phanerochaetaceae) una especie hidnoide poco conocida, nueva cita para el Paraguay. *Steviana* 9(1):25–30.

Campi, M., Y. Maubet & L. Britos. 2015a. Mycorrhizal fungi associated with plantations of *Pinus taeda* L. from the National University of Asunción, Paraguay. *Mycosphere* 6:486–492. https://doi.org/10.5943/mycosphere/6/4/10

Campi, M., Y. Maubet, E. Cristaldo, C. Mancuello & G. Robledo, G. 2022a. Funga del Parque Nacional Ybycuí: listado preliminar de especies de macrohongos y nuevas citas para Paraguay. *Revista Steviana* 13:78–100.

Campi, M. G., Azevedo-Olivera, C., Costa-Rezende, D., Maubet, Y. E., Morera, G., Urcelay, C., ... & Robledo, G. L. (2022b). What are the Laetiporus species present in southern South America? Lilloa 59 (Suplement): 193–218. https://doi.org/10.30550/j.lil/2022.59.S/2022.09.19

Campi, M., Y. Maubet, E. Grassi, E. Cristaldo & N. Menolli. 2019. *Pluteus* Fr. (Pluteaceae, Agaricales) in Paraguay: Morphological studies and new records. *Check List* 15:313–322.

Campi, M., Y. Maubet, E. Grassi, N. Niveiro & L. Guzmán-Dávalos. 2021b. First contribution to the genus *Gymnopilus* (Agaricales, Strophariaceae) in Paraguay. *Rodriguésia* 72:e00752019.

Campi, M., C. Mancuello, Y. Maubet, E. Cristaldo, B. Veloso, F. Ferreira & G. Robledo. 2023a. Biochemical, nutritional, and toxicological properties of the species *Phlebopus beniensis* with ethnomycological notes from Paraguay. *Brazilian Journal of Food Technology* 26:1–18.

Campi, M., Mancuello, C., Ferreira, F., Maubet, Y., Cristaldo, E. L., Gayozo, E., & G. Robledo. (2023). Does the source matter? Phenolic compounds and antioxidant activity from mycelium in liquid medium, wild and cultivated fruiting bodies of the neotropical species Ganoderma tuberculosum. *Journal of Microbiology, Biotechnology and Food Sciences* 13(1):e6148. https://doi.org/10.55251/jmbfs.6148

Campi, M., C. Mancuello, Y. Maubet & N. Niveiro. 2017f. *Laccaria fraterna* (Cooke & Mass:: Sacc.) pegler, 1965 (agaricales, basidiomycota) associated with exotic *Eucalyptus* sp. in northern Argentina and Paraguay. *CheckList* 13(4):87–90. https://doi.org/10.15560/13.4.87

Campi, M., Y. Maubet, B. Miranda, J. Armoa & E. Cristaldo. 2017c. Dos nuevas citas de Mycenaceas para el Paraguay: *Xeromphalina tenuipes* & *Filoboletus gracilis*, un interesante agarical poroide. *Steviana* 9(1):16–24.

Campi, M., Y. Maubet & L. Trierveiler-Pereira. 2018a. "Estrellas de tierra" *Geastrum* (Geastraceae, Basidiomycota): nuevas citas para el Pantanal paraguayo. Steviana 10(2):3–16.

Campi, M., Y. Maubet & L. Trierveiler-Pereira. 2021a. Rediscovery of *Clathrus argentinus* and new contributions to the gasteroid mycobiota of Paraguay. *Current Research in Environmental & Applied Mycology* 11:90–111.

Campi, M., Y. Maubet, L. Trierveiler-Pereira & G. Moreno. 2018b. Contribución a la micobiota gasteroide de Paraguay. Nuevas citas. *Sociedad Micologica de Madrid. Boletín* 42:115–124.

Campi, M., Y. Maubet, B. Veloso, E. Cristaldo & C. Brehm. 2025. Generando cultura fúngica: hongos comestibles en el Paraguay. *Lilloa* (Suplemento 1 [Hongos comestibles en Latinoamérica]):163–179. https://doi.org/10.30550/j.lil/1816

Campi, M., Y. Maubet & A. Weiler. 2015b. *Podaxis pistillaris* (Basidiomycota, Agaricomycetes) nuevo para el chaco paraguayo. *Boletín de la Sociedad Micologica de Madrid* 39:77–81.

Campi, M., Y. Maubet, A. Weiler & G. Moreno. 2017e. *Battarrea phalloides* (Basidiomycota, Agaricomycetes) nuevo para el chaco paraguayo. *Boletín de la Sociedad Micologica de Madrid* 41:47–51.

Campi, M., Y. Maubet & L. Trierveiler Pereira. 2017d. Contributions to the Paraguayan mycota: New records of Phallales (Agaricomycetes, Basidiomycota). *Mycotaxon* 132:361–372.

Campi, M., G. Robledo, Y. Maubet & E. Grassi, E. 2017b. *Amylosporus guaraniticus* sp. nov. (Wrightoporiaceae—Russulales) a new neotropical species from Paraguay. *Mycosphere* 8:1060–1070.

Campi, M., L. Trierveiler-Pereira & Y. Maubet. 2017a. New records of Phallales from Paraguay. *Mycotaxon* 132(2):361–372.

Campi, M. & B. Veloso. 2025. *Hongos comestibles y medicinales del Paraguay*. Ediciones de la Universidad Nacional de Asunción. San Lorenzo, Paraguay.

Campi, M. G., Maubet, Y., Veloso, B., Cristaldo, E., & Brehm, C. (2025). Generando cultura fúngica: hongos comestibles en el Paraguay. Lilloa 62(2): 163–179. https://doi.org/10.30550/j.lil/2022.59.S/2022.09.19

Cristaldo, E.L., T. Kossmann, M. Campi, Y. Maubet, D. Costa Rezende, R. Drechsler Santos & G.L. Robled. 2022. Neotropical Daedalea (Basidiomycota, Fomitopsidaceae) revisited: Daedalea rajchenbergiana sp. nov. from Brazil. *Lilloa* 59 (Suplemento):273–289. http://dx.doi.org/10.30550/j.lil/2022.59.S/2022.09.16

de Battini, B.E.V. & L.M. Kornfeld. 1980. *Cuentos y leyendas populares de la Argentina (Vol. 1)*. Ediciones Culturales Argentinas, Secretaría de Estado de Cultura, Ministerio de Cultura y Educación. Buenos Aires, Argentina.

Dirección General de Estadística, Encuestas y Censos (DGEEC). 2013. *Anuario Estadistico del Paraguay*. https://www.ine.gov.py/publicationsingle/php?codec=87

do Prado, C.M., E. Razzolini, G. Santacruz, L. Ojeda, M.R. Geraldo, N. Segovia, J. Pereira Brunelli, V.A. Vicente, W.K. Svoboda & F. Queiroz-Telles. 2023. First cases of feline sporotrichosis caused by *Sporothrix brasiliensis* in Paraguay. *Journal of Fungi* 9. https://doi.org/10.3390/JOF9100972

Flecha, A., B. de Madrignac & M. Campi. 2014. El género *Pleurotus* (Pleurotaceae-Basidiomycota) en Paraguay. *Steviana* 6:70–79.

García Duarte, J.M., V.R. Wattiez Acosta, P.M.L. Fornerón Viera, A. Aldama Caballero, G.A. Gorostiaga Matiauda, V.B. Rivelli de Oddone & J.G. Pereira Brunell. 2017. Esporotricosis trasmitida por gato doméstico. Reporte de un caso familiar. *Revista del Nacional* 9:67–76. https://doi.org/10.18004/RDN2017.0009.02.086-095

Hussein, J.M., D.D. Tibuhwa & S. Tibell. 2018. Phylogenetic position and taxonomy of *Kusaghiporia usambarensis* gen. et sp. nov.(Polyporales). *Mycology* 9(2):136–144.

Jiménez, D.A.A. 2023. Personajes mitológicos conocidos en Paraguay. Un estado de la cuestión a partir de un sondeo en línea. *Revista del Museo de Antropología* 16:155–168.

Laspina, F., M. Samudio, D. Cibils, C.N. Ta, N. Fariña, R. Sanabria R, Klauß & H.M. de Kaspar 2004. Epidemiological characteristics of microbiological results on patients with infectious corneal ulcers: A 13-year survey in Paraguay. *Graefe's Archive for Clinical and Experimental Ophthalmology* 242:204–209. https://doi.org/10.1007/s00417-003-0808-4

Lehner, B. 2005. Los pueblos indígenas del Paraguay una breve introducción. *Debate Teórico Paraguayo Legislación Antidiscriminatoria* 95.

Lofgren, L.A. & J.E. Stajich. 2021. Fungal biodiversity and conservation mycology in light of new technology, big data, and changing attitudes. *Current Biology* 31:R1312–R1325. https://doi.org/10.1016/J.CUB.2021.06.083

Madrignac, B., A. Flecha, M. Campi & A. Ortellado. 2013. Nuevos Registros del género Marasmius (Basidiomycota-Marasmiaceae) para la Región de la Reserva Natural Laguna Blanca, San Pedros-Paraguay. *Reportes Científicos de la FaCEN* 4:5–10.

Mancuello, C. R., Benítez, D., Maubet, Y., Cristaldo, E., Veloso, B., Ferreira, F., & Campi, M. (2022). Explorando las propiedades de la funga neotropical: perfil químico, actividades antioxidantes y antimicrobianas de Stiptophyllum erubescens (Berk.) Ryvarden. *Steviana* 14(2):5–16.

Mancuello Chaparro, C.R., Y.E., Maubet Cano, E. Cristaldo, B.A. Veloso Paredes, G.L. Robledo, A. Traba Daviccino, . . . & M.G. Campi Gaona. 2024. *Oudemansiella cubensis* an edible mushroom from the Neotropics with biological and nutritional benefits. *Natural Resources for Human Health* 4(3):257–268. https://doi.org/10.53365/nrfhh/189170

Martínez, M., C.A. Salvador-Montoya, A. de Errasti, O.F. Popoff & M. Rajchenberg. 2023. *Fulvifomes wrightii* (Hymenochaetales), a new species related to *F. robiniae* from Argentina and Paraguay. *Fungal Systematics and Evolution* 12:47.

Maubet, Y. 2023. *Geastrum violaceum. The IUCN Red List of Threatened Species* 2023:e. T238218011 A245223763. https://dx.doi.org/10.2305/IUCN.UK.2023-1.RLTS.T238218011A245223763.en

Maubet, Y., M. Campi, J. Armoa & E. Cristaldo. 2017. Nuevas citas de Cyathus Haller (Agaricaceae, Basidiomycetes) para Paraguay y ampliación de la distribución del género. *Steviana* 9:31–39.

Maubet, Y., M. Campi, E. Cristaldo & G.L. Robledo. 2022. Introducing a new member of the genus *Chlorophyllum*: *Chlorophyllum sapukai* sp. nov., and new records from Paraguay. *Plant and Fungal Systematics* 67:5–85. https://doi.org/10.35535/pfsyst-2022-0009

Maubet, Y., M.G. Campi, N. Niveiro, D. Pelissero, R. Sánchez, C. Ranieri, M.L.H. Caffot, G. Robledo & D. Torres. 2023a. *Phellinotus xerophyticus. The IUCN Red List of Threatened Species* 2023:e.T238221155A245223535. https://dx.doi.org/10.2305/IUCN.UK.2023-1.RLTS.T238221155A245223535.en

Maubet, Y., M.G. Campi, N. Niveiro, D. Pelissero, R. Sánchez, C. Ranieri, M.L.H. Caffot, G. Robledo & D. Torres. 2023b. *Clathrus argentinus. The IUCN Red List of Threatened Species* 2023:e.T238217921A245224007. https://dx.doi.org/10.2305/IUCN.UK.2023-1.RLTS.T238217921A245224007.en

Maubet, Y., M. Campi & G.L. Robledo. 2020. Hymenochaetaceae from Paraguay: Revision of the family and new records. *Current Research in Environmental & Applied Mycology* 10:242–261.

Maubet, Y., G. Robledo, M.G. Campi, M. Monteiro, S. Martinez & O. Popoff. 2023c. *Xylobolus gongylodes. The IUCN Red List of Threatened Species* 2023: e.T245800846A245801554. https://dx.doi.org/10.2305/IUCN.UK.2023-1.RLTS.T245800846A245801554.en

Medina, N., A. Alastruey-Izquierdo, D. Mercado, D.W. Denning, E. Arathoon & J.L. Rodriguez-Tudela. 2022. Diagnostic mycology laboratories should have a central role for the management of fungal disease. *Journal of Fungi* 8:1285. https://doi.org/10.3390/JOF8121285

Moura-Mendes, J., C.C. Cazal-Martínez, C. Rojas, F. Ferreira, P. Pérez-Estigarribia, N. Dias, P. Godoy, J. Costa, C. Santos & A. Arrua. 2023. Species identification and mycotoxigenic potential of *Aspergillus* Section Flavi isolated from maize marketed in the metropolitan region of Asunción, Paraguay. *Microorganisms* 11:1879. https://doi.org/10.3390/MICROORGANISMS11081879/S1

Mueller, G.M., K.M. Cunha, T.W. May, J.L. Allen, J.R. Westrip, C. Canteiro & A. Dahlberg, A. 2022. What do the first 597 global fungal red list assessments tell us about the threat status of fungi? *Diversity* 14:736.

Mueller, G.M., A. Dahlberg & M. Krikorev, M. 2014. Bringing fungi into the conservation conversation: The Global Fungal Red List Initiative. *Fungal Conservation* 4:12–16.

Nucci, M., F. Queiroz-Telles, A.M. Tobón, A. Restrepo & A.L. Colombo. 2010. Epidemiology of opportunistic fungal infections in Latin America. *Clinical Infectious Diseases* 51:561–570. https://doi.org/10.1086/655683/2/M_51-5-561-FIG001.GIF

Paul, R., A. Chakrabarti, H. Kaur, A. Ghosh, R. Agarwal, N. Panda, S.-A. Rudramurthy, T. Samudio, G. Aguilar, O. Sued. et al. 2022. P258 Incidence of histoplasmosis, cryptococcosis, and TB among people living with HIV in Paraguay-preliminary report. *Medical Mycology* 60(Supplement 1). https://doi.org/10.1093/MMY/MYAC072.P258

Popoff, O. & J. Wright. 1994. Two new corticioid fungi (Aphyllophorales) from NE Argentina and Paraguay. *Mycotaxon* 51:317–324.

Popoff, O. & J. Wright. 1998. Preliminary check-list of fungi of Paraguay. I. Wood-inhabiting Polypores (Aphyllophorales, Basidiomycota). *Mycotaxon* 67:323–340.

Ramirez-Garcia, A., A. Pellon, A. Rementeria, I. Buldain, F. Barreto-Bergter, R. Rollin-Pinheiro, J.V. De Meirelles, M.I.D.S. Xisto, S. Ranque, V. Havlicek, et al. 2018. *Scedosporium* and *Lomentospora*: An updated overview of underrated opportunists. *Medical Mycology* 56 (Supplement 1):S102–S125. https://doi.org/10.1093/MMY/MYX113

Robledo, G. & C. Urcelay. 2009. *Hongos de la madera en árboles nativos del centro de Argentina*. Editorial de la Universidad Nacional de Córdoba. Córdoba, Argentina.

Rodrigues, M.L. & J.D. Nosanchuk. 2020. Fungal diseases as neglected pathogens: A wake-up call to public health officials. *PLoS Neglected Tropical Diseases* 14:e0007964. https://doi.org/10.1371/JOURNAL.PNTD.0007964

Rossow, J.A., F. Queiroz-Telles, D.H. Caceres, K.D. Beer, B.R. Jackson, J.G. Pereira, I.D.F. Gremião & S.A. Pereira. 2020. A one health approach to combatting *Sporothrix brasiliensis*: Narrative review of an emerging zoonotic fungal pathogen in South America. *Journal of Fungi* 6:247. https://doi.org/10.3390/JOF6040247

Servián, A. & M. Campi. 2020. *Hongos silvestres y comestibles de Paraguay y algunas recetas*. Editorial FungiParaguay. Asunción, Paraguay.

Silvero, M.G.S., C.M. do Prado, B. Spruijtenburg, F.A.L. Codas, M.L. Ojeda, B.J.F. de Souza Lima, N.S. Coronel, J.P. Brunelli, V.A. Vicente & T. de Groot. 2025. The first autochthonous human case of sporotrichosis by *Sporothrix brasiliensis* in Paraguay. *Journal of Medical Mycology* 35:101536. https://doi.org/10.1016/j.mycmed.2025.101536

Sonego-Krone, S., D.S.-D. Martino, R. Ayala-Lugo, G. Torres-Alvariza, C.N. Ta, L. Barbosa & H.M. de Kaspar. 2006. Clinical results of topical fluconazole for the treatment of filamentous fungal keratitis. *Graefe's Archive for Clinical and Experimental Ophthalmology* 244:782–787. https://doi.org/10.1007/S00417-005-1127-8

Spegazzini, C. 1891. Fungi guaranitici nonnulli novi vel critici. *Revista Argentina de Historia Natural* 1:398–432

Spegazzini, C. 1922. Fungi Paraguayenses. *Museo Nacional de Historia Natural de Buenos Aires* 31:355–450.

Staicu, M.L., M.R. Laguio-Vila, J. Uzoma, R. Vansice, R.G. Hospital, N. Rochester, R.R. Health, N.Y. Rochester, F.M. Núñez & L. Woc-Colburn. 2020. Clinical and epidemiological characteristics of patients with paracoccidioidomycosis in Asuncion, Paraguay. *Open Forum Infectious Diseases* 7(Supplement 1):S604–S604. https://doi.org/10.1093/OFID/OFAA439.1341

Tonon, S.A., R.S. Marucci, G. Jerke & A. García. 1997. Mycoflora of paddy and milled rice produced in the region of northeastern Argentina and southern Paraguay. *International Journal of Food Microbiology* 37:231–235. https://doi.org/10.1016/S0168-1605(97)00066-4

Violeta Araújo López, P., G. Aguilar Fernández, R. Zancopé-Oliveira, E.R. del Ambiente, J.M. Boettner, P. Revisor, M.F.M. Mora & G. Giusiano. 2024. Pasado y presente de la Paracoccidioidomicosis en Paraguay. *Revista del Nacional* 16:167–196. https://doi.org/10.18004/RDN2024.DIC.03.167.196

Xavier, M.O., V.R. Poester, M.R. Trápaga & D.A. Stevens. 2023. *Sporothrix brasiliensis*: Epidemiology, therapy, and recent developments. *Journal of Fungi* 9:921. https://doi.org/10.3390/JOF9090921

Index

Note: Page numbers in *italics* indicate a figure and page numbers in **bold** indicate a table on the corresponding page.

D

E

F

N

O

P

T

U

V

W

X

Y

For Product Safety Concerns and Information please contact our
EU representative GPSR@taylorandfrancis.com Taylor & Francis
Verlag GmbH, Kaufingerstraße 24, 80331 München, Germany